Lecture Notes in Artificial Intelligence 5253

Edited by R. Goebel, J. Siekmann, and W. Wahlster

Subseries of Lecture Notes in Computer Science

Danail Dochev   Marco Pistore
Paolo Traverso (Eds.)

Artificial Intelligence: Methodology, Systems, and Applications

13th International Conference, AIMSA 2008
Varna, Bulgaria, September 4-6, 2008
Proceedings

 Springer

Series Editors

Randy Goebel, University of Alberta, Edmonton, Canada
Jörg Siekmann, University of Saarland, Saarbrücken, Germany
Wolfgang Wahlster, DFKI and University of Saarland, Saarbrücken, Germany

Volume Editors

Danail Dochev
Bulgarian Academy of Sciences
Institute of Information Technologies
Acad. G. Bonchev 29 A
1113 Sofia Bulgaria
E-mail: dochev@iinf.bas.bg

Marco Pistore
Paolo Traverso
Fondazione Bruno Kessler - IRST
Center for Information Technology
Via Sommarive 18, Povo
38100 Trento, Italy
E-mail: {pistore, traverso}@fbk.eu

Library of Congress Control Number: Applied for

CR Subject Classification (1998): I.2, H.4, F.1, H.3, I.5

LNCS Sublibrary: SL 7 – Artificial Intelligence

ISSN   0302-9743
ISBN-10   3-540-85775-3 Springer Berlin Heidelberg New York
ISBN-13   978-3-540-85775-4 Springer Berlin Heidelberg New York

Springer is a part of Springer Science+Business Media

springer.com

© Springer-Verlag Berlin Heidelberg 2008

Typesetting: Camera-ready by author, data conversion by Scientific Publishing Services, Chennai, India
Printed on acid-free paper   SPIN: 12519842   06/3180   5 4 3 2 1 0

Preface

The 13th Conference on Artificial Intelligence: Methodology, Systems, Applications (AIMSA 2008) was held in Varna, Bulgaria, on September 4–6, 2008. The AIMSA conference series has provided a biennial forum for the presentation of artificial intelligence research and development since 1984. The conference covered the full range of topics in artificial intelligence and related disciplines and provided an ideal forum for international scientific exchange between Central/Eastern Europe and the rest of the world.

For AIMSA 2008, we decided to place special emphasis on the works on theories, experimentations, applications, and case studies that demonstrate in practice the added value and the impact of AI science, methodologies and techniques for the market, society, and culture. We believe that this vision of "AI@WORK" truly reflects the original intention of AIMSA, a conference dedicated to artificial intelligence in its entirety but with a special emphasis on "methodology, systems, and applications", as explicitly stated in its name. According to this vision, we have also paid special attention to works that "cross the boundaries of AI", i.e., works that demonstrate the influence of AI on different fields of research and different scientific communities, such as software engineering, data bases, distributed systems, information systems, information retrieval, etc. Both of these aspects were covered by the papers that were submitted and by those that were accepted for publication.

A total of 109 papers were submitted to the conference, and the program committee accepted 30 full papers and 10 posters. Every submission received at least three reviews. Critical submissions were discussed by the program committee, and when necessary, additional reviews were collected. Poster authors were able to present their papers with a short talk and exhibit and discuss their work during a poster session. Among the accepted papers, 6 out of the 30 full papers describe applications of AI research to different fields. The proceedings also report on work crossing the boundaries of AI towards areas such as information retrieval and distributed systems. In addition to the selected papers, the AIMSA conference featured a workshop on "Cross-Media and Personalized Learning Applications with Intelligent Content". AIMSA 2008 had two invited speakers whose research is extremely relevant for the special theme of the conference ("AI@WORK"). Their keynotes provided an exciting occasion for a lively discussion at the conference. Marco Gori, from the University of Siena, addressed the problem of information retrieval by means of machine learning techniques, while Massimo Paolucci, from DoCoMo Euro-labs, discussed the application of semantic web technologies to mobile phones.

September 2008

Marco Pistore
Paolo Traverso

Conference Organization

Program Chairs

Marco Pistore
Paolo Traverso

Program Committee

Gennady Agre
Annalisa Appice
Franz Baader
Roman Barták
Daniel Borrajo
Luca Console
Silvia Coradeschi
Marie-Odile Cordier
Ulises Cortes
Dan Cristea
Christo Dichev
Danail Dochev
John Domingue
Stefan Edelkamp
Thomas Eiter
Floriana Esposito
Jérôme Euzenat
Michael Fisher
Hector Geffner
Malik Ghallab
Chiara Ghidini
Enrico Giunchiglia
Marco Gori
Peter Haase
Joachim Hertzberg
Andreas Herzig

Joerg Hoffmann
Werner Horn
Felix Ingrand
Irena Koprinska
Oliver Kutz
Ruben Lara
Derek Long
Bernardo Magnini
Pedro Meseguer
Michela Milano
Bernhard Nebel
Radoslav Pavlov
Terry Payne
Enric Plaza
Axel Polleres
Ioannis Refanidis
Francesca Rossi
Paolo Rosso
Luciano Serafini
Barry Smyth
Doina Tatar
Dan Tufis
Matthias Wagner
Toby Walsh
Annette ten Teije

Local Organization Chair

Danail Dochev

External Reviewers

Paolo Avesani
Piergiorgio Bertoli
Sami Bhiri
Ismel Brito
Francesca Carmagnola
Costantina Caruso
Giovanna Castellano
Federica Cena
Silvia Craft
Jordi Dalmau
Roxana Danger
Jérôme David
Nicola Di Mauro
Tommaso Di Noia
Stefan Dietze
Dimo Dimov
Francesco Donini
Christian Dornhege
Fernando Fernandez
Rocio Garcia
Angel Garcia-Olaya
Marco Gavanelli
Alessio Gugliotta
Philip Kilby
Sebastian Kupferschmid
Oswald Lanz
Chan Le Duc
Achim Lewandowski
Corrado Loglisci
Ilaria Lombardi
Dominik Luecke
Thomas Lukasiewicz

Ines Lynce
Marie-Laure Mugnier
Annapaola Marconi
Corrado Mencar
David Millan
Amy Murphy
Nina Narodytska
Barry Norton
Claudia Picardi
Ionut Pistol
Marco Ragni
Marius Raschip
Roman Rosipal
Gabriele Röger
Francois Scharffe
James Scicluna
Meinolf Sellman
Dmitry Shaparau
Selmar Smit
Holger Stenzhorn
Olga Symonova
Vlad Tanasescu
Hans Tompits
Paolo Tonella
Ilaria Torre
Stefan Trausan-Matu
Vassil Vassilev
Matthias Wagner
Shenghui Wang
Fang Wei
Makoto Yokoo
Massimo Zancanaro

Table of Contents

4   Knowledge Representation and Reasoning

5   Constraints, Heuristics, and Search

6 Applications

7 Posters

Context Model for Multi-Agent System Reconfiguration

Krzysztof Stebel and Dariusz Choiński

Faculty of Automatic Control, Electronics and Computer Science,
Silesian University of Technology,
ul. Akademicka 16, 44-100 Gliwice, Poland
{krzysztof.stebel,dariusz.choinski}@polsl.pl

Abstract. This paper presents a reconfigurable multi-agent system (MAS) applied to distributed control system (DCS). Agents co-operate, evaluating the system on different levels of information abstraction. Context-based statistical remote expert agent or local supervisory agent are used to evaluate control agent performance. Using expert-selected period of good performance the reference distribution function is proposed. Periodically, for any monitoring period, a sample of observations is taken for local or remote system performance monitoring. Because evaluation may also be carried out remotely two cases should be considered. Remote expert observes changes of parameters that come from process performance degradation. Second case refers to communication problems when data transmission is corrupted and can not be used for system evaluation. Because of that application, context model is necessary to inform the remote expert about transmission quality. For evaluation of transmission channel, the idea of a context tree is utilised. Number of nodes and leaves taken into considerations depends on the expert's knowledge.

Keywords: Multi-Agent System, context model, system reconfiguration, Kullback–Leibler divergence.

1 Introduction

Distributed control systems (DCS) are difficult to control and exhibit many challenges for engineers. It can be considered as an advanced hierarchical control system where one of the possibilities to optimise and maintain such system is application of Multi-Agent System (MAS). In case of large plants, distributed agent systems are used to simultaneously control hundreds of system variables. Due to the growing number of instruments and the increasing number of various possible faults, the system reliability and efficiency can be seriously degraded. Increasing plant efficiency impacts the improvement of economic factors. This can be done through fast detection of anomalies and proper maintenance [1]. Evaluation of equipment performance is very important and becomes a common practice. However, it is time demanding. Therefore, it is desired to develop a local or remote agent to flag a poorly performing system and even to attempt improving it [2], [3]. The main problem is to find suitable criterion for

D. Dochev, M. Pistore, and P. Traverso (Eds.): AIMSA 2008, LNAI 5253, pp. 1–11, 2008.
© Springer-Verlag Berlin Heidelberg 2008

obtained data analysis because of usually unknown distribution function. In practice, a common way is utilisation of expert knowledge that can propose a reference distribution function, which characterises a well performing system. Next, two distributions are compared using statistic tests, one coming from current data and the second being a reference distribution [4]. The proposed reference distribution is created from the real data coming from a user-selected good performance period, and therefore the reference distribution usually does not follow a known specific distribution. Hence, without any assumption about the underlying distribution, the chi-square goodness-of-fit test is applicable and allows deciding whether a taken sample is from a population with the same distribution as proposed by the user [2]. The idea described above was used in previous paper [5]. Another way of poor instrument performance detection is connected with an assumption that a complete binary tree may describe this class of process, where the states are equated with the leaves. The leaves refer to context, and each leaf may be interpreted as a different class of states. The context can be taken as any ordered set of observations [6]. At the beginning, this idea was applied in data compression systems [7]. Procedures based on context-tree were developed and evaluated and claimed to be very efficient and promising [8], [9]. An interesting example is inspection operations, that can often be treated as procedures used to classify the obtained product according to its conformance to the assumed set of requirements and to assign those products to possible classes. Another example is a transmitted message from a source to a receiver by any communication channel where similar errors can be observed. As a result, the received information context may differ from the initially transmitted information [10]. The idea of context tree may be extended by application of optimisation procedures for context selection. Optimal context is defined for each element in the data string, and gives the probability of the element eliminating redundancy in the tree [11].

In this paper a context-based statistical model for remote system performance evaluation over TCP/IP network is presented. Considering expert-selected good performance period the reference context tree is proposed. Then, for any monitoring time period, a sample of sequenced observations is used for system monitoring. The obtained distribution function is compared with reference distribution using the chi-square goodness-of-fit test. In case of remote expert system the evaluation is performed under the assumption of partially accessible information. Several reasons can cause partial information accessibility about the process. Hence, before process evaluation the information about it must also be evaluated. Because of that, the remote expert agent uses the context model to inform about possible communication problems.

Tasks for monitoring the system can be listed as follows:

-   sufficient network channel capacity verification
-   optimisation of channel capacity utilisation

Those tasks should be realised under the assumption that statistical parameters are unknown. To detect improper data transmission in case of insufficient network capacity, the context-tree idea is utilised. The context-tree is constructed on the front-end and back-end of communication channel. Using statistical measure given by Kullback [12] chosen nodes of obtained context-trees are compared. Because only one value per five measured values is send to the remote agent, the distance between obtained trees is nonzero.

The presented paper highlights the need of connection between Multi-Agent system idea described in section 2 with the necessity of data exchange verification. To verify exchanged data the context-based statistical model is used and is described in section 3.

2 Multi-Agent System for Hierarchical Hybrid Control System with Context Model

Practice of biotechnological processes automation [13], [14] shows that the automated plant may be represented as hybrid state machine [15]. Such hybrid system is described with two sets:

- Ω – set of continuous state variables. Range of those variables depends on the features of the considered system i.e. system constraints, measurements, activator's capabilities, etc.;
- Φ – set of events conditions enabling transitions between states. The Φ set can be divided into two subsets:
  - o Φu – subset of uncontrollable events – transitions caused by internal or external process behaviour but the control system can not prevent them;
  - o Φc – subset of controllable events – transitions which can be initiated and/or prevented by the control system, so those transitions can be used to obtain desired process state.

The state can change in two ways:

- instantaneously by discrete transition described by the sequences S of actions from source state to target state
- in a time pass according to a trajectory of the state variables change as a function f of input and output variables

Agents are responsible for transitions, therefore Φ describes the agent sensors and Ω describes common agent effectors. Apart from effectors used commonly by all agents, each agent possesses other effectors that are characteristic and dependant on the knowledge source.

The type of agent depends on the way the state changes and on the agent's knowledge. The agents are specified on a design level in an iterative process in addition to knowledge needed. Changes in system can be performed as discrete events when certain conditions defined in Φ are fulfilled. Control of a hybrid automaton is a complex task, which can be achieved with Multi-agent Systems (MAS). The proposed structure of control agents is hierarchical and consists of three layers: control, supervisory and expert.

The lowest layer consists of Control Agents. It implements all the required control algorithms for direct influence on process. Agents in control layer work in time-driven mode to be able to maintain time determined control algorithms. Uncontrollable events are treated as disturbances for Control Agent. The layer in the middle of the control system is formed by Supervisory Agent. Each agent may supervise a set of Control

Agents and monitor system performance quality. Supervisory Agent, partially utilising expert knowledge, is capable of more advanced state recognition and trajectory planning, and may decide when to switch the system into some Ωi state (or through some $\Omega i, \ldots, \Omega n$ states) to fulfil a given goal. The plan is then carried out by the specific Control Agent as a proxy. The top layer contains an Expert Agent. The Expert Agent's task is to supply Supervisory Agents with additional knowledge about the system. It is expected for them to work on abstract knowledge, not the specific realisation. Both supervisory layer and expert layer work in a event-driven mode, which in consequence means, that the multi-agent layered structure decomposes control system into two parts: time-driven (Control Agents) and event-driven (Supervisory and Expert Agents). Such decomposition allows for separate analysis and decreases complexity level of the control system.

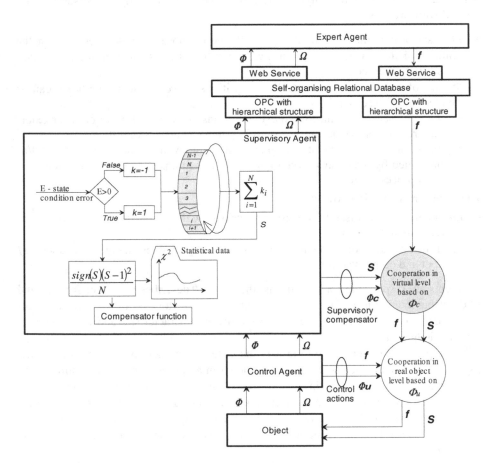

Fig. 1. The expert agent connected to hierarchical hybrid system by OPC server technology (www.opcfoundation.org)

As an example neutralisation process is considered. Usually, the system output fluctuates around a desired value. Statistically, it can be expected that the difference between time when system output is above or below the desired value is close to zero with some variance. Hence chi-square test (eq. 1) with one degree of freedom was applied to monitor difference of this time (S). N is window time length when the system is observed.

$$\chi^2 = \frac{(|S|-1)^2}{N} \tag{1}$$

It is assumed that the null hypothesis can be rejected when current chi-square value is greater than critical chi-square value with level $\alpha=0.1$ of significance. In case when null hypothesis is rejected, the Control Agent's goal is adjusted (fig 1). The goal is changed if necessary only once at the end of each time window allowing Control Agent and process to react.

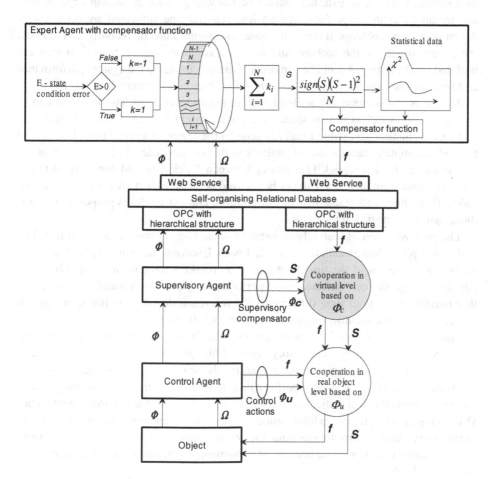

Fig. 2. Reconfigured hierarchical hybrid system

The idea of hierarchical hybrid system, shown in figure 1, previously described in [13], was extended by a compensator function based on statistical data analyses. The Control Agent directly influences the system to obtain the desired goal. Events from set **Φu** are uncontrollable from the Control Agent's point of view. However, because of cooperation in real object level some events from **Φu** set might be transferred to **Φc** set. It is possible because the Supervisory Agent is capable of more advanced state recognition and trajectory planning, hence it can switch the Control Agent to another state **Ωi**. When decision is made the compensator function chooses the **Ωi** state for Control Agent. It leads to better system performance in presence of uncontrollable events from lower control level point of view.

Some events might belong also to uncontrollable events subset **Φu** even from the Supervisory Agent's point of view. To avoid such situation, the Expert Agent is incorporated as the upper layer of hierarchical hybrid control system. The Expert Agent as a system interface to external sources of knowledge such as human experts. It is not meant to evaluate performance on-line but may be informed by Supervisory Agent about the problems. If this is the case, the system must be reconfigured to allow the expert to analyse the problem and to find a solution (fig. 2). Statistical tests are initiated on the Expert Agent level in order to deeply evaluate the system performance and for example to propose a new reference distribution for Supervisory Agent.

The proposed solution is possible if communication channel has sufficient capacity. In case of wireless transmission its capacity might be very limited, hence only part of information accessible locally is send to the remote Expert Agent. It is agreed to send incomplete data for expert analysis with the certain level of relative information distance, in a Kullback [12] sense, between local data and remotely obtained data. The maximum distance can not be crossed because expert analysis might not be valid. To monitor acceptable loss of information, context model is proposed which is discussed in the next chapter.

The proposed idea of hierarchical hybrid system (fig. 1) was applied and tested on real world pilot plant installation, sampled with 1-second interval. At the beginning, statistical properties of the time between zero crossings have been tested. Using expert knowledge the user-selected good performance period was found, and based on that maximum value of time differences (S) reference distribution function was set. Hence, finally time window length was chosen as 310 samples.

Figures 3 shows Control Agent state transitions **Ωi** (fig 3b) and corresponding chi-square values obtained by Supervisory Agent. Chi-square fluctuates around a value of zero and is a measure of performance quality; the smaller chi-square value obtained the better the quality of system performance (fig. 3a). After each 310 samples chi-square is checked and if critical chi-square value is crossed, the Control Agent's state **Ω** is changed. The process exhibits some nonstationary behaviour and in this case the Supervisory Agent helps to maintain the system output close to the desired value more efficiently. It is possible because of permanent cooperation between Supervisory and Control Agents.

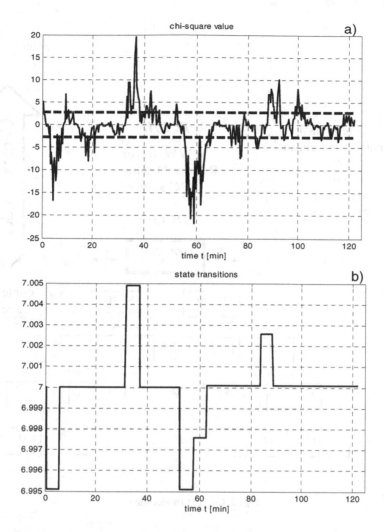

Fig. 3. Chi-square coefficients variation for Control Agent with activated Expert Agent (a), state transitions set for Control Agent (b)

3 Context-Based Statistical Model

Disturbance in communication might be an important problem especially when communication channel is overloaded or wireless communication is used. It is necessary for the proposed context model to measure distance between information on both ends of communication channel. To achieve this goal, communication can be described by a context tree and can be useful for transmission error detection, as was proposed e.g. by [10].

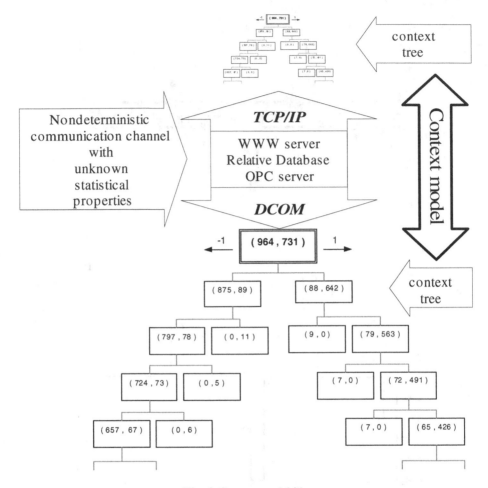

Fig. 4. Context model idea

The sample of N sequenced observations (X) is taken for monitored tree generation (fig. 4). In this case the transmitted symbol can be 1 or -1 hence each node of the tree contains two counters, one counter for each symbol. Constructing the tree recursively, next symbol is read from the sample going along the path defined by previously read symbols. For each node visited along the path, the counter value of the symbol $X(i)$ is incremented. The symbol can be read in context of the same symbol and as long as algorithm read the same element it goes deeper. When different symbol is obtained, going down is finished and in the next step the algorithm starts from the root. Depending on the read symbol the algorithm goes to the left or to the right. Values of counters in each node can be interpreted as frequencies of symbol occurrences for given context defined by nodes between root and the considered node. Obtained symbol frequencies for given context can easily be recalculated to probability [7]. Having two context-trees of probabilities, that describe the sample of sequenced observations at both ends of the communication channel, statistical measure can be used.

Commonly used measure for the relative information distance or the discrimination between two probability functions Q(x) and Q_0(x) (eq. 2) was given by Kullback [12].

$$KL\big(Q(x),Q_0(x)\big)=\sum_{x\in X}Q(x)\log\frac{Q(x)}{Q_0(x)}\geq 0 \qquad (2)$$

Kullback [12] has shown that the KL distance between a d-category multinomial distribution $Q_0(x)$ and $Q(x)$ is asymptotically chi-squared distributed with *d-1* degrees of freedom.

Usually unknown $Q(x)$ is estimated by $\hat{Q}(x)$ obtained by dividing counter of symbol in a context tree node by the total counter of symbol in the whole sample. In this case *d* refers to the number of context tree nodes taken into account increased by one.

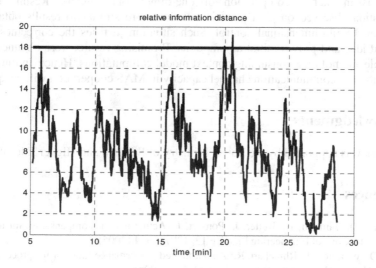

Fig. 5. Relative information distance values with presence of communication disturbances

Over wireless network only one measurement per five samples of data was send. Because of that, even if there is no communication problem there is a distance between local and remote information (fig 5). Because only a set of N measurements was taken into account, the relative information distance is multiplicatively proportional to this number of measurements.

Context model allows validating the coming data and flags error if the acceptable value of distant measure is not satisfied. The boundary value was set to 18 according to Expert Agent knowledge and it can be interpreted as a certain level of information significance based on chi-square probability density function. Context model flags error for the process time around 20min. This is information to Expert Agent that this data can not be considered and any correction function should not be taken because it observes communication disturbances as process fluctuations.

4 Conclusions

This paper presents Multi-Agent System for hierarchical hybrid control system for the local or remote monitoring and maintenance of continuous process. Using simple chi-square test it is possible to monitor quality of the process performance. When the obtained chi-square value is out of the assumed boundary condition, the system is influenced by change in Control Agent state. In order to avoid the influence of the communication channel, capacity or transfer data delay on remote side the context model is used. If distance between front-end and back-end system information is too large the error is flagged. It means that the obtained data is disturbed too much and using this data for any process modification should not be allowed. The proposed system was applied and tested on laboratory installation. Reconfiguration of Multi-Agent System is in fact a co-operation of one-dimensional agents. Results of such co-operation observed on pilot plant installation were similar to results obtained by complicated multidimensional control. Such situation justifies the correctness of the proposed idea and proves initial assumptions. Promising results were obtained due to the simple and not time consuming context model computations. Hence, in future work optimisation of communication channel capacity in MAS cooperation will be applied.

Acknowledgments

This work was supported by the Polish Ministry of Science and Higher Education.

References

1. Sauter, D., Jamouli, H., Keller, J., Ponsart, J.: Actuator fault compensation for a winding machine. Control Engineering Practice 13, 1307–1314 (2005)
2. Li, Q., Whiteley, J., Rhinehart, R.: An automated performance monitor for process controllers. Control Engineering Practice 12, 537–553 (2004)
3. Choinski, D., Nocon, W., Metzger, M.: Application of the Holonic Approach in Distributed Control Systems Design. In: Mařík, V., Vyatkin, V., Colombo, A.W. (eds.) HoloMAS 2007. LNCS (LNAI), vol. 4659, pp. 257–286. Springer, Heidelberg (2007)
4. Taber, R., Yager, R., Helgason, C.: Quantization effects on the equilibrium behaviour of combined fuzzy cognitive maps. International Journal of Intelligent systems 22, 181–202 (2007)
5. Stebel, K., Choiński, D.: Programmable on-line performance monitor integrated with plant by OPC standard Programmable Devices and Systems PDeS 2006, Brno, Czech Republic (2006)
6. Rissanen, J.: Fast Universal Coding with Context Models. IEEE Transaction on Information Theory 45(4), 1065–1071 (1999)
7. Rissanen, J.: A universal data compression system. IEEE Transactions on Information Theory IT-29(5), 656–664 (1983)
8. Weinberger, M., Rissanen, J., Feder, M.: A universal finite memory source. IEEE Transactions on Information Theory 41(3), 643–652 (1995)
9. Willems, F., Shtarkov, Y., Tjalkens, T.: The context-tree weighting method: basic properties. IEEE Transactions on Information Theory 41(3), 653–659 (1995)

10. Ben-Gal, I., Herer, Y., Raz, T.: Self-correcting inspection procedure under inspection errors. IIE Transactions 34, 529–540 (2002)
11. Ben-Gal, I., Morg, G., Shmilovici, A.: Context-Based Statistical Process Control: A Monitoring Procedure for State-Dependent Processes. Technometrics 45(4), 293–311 (2003)
12. Kullback, S.: Information Theory and Statistics. Wiley, New York (1959)
13. Choiński, D., Nocoń, W., Metzger, M.: Multi-agent System for Hierarchical Control with Self-organising Database. In: Nguyen, N.T., Grzech, A., Howlett, R.J., Jain, L.C. (eds.) KES-AMSTA 2007. LNCS (LNAI), vol. 4496, pp. 655–664. Springer, Heidelberg (2007)
14. Davidsson, P., Wernstedt, F.: Software agents for bioprocess monitoring and control. Journal of Chemical Technology and Biotechnology 77, 761–766 (2002)
15. Lynch, N., Segala, R., Vaandrager, F.: Hybrid I/O Automata. Inf. and Comp. 185, 105–157 (2003)

NN-Based Multiagent Platform for Testing Investing Strategies

Darius Plikynas

Vilnius Management School, J. Basanaviciaus 29a, Vilnius, Lithuania
darius.plikynas@vva.lt

Abstract. The proposed framework constructs an intelligent agent composed of a few trained neural networks, each specialized to make investment decisions according to the assigned investment strategies. The artificial investing agents are not only forecasting but also managing their portfolios. A novel simulation platform designs a multi-agent system, which is composed from those intelligent agents. Agents are trained to follow different investing strategies in order to optimize their portfolios. The proposed multi-agent portfolio simulation platform gives us an opportunity to display the multi-agent competitive system, and find the best profit-risk performing agents, i.e. investing strategies. In sum, simulations have shown that the proposed NN-based multi-agent system produces similar results in terms of e.g. wealth distribution compared with empirical evidence from the actual investment markets' behavior.

Keywords: Neural Networks, Investing Strategies, Multi-agent systems.

1 Introduction

In the economy in which agents have heterogeneous beliefs or heterogeneous learning rules, structure on long-run prices arises from the forces of market selection. It has long been assumed that those with better beliefs will make better decisions, driving out those with worse beliefs, and thus determining long run assets prices and their own welfare [12], [5].

In the current approach, we hold that a very effective way to employ a passive strategy with common stocks is to invest in an indexed portfolio. An increasing amount of mutual fund and pensions fund assets can be described as passive equity investments [3]. Using index funds, these asset pools are designed to duplicate, as precisely as possible, the performance of some market indexes.

Regarding the methodical aspect of the proposed novel technique (see the following sections), reinforced learning seems to be a close match, as it is a sub-area of machine learning, concerned with how an agent should act in an environment so as to maximize some notion of long-term reward.

Recent extensive analysis of short-term returns has shown that price differences, log-returns, and rates of return are described extremely well by a truncated Levy distribution [1]. In this respect, proposed research is also significant, since it provides a simulation framework within which empirical laws like power-law and Levy distribution can be observed and studied within the parameter's space of the simulated model [9], [12], [10].

D. Dochev, M. Pistore, and P. Traverso (Eds.): AIMSA 2008, LNAI 5253, pp. 12–21, 2008.
© Springer-Verlag Berlin Heidelberg 2008

In fact, almost all proposed systems are either employed for various technological or robotics applications and very few of them like the Multilayered Multi Agent Situated System [14] (MMASS), Agent-Based Computational Demography (ABCD) or Agent-Based Computational Economics (ACE Trading World application: simulation of economic phenomena as complex dynamic systems using large numbers of economic agents involved in distributed local interactions) are suitable for programmable simulations of social phenomena.

The paper is organized as follows. Section 2 gives a brief understanding of the application domain, whereas, section 3 addresses the experimental setup, giving us more details about the design and implementation of the proposed approach. Besides, it offers an overview of operational procedures, i.e. how the agent operates in the investment environment. Section 4 summarizes the findings and discusses further implications.

2 Application Domain

According to the above discussion, this section discusses research issues related with the investment market, which is represented using a real data set from S&P500, Dow Jones Composite and US Treasury Bonds indices (value, open/close values, trade volumes, volatility etc). Each agent, according to his investment strategy, is optimized for the best possible forecasting of index fund fluctuations. This system of agents is targeted to find an optimal risk-profit assessment on the efficient frontier according to Markowitz theory [8].

The procedure of building a portfolio of financial assets is following certain steps. Specifically it identifies optimal risk-return combinations available from the set of index funds and non-risky assets, following Markowitz efficient frontier analysis.

In the proposed model, each agent is forming his portfolio composed from the three investment options (in the next version the number of portfolio assets will not be bound to only three assets): two index funds and a Treasury bond, see above. This is a minimum set of assets needed for the design of a profit-risk adjusted portfolio, where the Treasury bond is assumed as an almost risk free asset.

It is important to notice that during simulation, an agent can change one owned asset to another, having associated transaction costs, but no liquidity issues. The transfer of an owned asset is via the current selling market price, and the buying of another asset is via the current market buying price. Therefore, there is no need to design any auctioning mechanism, and address liquidity issues.

If our simulations are showing that proposed NN-based multi-agent system produces similar results in terms of wealth distribution comparing with empirical evidences from the real investment markets' behavior, as described by Pareto wealth and Levy returns distributions [6], then we can conclude, that the proposed model is close to reality.

So, Pareto discovered that at the high-wealth range (social agent's wealth is in the high-wealth range if it is over some typical value characteristic for that particular society. In the current case, W_0 was chosen as 20% of the wealthiest out of all population), wealth (and also income), are distributed according to a power-law distribution. This law became known as the Pareto distribution or Pareto law. The Pareto distribution is given by the following probability density function

$$P(W) = CW^{-(1+\alpha)} \quad for \ W \geq W_0, \tag{1}$$

where W stands for wealth, $P(W)$ is the density function, W_0 is the lower of the high-wealth range, C is a normalization constant, and α is known as the Pareto constant. Pareto's finding has been shown in numerous studies to provide an excellent fit to empirical wealth distribution in various countries [1].

Pareto distribution is related to wealth distribution, which is acquired by the cumulative process of price differences. Meanwhile, distribution of stock returns, especially when calculated for short time intervals, are not suited to the normal distribution. Rather, the distributions of stock returns are leptokurtic or "fat tailed". Mandelbrot [6] has suggested that log-returns are distributed according to the symmetrical Levy probability distribution defined by

$$L_{\alpha_L}^{\gamma}(x) \equiv \frac{1}{\pi} \int_0^{\infty} \exp(-\gamma\Delta t q^{\alpha_L}) \cos(qx) dq \tag{2}$$

Above, is estimated Levy probability density function at x; α_L is the characteristic exponent of the distribution; $\gamma\Delta t$ is a general scale factor; γ is the scale factor for $\Delta t=1$, and q is an integration variable [7]. Recent analysis of returns has shown that price differences, log-returns, and rates of return are described extremely well by a truncated Levy distribution [5], [6].

The next section outlines how the actual experimental setup is constructed. It outlines the main design and methodical issues, and also explains how the NN-based multi-agent system is configured to meet the intended empirical application goals.

3 Experimental Setup

In the proposed model, each investor is represented as an agent. Depending on the simulation parameter set, the total number of agents can be chosen, varying from a few to even thousands. In fact, one could simulate as many agents as possible in order to get power law distributions of their portfolios, but we are not intended to simulate the behavior of the whole market. Asset prices are given as exogenous factors, i.e. our subsystem of agents composes just a small fraction in the overall market, which doesn't significantly influence the overall market behavior and asset prices respectively.

A completely different approach for actualizing the market feedback via e.g. an auctioning mechanism, which would give price distribution involving market makers, matching demand and supply respectively [2]. At this stage of research we don't pretend to build the whole market mechanism.

In a methodical sense, agents are modeled using multilayer perceptron (MLP) backpropagation neural networks [NN] with fast Levenberg-Marquardt learning algorithm [4]. The application is compiled using MatLab environment and NN Toolbox in particular. To summarize, the virtual lab software platform is based on Condor, Tomcat server and Oracle XE DB. The Win2003 server is running on Intel Core2 Quad Q6600 2.4 GHz/1066FSB/8MB/RAID2*500GB HDD hardware server platform (integrated 4 processors).

Concerning the "internals" of an agent, each investment asset is analyzed by a dedicated NN (neural network). This means, that an agent is a modular entity composed from a number of different NNs, where separate NNs are dealing with different market data, e.g. time series analyses, forecasting, recognition (when and which investment strategy to employ) etc.

For the effective implementation of the model there is an additional algorithmic layer of procedures above NNs. It deals with agents' portfolios and operational decisions. Thus, profitable decision-making is not solely based on the NNs performance alone. Therefore, the second layer assists as procedural shell, which maintains market interventions in the operational sense, and also feedback, and agent's own capital estimates.

The employed connectionist approach is working rather like a reactive system, implementing direct mappings from perception to action that avoid expensive intermediate steps of representation and reasoning. This loosely follows the idea that intelligent systems should be constructed from large numbers of separate agents, each with percepts, actions, and goals of its own.

Now let's focus on the data setup. All real market data is collected from the publicly available yahoo/finance data source, where we got our S&P500, Dow Jones Composite and US Treasury Bonds indices data sets (opening, closing, volume, average price, etc). This data is checked for consistency, but is not de-trended as we definitely wanted all periodic and non-periodic fluctuation to be present for NNs to learn.

The reader should also be aware that the size of the data set plays a crucial part in the NN learning and testing stage (in the current case, over 6581 business day observations were collected beginning from the 1981-ies; 1581 observations were dedicated for the NN testing stage).

As a standard procedure, optimally designed NN went through the initial training and validation stage. Only afterwards were they ready for the simulation stage, i.e. forecasting and recognizing. NN learned to forecast the market behavior (index fluctuations, trade volume, market timing moments, specific patterns described by the technical analyses etc) and recognize investment strategies.

In general, the more data we have, the better we are in forecasting and recognizing if overfitting of NN is ruled out. In the model, the experimenter can generate a set of NNs with differing internal configurations (learning algorithms, topology, number of epochs, biases, etc), but for simplicity at this stage we were using only one type of optimized NN, which has been tailored for all agents later.

Once we have NN optimized for forecasting of the particular data set, we can make use of it for recognizing different investing strategies by simply teaching it to do so. In the current study, we applied seven investing strategies. Therefore, one and the same NN configuration has been trained with different investing strategies obtaining appropriate NNs, each tailored for a different type of investing strategy (naïv, buy and hold, bull trading, support&resistance, contrary, risk averse, filter rule).

Following the previous discussion, we are investigating how traditional strategies are operating in our multi agent model. As we can see from Fig. 1, all strategies are wrapped in the same principal simulating scheme.

Actually, the stage of forecasting and recognition usually go one after another, though, in this particular model, the NN learns forecasting and recognizing at the same time (the initial data set has been preprocessed to add needed technical indicators for the recognition of strategies). MSE-mean square error and R-square were used mainly as performance criteria, comparing with the standard multivariate regression. As a matter of fact, a novel type of NN has been designed, which has a unique input weight matrix designed [11].

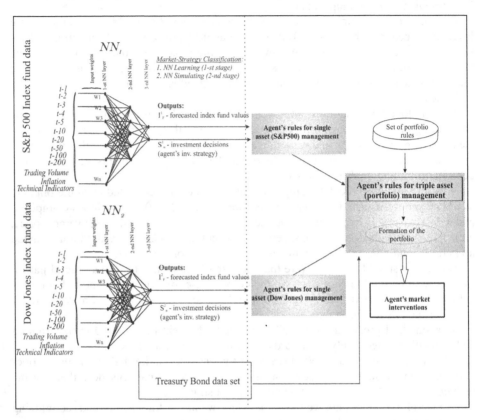

Fig. 1. Agent's principal simulating scheme

In conclusion, a model was designed that could relate the domain space (real investment market data) with solutions space (simulated agents' investment decisions). A modular NN-based approach was adopted for simulation of complex strategies existing in the investment markets. An agent as a backpropagation NN (each NN for an asset) is aimed at providing (i) forecasting of asset values, (ii) recognizing market situations, (iii) making timely market interventions.

In practical terms, each investing strategy has been facilitated by a set of appropriate technical indicators and procedures, which once satisfied, give incentives for the appropriate market interventions.

The leading experts in the field [3], [2], [13] have published a number of papers providing support for some basic technical indicators, e.g. MACD – moving average convergence/divergence, W%R – Williams' oscillator, MFI – money flow index, WMA – weighted moving average etc. Following their advice, in the presented model, we have adopted them as a benchmark.

Therefore, our model addresses not only trend issues, and selling/buying volumes, but also more sophisticated indicators like filtering rules etc. We also have included the estimates of dispersion σ, which is very important for some investing strategies, (mostly used by risk averse institutional market players).

To the author's knowledge, there is no experiment in the field that focuses on the market mechanisms in such a virtual setting. The management of the agents' operational procedures thus gives us an insightful breakthrough, and unique tool for the simulation of investment strategies. Simulations' results are discussed next.

4 Results and Conclusions

Hence, our agents simulate the above mentioned investment strategies as we can see in Fig. 2, where the blue line represents the NN performance during simulation of investment decisions following a particular investment strategy (S_2). Even from the visual impression (see lower testing graphs in Fig. 2), it is clear that NNs are more or less following real investors' behaviors, and can be delegated to perform routine trading actions in the market.

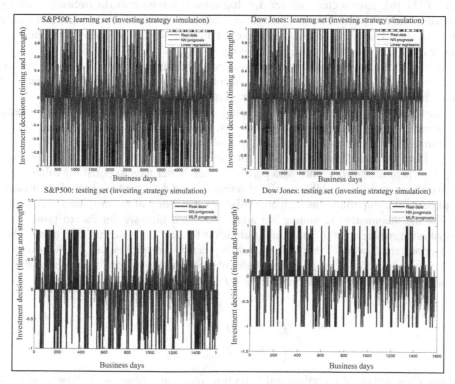

Fig. 2. Investment strategy (S_2) simulation during the NN learning and testing stages (agent's decisions vary between [-1;1], where 1 means all funds reallocation to this specific asset, and -1 means selling all funds allocated in this specific asset)

The ultimate long-term goal for each agent differs, but in general each of them is looking for the initial capital maximization (initial capital equals 1 for each agent). The model is enriched with some heuristics, based on the nonlinear properties and generalization features of neural networks, which, as numerous studies have shown [5], [11], [4], have systematically outperformed the traditional regression methods so popular in finance.

What is next? Now we are ready to go into the real market and test our simulated investment strategies. We generate a chosen number of agents for each investment strategy, allocate them initial capital (initial capital is shared among the three assets proportionally 0.33:0.33:0.33) and start simulation.

Certainly, before doing that we have to parameterize the simulation model itself. After parameters are set and simulation has started, agents are reallocating their funds, i.e. buying and selling assets at the market price (there are no liquidity considerations involved). In this way, they are designing their own portfolio. The final goal is to find out which particular strategies have led to the best performance in terms of accumulated wealth and risk performance.

Empirical tests, whether simulated log-returns are distributed according to the symmetrical Levy probability distribution defined by Eq. 2 or truncated Levy distribution [5], [6], gave a clear answer, i.e. log-returns are described extremely well by the truncated Levy distribution. Though, that is more obvious for portfolio than single asset returns. Empirical results also gave an insight which simulated strategies are closer to the real investment market behavior.

Consequently, aggregated returns estimated for all agents give us portfolio capital distributions, as it is depicted in Fig. 3. According to numerical estimates, which are given below, it perfectly resembles real market behavior.

Starting simulation from the initial capital 1, at the end of simulation we have peak frequencies in the range 1.36±0.3 (conf. int. 95%) of initial portfolio value, which well matches with the real market participants' performance [1-3].

In fact, simulated portfolios are following Pareto wealth distribution law (see Eq. 1). There are some agents who perform much better having portfolio values well above the average. The ability of the simulating model to produce such distributions is of particular importance as it clearly validates feasibility of the proposed approach.

For the quantitative reasoning we adapt truncated high-wealth Pareto power law distribution, where $w_0 = 0.7$ (the minimum limit for the "high-wealth" end of distribution), constant $C = \alpha \, w_0^{\alpha}$ (because probability density integral equals 1). We assume, that obtained portfolios are distributed according to the Pareto law, if simulated distribution F_D matches Pareto distribution $F_D \sim Pareto(C, w_0)$.

The null hypothesis:

$$\begin{cases} H_0 : F_D = F_{Pareto} \\ H_1 : F_D \neq F_{Pareto} \end{cases}, \tag{3}$$

where F_D – truncated (simulated) distribution function (values with "low-wealth" < w_0 are filtered out). We adapt Kolmogorov-Smirnov test with standard significance level (5%)

The main question addressed by practitioners is, whether particular investment strategy is dealing well in the particular market. They are not only interested in terms of returns (profits), but also in terms of associated risk, as it is described by CAPM (Capital Asset Pricing Model). Proposed simulation environment gives such an opportunity. It estimates risks (MSE, VAR, CVAR etc) associated with each investment strategy, as an example, see Fig. 3.

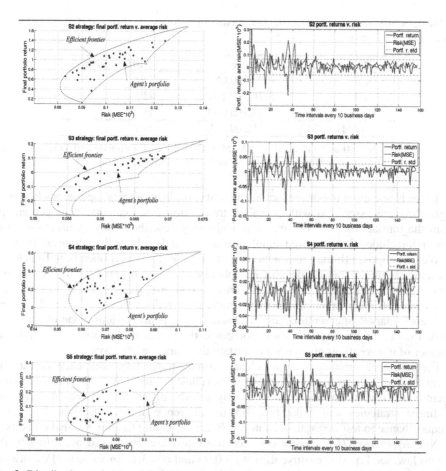

Fig. 3. Distributions of risk-profit portfolios due to investment strategies S2-S5 (see left side) and time dependant correlations of returns v. risk for the sample agents chosen from the appropriate strategy set (see right side). Final portfolio returns (profit) = final value – initial value.

As we can see from the Fig. 3 (see left side), simulation model clearly demonstrates how well each investment agent is doing in terms of risk-profit performance (blue dots represent individual agents). An investment analyst may clearly see the difference between different investment strategies and make probabilistic inference about outcomes. It gives a unique chance to test investment strategies before applying them in the real markets.

From the results above, see Table 1, we can infer that simulated portfolio distribution fits truncated Pareto power law distribution [1], [3], [5], [6]. Though, the truncated portfolio data set is relatively small and p values are not very high. Therefore, for stronger evidence we need at least 10,000 or bigger portfolio set.

Table 1. Null hypothesis: testing whether obtained portfolios are distributed according to the Pareto law (see Fig.3)

Index*	P-value	Null hypothesis
SP	0.937±0.062	H0
DJ	0.767±0.191	H0
TB	0.985±0.003	H0
SP+DJ+TB	0.552±0.250	H0

* - SP, DJ and TB stand for S&P500, Dow Jones Composite and Treasury Bonds (30 year) accordingly; whereas SP+DJ+TB – portfolio values.

There is one more important factor. In reality, ordinary agents' operations like buying/selling are associated with so called transaction costs calculated as percentage from the transaction sum. For the empirical study described above transaction costs were not included, but proposed simulation approach does have this opportunity.

We can impose arbitrary transaction cost rate for each market operation. Hence, the implementation of transaction cost completely changes the profitability levels and wealth distribution accordingly. In sum, research shows, that even very low transaction rates (>0.3% market deal price) significantly reduce investors' wealth if high frequency market interventions are involved. There is an obvious trade-off between market interventions rate and transaction costs: the higher transaction costs, the lower number of interventions are needed to sustain the same wealth level.

For instance, simulation model clearly demonstrates how well each investment agent is doing in terms of risk-profit performance (blue dots represent individual agents). An investment analyst may clearly see the difference between different investment strategies and make probabilistic inference about outcomes. It gives a unique chance to test investment strategies before applying them in the real markets.

Null hypothesis H_0 (whether obtained portfolios are distributed according to the Pareto law, see Eq. 5) is positive then $W_0 = 0.95$ and $\sigma = 3.17$ (p-value 0.91). Notwithstanding obtained results, additional research needs to be done to examine more representative samples of investment strategies and bigger set of portfolios accordingly.

Simulation gives us an opportunity to display the multi-agent competitive system, and find the best profit-risk performing agents, i.e. investing strategies. This gives an edge over costly and time consuming manual experimentation in the diversified investment markets with real-life index funds (or stocks) and non-risky financial assets. Unfortunately, the limited space available for this paper does not allow to demonstrate plenty of validation and robustness results. The first version of the model is already available via the web, using our virtual lab environment, which is intended to enable others to access the model and experiment with it (http://vlab.vva.lt/; Investment Simulation_v1).

Acknowledgments

The author is grateful for the collaborative help and advice from numerous colleagues associated with the other stages of the current project. This work is a part of the project identification No. BPD2004-ERPF-3.1.7-06-06/0014.

References

1. Blume, L.E., Durlauf, S.N. (eds.): The Economy as an Evolving Complex System III. The Santa Fe Institute in the sciences of complexity. Oxford University Press, New York (2006)
2. Darley, V., Outkin, A.V.: A NASDAQ Market Simulation: Insights on a Major Market from the Science of Complex Adaptive Systems, p. 152. World Scientific, Singapore (2007)
3. Jones, C.P.: Investments: Analysis and Management, 8th edn. John Wiley & Sons, Chichester (2002)
4. Haykin, S.: Neural Networks: A Comprehensive Foundation, 2nd edn. Prentice-Hall, New Jersey (1999)
5. Hoffmann, A.O.I., Jager, W., Von Eije, J.H.: Social Simulation of Stock Markets: Taking It to the Next Level. Journal of Artificial Societies and Social Simulation (JASSS) 10(2), 7 (2007), http://jasss.soc.surrey.ac.uk/10/2/7.html
6. Levy, M., Solomon, S.: New Evidence for the Power-Law Distribution of Wealth. Physica A 242 (1997)
7. Mantegna, R.N., Staney, H.E.: An Introduction to Econophysics: Correlations and Complexity in Finance. Cambridge University Press, Cambridge (2000)
8. Markowitz, H.M.: Portfolio Selection:Efficient diversification of investments. Blackwell Publishers, Malden (1991)
9. McCauley, J.L.: Dynamics of Markets: Econophysics and Finance. Cambridge University Press, Cambridge (2004)
10. Moss, S., Edmonds, B.: Towards Good Social Science. Journal of Artificial Societies and Social Simulation 8(4) (2005), http://jasss.soc.surrey.ac.uk/8/4/13.html
11. Plikynas, D.: Proceedings of the First World Congress on Social Simulation, WCSS 2006, August 21-25, Kyoto University, Kyoto, Japan (2006)
12. Sapienza, M.: Do Real Options perform better than Net Present Value? Testing in an artificial financial market. Journal of Artificial Societies and Social Simulation 6(3) (2003), http://jasss.soc.surrey.ac.uk/6/3/4.html
13. Sharpe, W.F.: Capital asset prices: A theory of market equilibrium under conditions of risk. Journal of Finance 19(3), 425–442 (1964)
14. Testafsion, L., Judd, K.: Handbook of computational economics: agent- based computational economics, vol. 2, p. 828. North Holland, Amsterdam (2006)
15. Wooldridge, M.: Multiagent Systems, p. 346. John Wiley & Sons Ltd, Chichester (2002)

Towards Agent-Oriented Relevant Information

Laurence Cholvy[1] and Stéphanie Roussel[1,2]

[1] ONERA Centre de Toulouse, 2 avenue Edouard Belin,
31055 Toulouse, France
[2] ISAE, 10 avenue Edouard Belin,
31055 Toulouse, France

Abstract. In this paper, we give some elements about agent-oriented relevance. For that, we define a new modal operator R_a^Q so that $R_a^Q \varphi$ means that φ is relevant for agent a concerning the request Q. We discuss properties of this new operator, as well as its extensions and limits.

Keywords: information relevance, multi-agent systems, information need.

1   Introduction

The general context of this work[1] is modelling multi-agent systems i.e, systems in which some entities have to cooperate in order to achieve a given task that none of them would be able to achieve alone. In such a context, these cooperating entities (henceforth called agents) have to communicate i.e exchange information, in particular to share a common view of the current environment. However, in most systems, communication channels saturate if any agent sends to others all pieces of information he possesses. Thus, in order to be efficient, only a few information should be exchanged within agents. More precisely, the information exchanged should be the very one needed by agents in order to achieve their goals, that means *relevant information.*

Characterization of relevant information in the context of information exchange is the main object of this paper.

Following Borlund [14], relevance concept can be addressed following two different kinds of approaches: System-oriented approaches and Agent-oriented ones. System-oriented approaches analyse relevance in terms of topicality, aboutness, degrees of matching between a piece of information and a request, or independence. System-oriented approaches are the basis of Information Retrieval area ([8,6,12]) and Relevant Logics ([15,4]) domain. They do not try to define a relation between some information and an informee like Agent-oriented approaches which, on the other hand, analyse relevance in terms of agent's utility, informativeness for the agent... aiming at defining relevant information according to agent's information need. According to Floridi [9], Agent-oriented Relevance lacks definition: *"The current situation can be summarised thus : some philosophical work has been done on several formal aspects of system-based or causal relevance, but the key question, namely what it means for some*

---

[1] This work has been realized during PhD studies granted by DGA (Direction Générale de l'Armement).

D. Dochev, M. Pistore, and P. Traverso (Eds.): AIMSA 2008, LNAI 5253, pp. 22–31, 2008.

information to be relevant to some informee, still needs to be answered. We lack a foundational theory of agent-oriented or epistemic relevance. The warming up is over. The time has come to roll up our sleeves. ". Floridi has developed an interpretation of epistemic relevance. It is based on an analysis of the degree of relevance of some semantic information to a rational informee as a function of the accuracy of the information understood as an answer to a question, given the probability that this question might be asked by the informee.

In this present paper, our aim is to contribute, like Floridi, to the study of agent-oriented relevance. For doing so, we use a widely respected model of agents, the *belief-desire-intention* model (BDI) [13]. This model assumes that an agent is characterized by its mental attitudes, mainly belief, desire and intention. Most formal models based on BDI are modal logics whose modal operators are used to represent the different mental attitudes. The semantic of those of operators are generally given by the possible world semantics [5]. Using this kind of formalism, we aim at defining a new modal operator to represent agent-oriented information relevance.

This paper is organized as follows. Section 2 presents the formal framework we base our work on. Section 3 deals with relevance defined according to an agent's information need. In section 4, we consider a multi-agent context. Section 5 addresses a more general case of information need. Finally section 6 concludes this paper.

2 Formal Framework

The formal framework on which our work is based on is the one defined in [1]. It is a propositional multimodal logic[2] whose modal operators are belief and intention.

Let $\mathcal{A}$ be the set of agents.

- **Belief.** A modal operator of belief B_a is associated to every agent a in $\mathcal{A}$. The formula $B_a A$ is read "agent a believes that A". We will write $Bif_a A$ instead of $B_a A \vee B_a \neg A$. $Bif_a A$ is read "agent a knows[3] if A is true or false". Belief operator is ruled by KD45 axioms. Thus, in this framework, agents do not have inconsistent beliefs and that they are aware of their beliefs and of their disbeliefs.
- **Intention.** In [1], intention is a primitive operator. A modal operator I_a is associated to every agent of $\mathcal{A}$. $I_a A$ is read "agent a intends that A". For this operator, nothing is supposed about logical consequence, conjunction or material implication. It is only postulated that: $\frac{A \leftrightarrow B}{I_a A \leftrightarrow I_a B}$.
- **Relation between belief and intention.** As in [1], we suppose strong realism, that means we consider that one does not intend a proposition if one does not believe that this proposition is false. This is expressed by : $I_a A \rightarrow B_a \neg A$. From this postulate, it has been proven that: $I_a A \rightarrow \neg B_a A$: an agent does not intend A as soon as he believes A (weak realism); $I_a A \rightarrow \neg I_a \neg A$; $B_a A \rightarrow \neg I_a A$; $\neg B_a A \rightarrow \neg I_a \neg B_a A$.

    Moreover, positive and negative intention introspection are supposed. I.e, $I_a A \rightarrow B_a I_a A$ and $\neg I_a A \rightarrow B_a \neg I_a A$.

---

[2] The logic used in [1] was a first-order multimodal logic. We do not need the first-order here.

[3] "Knowledge" is here used instead of "belief" because in natural language, we don't say that an agent wants to belief if something is true or false.

3   Relevance

3.1   Informal Definition

In this part, we want to characterize what makes an information relevant.

First of all, we consider that a piece of information is relevant according to an agent. We do not conceive a piece of information to be relevant without anyone to whom it will be useful. Secondly, we consider that relevance depends on an information need this agent has and we suppose that a request models this information need[4].

Considering these two assumptions only, we get what we call *intrinsic relevance*[5]. According to this kind of relevance, any piece of information which contributes to information need of the agent is relevant. Even false information can be relevant, which is questionable in the context of multi-agent systems.

If we add that a relevant information has to be true, then we get what we call *contingent relevance*. The term "contingent" is justified by the fact that this notion of relevance here highly depends on the current situation.

In this paper, we only focus on contingent relevance for a given agent concerning a particular information need. We will henceforth say "relevance". Thus, an information is relevant concerning an information need if the agent actually has got this information need, if the relevant information helps the agent answering his information need and if the information is true. This shows that some elements have to be seen in relevance definition: 1.the agent information need, 2.the agent beliefs (that allows him to answer his information need from the relevant piece of information) and 3. the truth value of the relevant piece of information.

3.2   Information Need

In what follows, a *request* is a formula without modal operators.

We consider an agent a that needs information. For a first approach, we consider that this need is quite simple and can be expressed the following way: "agent a wants to know if Q or if $\neg Q$", where Q is a request. Thus, an agent's information need can be expressed by $I_a Bif_a Q$. This formula is read "agent a intends to know if Q".

A few comments on this information need:

- $I_a Bif_a \varphi \rightarrow \neg B_a \varphi \wedge \neg B_a \neg \varphi$ is a theorem, i.e, an agent does not intend to know if φ is true if he already believes φ or $\neg \varphi$. To say it differently, this means that an information need corresponds to a lack of information ([16]).
- If φ is a tautology and ψ a contradiction, then $\neg I_a Bif_a \varphi$ and $\neg I_a Bif_a \psi$ are theorems. It means requests are not tautologies nor contradictions.
- To know if Q comes to know if $\neg Q$. In fact, $I_a Bif_a Q \leftrightarrow I_a Bif_a \neg Q$ is a theorem.

---

[4] This request may be explicitly asked by the agent or not.

[5] Thanks to Robert Demolombe for proposing the two terms contingent relevance and intrinsic relevance.

3.3   Formal Definition of Relevance

In this part, we give a definition for a new modal operator R_a^Q so that $R_a^Q \varphi$ means that information φ is relevant for agent a concerning request Q. When there is no ambiguity, we will say that φ is relevant.

In what follows, $\otimes$ is the exclusive disjunction generalized to n formulas i.e. if A_1, ... A_n are n formulas then $A_1 \otimes ... \otimes A_n$ is true if and only if $A_1 \vee ... \vee A_n$ is true and $\forall i, j \; \neg(A_i \wedge A_j)$ is true.

Definition 1. *Let a be an agent of $\mathcal{A}$, φ a formula, Q a request.*

$$R_a^Q \varphi \equiv I_a Bif_a Q \wedge (B_a(\varphi \to Q) \otimes B_a(\varphi \to \neg Q)) \wedge \varphi$$

In this definition, the three elements that have been given in the informal definition of relevance appear:

- **agent's information need:** $I_a Bif_a Q$. Relevance assumes an information need.
- **agent's beliefs:** $B_a(\varphi \to Q) \otimes B_a(\varphi \to \neg Q)$. From what he knows and with a relevant information φ, the agent can answer his information need Q, that means he can deduce Q or he can deduce $\neg Q$. Here, we have chosen the simplest way to represent this deduction: logical implication.
  Two points are important here:
  - First it is important that the agent uses his beliefs to answer his information need.

    For example, let's consider a doctor a that needs to know if his patient has disease d or not. This information need is formalized by $I_a Bif_a d$. Let's suppose that if the blood formula is normal (modelled by n), then the patient does not have the disease d.

    Even if the patient has got a normal blood formula, if the doctor does not know that $n \to \neg d$, then n is totally useless to answer $\neg d$. n is relevant only if the doctor knows that $n \to \neg d$.
  - $\otimes$: if an agent, from a formula φ, can deduce Q and $\neg Q$, it means that this formula does not really answer the request as it does not allow the agent to know which one of Q or $\neg Q$ is his answer. With the $\otimes$ operator, we prevent this case to happen.
- **truth value of φ:** only true information is relevant as we deal with contingent relevance.

3.4   Properties

In this part, let us take a an agent of $\mathcal{A}$, Q, Q_1 and Q_2 some requests, φ, φ_1, φ_2 some formulas. The following propositions are theorems of our logic.

Proposition 1

$$R_a^Q \varphi \to \neg B_a \varphi$$

If information φ is relevant, then the agent does not know it (otherwise, he would already be able to answer his own request).

Proposition 2

$$R_a^Q \varphi \rightarrow \neg B_a \neg \varphi$$

If information φ is relevant, then the agent does not believe that it is false. Indeed, if a believes $\neg\varphi$, his belief set is inconsistent.

Proposition 3. *Let $*$ be an operator of belief revision satisfying AGM postulates [2]. Let Bel_a be agent a's belief set and $Bel_a * \varphi$ be a's belief set after revising Bel_a with φ by $*$. Then*

$$R_a^Q \varphi \rightarrow ((Bel_a * \varphi) \rightarrow Q) \otimes ((Bel_a * \varphi) \rightarrow \neg Q)$$

This proposition shows that the deduction operator that we have chosen, the implication, corresponds to a "basic" belief set revision operator. If a piece of information is relevant concerning a request Q, then the agent, by adding this information to his current belief set can deduce either Q or $\neg Q$.

Proposition 4

$$I_a Bif_a Q \rightarrow R_a^Q Q \otimes R_a^Q \neg Q$$

If agent a has a request Q then either Q or $\neg Q$ is relevant.

Proposition 5

$$(Q_1 \leftrightarrow Q_2) \rightarrow (R_a^{Q_1} \varphi \leftrightarrow R_a^{Q_2} \varphi)$$

For a given agent, two requests equivalent to each other have the same relevant information.

Proposition 6

$$R_a^Q \varphi \leftrightarrow R_a^{\neg Q} \varphi$$

The two requests Q and $\neg Q$ are equivalent for relevant information research.

Proposition 7

$$\neg(\varphi_1 \wedge \varphi_2) \rightarrow \neg(R_a^{Q_1} \varphi_1 \wedge R_a^{Q_2} \varphi_2)$$

Two contradictory information are not both relevant for an agent concerning some requests.

Proposition 8

$$R_a^Q \varphi \rightarrow \neg R_a^Q \neg \varphi$$

If φ is relevant for an agent concerning a request then $\neg\varphi$ is not relevant for this agent concerning this request.

Proposition 9

$$R_a^Q \varphi \rightarrow \neg B_a R_a^Q \varphi$$

If φ is relevant for an agent concerning a request then the agent does not know it. In fact, if a knows that an information is relevant then he knows that it is true. If he knows that the information is true, then he can answer his request. And as soon as he can answer his request, he does not have his information need anymore.

Notation. In what follows, we will note $B_a(\varphi_1, \varphi_2/Q)$ instead of $\neg(B_a(\varphi_1 \to Q) \wedge B_a(\varphi_2 \to \neg Q)) \wedge \neg(B_a(\varphi_1 \to \neg Q) \wedge B_a(\varphi_2 \to Q))$. This formula means that a believes that φ_1 and φ_2 do not allow to deduce contradictions concerning Q.

Proposition 10

$$B_a(\varphi_1, \varphi_2/Q) \to (\varphi_2 \wedge R_a^Q \varphi_1 \to R_a^Q(\varphi_1 \wedge \varphi_2))$$

Proposition 11

$$B_a(\varphi_1, \varphi_2/Q) \to (R_a^Q \varphi_1 \wedge R_a^Q \varphi_2 \to R_a^Q(\varphi_1 \vee \varphi_2))$$

Those two propositions show the previous definition of relevance characterizes too many relevant information.

For example, supposing that the doctor knows that $n \to d$ we have seen that n is a relevant information for doctor a who has a request d. If we consider that r is true (for instance r means that it is raining) then $n \wedge r$ is relevant for a concerning d. This is true because $n \wedge r$ contains the relevant element that really answers the doctor's request. However, we would like to say that n is more relevant than $n \wedge r$, because r is not needed to answer the doctor's need. Thus, the problem we face is to consider a hierarchy in relevance in order to characterize the most relevant information. The next section addresses this point.

3.5   Hierarchy in Relevance

Let $\mathcal{R}_a^Q$ be the set of relevant formulas. We consider successively the case when such formulas are clauses, cubes, or more general formulas.

Clauses
Let suppose that $\mathcal{R}_a^Q$ is a set of clauses.

Definition 2. *Let φ_1 and φ_2 be two clauses. We define $\varphi_1 \leq_{Cl} \varphi_2$ iff $\vdash \varphi_2 \to \varphi_1$.*

Proposition 12. $\leq_{Cl}$ *is a preorder.*

If we consider that relevant information are clauses, the most relevant are the maxima with this preorder. Indeed, the maxima with this preorder are the clauses that are not subsumed by any other. Those maxima formulas are the most precise. Relevance degrees can also be defined by taking successive maxima.

For example, let us suppose that $\mathcal{R}_a^Q = \{p, q, p \vee q, r \vee s, p \vee q \vee r\}$. Thus, the most relevant information set is $\{p, q, r \vee s\}$. Then, we have $\{p \vee q\}$ and $\{p \vee q \vee r\}$.

Cubes
Let suppose that $\mathcal{R}_a^Q$ is a set of cubes (conjunction of literals).

Definition 3. *Let φ_1 and φ_2 be two cubes (literals conjunction). We define: $\varphi_1 \leq_{Cu} \varphi_2$ iff $\vdash \varphi_1 \rightarrow \varphi_2$*

Proposition 13. $\leq_{Cu}$ *is a preorder.*

If we consider that relevant information are cubes, the most relevant are the maxima with that preorder. Indeed, the maxima of this preorder are prime implicants [6] [7] of Q and $\neg Q$. This corresponds to information that contain only necessary elements to answer the information need. Relevance degrees can also be defined by taking successive maxima.

For example, let us suppose that $\mathcal{R}_a^Q = \{p, q, p \wedge q, r \wedge s, p \wedge q \wedge r\}$. Thus, the most relevant information set is $\{p, q, r \wedge s\}$. Then, we have $\{p \wedge q\}$ and $\{p \wedge q \wedge r\}\}$.

General formulas

For general formulas, we consider them under a particular form.

Definition 4. *A formula is CNF- iff it is written under conjunctive normal form in which clauses which are tautologies have been deleted as well as subsumed clauses.*

Then we suggest the following algorithm to retrieve the most relevant formulas.

Definition 5. *Let φ_1 and φ_2 be two formulas of a set $\mathcal{E}$ which are CNF-. We consider $\mathfrak{R}_S$ a binary relation on $\mathcal{E} \times \mathcal{E}$ defined by: $\varphi_1 \mathfrak{R}_S \varphi_2$ iff $Cl(\varphi_1, \varphi_2) \geq Cl(\varphi_2, \varphi_1)$ with $Cl(\varphi_1, \varphi_2)$ a function of $\mathcal{E} \times \mathcal{E}$ in $\mathbb{N}$ and that sends back the number of clauses of φ_1 that subsume clauses of φ_2.*

Proposition 14. $\mathfrak{R}_S$ *is an ordinal preference relation.*

A formula φ_1 is preferred to φ_2 if it subsumes "globally" more φ_2 than φ_2 subsumes φ_1. For example, $Cl(a, a \vee p) = 1$ and $Cl(a \vee p, a) = 0$ so $a\mathfrak{R}_S(a \vee p)$.

>From this preference, we can define an indifference relation $\mathfrak{R}_I$.

Definition 6. *Let φ_1 and φ_2 be two formulas of $\mathcal{E}$. $\varphi_1\mathfrak{R}_I\varphi_2$ iff $\varphi_1\mathfrak{R}_S\varphi_2$ and $\varphi_2\mathfrak{R}_S\varphi_1$.*

For example, $a\mathfrak{R}_I p$, $a \wedge (p \vee q)\mathfrak{R}_I(a \vee b) \wedge p$.

With that indifference relation, we can create classes of indifferent formulas. In particular, we have $max_{\mathfrak{R}_S} \mathcal{R}_a^Q$, i.e. relevant formulas preferred by $\mathfrak{R}_S$.

For example, if $\mathcal{R}_a^Q = \{a, a \vee p, (a \vee r) \wedge p, r \vee s, a \wedge p, p, a \wedge (t \vee s)\}$, then $max_{\mathfrak{R}_S} \mathcal{R}_a^Q = \{a, p, r \vee s, a \wedge p, a \wedge (t \vee s)\}$. We can see that there are still elements that we had removed with cubes such as $a \wedge p$. Thus, we have to use a preorder on the obtained set. We extend to general formulas the preorder used for cubes.

Definition 7. *Let φ_1 and φ_2 two formulas of a set $\mathcal{E}$. $\geq$ is a relation on $\mathcal{E} \times \mathcal{E}$ defined by: $\varphi_1 \geq \varphi_2$ iff $\vdash \varphi_2 \rightarrow \varphi_1$.*

Proposition 15. $\geq$ *is a preorder.*

---

[6] We remind that an implicant of φ is a conjunction of literals and that an implicant α of φ is prime iff it ceases to be an implicant upon the deletion of any of his literals.

What interests us here is the set $max_{\geq} max_{\mathfrak{R}_s} \mathcal{R}_a^Q$ that we will note $\mathfrak{R}m_a^Q$.

For example, if we take back $\mathcal{R}_a^Q = \{a, a \vee p, (a \vee r) \wedge p, r \vee s, a \wedge p, p, a \wedge (t \vee s)\}$, we have $\mathfrak{R}m_a^Q = \{a, p, r \vee s\}$.

We write $Rm_a^Q \varphi$ instead of "$R_a^Q \varphi$ and $\varphi \in \mathfrak{R}m_a^Q$". $Rm_a^Q \varphi$ is read φ *is maximal relevant for a concerning Q*.

This algorithm generalizes the two previous cases : for clauses, the set of most relevant information obtained by this algorithm is the set of most relevant information with the preorder $\leq_{Cl}$. Likewise, for cubes, the set of most relevant information obtained by this algorithm is the set of most relevant information with the preorder $\leq_{Cu}$.

The problem with maximal relevance defined this way is that we do not, for the moment, have a semantic characterization of formulas obtained with this algorithm. Moreover, contrary to cubes and clauses, we do not have a stratification of relevant information but we get only the set of most relevant formulas.

4 Multi-agent Case

Let us now extend the previous definition to several agents. Let a and b be two agents of $\mathcal{A}$ and φ a formula. Let's suppose that a has an information need modelled by request Q. From relevance definition, we get:

$$B_b R_a^Q \varphi \leftrightarrow B_b(I_a Bif Q) \wedge B_b \varphi \wedge B_b(B_a(\varphi \to Q) \otimes B_a(\varphi \to \neg Q))$$

That means that agent b believes that φ is relevant for a concerning Q iff b believes that a has an information need Q, b believes that φ is true and b believes that a, from his own beliefs and from φ, can deduce Q or (exclusive) deduce $\neg Q$. Thus, b can believe that an information is relevant for another agent a concerning Q, only if b knows about a's information needs.

Moreover, it can happen that b thinks that an information is relevant for a concerning an information need while it is not. This happens when b's beliefs are wrong i.e when b believes that a has an information need and a does not have this information need, or when b believes that φ is true and φ is not, or when b believes that a, from his beliefs and from φ, can deduce Q or (exclusive) deduce $\neg Q$ and a does not have enough knowledge.

5 Information Need Generalized

Information need considered until now was *a wants to know if Q or if $\neg Q$* . This information need can be extended in the following way *a wants to know if q_1 or if q_2, ... or if q_n* , q_i being requests and being mutually exclusives to each other.

Let $\mathcal{Q}$ be the set of alternative answers to information need. We suppose that $\mathcal{Q}$ is finite and countable and that all formulas q_i of $\mathcal{Q}$ are mutually exclusives to each other. We can model the information need the following way:

$$I_a Bif_a \mathcal{Q} \leftrightarrow I_a(B_a q_1 \otimes B_a q_2 \otimes ... \otimes B_a q_n)$$

For example, information need for a to know if the result of an exam is *good*, *average* or *bad* is $I_a Bif_a \mathcal{Q}$, with $\mathcal{Q} = \{good, average, bad\}$.

Thus, we can extend relevance the following way:

Definition 8. *Let $\mathcal{Q}$ be a set of n requests exclusive to each other.*

$$R_a^{\mathcal{Q}}\varphi \equiv I_a Bif_a\mathcal{Q} \wedge (B_a(\varphi \rightarrow q_1) \otimes B_a(\varphi \rightarrow q_2) \otimes ... \otimes B_a(\varphi \rightarrow q_n)) \wedge \varphi$$

$R_a^{\mathcal{Q}}\varphi$ *is read φ is relevant concerning $\mathcal{Q}$ for a.*

Let's take our exam example. We will note ex_1 and ex_2 the two literals representing the facts that exercises number 1 and 2 have been succeed. Let a's set of beliefs be : $\{ex_1 \wedge ex_2 \rightarrow good, ex_1 \otimes ex_2 \rightarrow average, \neg ex_1 \wedge \neg ex_2 \rightarrow bad\}$, i.e. the exam is good if exercises 1 and 2 have been succeed, the exam is average if only one of the two exercises has been succeed and the exam is bad if no exercise has been succeed. In that case, $R_a^{\mathcal{Q}}(ex_1 \wedge ex_2), R_a^{\mathcal{Q}}(ex_1 \otimes ex_2), R_a^{\mathcal{Q}}(\neg ex_1 \wedge \neg ex_2)$.

Properties. We have the same properties than before.

Proposition 16
$$R_a^{\mathcal{Q}}\varphi \rightarrow \neg B_a\neg\varphi$$

Proposition 17
$$R_a^{\mathcal{Q}}\varphi \rightarrow \neg B_a\varphi$$

Proposition 18
$$R_a^{\mathcal{Q}}\varphi \rightarrow ((Bel_a * \varphi) \rightarrow q_1) \otimes ... \otimes ((Bel_a * \varphi) \rightarrow q_n)$$

Proposition 19
$$(q_1 \otimes ... \otimes q_n) \wedge I_a Bif_a\mathcal{Q} \rightarrow R_a^{\mathcal{Q}} q_1 \otimes ... \otimes R_a^{\mathcal{Q}} q_n$$

The hypothesis $q_1 \otimes ... \otimes q_n$ is necessary to make sure than at least one request q_i is true. Without that, relevance is impossible.

6 Conclusion

The main contribution of this paper is a definition of agent-oriented relevance. In the framework of BDI models, we defined a new modal operator that represents the fact that a piece of information is relevant for an agent concerning an information need. We have shown how this definition can be applied in the multi-agent case. However, formulas which are considered as relevant are too many. So, we have given a method selecting the most relevant ones.

There are several extensions to this work.

First, event if it is done in the case of clauses or cubes, giving a semantic characterization of the most relevant information remains to be done for general formulas. Studying the properties of these formulas will be possible only when this is done.

Secondly, in the case of multiple choice requests, it would be interesting to define degrees of relevance. Thus, a piece of information would be partially relevant if it would allow to eliminate some requests among all possible requests of multiple choice.

Dealing with first order logic and open requests also remains to be done.

It would also be very interesting to address other needs than information need as well as or to deal with more complicated requests. For example, an agent may have a *verification need* that means he needs to verify that his beliefs are true. In this case, relevant information are the ones reinforcing or contradicting his beliefs. Symmetrically, an the agent may have a *need for completion* that means that he wants to be aware of any information which are true (or any information in a given domain).

Finally, a future work will consist in drawing a parallel between the notion of relevance developped in this paper and the different varieties of relevance and irrelevance suggested in Artificial Intelligence Journal, special issue on Relevance ([10], [11], [3]).

References

1. Herzig, A., Longin, D.: A logic of intention with cooperation principles and with assertive speech acts as communication primitives. In: Castelfranchi, C., Johnson, W.L. (eds.) AAMAS 2000 and AAMAS 2002, pp. 920–927. ACM Press, New York (2002)
2. Alchourrón, C., Gärdenfors, P., Makinson, D.: On the logic of theory change: partial meet contraction and revision functions. Journal of Symbolic Logic 50, 510–530 (1985)
3. Darwiche, A.: A logical notion of conditional independence: properties and applications. Artificial Intelligence, 45–82 (December 1997)
4. Voishvillo, E.K.: Philosophical-methodological aspects of relevant logic. Bulletin of Section of Logic 18/4, 146–151 (1989)
5. Chellas, B.F.: Modal logic: An introduction. Cambridge University Press, Cambridge (1980)
6. Crestani, F., Lalmas, M.: Logic and uncertainty in information retrieval, pp. 179–206 (2001)
7. Enderton, H.B.: A Mathematical Introduction to Logic. Academic Press, London (1972)
8. Chevallet, J.P.: Modélisation logique pour la recherche d'information, Hermes, pp. 105–138 (2004)
9. Floridi, L.: Understanding epistemic relevance. Erkenntniss (2007)
10. Lakemeyer, G.: Relevance from an epistemic perspective. Artif. Intell. 97(1-2), 137–167 (1997)
11. Levy, A.Y., Fikes, R.E., Sagiv, Y.: Speeding up inferences using relevance reasoning: a formalism and algorithms. Artif. Intell. 97(1-2), 83–136 (1997)
12. Lalmas, M., van Rijsbergen, C.J.: A model of information retrieval system based on situation theory and dempster-shafer theory of evidence. In: Proceedings of the 1st Workshop on Incompleteness and Uncertainty in Information Systems, pp. 62–67 (1993)
13. Wooldridge, M.: Reasoning about rational agents. MIT Press, Cambridge (2000)
14. Borlund, P.: The Concept of Relevance in IR. Journal of the American Society for Information Science and Technology 54(10), 913–925 (2003)
15. Tagawa, T., Cheng, J.: Deontic relevant logic: A strong relevant logic approach to removing paradoxes from deontic logic. In: Ishizuka, M., Sattar, A. (eds.) PRICAI 2002. LNCS (LNAI), vol. 2417, pp. 39–48. Springer, Heidelberg (2002)
16. Wilson, T.D.: Models in information behaviour research. Journal of Documentation 55(3), 249–270 (1999)

The Agent Modeling Language (AMOLA)

Nikolaos Spanoudakis[1,2] and Pavlos Moraitis[2]

[1] Technical University of Crete, Department of Sciences,
University Campus, 73100, Kounoupidiana, Greece
`nikos@science.tuc.gr`
[2] Paris Descartes University, Department of Mathematics and Computer Science,
45, rue des Saints-Pères, 75270 Paris Cedex 06, France
`{nikos,pavlos}@math-info.univ-paris5.fr`

Abstract. This paper presents the Agent MOdeling LAnguage (AMOLA). This language provides the syntax and semantics for creating models of multi-agent systems covering the analysis and design phases of the software development process. It supports a modular agent design approach and introduces the concepts of intra-and inter-agent control. The first defines the agent's lifecycle by coordinating the different modules that implement his capabilities, while the latter defines the protocols that govern the coordination of the society of the agents. The modeling of the intra and inter-agent control is based on statecharts. The analysis phase builds on the concepts of capability and functionality. AMOLA deals with both the individual and societal aspect of the agents. However, in this paper we focus in presenting only the individual agent development process. AMOLA is used by ASEME, a general agent systems development methodology.

Keywords: Multi-agent systems, Tools and methodologies for multi-agent software systems.

1 Introduction

Agent oriented development emerges as the modern way to create software. Its main advantage – as referred to by the literature – is to enable intelligent, social and autonomous software development. In our understanding it aims to provide the system developers with adequate engineering concepts that abstract away the complexity that is inherent to such systems. Moreover, it should allow for modular development so that successful practices can easily be incorporated in new systems. Finally, it should cater for model transformation between the different software development phases so that the process can be automated.

In the past, we introduced the Gaia2JADE process for Multi-agent Systems (MAS) development ([12]) and since then it has been used for the successful development of a number of MAS, see e.g. [8]. However, our more recent work (i.e. [11]) was about a more complex MAS that called for modularity, abstraction and support for iterative development. At the same time, we observed that the "services model" of Gaia [19] didn't apply to modern agents who provide services through agent interaction protocols. Moreover, we had no specific requirements analysis models that would be

D. Dochev, M. Pistore, and P. Traverso (Eds.): AIMSA 2008, LNAI 5253, pp. 32–44, 2008.

transformed to analysis models. Furthermore, the protocol model of Gaia did not provide the semantics to define complex protocols and the Gaia2JADE process additions remedied this situation only for simple protocols.

In order to address these issues, we used method fragments [4] from other methodologies and introduced a new language for the analysis and design phases of the software development process, namely the *Agent MOdeling LAnguage (AMOLA)*. All these changes led to the proposal of a new methodology, ASEME, an acronym of the full title *Agent SystEms Methodology*. Through this methodology, we were given an opportunity to express our point of view on modular agent architectures (see e.g. [9]) that we have been supporting several years now.

AMOLA not only modernizes the Gaia2JADE process but offers to the system developer new possibilities compared to other works proposed in the literature (a large set of which can be found in [7]). It allows models transformation from the requirements analysis phase to implementation. It defines three levels of abstraction, namely the society level, the agent level and the capability level, in each software development phase, thus defining a top-down analysis and design approach. Finally, using in an original way the statecharts and their orthogonality feature, it defines the inter- and intra-agent control models, the first for coordinating a group of agents and the second for coordinating an agent's modules.

In this paper, we present AMOLA focusing in the individual agent development issues, leaving aside, for the moment, the society issue. However, the reader will get an idea of how this is accomplished. Firstly, we discuss AMOLA's basic characteristics in section two, followed by the analysis and design phase models, in sections three and four respectively. For demonstrating the different elements of AMOLA we use a real-world system development case-study. Then, we discuss related work in section five and, finally, section six concludes.

2 The Basic Characteristics of AMOLA

The Agent Modeling Language (AMOLA) describes both an agent and a multi-agent system. Before presenting the language itself we identify some key concepts. Thus, we define the concept of *functionality* to represent the thinking, thought and senses characteristics of an agent. Then, we define the concept of *capability* as the ability to achieve specific tasks that require the use of one or more functionalities. The agent is an entity with certain capabilities, also capable for inter and intra-agent communication. Each of the capabilities requires certain functionalities and can be defined separately from the other capabilities. The capabilities are the modules that are integrated using the *intra-agent control* concept to define an agent. Each agent is considered a part of a community of agents. Thus, the community's modules are the agents and they are integrated into it using the *inter-agent control* concept.

The originality in this work is the intra-agent control concept that allows for the assembly of an agent by coordinating a set of modules, which are themselves implementations of capabilities that are based on functionalities. Here, the concepts of capability and functionality are distinct and complementary, in contrast to other works where they refer to the same thing but at different stages of development (e.g. Prometheus [14]). The agent developer can use the same modules but different assembling strategies, proposing a different ordering of the modules execution producing in that way different profiles of an agent (like in the case of the KGP agent [1]). Using this approach, we can design an agent with the *reasoning capability* that is based on the *argumentation based decision making functionality*. Another implementation of the same capability could be based on a different functionality, e.g. abductive reasoning.

Then, in order to represent our designs, AMOLA is based on statecharts, a well-known and general language and does not make any assumptions on the ontology, communication model, reasoning process or the mental attitudes (e.g. belief-desire-intentions) of the agents giving this freedom to the designer. Other methodologies impose (like Prometheus [14] or Ingenias [16]) or strongly imply (like Tropos [3]) the agent meta-models (see [7] for more details). Of course, there are some developers who want to have all these things ready for them, but there are others that want to use different agent paradigms according to their expertise. For example, one can use AMOLA for defining Belief-Desire-Intentions based agents, while another for defining procedural agents.

The AMOLA models related to the analysis and design phases of the software development process are shown in Figure 1. These models are part of a more general methodology for developing multi-agent systems, ASEME (Agent Systems Methodology, a preliminary report of which can be found in [17]). ASEME is a model-driven engineering (MDE) methodology. MDE is the systematic use of models as primary engineering artifacts throughout the engineering lifecycle. It is compatible with the recently emerging Model Driven Architecture (MDA[1], [10]) paradigm. MDA's strong point is that it strives for portability, interoperability and reusability, three non-functional requirements that are deemed as very important for modern systems design. MDA defines three models:

- A computation independent model (CIM) is a view of a system that does not show details of the structure of systems. It uses a vocabulary that is familiar to the practitioners of the domain in question as it is used for system specification.
- A platform independent model (PIM) is a view of a system that on one hand provides a specific technical specification of the system, but on the other hand exhibits a specified degree of platform independence so as to be suitable for use with a number of different platforms.
- A platform specific model (PSM) is a view of a system combining the specifications in the PIM with the details that specify how that system uses a particular type of platform.

The system is described in platform independent format at the end of the design phase. We will provide guidelines for implementing the AMOLA models using JADE or STATEMATE (two different platform specific models).

We define three levels of abstraction in each phase. The first is the *societal level*. There, the whole agent community system functionality is modeled. Then, in the *agent level*, we model (zoom in) each part of the society, the agent. Finally, we focus in the details that compose each of the agent's parts in the *capability level*. In the first three phases the process is top-down, while in the last three phases it is bottom-up.

---

[1] The Model Driven Architecture (MDA) is an Object Management Group (OMG) specification for model driven engineering, http://www.omg.org/mda/

AMOLA is concerned with the first two levels assuming that the analysis and design in the capability level can be achieved using classical software engineering techniques.

Development Phase	Levels of Abstraction		
	Society Level	Agent Level	Capability Level
Analysis Phase AMOLA Models	Roles and Protocols Use case Diagram Agent Interaction Protocols	Capabilities Roles Model	Functionalities
Platform Independent Model — **Design Phase** AMOLA Models	Agent Types And Organization Inter-agent control model	Modules Intra-agent control model	Components

Fig. 1. The Agent Systems Methodology (ASEME) phases in three levels of abstraction and the AMOLA models related to each level

For the AMOLA models demonstration we present the analysis and design models of an agent participating in a real-world system that we developed. Our requirements were to develop a system that allows a user to access a variety of location-based services (LBS) that are supported by an existing brokering system, using a simple service request protocol based on the FIPA Agent Communication Language (ACL). The system should learn the habits of the user and support him while on the move. It should connect to an OSGi[2] service for getting the user's coordinates using a GPS device. It should also handle dangerous situations for the user by reading a heart rate sensor (again an OSGi service) and call for help. A non-functional requirement for the system is to execute on any mobile device with the OSGi service architecture. For more details about the real-world system the reader can refer to [11].

3 The Analysis Phase Models

The main models associated with this phase are the *use case model* and the *roles model*. The former is an extended UML *use case diagram* and the latter is mainly inspired by the Gaia methodology [19] (a Gaia roles model method fragment can be used with minimal transformation effort).

The use case diagram helps to visualize the system including its interaction with external entities, be they humans or other systems. No new elements are needed other than those proposed by UML. However, the semantics change. Firstly, the actor "enters" the system and assumes a role. *Agents* are modeled as roles, either within the system box (for the agents that are to be developed) or outside the system box (for existing agents in the environment). Human actors are also represented as roles outside the system box (like in traditional UML use case diagrams). We distinguish the human roles by their name that is written in italics. This approach aims to show the concept that we are modeling artificial agents interacting with other artificial agents or human agents. Secondly, the different use cases must be directly related to at least

---

[2] The Open Services Gateway initiative (OSGi) alliance is a worldwide consortium of technology innovators defining a component integration platform, http://www.osgi.org

one artificial agent role. These general use cases can be decomposed to simpler ones using the *include* use case relationship. Based on the use case diagram the system modeler can define the roles model. A use case that connects two or more (agent) roles implies the definition of a special capability type: the participation of the agent in an interaction protocol (e.g. negotiation). A use case that connects a human and an artificial agent implies the need for defining a human-machine interface (HMI). The latter is modeled as another agent functionality. A use case can *include* a second one showing that its successful completion requires that the second also takes place.

Referring now to our case study, in the agent level, we define the agent's capabilities as the use cases that correspond to the goals of the requirements analysis phase. The relevant actor diagram that was the result of the previous phase is presented in Figure 2.

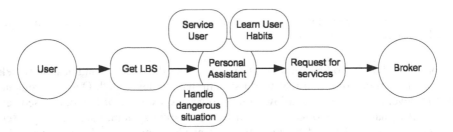

Fig. 2. Actor diagram. The circles represent the identified actors and the rounded rectangles their goals. It is the output of the requirements analysis phase.

In the Analysis phase, the actor diagram is transformed to the use case diagram presented in Figure 3. The activities that will be contained in each capability are the use cases that the capability includes. Then we define the role model for each agent role (see Figure 4). Firstly we add the interaction protocols that this agent will be able to participate in. In this case it is a simple service protocol with a requester (our agent) and a responder. We continue with the definition of the liveness model inside the roles model. The liveness model has a formula at the first line (*root formula*) where we can add activities or capabilities. A capability must be decomposed to activities in the following line.

In the capability abstraction level the activities are associated to generic functionalities. The latter must clearly imply the technology needed for realizing them (see Figure 5). In this case, returning to our running example, the requirement for functioning on any mobile device running OSGi services reveals that such a device must at least support the Java Mobile Information Device Profile (MIDP), which offers a specific record for storing data. Therefore, the activities that want to store or read data from a file must use the MIDP record technology. Finally, the reader should note that a special capability not included in the use–case diagram named *communicate* appears. This capability includes the **send message** and **receive message** activities and is shared by all agents and is defined separately because its implementation is relative to the functionality provided by the agent development platform, e.g. JADE[3].

---

[3] The Java Agent Development Environment (JADE) is an open source framework that adheres to the FIPA standards for agents development, http://jade.tilab.com

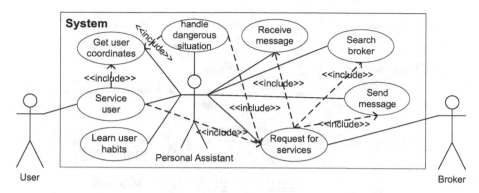

Fig. 3. Use case diagram

Role: Personal Assistant
Protocols: Service protocol: initiator
Liveness:
personal assistant = (service user)$^{\omega}$ || (handle dangerous situation)$^{\omega}$
service user = get user order. get user coordinates. get user preferences. request for
 services. present information to the user. learn user habits.
handle dangerous situation = invoke heart rate service. determine user condition. [get
 user coordinates. request for services]
request for services = search broker. [send message. receive message]
learn user habits = learn user preference. update user preferences.

Fig. 4. The role model, including five liveness formulas

4 The Design Phase Models

The models associated with the *Design phase* are the *inter-agent control* and *intra-agent control*. They define the functional and behavioral aspects of the multi-agent system. In the past, the MaSE methodology [5] defined agent behavior as a set of concurrent tasks, each specifying a single thread of control that integrates inter-agent as well as intra-agent interactions. Our model goes one step further by modeling the interaction among the capabilities of an agent, i.e. what they call the different threads of control, but also their execution cycle. The model associated to the first level of this phase is the inter-agent control, which defines interaction protocols by defining the necessary roles and the interaction among them. The implementation of the inter-agent control is done at the agent level via the capabilities and their appropriate interaction defined via the intra-agent control. Finally, in the third level each capability is defined with regard to its functionality, what technology is used, how it is parameterized, what data structures and algorithms should be implemented. The models defined in this phase are the Platform Independent Model (PIM).

For the Design Phase models we use the language of statecharts as it is defined in [6]. Statecharts are used for modeling systems. They are based on an activity-chart that is a hierarchical data-flow diagram, where the functional capabilities of the

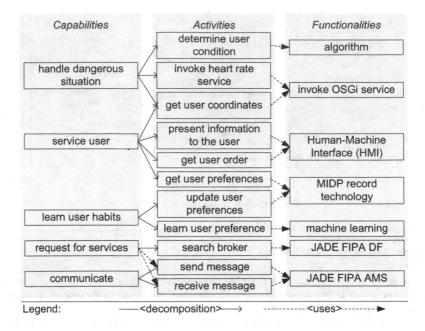

Fig. 5. Capabilities, activities and functionalities

system are captured by activities and the data elements and signals that can flow between them. The behavioral aspects of these activities (what activity, when and under what conditions it will be active) are specified in statecharts. There are three types of states in a statechart, i.e. OR-states, AND-states, and basic states. OR-states have substates that are related to each other by "exclusive-or", and AND-states have orthogonal components that are related by "and" (execute in parallel). Basic states are those at the bottom of the state hierarchy, i.e., those that have no substates. The state at the highest level, i.e., the one with no parent state, is called the root. Each transition from one state (source) to another (target) is labeled by an expression, whose general syntax is e[c]/a, where e is the event that triggers the transition; c is a condition that must be true in order for the transition to be taken when e occurs; and a is an action that takes place when the transition is taken. All elements of the transition expression are optional. Action a can be the special action start(P) that causes the activity P to start. The scope of a transition is the lowest OR-state in the hierarchy of states. Multiple concurrently active statecharts are considered to be orthogonal components at the highest level of a single statechart. If one of the statecharts becomes non-active (e.g. when the activity it controls is stopped) the other charts continue to be active and that statechart enters an idle state until it is restarted.

In the past, statecharts have been used for modeling agent behaviors in MaSE [5]. In our work we use statecharts to model *intra-agent control*. As we said before, it corresponds to modeling the interaction between different capabilities, defining the behavior of the agent. This interaction defines the interrelation in a recursive way between capabilities and also between activities of the same capability that can imply concurrent or sequential execution. This is the basic and main difference with the way that statecharts have been used in the past. Moreover, we use statecharts in order to model agent interaction, thus using the same formalism for modeling inter and intra-agent control, which is also a novelty. However, the use of statecharts for the inter-agent control is out of the scope of this paper.

In the agent level, we define the intra-agent control by transforming the liveness model of the role to a state diagram. We achieve that, by interpreting the Gaia operators in the way described in Table 1. The reader should note that we have defined a new operator, the $|x^{\omega}|^n$, with which we can define an activity that can be concurrently instantiated and executed more than one times (n times). Initially, the statechart has only one state named after the left-hand side of the first liveness formula of the role model (probably named after the agent type). Then, this state acquires substates. The latter are constructed reading the right hand side of the liveness formula from left to right, and substituting the operator found there with the relevant template in Table 1. If one of the states is further refined in a next formula, then new substates are defined for it in a recursive way.

Table 1. Templates of extended Gaia operators (Op.) for Statechart generation

At this stage, the activities that have been defined in the roles model are assigned to the states with the same name in the statechart. An agent percept, a monitored for environmental effect, an event generated by any other executing agent activity, or the ending of the executing state activity can cause a transition from one state to another.

In Figure 6 we present the statechart that is derived from the liveness model of our example presented in Figure 4. At this point, we need to define the events that cause transitions, their conditions and also the data elements that will be used for the statechart. These events can be inter-agent messages, or other kinds of events generated by the execution of the agent activities.

Finally, the designer defines the modules that will be used for the agent. The modules are typically as many as the agent capabilities. This allows for a modular representation of the agent's architecture and defines the right level of decomposition of an agent. Thus, it allows for the reusability of the modules as independent software components in different types of agents, having common capabilities. This is also a main difference with other methodologies. The agent implements the root formula of the statechart. The substates are implemented in the relevant modules. The modules are ready for development by transforming the statecharts to code, not restricted to JADE development like in [12], but using any tool that transforms statecharts to code, e.g. STATEMATE [6] for object oriented languages. In order to transform the models

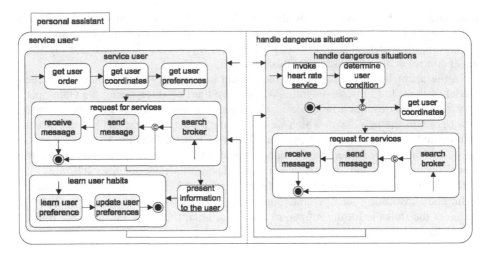

Fig. 6. The intra-agent control model

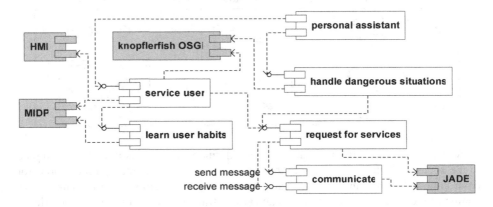

Fig. 7. The agent modules

to JADE code the developer should transform complex states to instances of the *FSMBehaviour* class and the simple states to *SimpleBehaviour* instances in a fashion relevant to the one in [12].

In the next ASEME phase the modeler firstly creates the platform specific models and then implements the system. The designer defines the modules that will be used for the agent. The modules are presented in a UML component diagram; Figure 7 shows the modules of our example. The modules are typically as many as the agent capabilities. The aggregation of these modules leads to a new module, namely the agent. The agent module only implements the root formula of the statechart. The substates are implemented in the relevant modules. All these modules are now concrete components and could be reused in the future by another agent. The grey components in Figure 7 are the used functionalities. The reader will notice that the abstract functionalities like *algorithm* are not shown in this diagram as they do not refer to an existing software component. They are analyzed, designed and programmed like any other software component using an existing method, e.g. UML.

5 Related Work

Comparing AMOLA with the Gaia methodology [19] we first notice that the latter does not support the requirements analysis phase and its agent design models do not lead in a straightforward way to implementation. For example, the services model isn't concrete – does not relate to code. In the past, Gaia has been modified in order to cover the implementation phase [12], but certain aspects proved difficult to deal with, such as the definition of complex agent interaction protocols or the way to merge two roles in one agent. In [12] we offered some extensions but they were in rather practical than conceptual level. These extensions allowed for easily conceiving and implementing relatively simple agents. Finally, its models cannot be used for simulation-optimization. AMOLA can be connected with successful method fragments in the requirements analysis phase (such as the actor diagram of Tropos) and its models are statecharts. The latter is a well known language for which there are numerous tools for code generation and simulation/optimization.

TROPOS [3] provides a formal language and semantics that greatly aid the requirements analysis phase. It can also lead to successful requirements verification for a system. However, the user must come down to attributes definitions (extremely detailed design including data types) in order to use simulation. It is a process centric design approach, not a module based one, like AMOLA. We believe that the module based approach proposes the right level of decomposition of an agent because it allows for the reusability of the modules as independent software components in different types of agents, having some common capabilities. Moreover, the detailed design phase of TROPOS proposes the use of AUML [13] (as well as in the work presented in [14]). However, AUML has specific shortcomings when it comes to defining complex protocols (the reader can refer to [15] for an extensive list). Finally, Tropos has been applied for modeling relatively simple agents, not complex ones [7].

AUML has been proposed as a language for modeling multi-agent systems. However, it does not come along with a methodology or a complete process for software development. Thus, many methodologies just use some of its models, mainly the agent interaction protocol (AIP) model that has been defined as an extension to the UML sequence diagram. The Layered approach to protocols provides a mechanism for specifying the program that implements a protocol but does not specify how it is integrated with other such programs (other protocols), or how to integrate it with the other agent capabilities. AMOLA caters for this issue by using the same formalism (statecharts) for modeling inter and intra-agent control.

MaSE [5] defines a system goal oriented MAS development methodology. They define for the first time inter and intra-agent interactions that must be integrated. However, in their models they fail to provide a modeling technique for analyzing the systems and allowing for model transformation between the analysis and design phases. Their concurrent tasks model derives from the goal hierarchy tree and from sequence diagrams in a way that cannot be automated. In our work the model transformation process is straightforward. For example, we provide simple rules for obtaining the design phase intra-agent control from the analysis phase liveness model. Moreover, we distinguish independent modules that are integrated for developing an agent, which can be reusable components. We define agent types that originate from actors of the requirements phase, while the agents in MaSE are related to system goals. This restricts the definition of autonomous agents. Finally, in MaSE agents are implemented using AgentTool while in AMOLA more implementation possibilities are allowed.

In Prometheus [14] the authors use the terms of functionality and capability. However, they correspond to different concepts compared to our work. In fact, functionalities and capabilities refer to same concept as it evolves through the development phases (i.e. the abilities that the system needs to have in order to meet its design objectives). In our work capabilities refer to a specific goal and functionality is related to the used technologies that are application independent (e.g. argumentation, abduction, induction for reasoning mechanism implementation). Moreover, in our approach, with the proposal of the intra-agent control we are able to model in a recursive way the dynamic interaction between capabilities and between activities of the same capability. The support for implementation, testing and debugging of Prometheus models is limited and it has less focus on early requirements and analysis of business processes [7]. AMOLA's capability to be integrated with method fragments and the fact that its design models are statecharts overcomes these issues.

One interesting approach that is based on UML is Ingenias [16], which proposes several meta-models that define specific structures for different concepts like *agent*, *role*, *resource*, etc. For some developers it could be useful, but for others it could be considered as a constraint, as it imposes a particular structure for an agent and agent organizations. Moreover, Ingenias does not offer the convenience of gradually modeling a multi-agent system by considering it at three levels of abstraction.

The authors of [2] proposed a capability concept for BDI agents. In their view, capability is "a cluster of plans, beliefs, events and scoping rules over them". Capabilities can contain sub-capabilities and have at most one parent capability. Finally, the agent concept is defined as an extension of the capability concept aggregating capabilities. The differences of our work in comparison to this one is the fact that the capability concept for us is more general (not limited to BDI agents) and it leads to the definition of the module that can be a reusable software component.

Capability in AML [18] is used to model an abstraction of a behavior in terms of its inputs, outputs, pre-conditions, and post-conditions. A behavior is the software component and its capabilities are the signatures of the methods that the behavior realizes accompanied by pre-conditions for the execution of a method and post-conditions (what must hold after the method's execution). However, in AMOLA the concept of capability is more abstract and is used for modeling an agent's abilities that are more general than method signatures. The latter are defined as functionalities and the activity within the capability defines when and how the functionalities are used by the agent.

6 Conclusion

Concluding, in this paper we defined AMOLA, a language for modeling agent systems that has many qualities compared to other relevant methodologies (e.g. the Prometheus, Gaia, Tropos, PASSI and MaSE discussed in [7]):

- The intra-agent control, whose novelty is to allow the modeling of interactions between the different capabilities of an agent. For this purpose we use statecharts and their orthogonality concept in an original way
- The inter-agent control that corresponds to the agent interaction protocol. This part of the methodology is out of the scope of this paper but that which is important is the use of statecharts like in the intra-agent control, thus simplifying the designer's task by using the same formalism
- There is a straightforward transformation process between the models of the analysis phase to those of the design phase and then to an implementation platform
- It defines three abstraction levels (the society, agent and capability levels), thus supporting the development of large-scale systems.
- The models of AMOLA can lead to agent development without imposing constraints on how the mental model of the agent will be defined (e.g. like in Ingenias [16] and Prometheus [14])
- We define the terms of capability and functionality that have been used with different meanings in the past in order to provide new concepts for modeling agent-based systems with relation to previous methodologies like, e.g. for object-oriented development.

Currently we are working on the society level using statecharts in order to model agent interaction protocols. Moreover, we are working on the way that these models will be integrated and implemented through the agent capabilities.

References

1. Bracciali, A., Demetriou, N., Endriss, U., Kakas, A., Lu, W., Stathis, K.: Crafting the Mind of a PROSOCS Agent. Applied Artificial Intelligence 20(4-5) (April 2006)
2. Braubach, L., Pokahr, A., Lamersdorf, W.: Extending the Capability Concept for Flexible BDI Agent Modularization. In: Bordini, R.H., Dastani, M., Dix, J., El Fallah Seghrouchni, A. (eds.) PROMAS 2005. LNCS (LNAI), vol. 3862. Springer, Heidelberg (2006)
3. Bresciani, P., Giorgini, P., Giunchiglia, F., Mylopoulos, J., Perini, A.: TROPOS: An Agent-Oriented Software Development Methodology. Journal of Autonomous Agents and Multi-Agent Systems (2004)
4. Cossentino, M., Gaglio, S., Garro, A., Seidita, V.: Method fragments for agent design methodologies: from standardisation to research. Int. Journal of Agent-Oriented Software Engineering 1(1), 91–121 (2007)
5. Deloach, S.A., Wood, M.F., Sparkman, C.H.: Multiagent Systems Engineering. Int. Journal of Software Engineering and Knowledge Eng. 11(3), 231–258 (2001)
6. Harel, D., Naamad, A.: The STATEMATE Semantics of Statecharts. ACM Transactions on Software Engineering and Methodology 5(4), 293–333 (1996)

7. Henderson-Sellers, B., Giorgini, P. (eds.): Agent-Oriented Methodologies. Idea Group Publishing (2005)
8. Karacapilidis, N., Lazanas, A., Megalokonomos, G., Moraitis, P.: On the Development of a Web-based System for Transportations Services. Information Sciences 176(13), 1801–1828 (2006)
9. Karacapilidis, N., Moraitis, P.: Intelligent Agents for an Artificial Market System. In: Proc. fifth Int. Conf. on Autonomous Agents (AGENTS 2001), Montreal, Canada, pp. 592–599 (2001)
10. Kleppe, A., Warmer, S., Bast, W.: MDA Explained. The Model Driven Architecture: Practice and Promise. Addison-Wesley, Reading (2003)
11. Moraitis, P., Spanoudakis, N.: Argumentation-based Agent Interaction in an Ambient Intelligence Context. IEEE Intelligent Systems 22(6), 84–93 (2007)
12. Moraitis, P., Spanoudakis, N.: The Gaia2JADE Process for Multi-Agent Systems Development. Applied Artificial Intelligence Journal 20(4-5) (2006)
13. Odell, J., Parunak, H.V.D., Bauer, B.: Representing Agent Interaction Protocols in UML. In: Ciancarini, P., Wooldridge, M.J. (eds.) AOSE 2000. LNCS, vol. 1957. Springer, Heidelberg (2001)
14. Padgham, L., Winikoff, M.: Developing Intelligent Agent Systems: A Practical Guide. Wiley, Chichester (2004)
15. Paurobally, S., Cunningham, R., Jennings, N.R.: Developing agent interaction protocols using graphical and logical methodologies. In: Dastani, M., Dix, J., El Fallah-Seghrouchni, A. (eds.) PROMAS 2003. LNCS (LNAI), vol. 3067. Springer, Heidelberg (2004)
16. Pavón, J., Gómez-Sanz, J.J., Fuentes, R.: The INGENIAS Methodology and Tools. In: Henderson-Sellers, B., Giorgini, P. (eds.) Agent-Oriented Methodologies, pp. 236–276. Idea Group Publishing (2005)
17. Spanoudakis, N., Moraitis, P.: The Agent Systems Methodology (ASEME): A Preliminary Report. In: Proceedings of the 5th European Workshop on Multi-Agent Systems (EUMAS 2007), Hammamet, Tunisia, December 13 - 14 (2007)
18. Trencansky, I., Cervenka, R.: Agent Modelling Language (AML): A comprehensive approach to modelling MAS. Informatica 29(4), 391–400 (2005)
19. Zambonelli, F., Jennings, N.R., Wooldridge, M.: Developing multiagent systems: the Gaia Methodology. ACM Trans. on Software Eng. and Methodology 12(3), 317–370 (2003)

Tailoring the Interpretation of Spatial Utterances for Playing a Board Game

Andrea Corradini

Institute of Business Communication and Information Science
University of Southern Denmark
6000 Kolding, Denmark
andrea@sitkom.sdu.dk

Abstract. In order to build an intelligent system that allows human beings to cooperate with a computing machine to perform a given task it is important to account for the individual characteristics of each user. In this work, we describe a flexible and adaptive user interface that is capable of interpreting spatial linguistic expressions according to the peculiar preferences and use of language of the user. As a proof of concept, we have implemented a graphics board game in which players can move, remove and spatially manipulate in 2D various game pieces on the computer screen by issuing speech commands during a dialogue session with the application. After an initial guided interaction, the system analyzes on the fly subjective interpretations of several linguistic relationships that form the basis for spatial expressions to tune a set of system's parameters. We tested the system with different human subjects. Automatic judgment of spatial expressions based on each user's interaction behavior over-performed sentence interpretation of a model that we previously created to tailor the characteristics of some abstract typical player. Post interaction informal talks also indicated a higher level of user satisfaction in subjects playing with the customized application rather than the more general one.

1 Introduction

One of the results of the growth in information and communication technology research is the large quantity of electronic devices available nowadays at a reduced cost. Low-cost digital equipment has boosted the number of people adopting new technologies, thereby causing an increased diversification of end users. The inherent user variety poses several issues for designers and developers who want to create robust, reliable and usable systems that accommodate to the individual needs and distinct characteristics of a large number of people. The demand for personalized applications also gives rise to a need to expand the range of tasks and the physical contexts in which perform them.

The standard dyadic model of human-computer interaction for knowledge-based systems assumes that machines process user input according to a kernel of knowledge [10]. This body of knowledge ranges from information about the application domain to constrain user actions and available system operations [14], to information regarding the communication process to determine whether, how, and when to assist the

D. Dochev, M. Pistore, and P. Traverso (Eds.): AIMSA 2008, LNAI 5253, pp. 45–57, 2008.
© Springer-Verlag Berlin Heidelberg 2008

user [15], and finally to knowledge about the current communication partner for the system to interact with the user in a cooperative way. Any successful man-machine interaction strategy depends on both the context at hand and the current user for there is no typical system user. Existing research has thus attempted to address the challenges posed by adaptive application by focusing on how content can be customized on the basis of user models and interaction history.

In this paper, we present a multimodal interface in which speech, typing, and mouse clicking can be used to control a run-time board game system. We exploit the information from the different input modes to interpret user's intention and solve ambiguities in spatial utterances containing topological relations like "*next to*", projective relations like "*under*", spatial relations of the kind "*between*", and absolute indication of spatial references such as "*upper rightmost corner*". We exploited the behavioral data collected from players interacting with the game to infer a set of system parameters to model the average player. In an optional introductive session, new players are asked to follow a protocol to learn how to interact with the game. During such a session, the system accommodates for each user's playing behaviors by adjusting its internal parameters while building a new one that is personalized to the subject currently playing.

In the rest of this paper, we start with a review of works that relate to our approach. We then outline the main features of our systems. We continue by presenting the results of a usability study accomplished with human subjects to check the feasibility of our method. Finally, we conclude with a brief discussion and notes on issues of user modeling in our context.

2 Related Work

The end users' diversification is tautological to say that there is no universal interface and thus calls for user applications that can accommodate to the given user, task and context at hand. Over the last years, several interfaces have been proposed to take advantage of the individual user differences in education, environment, physical impairments, cognitive abilities, social background, age, etc. The search of a strategy that helps determining recurrent patterns in user behaviors and further exploits them to generate an appropriate contextualized response is a central task [24] common to most of these systems.

In early works, *adaptable* interfaces were put forward to achieve a certain degree of system personalization. This was accomplished typically by requiring the user to either set some preferences in advance or fill in some form from which a designer could infer a set of preferences [1]. State-of-the-art *adaptive* systems require approaches that are more complex. The aim of those systems is to ferret out what is in the head of the user automatically from the analysis of off-line interactive sessions i.e. with data collected in advance and/or on the fly as the user interaction proceeds [9]. A detailed review of the development of user modeling systems is given in [18]. [31] presents a study of issues in user modeling with focus on standard machine learning techniques.

As pointed out in [3], sometimes the terms context-awareness and adaptation are used interchangeably. However, they denote distinct concepts where the former one refs to the digital device capability of accommodating to the physical-virtual context in which the human user is situated. This tightly relates to the notion of location and environment and is relevant for interactions with electronic equipments that operate in immersive environments and/or allow the user to be mobile [7, 13, 25]. Both user and device models must be accounted for to achieve customized content, presentation, and response.

Voice user interfaces offer a great potential for enhancing the interaction with computing machines. Due to the characteristics of human language, speech based applications are an ideal arena for experimenting with the concepts of adaptation and personalization. Over the last two decades, there have been a variety of practical task oriented spoken conversational systems for applications in limited domains [12, 16, 20, 27, 28]. These prototypes have provided good results in laboratory settings but have become commercial applications only in a few cases [22], notably in customer service centers and telephone-based information systems. Current technology limitations in addressing human language ambiguities and the difficulties to achieve user modeling in natural language applications [19, 32] are among the main reasons of this reduced performance in comparison with that of human beings. To this extent, some researchers have tried to improve the acceptance of dialogue systems by adapting the application grammar to the user's use of language either by adding new words and assigning a semantic meaning to them [8, 17] or by allowing users to expand it at run time with new rules [11].

Systems that exploit complementary information from other modalities have been also proposed to enhance spoken communication through user adaptation [16, 21]. Such multimodal spoken and dialogue systems offer more alternatives for adapting to user preferences and needs and indeed have been shown to achieve a higher degree of satisfaction in their users [26]. More work has focused on how to customize content based on user models and interaction history [2, 6, 26] similarly to early research on adaptation in hypermedia.

All the systems mentioned so far are difficult to compare with each other for the lack of standardized evaluation procedures. Despite a methodology to test multimodal output generated in response to a specific set of user, device, and situation constraints has been presented in [23], a thorough evaluation of an adaptable/adaptive multimodal systems remains a difficult and challenging task.

3 The Adaptive Game System

3.1 The Setting

As a test-bed for our approach we chose the digital version of a math puzzle game called Pentomino. In such a game, each piece is composed of five congruent unit squares, connected orthogonally. Reflectional and rotational symmetries of Pentomino pieces do not count. Hence, there is a set of only twelve valid different arrangements that may be used to play. They are usually named after the letters of the Latin alphabet that they resemble. Solving a Pentomino puzzle consists of tiling a two dimensional game board (see e.g. Figure 1), i.e. covering and filling a designated region without overlaps and without gaps using up the available pieces. Before placement, pieces may be rotated, shifted, flipped, and/or mirrored to make them fit onto the intended board location.

Early studies that we carried out on human-human communication to play the game show that subjects resort extensively to localization expressions when they collaboratively play towards the resolution of a puzzle. This makes the game an excellent scenario for testing ambiguities in the use understanding and interpretation of spatial language.

3.2 Resolution of Spatial Expressions for the 'Average' Player

In [4], we presented an approach for the creation of a computational model capable of interpreting a set of linguistic spatial propositions in the restricted domain of Pentomino. In an experiment with 38 subjects, we analyzed human judgment of spatial expressions that allowed us to come up with a set of criteria that explain human preferences for certain interpretations over others.

For each of these criteria, we defined a metric that combines the semantic and pragmatic contextual information regarding the game as well as the utterance being resolved. Each metric gives rise to a potential field that characterizes the degree of likelihood for carrying out the instruction at a specific hypothesized location.

We then resorted to multivariate linear regression techniques to determine a model that incorporates and summarizes all spatial interpretations given by the subjects during the experiment. By matching this model with the metric values calculated for an unknown spatial relationship, we could then resolve this latter to a specific location. The created model could be used to correctly explain spatial sentences of any hypothetical average user for 2.5 spatial sentences out of 4 on average.

This number indicates that there is still room for improvement.

3.3 Adaptation of the Resolution of Spatial Expressions to the Player

The analysis of data collected from observing human beings resolving spatial utterances while playing Pentomino, made it possible for us to come up with a set of nine criteria [4] that give helpful insights on the underlying strategy that people adopt to interpret situated communication in the game world. Each criterion reflects a specific behavioral pattern, which is common to the great majority of subjects. At the same time, each criterion incorporates some implicit parameters, which vary in quantity from individual to individual.

For instance, we noticed that users tend to place a piece *Obj* as closer as possible to the reference piece *Ref* to resolve sentences like "*drop the green piece left to the blue one*" (here being *Obj* the green piece and *Ref* the blue one). This *proximity* criterion generalizes very well over all subject population yet it tells us little about how we can quantify the notion of closeness in terms e.g. of number of units on the game board. For the average user, we can exploit the finding that over 97% of the subjects considered locations on the board grid that are within a distance of up to three units far away from the referent. This information restricts the search area for the resolution of the spatial relation under investigation since it indirectly defines an area of possible candidate locations on the board.

We refer to this area as the *region of interest*, *RoI* in short. Eventually, any ambiguous utterance involving spatial relations resolves to a location within the *RoI*.

If we further focus on each specific user, we can restrict such region and make it fit even more to the user's own mental idea of closeness. Hence, the next step consists in defining a maximum and minimum value for the extension of the *RoI* that can quantitatively express the notion of closeness as intended by each specific user. In the quest for such a peculiar value, we have to factor in also other criteria that usually co-occur in the characterization of human interpretation of spatial utterances. For instance, we need to account for findings that indicates that users tend to place a piece *Obj* at positions that maximize the number of physical contacts with other game entities like e.g. board edges or other pieces onboard (*adherence* criterion). Moreover, we must take into account the evidence that preferred positions for the placement of *Obj* are those that minimize the distance between the centers of mass of piece *Obj* and the referent *Ref*, respectively (*center of mass* requirement). Other factors that have an influence on a personalized choice of the *RoI* can be inferred from the analysis of the game history. We observed uniformity in the way users play the game: they start building up the puzzle from a certain region and then incrementally make it bigger and bigger. They seldom jump from one area of the board to another one farther away from it (*uniformity* criteria). Players that want to pursue the game objective do not place pieces on the board in a random way except in very specific cases (Figure 1). The analysis of the game history helps restricting the possible locations to include in the *RoI* thus ultimately accommodating for the playing style of the user. Finally yet importantly, an experienced player knows that piece placement must not give rise to enclosed regions on board that cannot be filled up with the remaining pieces since this would make the puzzle unsolvable (*solvability* criteria). This criterion relies on contextual game conditions as well as game semantics and thus relates directly to the user experience and skills. This becomes more and more important as the player improves and becomes an expert player.

In order to determine a user-specific *RoI* underlying the criteria that we briefly touched upon previously, we guide new users through an interactive session where they are asked by the application to perform a sequence of specific game moves. For this task, we created approximately 500 different configurations to provide a broad comprehensive coverage of use cases for each criterion as well as for situations where more criteria are potentially in conflict with each other (Figure 1).

In the same way as outlined in [4], we then resorted to multivariate linear analysis to determine the parameters that act as weighting factors for each of the criterion. These parameters determined by this standard machine learning technique are tailored to the playing style, behavioral patterns, knowledge of the game, and skills of the current user ultimately constitute an average user model. At the same time, the analysis of the response of the subjects to the different tasks assigned to them, allows us to determine a set of regions of interests for each kind of spatial relationship personalized to the particular individual. The determination of a single *RoI* to explain any given spatial relation class occurs then by inter-class intersection of each of the *RoI*s as determined during the guided session. Spatial ambiguity is eventually resolved by matching each game move with the average user model as explained in detail in [4].

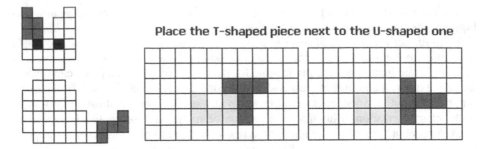

Fig. 1. (*left*) A cat-shaped board "naturally" lures players into placing specific pieces at particular locations like the upper or lower-right corners thus making the game history look not uniform in terms of spatial locations; (*middle and right*) this does not generally happen with standard rectangular boards; however other issues do still occur using these boards like the T-piece placement near the U that gives rise to conflicts between proximity and adherence criteria

3.4 Evaluation

We evaluated the adaptation capabilities of the Pentomino puzzle system on a standard computer with 2.20 GHz Centrino Pro CPU and 1.96 GB memory under the MS Windows XP operating system. In addition to the usual display, loudspeakers, keyboard and mouse, users were equipped with a microphone to enter spoken commands. The application software is written in JAVA and Prolog. We use a context-free grammar compliant to the Java Speech Grammar Format. We utilize the open-source Sphinx-4 [30] as speech recognition engine and FreeTTS [29] as speech synthesizer. The application dictionary contains approximately 500 words.

In order to test and evaluate the system, we collected the data from 13 participants; seven subjects were female and six were male with ages ranging from 19 to 37. Among the subjects there were 3 native Italian speakers, 2 native English speakers, 6 native German speakers, and 2 native Danish speakers. All subjects reported to be familiar with IT and computer technology in general. Each experiment session lasted approximately 45 minutes plus an additional 30 minutes for both a pre session system explanation and a post session informal talk.

We did not give the subjects any specific task. We just asked them to freely play the game (the one for the average user) to get acquainted with the application and the game rules. Requiring the users to play following a specific sequence of operations would have forced them to adapt their language to fit the assigned task i.e. the very opposite of the goal of our investigation. We provided the users with a digital button that we asked them to push on anytime the spoken command issued was not correctly interpreted by the system as they intended. Each interactive session was logged.

In a second experimental session, we invited the same players one more time. This time though, we asked them to first play the game in guided mode i.e. by following instructions as they appeared on screen. We did so for the system to collect user data and use it to build a model for the playing subject. Each guided session lasted some 45 minutes. We further allocated 45 more minutes for the user to play with the adapted system and 15 minutes for post-session discussion. As in the first experimental session, we equipped the participants with a digital button to signal system interpretation errors.

Participants were not told anything about system adaptation and the goal of our experiment. They were told that they were using another version of the system and that we were testing it against the older one for us to find out which one performs best.

Evaluation occurred by analyzing the log files from both the two interactive sessions and the informal talks that followed each interactive session.

Figures 2 through 4 provide a detailed overview of the performance of both the adaptive system and the more general non-adaptive system for the interpretation of each of the three classes of spatial relationships under consideration.

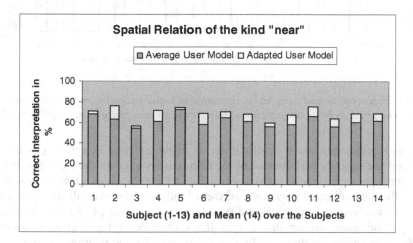

Fig. 2. User adaptation results for the interpretation of topological relations of the kind "near" by 13 subjects; the adaptive system over performs in 5.47% of the cases

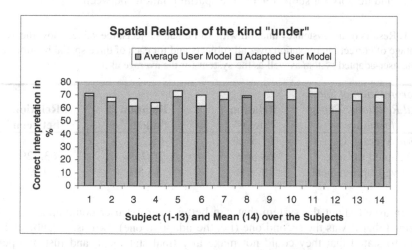

Fig. 3. User adaptation results for the interpretation of projective relations of the kind "under" by 13 subjects; the adaptive systems over performs in 7.3% of the cases

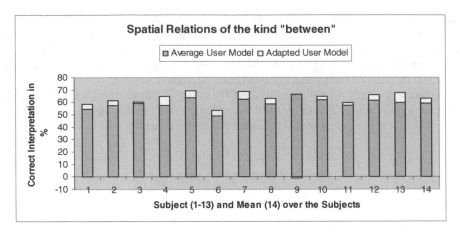

Fig. 4. User adaptation results for the interpretation of topological relations like "between" that involve two referents by 13 subjects; the adaptive systems over performs in 4.12% of the cases

Table 1 summarizes these results in terms of the average interpretation error and highlights the superior performance of the adaptive system over the more general one. The figures presented can be inferred from Table 2 that provides a tabular representation of the data collected from each of the 13 subjects and over all the spatial relations. The adaptive system achieves a relative error reduction (RER) compared to the more general game system that ranges from 18.97% for topological relations, to 15.5% for projective relations, to finally nearly 10% for the relation "between". The system created for the average user performs slightly better than the one tailored to the specific needs and habits of the user only in one single case (highlighted in Table 2) and notably for subject 9 and the spatial relation "between".

Table 1. Results of the system evaluation with 13 human subjects; the values show the average percentage of correct interpretation over all subjects and for each of three spatial relations using both the user-adapted system as well as the system for the average user

Interpretation of Spatial Relations: Performance	Relation "near"	Relation "under"	Relation "between"
For Average User	61.56%	65.28%	59.1%
Adapted to User	68.86%	70.75%	63.22%
Relative Improvement	18.97%	15.5%	10%

When asked about the system that can better interpret user commands, ten users confirmed that it was the second one (i.e. the adaptive one), two users (subject 1 and subject 9) stated that they could not make any final statement, and just one person (subject 8) indicated that the first system (i.e. the non adapted system) was the better one, despite it actually was not.

Table 2. Detailed analysis over all subjects and all spatial relations of the relative error reduction (RER) which is the percentage of increase/decrease of the correct interpretation of the user-adapted system with respect to the general one designated for the average user

Relative Error Reduction (RER)			
Spatial Relation Subject ID	"near"	"under"	"between"
1	10.38%	6.21%	9.21%
2	35.25%	9.77%	8.49%
3	5.26%	15.54%	2.91%
4	29.34%	11.39%	17.37%
5	7.64%	13.96%	16.44%
6	26.32%	22.08%	9.38%
7	15.58%	18.07%	16.53%
8	18.16%	4.73%	10.36%
9	8.24%	22.72%	-3.58%
10	22.84%	24.18%	6.84%
11	27.16%	15.44%	5.87%
12	17.87%	22.01%	10.94%
13	22.61%	15.43%	19.46%
Mean RER	18.97%	15.5%	10%
Standard Deviation	9.31%	6.29%	6.39%

It should be noticed that, despite our game system has an operative dialogue manager module that can engage the user in clarification requests to disambiguate incomplete, unclear utterances as well as misrecognition errors from the speech recognizer [5], we disabled such an agent during the evaluation sessions. Our decision is based on the fact that while a working dialogue manager is not necessarily relevant to the analysis of the phenomena we were investigating, it simultaneously does sometimes make wrong decisions thus introducing an additional complexity layer between the system and the user. The absence of a fully functional dialogue manager had the immediate consequence that during evaluation not all user instructions were executed in the desired (or undesired) way by the system. In case of a user instruction that did not directly translate into a game command, the system simply played back a text asking for a repetition of the utterance or an equivalent message.

4 Discussion and Conclusion

Traditionally, user interfaces are built incrementally as an iterative development process that involves the analysis of several factors such as modularity, reliability, efficiency, usability, and goal achievement. The usability aspects require specifically the collection of data about system use in various environments and on different tasks in order to improve the performance of the hypothetic average person representing the final user. The implicit assumption behind this strategy is that the subjects recruited to collect data constitute a representative homogeneous set. Yet, despite such interfaces work fine with the majority of the people in most situations and for a given task, a better system is tailored to the individual user rather than to an abstract typical person.

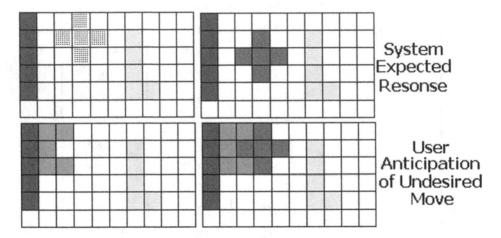

Fig. 5. (*1st row, left*) Given this game configuration, the dotted area indicates where the user intended to drop the cross-shaped piece for sentence "*place the cross between the pieces on-board*"; (*1st row, right*) Expected system response to the user's sentence; the player however does not issues that command as he anticipates the undesired result based on his mental model of the game; (*2nd row, left*) instead he changes his strategy and first places the U-shaped piece in the upper left region and then (*2nd row, right*) he drops the cross next to it (a command semantically equivalent to the originally intended one would have worked as well)

We described an extension of a computational model that approximates human interpretation and judgment of situated language while also accommodating for individual variability and ambiguity in spatial utterances. A proof-of-concept is made by incorporating our approach in the Pentomino board game involving speech user interaction. Observations of the user's behavior provided interaction data that we used to form a model designed to predict future systems actions in response to specific inputs. In the current development, a customized model takes into account playing patterns, level of experience, skill, etc. of the players for the interpretation of spatial expressions that occur while playing. This adaptive approach shows a relevant improvement of the performance with respect to the previous non-adaptive system.

The user model extrapolated by our application during the introductive session should not be confused with the mental model that each user has of the system itself. These two models are intertwined and contribute to the creation of each other. On the one hand, the user model is built by the system during a guided game session and thus is, almost by definition, based on the user's mental model of the game. On the other one, the opposite is true as well. In fact, we noticed that sometimes players anticipated expected, yet unwanted, system responses to a specific input command by changing the sequence of moves to obtain a desired game configuration (Figure 5). Such behaviors are a mere trick to prevent game moves and do not reflect the user interpretation of specific spatial relationships. Hence, they are potentially damaging to our application since they undermine the collection of consistent training data aimed at reflecting the player's intentions and game style.

It is worth noticing, that learning based on the user's interaction behavior may actually misinterpret the user's intention, when the user is making unexpected mistakes. It is therefore important to assist each user during the guided session to avoid misunderstandings. This fact needs to be considered particularly when an interface is to be used also by people with disabilities or by those who are prone to enter wrong inputs due to their physical limitations.

Due to the way we run the evaluation sessions, our system is little vulnerable to wrong interpretations of user input used as training data. Nonetheless, if an unknown user wants to play the game we need either a preliminary training session to create the user model or to collect training data on the fly while the user is playing. While the former is not always a feasible solution, the latter strategy is however prone to inevitable data interpretation errors. Indeed, erroneous training data is a major problem for any automated learning system, and usually a "smoothing" method is provided to partially account for it. At this point, we do not employ any such method but this is on top of our to-do list.

Our approach can be extended to other domains such as e.g. two-dimensional map based applications, other grid-like scenarios and geographic information systems also in the three dimensional case. A useful augmentation of the current system would be the resolution of spatial relations applied to groups of objects. For instance, an utterance like *"select the red pieces under the blue one(s)"* requires the selection of a subset of red pieces that are located underneath a specified referent which, in this case, can be itself a subgroup of the set of blue colored pieces. It remains to investigate user variability in the use of referential expressions involving groups of elements.

Acknowledgments: We are grateful to the EU Marie Curie Transfer of Knowledge Programme under Grant #FP6-2002-Mobility-3014491 for partial financial support.

References

1. Ardito, C., Pederson, T., Costabile, M.F.: CHAT - Towards a General-purpose Infrastructure for Multimodal Situation-adaptive User Assistance. In: Proc. ACM MobileHCI Workshop on Modelling and Designing User Assistance in Intelligent Envs, Espoo, Finland (2006)
2. Benyon, D.R.: Accommodating Individual Differences through an Adaptive User Interface. In: Schneider-Hufschmidt, M., Kühme, T., Malinowski, U. (eds.) Adaptive User Interfaces - Results and Prospects. Elsevier Science Publications, Amsterdam (1993)

3. Celentano, A., Gaggi, O.: Context-aware Design of Adaptable Multimodal Documents. Multimedia Tools Applications 29(1), 7–28 (2006)
4. Corradini, A.: A Computational Model for Spatial Expression Resolution. In: Proceedings of the Ninth International Conference on Multimodal Interfaces, Nagoya, Japan, pp. 67–70 (2007)
5. Corradini, A., Samuelsson, C.: A Generic Spoken Dialogue Manager Applied to an Interactive 2D Game. In: André, E., Dybkjær, L., Minker, W., Neumann, H., Pieraccini, R., Weber, M. (eds.) PIT 2008. LNCS (LNAI), vol. 5078, pp. 3–13. Springer, Heidelberg (2008)
6. De Bra, P., Brusilovsky, P., Houben, G.J.: Adaptive Hypermedia: From Systems to Framework. ACM Computing Surveys 31, 1–6 (1999)
7. Dey, A.K.: Understanding and Using Context. Personal Ubiquitous Computing 5(1), 4–7 (2001)
8. Dusan, S., Flanagan, J.: Adaptive Dialog Based upon Multimodal Language Acquisition. In: Proceedings of the Fourth IEEE International Conference on Multimodal Interfaces, Pittsburgh, PA, USA, pp. 135–140 (2002)
9. Fink, J., Kobsa, A., Nill, A.: Adaptable and Adaptive Information Provision for All Users, including Disabled and Elderly People. The New Review of Hypermedia e Multimedia 4, 163–188 (1998)
10. Fischer, G.: User Modeling in Human-Computer Interaction. In: User Modeling and User-Adapted Interaction, vol. 11, pp. 65–86. Kluwer Academic Publishers, Dordrecht (2001)
11. Gavalda, M., Waibel, A.: Growing Semantic Grammars. In: Proceedings of the ACL International Conference on Computational Linguistics, Montreal, Canada (1998)
12. Gorin, A.L., Riccardi, G., Wright, J.H.: How I Help You? Speech Communication. Speech Communication 23, 113–127 (1997)
13. Harter, A., Hopper, A., Steggles, P., Ward, A., Webster, P.: The Anatomy of a Context-aware Application. In: Proceedings of the 5th Annual ACM/IEEE International Conference on Mobile Computing and Networking, Seattle, WA, USA, pp. 59–68 (1999)
14. Horvitz, E., Breese, J., Heckerman, D., Hovel, D., Rommelse, K.: The Lumiere Project: Bayesian User Modeling for Inferring the Goals and the Needs of Software Users. In: Proceedings of the 14th Conference on Uncertainty in Artificial Intelligence, Madison, WI, USA, pp. 256–265 (1998)
15. Horvitz, E.: Users are Individuals: Individualizing User Models. In: Proceedings of the ACM CHI, Pittsburgh, PA, USA, pp. 159–166 (1999)
16. Johnston, M., Bangalore, S., Vasireddy, G., Stent, A., Ehlen, P., Walker, M., Whittaker, S., Maloor, P.: MATCH: An Architecture for Multimodal Dialogue Systems. In: Proc. of Annual Meeting of Association for Computational Linguistics, Philadelphia, PA, USA (2002)
17. Kaiser, E.: Using redundant speech and handwriting for learning new vocabulary and understanding abbreviations. In: Proceedings of the International Conference on Multimodal Interfaces, Banff, Alberta, Canada, pp. 347–356 (2006)
18. Kobsa, A.: Generic User Modeling Systems. In: User Modeling and User-Adapted Interaction, pp. 49–63. Kluwer Academic Publishers, Dordrecht (2001)
19. Kobsa, A., Wahlster, W.: User Models in Dialog Systems. Springer, New York (1989)
20. Natarajan, P., Prasad, R., Suhm, B., McCarthy, D.: Speech Enabled Natural Language Call Routing: BBN Call Director. In: Proceedings of the International Conference on Spoken Language Processing, Denver, CO, USA (2002)
21. Oviatt, S.L., Darrell, T., Flickner, M.: Multimodal Interfaces that Flex, Adapt, and Persist. Communications of the ACM, 30–33 (2004)

22. Pieraccini, R., Lubensky, D.: Spoken Language Communication with Machines: The Long and Winding Road from Research to Business. In: Proceedings of the 18th International Conference on Industrial and Engineering Applications of Artificial Intelligence and Expert Systems, Bari, Italy, pp. 6–15 (2005)
23. Panttaja, E., Reitter, D., Cummins, F.: The evaluation of adaptable multimodal system outputs. In: Proceedings of the DUMAS Workshop on Robust and Adaptive Information Processing for Mobile Speech Interfaces (2004)
24. Rich, E.: Principles of Mixed-Initiative User Interfaces. International Journal of Man-Machine Sciences 18, 199–214 (1981)
25. Schmidt, A., Beigl, M., Gellersen, H.-W.: There is more to Context than Location. Computers and Graphics 23(6), 893–901 (1999)
26. Stent, A., Walker, M., Whittaker, S., Maloor, P.: User-Tailored Generation for Spoken Dialogue: An Experiment. In: Proceedings of International Conference on Spoken Language Processing, Denver, CO, USA, pp. 1281–1284 (2002)
27. Strik, H., Russel, A., van den Heuvel, H., Cucchiarini, C., Boves, L.: A Spoken Dialogue System for the Dutch Public Transport Information Service. International Journal on Speech Technology 2(2), 119–129 (1997)
28. Walker, M., Rudnicky, A., Prasad, R., Aberdeen, J., Bratt, E., Garofolo, J., Hastie, H., Le, A., Pellom, B., Potamianos, A., Passonneau, R., Roukos, S., Sanders, G., Seneff, S., Stallard, D.: Darpa Communicator: Cross-system Results for the 2001. In: Proceedings of the International Conference on Spoken Language Processing, Denver, CO, USA (2002)
29. Walker, W., Lamere, P., Kwok, P.: FreeTTS – A Performance Case Study. Technical Report SMLI TR-2002-114, Sun Microsystems Inc. (2002)
30. Walker, W., Lamere, P., Kwok, P., Raj, B., Singh, R., Gouvea, E., Wolf, P., Woelfel, J.: Sphinx-4: A Flexible Open Source Framework for Speech Recognition. Technical Report SMLI TR-2004-0811, Sun Microsystems Inc. (2004)
31. Webb, G.I., Pazzani, M.J., Billsus, D.: Machine Learning for User Modeling. In: User Modeling and User-Adapted Interaction, vol. 11, pp. 19–29. Kluwer Academic Pubs., Dordrecht (2001)
32. Zukerman, I., Litman, D.: Natural Language Processing and User Modeling: Synergies and Limitations. User Modeling and User-Adapted Interaction 11(1-2), 129–158 (2001)

ASAP- An Advanced System for Assessing Chat Participants

Mihai Dascalu[1], Erol-Valeriu Chioasca[1], and Stefan Trausan-Matu[1,2]

[1] "Politehnica" University of Bucharest,
313, Splaiul Indepentei, 060042 Bucharest, Romania
[2] Romanian Academy Research Institute for Artificial Intelligence,
13, Calea 13 Septembrie
Bucharest, Romania
{mikedascalu,erolvaleriu}@yahoo.com, stefan.trausan@cs.pub.ro

Abstract. The paper presents a method and an implemented system for the assessment of the participants' competences in a collaborative environment based on an instant messenger conversation (chat). For each utterance in the chat, a score is computed that takes into account several features, specific to text mining (like the presence and the density of keywords, via synonymy), natural language pragmatics and to social networks. The total rating of the competence of a participant is computed considering the scores of utterances and inter-utterance factors. Within the frame of the developed system, special attention was given to multiple ways of visualizing the analysis' results. An annotation editor was also implemented and used in order to construct a "golden standard", which was further employed for the evaluation of the developed assessment tools.

Keywords: Computer Supported Collaborative Learning, Social Networks, Natural Language Processing, WordNet, Text Mining.

1 Introduction

In the last few years, at the same time as computers spread and developed, new communication environments have emerged. For example, instant messaging (chat) determined a change in the way collaborative work is regarded, becoming a viable alternative to classic learning: Computer Supported Collaborative Learning [5]. However, in this new context new problems have appeared related to the kind of computer support that may be provided. For instance, automatic assessment tools are required since the examination of the chat transcripts by a professor is an extremely time consuming task. The development of such assessment tools is essentially difficult because of natural language processing, constituting one of the most intricate domains of artificial intelligence.

In contrast with the classical, cognitive approach to artificial intelligence, Computer Supported Collaborative Learning moves towards a socio-cultural paradigm, focusing on the idea that knowledge is created collaboratively [5]. Therefore, the

D. Dochev, M. Pistore, and P. Traverso (Eds.): AIMSA 2008, LNAI 5253, pp. 58–68, 2008.
© Springer-Verlag Berlin Heidelberg 2008

implemented tools should also take into account the idea of this new paradigm. Consequently, in the research presented in this paper, analysis techniques from the social networks theory [1] were employed, in addition to classical statistical natural language processing [4].

Taking as a starting point the utterances of a chat, their sequencing and the social network of the participants, the paper presents and evaluates an automatic assessment tool which assigns a grade to each participant. In order to achieve this goal, several features are considered, starting from the simplest, such as the dimension of utterances, and ending with pragmatics issues such as speech acts and even social aspects.

The second section introduces some basic notions about evaluating chats using social networks analysis techniques. The third section presents the approach used for the assessment system. The evaluation of the performance of this system is discussed in the fourth section.

2   Analysis Factors

There are several analysis factors that we have considered for discovering the most competent user in a chat, ranging from very simple to complex. Many of these factors are based on social network analysis [1] and some of them on the content of the utterances.

The number of characters written by a user is probably the simplest feature to consider when searching for the most competent participant in a chat conversation. This feature provides an indicator for the contribution of each participant, but it is obviously naive, because the content, the quality of the utterances is far more important. Average efficiency should be regarded as a balance between the length of the intervention and the consistency of the information that it comprises of.

Therefore, in addition to the number of characters per utterance, features that reflect the content should influence the mark obtained by a participant to the chat. For this purpose, the analysis process should consider the keywords or the topics of the discussion and marks be given accordingly.

Moreover, besides quantity and quality inferred from the utterances, social factors should also be taken into account. Consequently, a graph is generated from the chat transcript in concordance with the utterances exchanged between the participants. From the graph theory's point of view, the first two measures taken into consideration for the subsequent processing are the in-degree (number of arcs that enter a graph's node) and out-degree (the number of arcs that leave a node) [2].

The social networks theory [1] analyzes the number and distribution of arcs (number of exchanged utterances) among the chat participants. A social network is a graph which considers the participants in the cooperative or collaborative activity as nodes and the interactions among them as arcs (see figure 1). Examples of arcs may be the emails that are sent between the members of a group or chat utterances that refer previous ones using the explicit referencing facility of chat systems as ConcertChat [3] or the implicit links discovered by natural language processing techniques [5, 6].

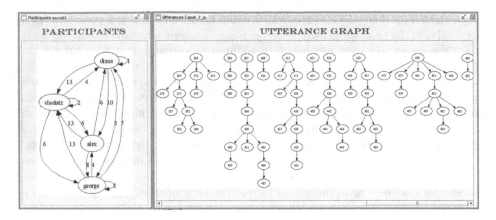

Fig. 1. Participants' social network and the utterance graph

As an example of the analysis of social interactions, let's start from the fact that in a chat environment every participant can fall under several categories, such as *knowledgeable*, *gregarious* or *passive*. From the point of view of the social network, a knowledgeable user is one who influences other users with whom he is linked (these neighbors have a limited number of links to other nodes). He is usually defined as a participant who receives a lot of questions and answers them correctly in a large proportion and therefore is more sought-after.

On the other hand, a gregarious user is a very active participant in the chat, enlarging the question pool. If one is to analyze the activity of the gregarious groups, the experts - usually marked as knowledgeable users - are discovered, thus their activity is crucial for the success of the chat and evaluation program.

A passive user generally responds only to questions specifically addressed to him. Because of the lack of involvement in active threads, although his remarks might have an important role to play in the overall evolution of the chat, the evaluation system might have difficulties estimating his overall performances, mostly from the perspective of social network analysis. This type must not be mistaken for an inactive user who has no direct impact on the ongoing discussion, posing no important questions, replies or remarks in the chat.

Social networks may be used for computing various metrics. One example is the centrality degree, defined as the extent to which a participant has a large in-degree or out-degree. This information is useful during the network analysis and particularly during the social relations analysis incorporated in it. Usually, this analysis is quantitative in character; however a qualitative one exists as well, and in this case the type of the utterance is determined, discovering what the items express.

Five types of centralities may be identified: closeness, graph centrality, betweenness, stress, and eigenvector [1]. Closeness is the measure of centrality of a node in a graph. It is proportional with the inverse of the minimal distance between the current node and all the other nodes in the graph.

In a social network, the graph centrality of a node is important for determining the relative closeness of the participants. It is computed as the maximal distance between a considered node and all the other nodes in the graph.

The Floyd-Warshall algorithm [2] can be used both for the graph's centrality as well as for the closeness, because it provides the minimum distance between each pair of nodes in an $O(n^3)$ complexity. Another measure of centrality in a graph is the betweenness centrality, which states that nodes with higher degrees can be located on more than one shortest path [1].

In the case of the stress factor, centrality is correlated with the sum of all shortest paths within the graph passing through the selected node. Centrality obtained through the measure of eigenvalues is another method of determining the importance of a participant from the network. This factor attaches relative marks to every node in the network on the following principle: a connection to a higher ranking node is more important than a set of connections to inferior ranked nodes [1].

3 The Assessment Tool

The first problem encountered in chat processing was the correction of spelling errors. Unfortunately, due to the nature of on-line keyboard interfacing in chats, the rate of occurrence of these errors is rather high. Other problems encountered are abbreviations and emoticons.

In order to address spellchecking, the Jazzy library is used (http://jazzy.sourceforge.net/, http://www.ibm.com/developerworks/java/library/j-jazzy/), which is based on the Aspell algorithm (http://aspell.net/), combining techniques from several previous spellchecking algorithms, such as phonetic matching and sequence comparison.

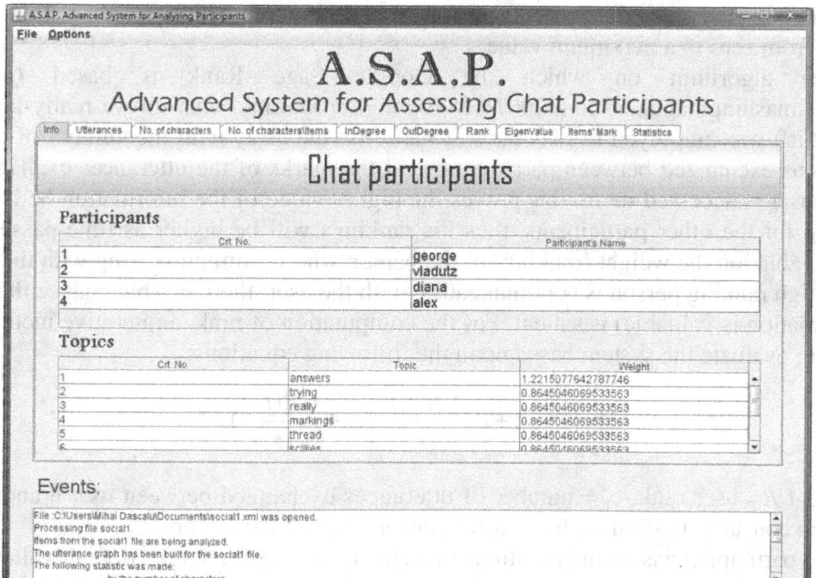

Fig. 2. The ASAP interface

Considering the peculiarities of chats, several improvements have been made:

- Jazzy does not detect a frequent error found in chats, the missing space between two words. Therefore, unknown words are divided in two and spellchecking is performed on each. Afterwards, based on the Levenshtein distance, it is verified if the initial word is a concatenation of two words. The considering of possible positions of the missing space follows the observed frequency of the occurrences of this error.
- A graph is built starting from a dictionary in which the occurrence frequency of each word is stored. In the case of a misspelling, these frequencies are also considered for selecting the best suggestion provided by Jazzy (that takes into account both phonetic similarities and Levenshtein distance),

The next chat processing steps are the elimination of stop words (very frequent words that bring no content, as prepositions, conjunctions, determiners, etc.) and stemming. The Snowball stemmer (http://snowball.tartarus.org/) was chosen because, in the context of the prior usage of a spellchecker, it offered better results, as compared to Porter (http://tartarus.org/~martin/PorterStemmer/) or Lovin stemmers (http://sourceforge.net/project/showfiles.php?group_id=24260). Moreover, Snowball is recommended even on Porter's web page (http://tartarus.org/~martin/PorterStemmer/).

To compute the percentage of each participant's eigenvalue in a chat, the following assumption is made: if all values are negative, the absolute values are considered; if both positive and negative values exist, then the percentage is distributed between the highest positive value and the lowest negative number; in the case that all eigenvalues are greater than zero, the percentage values will be computed as a distribution on a scale from zero to a maximum value.

The algorithm on which the Google Page Rank is based (http://www.smashingmagazine.com/2007/06/05/google-pagerank-what-do-we-really-know-about-it/) was employed to rank an user based on the array with the number of interventions exchanged between participants and the marks of the utterances exchanged. If an user is accessed more, this proves the higher value of the information he is providing for the other participants, thus his rankings will be higher as time passes. In this evaluation the weight (ranking) of the person who is communicating with the user (if a high ranking person is communicating with the user, there is a big chance that the information is valuable) is salient. For the computation of rank, an iterative method is used to evaluate the system based upon the following equations:

$$UR(A) = (1-d) + d\left(\frac{UR(t_1)}{c(t_1)} + ... + \frac{UR(t_n)}{c(t_n)}\right),$$ (1)

where UR = user rank; c_i = number of utterances exchanged between user t_i and user A; d = a constant (0.85) used for a faster convergence of the system.

If a participant has no interventions in a chat his rank is 0, not the default value 1.

Furthermore, taking quality, quantity and efficiency into account, the mark of the intervention is based on both the words used and the length of the utterance. By words we understand only those terms that are relevant to speech, and not the stop words, which are eliminated from the analysis.

In the evaluation of an intervention, besides the length, the following factors are considered:

- The number of keywords which remain after the spell-check, stemming and elimination of stop words - *MClength*.
- The number of times that any given word is repeated during a sentence *no_ocuurences*.
- The level at which the participant's intervention can be found in the utterance hierarchy (see figure 1) constructed along the referencing facility [3].

Thus the following formula is obtained:

$$mark_{empiric} = \frac{\frac{length}{6} + \frac{MClength*5}{6}}{1 + \frac{1}{level}} * \left(1 + \frac{1}{10} * \sum_k \log_2(no\_occurances_k + 1)\right) \qquad (2)$$

Based upon this empiric mark, the final mark of the utterance is obtained:

$$mark_{final} = mark_{previous\_utterance} + coefficient * mark_{empiric} , \qquad (3)$$

where the coefficient is determined by the type of the current utterance and the utterance to which it is tied.

The in-degree, out-degree, centrality, eigenvalues and rank factors are applied on the array with the number of interchanged utterances and the array which takes into consideration the empiric mark (2) of an utterance instead of the default value of 1 (an utterance has been changed between users), therefore quality, not quantity is important (an element [i,j] equals the sum of $mark_{empiric}$ for each utterance from participant i to participant j). This approach provides a deeper analysis of the chat by using a social network's approach based on a semantic utterance evaluation.

Each of the analysis factors (applied on both arrays, besides the number of written characters and the number of characters per utterance) is converted to a percentage (current grade/sum of all grades for each factor).

The final grade for each participant takes into consideration all these factors and the weight of each one:

$$final\_grade_i = \sum_k weight_k * percentage_{k,i} , \qquad (4)$$

where k is a factor used in the final evaluation of the participant i and the weight of each factor is read from a configuration file.

Equation (3) is important also for the determination of other important information, like collaboration regions (sequences of utterances where collaborative moments [5] appear or, in other words, when the participants inter-animate [7]). These regions are detected starting from the idea that a thread or a set of threads in which the final mark increases is likely to be part of a zone in which important aspects are being treated.

An important role in choosing the coefficient is played by the identification of speech acts in every utterance [6]. For this aim, the verbs as well as the presence of some punctuation signs and certain keywords in the intervention are inspected.

It is in this way that a list of features per utterance is obtained, and the coefficient is determined depending on the list of attributes of the previous utterance (with which the current utterance is linked). In the current implementation; an array containing these values exists, and the noted attributes are: negations, confirmations, questions and affirmations.

In the evaluation of an utterance, a semantic analysis of the social network may be performed and a filtering of the utterances will be made, from the point of view of the likelihood of the terms used through a list of words introduced by the user or through a list of keywords of the chat, statistically determined.

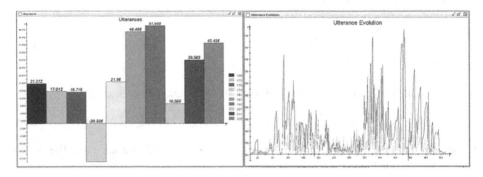

Fig. 3. Charts generated by A.S.A.P. representing: 1. Utterances' evolution in a single thread in chat; 2. Utterances' evolution in the whole chat;

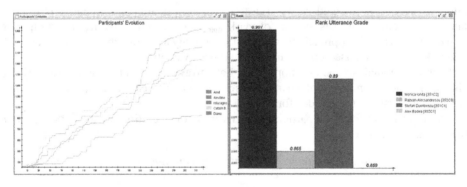

Fig. 4. 1. Overall participants' evolution; 2. Comparative results for a given factor

The lexical ontology WordNet (http://wordnet.princeton.edu) was used due to its facilities for finding semantic similarities. It was also used in detecting speech acts.

The grade of a discussion thread may be raised or lowered by each utterance (see figure 3). Therefore, depending on the type of an utterance and the speech act, the grade may have a positive or negative weight. The resulted marks are used for building a top of chat participants and their evolution (see figure 4). These calculations will be improved in the next version of the system by using an utterance type table containing weights for each type.

The application has been developed in JAVA 1.6 and graphics was implemented in SWING. JopenChart Drawer (http://jopenchart.sourceforge.net/) was used for generating the graph. Centrality and singular values were computed with the JAMA algebraic calculus library (http://math.nist.gov/javanumerics/jama/).

4 The Evaluation of the Assessment Tool

Annotated corpora are needed in order to run machine learning algorithms that train the analysis modules and to evaluate and tune the assessment tool, presented above,. Therefore, in order to facilitate the annotation, an editing program was developed. The application allows the chat annotation in an XML format with the following data:

- Comments and grades for an individual utterance.
- Grades for a participant's 20 successive utterances, reflecting his contribution to that segment of the conversation.
- Global grades for each participant in the chat.
- Two global grades on the chat as a whole: the degree to which the collaboration in the chat was successful and the degree ton which the chat follows the topics of the discussion.
- Implicit links between utterances, including their type and associated patterns.
- The main topics (keywords) of the chat.
- Segments of successful collaboration.

The annotation editor was used by more than 250 students in two courses at the "Politehnica" University of Bucharest: Human-Computer Interaction and Natural Language Processing. As a consequence, a golden standard was obtained for the evaluation of the performances of the system that grades the chat participants. In the future it will be used also for applying machine learning for developing the analysis tool.

In order to analyze the performance and the correctness of the method previously described, the evaluation process compares the results of the assessment tool with those of the golden standard. The overall task of evaluating each chat in the corpus is divided into smaller tasks (each chat has its own task) and all are inserted into a work pool.

A task consists of the following operations:

1. Perform the ASAP automatic evaluation on the current chat.
2. Obtain a certain grade (from 0 to 10) for each participant in the chat. Because ASAP provides a percentage for each participant, this is converted to a grade using a linear approximation. Therefore, the grade 10 is given to a participant with the maximum final percentage, and all other grades are calculated as the current grade divided by the maximum grade. Because grades in the evaluated corpus are generally high, a logarithmic approach will be taken into consideration.
3. Generate an XML file to save the current grades of the participants in the chat.

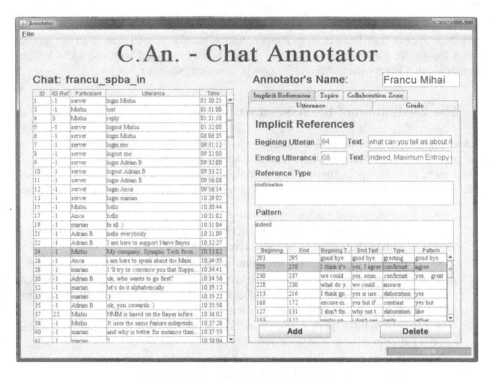

Fig. 5. The interface for the Chat Annotator

4. Compare the results obtained using the ASAP system and the grades given by an annotator (the "golden standard") obtained from the editor for corpus annotation and save them as a new XML file. Two measurements are taken into account: the absolute and the relative correctness for each participant's grade. Eventually, the final results – the average absolute and the relative correctness on all the chats considered – are displayed in the main window of the interface, as it can be seen in figure 6. The final results are also saved in an XML format, in order to centralize them.

Fig. 6. The main window of the Corpus Evaluation program

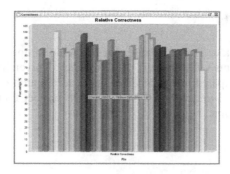

Fig. 7. Chart with relative correctness of all files in the corpus

Charts are also displayed in order to observe the evaluation results for each chat.

The average results obtained for the corpus are good (about 85% relative correctness and 75% absolute correctness in grading participants in each chat). Relative correctness and absolute correctness represent absolute/relative distances in a one-dimensional space, where the annotator's grade and the grade obtained automatically using the ASAP system are taken into consideration. The values for the average deviations suggest that the ASAP program is very close to really assessing students' capabilities within chat environments. We strongly believe that with further improvements and adjusted percentages for each factor, the results will improve. However, the human, subjective factor in the manual evaluation must be also taken into account.

5 Conclusions and Future Improvements

The first results allow us to conclude that the competence of a person in a chat can be automatically determined. In the future, the following improvements are going to be addressed:

- The discovery of implicit references between utterances using WordNet and LSA by finding the link between words in the same semantic field but in different utterances;
- Refining the intervention assessment method (the level of influence), by giving grades to utterances, based upon certain key elements: position of a specific word contained in that particular sentence, the actors involved and overall impact by correctly identifying speech acts to correlate interventions in a hierarchy.
- Identifying chat topics through the aid of LSA and WordNet, by narrowing down areas of discussion to common concepts (e.g. synonyms, words which have the same semantic field);
- After the creation of a broad collection of evaluated chats, reverse indexing to determine the most competent user overall;

Acknowledgements

We would like to thank to Traian Rebedea for his help in improving the quality of the text and to the students of Bucharest "Politehnica" University Computer Science Department that participated in our experiments. The research presented in this paper was partially performed under the FP7 EU STREP project LTfLL and the national CNCSIS project K-Teams.

References

1. Brandes, U.: A Faster Algorithm for Betweenness Centrality. Journal of Mathematical Sociology 25(2), 163–177 (2001)
2. Cormen, T., Leiserson, C., Rivest, R., Stein, C.: Introduction to Algorithms. MIT Press, Cambridge (2001)
3. Holmer, T., Kienle, A., Wessner, M.: Explicit Referencing in Learning Chats: Needs and Acceptance. In: Nejdl, W., Tochtermann, K. (eds.) EC-TEL 2006. LNCS, vol. 4227, pp. 170–184. Springer, Heidelberg (2006)
4. Manning, C., Schutze, H.: Foundations of Statistical Natural Language Processing. MIT Press, Cambridge (1999)
5. Stahl, G.: Group Cognition: Computer Support for Building Collaborative Knowledge. MIT Press, Cambridge (2006)
6. Trausan-Matu, S., Chiru, C., Bogdan, R.: Identificarea actelor de vorbire în dialogurile purtate pe chat, în Stefan Trăusan-Matu, Costin Pribeanu (Eds.), Interactiune Om-Calculator 2004, Editura Printech, Bucureşti, pp. 206–214 (2004)
7. Trausan-Matu, S., Stahl, G., Sarmiento, J.: Supporting Polyphonic Collaborative Learning. E-service Journal 6(1), 58–74 (2007)

Using Text Segmentation to Enhance the Cluster Hypothesis

Sylvain Lamprier*, Tassadit Amghar, Bernard Levrat, and Frédéric Saubion

LERIA - University of Angers, 2 Bd Lavoisier 49000 Angers, France
{lamprier,amghar,levrat,saubion}@info.univ-angers.fr

Abstract. An alternative way to tackle Information Retrieval, called Passage Retrieval, considers text fragments independently rather than assessing global relevance of documents. In such a context, the fact that relevant information is surrounded by parts of text deviating from the interesting topic does not penalize the document. In this paper, we propose to study the impact of the consideration of these text fragments on a document clustering process. The use of clustering in the field of Information Retrieval is mainly supported by the cluster hypothesis which states that relevant documents tend to be more similar one to each other than to non-relevant documents and hence a clustering process is likely to gather them. Previous experiments have shown that clustering the first retrieved documents as response to a user's query allows the Information Retrieval systems to improve their effectiveness. In the clustering process used in these studies, documents have been considered globally. Nevertheless, the assumption stating that a document can refer to more than one topic/concept may have also impacts on the document clustering process. Considering passages of the retrieved documents separately may allow to create more representative clusters of the addressed topics. Different approaches have been assessed and results show that using text fragments in the clustering process may turn out to be actually relevant.

Keywords: Information Retrieval, Text Segmentation, Cluster Hypothesis.

1 Introduction

An Information Retrieval (IR) system typically returns, as response to a user's query, a ranked list of documents. The documents in the list are ordered according to their likelyhood of being relevant to the request and the user has to examine each of them linearly, by beginning with the highest ranked. This organizational approach is adopted by most of existing search engines [1]. Nevertheless, there is a significant potential benefit in providing additional structures that could reduce the user's cognitive efforts. In this aim, several alternative organization approaches have been investigated over the recent years, most of them relying on document clustering [2,3,4]. Indeed, the emphasis of document

---

* With the support of Angers Loire Metropole.

D. Dochev, M. Pistore, and P. Traverso (Eds.): AIMSA 2008, LNAI 5253, pp. 69–82, 2008.
© Springer-Verlag Berlin Heidelberg 2008

relations may orient the user in its information search [5]. Document clustering techniques [6] have then been widely applied in IR in order to improve search and retrieval efficiency. Initially, clustering techniques were applied statically (i.e. over the whole document collection) [5,7]. However, more recent studies [8,9,10] have shown that better results can be obtained when the clustering is applied to the results of an IR system (i.e. a query-specific clustering). Since all of the retrieved documents are likely to fall in the same field, the resulting clustering may be more acute (moreover, the search area being smaller, the clustering process is less complex). The use of clustering in this context is mainly supported by the cluster hypothesis [8,11], which states that relevant documents tend to be more similar[1] one to each other than to non-relevant documents and hence are likely to appear in the same clusters. If this assumption holds for a given corpus of texts, relevant documents having been ranked low by the initial search engine, and thus having few chances to be examined by the user, might be grouped with other relevant documents, improving then the effectiveness of the IR system.

In all clustering studies reported above, clustering has been applied to entire documents. Nevertheless, some studies have shown that, in most cases, retrieving fragments of texts or using them to compute the proximity of a document with a query is more effective than considering entire documents [13,14]. Documents may contain fragments of texts related to different topics and considering them separately may allow to identify short blocks of relevant text inside of documents otherwise judged irrelevant. In passage retrieval approaches, where similarities are computed for each passage of the document set, a document in which only one passage is very similar to a query (others being really far) is more likely to be ranked at the top of the list since, according to the heuristic used, the relevant passage may be more weighted than others. Moreover, using short fragments of texts reintroduces a notion of locality of terms in the text since terms have to belong to the same fragment to be taken into account together. It limits biases induced by the classical 'bag of words' representation of documents.

The assumption stating that a document can refer to more than one topic may have also impacts on the document clustering process. Despite different thematic developments, documents may be strongly connected one to each other w.r.t. some particular parts of them. A distance measure depending on the whole documents is likely to judge them as dissimilar. Consequently, this paper first experiments an approach in which the similarity between two documents, rather than considering their whole sets of terms, is equal to the similarity of their closest thematic passages (resulting from a thematic segmentation process). In that case, the proximity of the documents may be more finely assessed, by widely increasing the similarity score of documents addressing strongly related topics. On the other hand, adding an heterogeneous document to a cluster may make deviating the cluster somewhat from its main topic because of parts of text which have nothing in common with the topic in concern. This heterogeneous

---

[1] Document similarities are usually evaluated thanks to measures depending on the term co-occurrences. Here, the terms, the indexing units, are meaningful words obtained after having removed stop words and applied a stemming process [12].

document may then attract documents far from the initial topic by some kind of 'transitivity'. This second observation led us to consider a query-oriented clustering of documents in which the proximity of a document with others only depends on the set of terms of its most similar passage to the query. In this more static process, the considered passage remains the same for each similarity computation and then biases induced by the heterogeneity of the documents are laid aside. Nevertheless, such an approach may lead to ignore some important aspects of the documents. In classical approaches, the benefit resulting from the document clustering arises from the fact that the identification of the relevant documents does not depend on the sole query terms but also considers the relationships between documents. Our approach may lead to ignore a part of a relevant document which could have allowed to group it with other relevant ones. Consequently to these observations, the main idea of this paper is to realize a preliminary clustering of homogeneous parts of text which could allow to produce clusters being more cohesive and representative of the collection topics. We extend so the cluster hypothesis itself, assuming that relevant fragments tend to be more similar one to each other than to non-relevant ones and hence are likely to appear in the same clusters. We expect the documents clusters, finally deduced from the passage clustering, to be more centered around specific topics, enhancing then the potential of the cluster hypothesis. Section 2 details our query-specific passage clustering approach and section 3 finally gives and interprets results of the experiments.

2   Query-Specific Passage Clustering

This section first presents the IR system used to obtain the initial ranked list of documents. Then, it describes the thematic segmentation process, the clustering method and the inter-document similarity measure used. At last, it addresses the different ways to deduce a document clustering from the segments clusters.

2.1   Initial Retrieval System

In [14], fragments of texts are defined at indexing time. They are then indexed as independent entities and the retrieval system returns a ranked list of fragments as response to the user's query. In order to evaluate their approach, authors reconstitute the ranked list of documents according to the ranked list of fragments. Here, the aim is not to assess the effectiveness of passage retrieval, which has been widely explored in the past [13,14], but to evaluate the impact of a segmentation of texts on a query-specific clustering of documents. We focus on the end of the retrieval process, the reorganization of the results provided by the initial system. Therefore, in our case, indexed and retrieved units are documents of the corpus, the segmentation of texts is applied afterward, on the retrieved set of documents.

The system used for this initial documents retrieval w.r.t. a user's query is based on the Cornell's Smart system [15] which relies on the well known vectorial model [16] where texts are encoded as vectors of weights w.r.t. their set of meaningful terms. The proximity between a query and a document can be computed as a Cosine measure:

$$Sim(D, Q) = \frac{\sum_{i=1}^{t} w_{D,i} \times w_{Q,i}}{\sqrt{\sum_{i=1}^{t} w_{D,i}^2 \times \sum_{i=1}^{t} w_{Q,i}^2}} \qquad (1)$$

where D represents the document whose similarity with the query Q has to be assessed, t the number of unique meaningful terms contained in $D \cup Q$, $w_{D,i}$ and $w_{Q,i}$ the weights of the term i in the document and the query respectively. Following the tf-idf weighting scheme, these weights are equal to:

$$w_{D,i} = 1 + \ln(tf_{D,i})$$
$$w_{Q,i} = (1 + \ln(tf_{Q,i})) \times \ln \frac{N}{n_i} \qquad (2)$$

where $tf_{D,i}$ and $tf_{Q,i}$ are the numbers of occurrences of the term i in the document D and the query Q respectively, N is the total number of documents in the collection and n_i is the number of documents in which the term i is present.

In [10], authors examined the effectiveness variation of a cluster-based IR system considering different numbers of top-ranked retrieved documents. Here, given the good results obtained with low numbers of documents in [10], experiments are realized with the fifty first documents retrieved.

2.2   Text Segmentation

Since the aim of the study is to assess the effectiveness of a system using a query-specific clustering of parts of texts to cluster documents, the first step to achieve is the segmentation of the documents. In [14], authors have shown that the best results of passage retrieval are obtained with arbitrary overlapping parts of text. However, this kind of passages cannot be used here because of the induced complexity. Moreover, since the aim is to cluster homogeneous parts of texts, arbitrary passages are not appropriate.

The purpose of thematic text segmentation is to identify the most important thematic breaks in a document in order to cut it into homogeneous units, disconnected from other adjacent parts [17]. More precisely, the segmentation process partitions a text by determining boundaries between contiguous segments related to different topics, defining so semantically coherent parts of text that are sufficiently large to expose some aspect of a given subject. Thematic segmentation of texts can also be seen as a grouping process of basic units (words, sentences, paragraphs...) in order to highlight local semantical coherences [18]. The granularity level of the segmentation depends on the size of the units gathered.

Many segmentation methods have been proposed, most of them relying on statistical approaches such as Text Tiling [19], C99 [20] or SegGen [21]. These approaches are based on an analysis of the distribution of the words in the text to locate the less thematically cohesive locations. SegGen [21], which appears to realize the most accurate segmentation of texts, uses a multiobjective evolutionary algorithm to maximize the internal segment cohesion and to minimize the similarity between adjacent segments. Contrarily to most of existing methods, which create boundaries sequentially, it evaluates segmentations of whole texts. This provides a more complete view of the layout of the text and allows to compute easier the measures of internal cohesion of segments and similarity between neighboring ones. This method is used in our experiments to segment the documents retrieved by the initial system. The total number of segments in each corpus is given in table 1.

2.3   Clustering Algorithm

Two types of clustering methods have been applied to IR: partitioning approaches and hierarchic ones [22]. Partitioning methods present the advantage of benefit from a low computational requirements [23] but are generally limited by their use of several arbitrarilyy determined parameters (such as the number of clusters to create). This observation makes partitioning methods inappropriate for query-specific clustering [10] since the number of topics that will be present in the retrieved document collection is unknown. Hierarchic methods, whose principle is to build a tree where each node is a cluster gathering two clusters of a lower level, appear to be, thanks to their low number of required parameters, the best suited for the clustering of responses furnished by an IR system. The search of a cluster satisfying a retrieval criterion in the hierarchy is independent from the clustering process [11]. At the beginning of the process, each object is assigned to a different cluster. Until all objects belong to a single cluster (the root of the tree), the algorithm merges the two closest clusters. While the distance between two singleton clusters is well-defined (the opposite of the similarity of both concerned objects), differences between methods rely on the heuristic of modification of the initial inter-cluster distance matrix after the merging of two clusters. In [10], the best results have been obtained with the Group Average method which defines the distance between two clusters as the average of the distances between every objects of both clusters normalized according to the clusters sizes. This method is the one used in our experiments.

As it is done for the similarity between a document and a query, the similarity $Sim(T, T')$ between two texts (documents or passages) is computed as a Cosine measure (formula 1). Term weights of both texts are computed as the weights of query terms:

$$w_{T,i} = (1 + \ln(tf_{T,i})) \times \ln \frac{N}{n_i}$$
$$w_{T',i} = (1 + \ln(tf_{T',i})) \times \ln \frac{N}{n_i}$$

In the case of segments clustering, passages of a same document, because of their common vocabulary or writing style, are likely to obtain high similarity scores and hence to be grouped in same clusters, what roughly comes down to a classical document clustering and leads to not really significant results differences. In order to prevent this side effect, the similarity of segments belonging to a same document is set to the average similarity between segments of different documents.

2.4   Documents Assignments

Two kinds of documents clustering can be deduced from the segments clustering: a soft one and a hard one. While a hard clustering refers to the assignment of a candidate document to a single cluster, a soft clustering allows the assignment of a same document to multiple clusters [24,25].

In recent years, soft clustering algorithms have been studied in document clustering contexts [24]. The fuzzy C-means algorithm [26] is a soft version of the K-means algorithm which uses a fuzzy membership function. A lot of approaches use this algorithm [27,24] but fuzzy C-means algorithms have the well-known problem of their dependency on initialization [25]. Another widely used approach for soft clustering is the Expectation-Maximization (EM) algorithm [28], which generates classification probabilities of documents. The classical EM algorithm produces good soft clustering of documents but encounters convergence difficulties when the search space is huge. To our knowledge, not any study has attempted to take profit from the observations made by passage retrieval approaches, i.e. using the locality of terms in texts and considering documents as a set of different topics. Moreover, no approaches have led to a hierarchical soft clustering of documents and the impact of soft clustering on the cluster hypothesis has not been investigated. In our case, the soft clustering of documents can be easily achieved from the clustering of their segments: the algorithm simply replaces, in clusters, each segment by the document it belongs to (if a document is assigned twice in a same cluster, a unique occurrence is kept). This process of substitution produces a soft clustering since clusters of a same level overlap each other. Each document finally appears in clusters representing the topics of its segments. Each cluster is likely to be centered around one topic, since clustering has been performed w.r.t. supposed homogeneous segments.

As regards hard document clustering methods, very few researchers have attempted to extract a document classification from the clustering of document fragments. In [25], authors experimented a clustering of legal documents fragments. This application does not require the use of any segmentation method since texts can be decomposed in legal cases. In other approaches, keywords are extracted, from documents [29] or from segments [30], to create a hierarchy of topics on which documents may be assigned [29]. However, the impact in IR, and especially on the cluster hypothesis, has not been assessed. Here, a hard document clustering can be deduced from the preliminary clustering of segments by assigning each document to its most similar topic/cluster (by computing its average similarity with segments of a given cluster).

3   Experimental Results

3.1   Document Collections

Four document collections from the TREC corpus are used: the ZIFF corpus includes computer science articles, copyrighted by Ziff-Davis Publishing Company, FR whole issues of the Federal Register, and AP and WSJ full articles from the

Table 1. Statistics of the collections

Document collection	ART	ZIFF	AP	WSJ	FR
Number of documents	21003	75180	84510	98732	25960
Terms per document	880	297	217	204	927
Unique terms per document	516	139	144	128	244
Terms of the 1000 shortest documents	397	19	14	11	49
Terms of the 1000 longest documents	1570	7915	512	1235	11304
Unique terms of the 1000 shortest documents	255	14	13	12	35
Unique terms of the 1000 longest documents	891	1503	323	623	1392
Number of segments	84510	138998	156147	179521	101326
Number of queries	34	31	33	39	13
Relevant documents per query	45.09	82.87	43.24	61.69	29
Relevant retrieved documents per query	9.44	12.19	9.75	23.04	4.23

Associated Press and the Wall Street Journal respectively. AP and WSJ, since being articles taken from multidisciplinary journals, present the greatest topical heterogeneity. An additional corpus, the ART corpus, has been artificially constituted. Each document of this corpus is made up of four articles of the AP corpus put end to end. This has been created in order to realize experiments on very heterogeneous documents. Moreover, fragments of text being articles, this allows to test the effectiveness of the optimal cluster without using a specific text segmentation method, which may induce some bias. Table 1 presents the statistics of these corpora. It should be noted that each corpus owns a great variety of document sizes. Whereas the ART and FR corpora contain a lot of very long documents, the AP collection contain rather short ones. ZIFF and FR corpora present the greatest document length heterogeneity. The same set of queries, the topics 1-50 of TREC, is used for each corpus. The ART corpus having been artificially built, no relevant document list is available. We then consider an ART document as relevant if it contains at least one AP article belonging to the relevant list. Table 1 gives the number of queries whose at least one document in the 50 first ones retrieved by the initial system described below is relevant. Same manner as the following experimental results, the two last statistics given by this table (i.e., the average number of relevant documents in each corpus and the average number of relevant documents in the 50 first retrieved documents) only consider these queries.

3.2   Optimal Cluster Evaluation

The evaluation of the standard IR systems is performed with well-known precision and recall measures. Systems relying on clustering techniques present a set of clusters rather than a ranked lists of documents as response to the user's query. The evaluation process of cluster-based IR systems is then mroe difficult: Unless rebuilding the documents ranked list w.r.t. clusters [8,9], process which may induce several biases, the computation of well-known precision and recall measures is not possible for cluster-based IR systems. In [11], authors proposed the evaluation function E which assesses the effectiveness of a cluster:

$$E = 1 - \frac{(\beta^2 + 1) \times P \times R}{(\beta^2 \times P) + R} \qquad (3)$$

where P and R correspond to the precision (proportion of documents being relevant in a given set) and recall (proportion of relevant documents occurring in the set)[2] measures applied on the set of documents of the concerned cluster and β is a parameter which determines the relative importance of the recall w.r.t. the precision[3]. In [11], authors introduce the notion of an 'optimal cluster search': a cluster-based IR system is evaluated, for each query, by the research of the best E score (the least) it is possible to find in the built hierarchy. Such a search, named MK1, presents the advantage that it isolates the cluster effectiveness from the bias induced by the search method employed. In order to compare this measure with the effectiveness of the simple ranked list, it would have been possible to capture the size of the optimal cluster and use this number of top-ranked documents to evaluate the effectiveness of the simple ranked list. Nevertheless, in [10], authors mentioned that this measure is unfair compared to MK1, since it does not consider the optimality of the ranked list (the number of documents taken into account is determined by the size of the optimal cluster). Therefore, they used another measure, the MK3 measure proposed in [11], which is more fair with the ranked list approach: it represents the best effectiveness (the least value E) that is reachable in the simple ranked list, by varying the number of first documents taken into account. These two measures, capturing the best effectiveness it is possible to reach according to the method employed, MK1 for the clustering-based approaches and MK3 for simple ranked list ones, are used in our experiments.

3.3 Initial and Passage Retrieval Effectiveness

Table 2 details results of experiments on two different ranked list approaches, in term of E criterion (E), number of relevant documents (R) and number of documents (N) taken into account by the MK3 measure (the optimal set of top documents). DR (Document retrieval) corresponds to the ranked list retrieved by the initial retrieval system (see section 2.1) and PR (passage retrieval) corresponds to the ranked list of documents established w.r.t. the similarity of their best passage. From this table, we can note that, except for the AP corpus where all documents are rather short, the Passage Retrieval approach obtained significantly better results than the classical Document Retrieval approach. The best improvements are recorded on the ART corpus, where passages of each document are very heterogeneous, and on the FR one, which contains a lot of very long documents.

---

[2] As realized in [10], the recall considers the whole set of relevant documents rather than the sole relevant documents retrieved in the 50 first ones by the initial system.

[3] $\beta = 1$ allocates equal importance to both measures, $\beta = 2$ attributes twice importance to recall and $\beta = 0.5$ twice importance to precision.

Table 2. Effectiveness of initial retrieval and passage retrieval

MK3 Ranked List		$\beta = 0.5$			$\beta = 1$			$\beta = 2$		
		E	R	N	E	R	N	E	R	N
ART	DR	0.716	6.618	22.06	0.745	8.588	33.06	0.728	9.088	38.32
	PR	0.631	6.765	16.29	0.684	8.735	25.50	0.682	9.088	29.59
FR	DR	0.645	3.923	11.00	0.695	4.000	11.23	0.692	4.154	12.69
	PR	0.602	3.923	9.462	0.638	4.077	10.85	0.643	4.154	11.46
ZIFF	DR	0.688	10.16	30.32	0.730	10.77	33.81	0.728	11.03	35.87
	PR	0.659	10.16	26.19	0.708	10.87	31.52	0.716	11.13	33.71
AP	DR	0.622	6.970	16.91	0.673	8.424	23.21	0.679	9.576	31.52
	PR	0.613	7.030	16.45	0.671	8.424	21.33	0.680	9.576	31.91
WSJ	DR	0.679	11.10	28.08	0.728	12.08	33.46	0.730	12.33	37.38
	PR	0.643	11.08	24.10	0.702	12.13	30.74	0.707	12.49	33.33

3.4 Cluster-Based Retrieval Effectiveness

The purpose of this section is to compare the following approaches in order to assess the benefit arisen from the consideration of the thematic segments in a cluster-based IR system:

- **D:** Classical query-specific document clustering presented in [10].
- **C:** Document clustering in which the similarity between two documents, rather than depending on both whole sets of terms, is equal to the similarity of their closest passages.
- **B:** Document clustering in which the similarity of a document with others, rather than depending on its whole set of terms, depends on the set of terms of its passage the most similar to the query.
- **S:** Soft document clustering deduced from a preliminary passage clustering.
- **H:** Hard document clustering deduced from a preliminary passage clustering.

First, the point is to compare the clustering tendencies of the different approaches, i.e. the degree at which the Cluster Hypothesis is likely to hold on the different corpora according to the approach used. This comparison is of impotance since, according to [11], the more a document collection is characterised by this hypothesis, the higher the effectiveness of a cluster-based IR system is likely to be. Two different methods have been proposed in order to test whether the Cluster Hypothesis holds on a given corpus: the Overlap test [11] and the Nearest Neighbour test [7]. The Overlap test considers, for each query, the relevant-relevant and relevant-nonrelevant inter-document similarity coefficients and checks how much the average coefficient of the former is larger than the latter. The lower the overlap of the two distributions of similarities is, the more the cluster hypothesis is likely to hold on the considered corpus. Nevertheless, it has been shown in Voorhees [7] that this test is biased by the relative frequency of relevant-nonrelevant pairs which is always greatly higher than the one of relevant-relevant pairs and causes then distortions in the

Table 3. Nearest Neighbours Test results

Clustering Tendencies	ART	FR	ZIFF	AP	WSJ
D	2.57	2.90	2.43	3.25	3.44
C	3.30	3.15	3.16	3.20	3.66
B	3.24	3.08	3.12	3.23	3.60

generated results. The Nearest Neighbour test, which has been designed to overcome the limitation of the overlap test, consists in counting the number of relevant documents belonging to the N nearest neighbours[4] (i.e. most similar documents) of each relevant document. The higher the average number of relevant document in the neighbourhood of each relevant document is, the more a given distribution of similarities is characterised by the cluster hypothesis.

Due to the absence of particular relevance judgments for the thematical segments, the application of the test on the two last approaches S and H, which depend on a preliminary thematical segments clustering, is not possible (since it is difficult to know which segment contains the relevant information). Nevertheless, the impact of the consideration of the segments on the clustering tendencies may be assessed by applying the test on the approaches C and B where the similarities depend on the thematical segments of the documents. Table 3 then reports the scores obtained with the Nearest Neighbour test for the approaches D, C and B on our five corpora[5]. According to the results, the cluster hypothesis is more likely to hold when the thematical segments of the documents are considered. Indeed, over all corpora (except on the AP corpus where the consideration of the passages has not led to any improvements of the ranked list effectiveness in the previous section), the scores obtained show that it is easier to reach relevant documents from other ones when the similarity between documents depends on their passages than on their whole sets of terms. This renders the better ability to gather all relevant documents in a single cluster with our approaches.

Table 4 reports the effectiveness of the different cluster-based approaches w.r.t. the accuracy of the optimal cluster found in their hierarchies by the MK1 measure, in term of E criterion (E), number of relevant documents (R) and total number of documents in the cluster (N). According to the results of this table, all of the clustering based approaches lead to a better effectiveness than the initial ranked lists of documents. The cluster hypothesis appears then to hold for all of the experimented corpora, the clustering of the retrieved documents enables to better identify the relevant ones. On each corpus where the ranked list based on passage retrieval has obtained better results than the initial ranked list of documents (i.e., ART, FR, WSJ and ZIFF), the use of thematic segments in the clustering process improves the results. The main assumption of this paper

---

[4] In our experiments, we take $N = 5$.

[5] The test only takes into account, for each query, only relevant documents having been retrieved in the first 50 documents by the initial IR system.

Table 4. Effectiveness of the query-specific clustering approaches

MK1		$\beta = 0.5$			$\beta = 1$			$\beta = 2$		
Group Average		E	R	N	E	R	N	E	R	N
ART	D	0.594	6.941	14.71	0.687	8.441	22.82	0.702	9.000	29.65
	C	0.528	7.206	12.03	0.621	8.794	19.50	0.643	9.147	25.97
	B	0.528	7.206	12.03	0.621	8.794	19.50	0.643	9.147	25.97
	S	0.532	7.235	12.24	0.627	8.824	20.32	0.643	9.088	24.76
	H	0.530	7.176	12.18	0.629	8.647	19.41	0.651	9.000	26.00
FR	D	0.484	3.846	5.308	0.570	4.000	5.615	0.607	4.077	6.462
	C	0.463	3.846	4.769	0.561	4.000	5.231	0.602	4.077	6.000
	B	0.470	3.846	4.923	0.565	4.000	5.385	0.603	4.077	6.154
	S	0.417	3.923	4.538	0.526	4.077	5.000	0.571	4.154	5.923
	H	0.413	3.846	4.308	0.525	4.077	4.923	0.583	4.077	4.923
ZIFF	D	0.600	9.742	22.10	0.684	10.71	27.65	0.704	11.10	31.81
	C	0.588	10.16	21.23	0.678	10.94	26.19	0.702	11.13	31.94
	B	0.589	10.16	21.35	0.676	11.03	26.65	0.700	11.16	32.29
	S	0.575	10.16	20.74	0.661	11.42	25.81	0.688	12.03	31.19
	H	0.569	10.06	18.35	0.662	10.87	23.32	0.695	11.29	31.55
AP	D	0.526	7.333	12.12	0.618	8.848	19.42	0.631	9.667	26.61
	C	0.525	7.333	12.06	0.618	8.788	19.09	0.632	9.667	27.21
	B	0.526	7.303	12.09	0.618	8.909	19.61	0.630	9.697	28.58
	S	0.524	7.394	12.18	0.616	9.152	20.67	0.629	9.727	28.88
	H	0.523	7.303	11.79	0.616	8.909	18.94	0.630	9.697	28.30
WSJ	D	0.593	10.95	19.13	0.680	12.05	23.33	0.711	12.28	27.44
	C	0.584	10.87	18.33	0.670	12.21	23.62	0.704	12.46	28.72
	B	0.586	10.90	18.67	0.668	12.26	24.74	0.701	12.51	29.90
	S	0.563	11.21	17.46	0.651	13.00	24.69	0.692	13.38	28.59
	H	0.561	11.10	17.13	0.654	12.49	22.70	0.698	12.87	26.31

appears to be verified, considering the passages of the documents allows to produce more accurate clusters.

On the ART corpus, the C and B methods obtain the best results. In this collection, passages of a same document being articles taken from a multidisciplinary journal, it is likely to have only one passage containing query terms in each document. Hence, the best similarity is recorded between passages related to the interesting topic and the information retrieval effectiveness is improved by the fact of ignoring disconnected passages. Nevertheless, on real-world documents corpora whose documents are more homogeneous, some documents adopting a same topic development may contain close irrelevant passages. In that case, the similarity of two documents in the C approach may depend on passages widely disconnected from the user's topic. The fact of adding a document in a cluster may then attract others having nothing in common with its initial subject. The B approach, which only considers the passage of each document the most similar to the query, does not obtain better results than the C one on these corpora: as expected in section 1, the fact of only considering the best passage of each document leads to ignore some important aspects of the documents which could have been useful to distinguish the different document groups.

As expected, approaches depending on a preliminary clustering of the passages (S and H) obtain the best effectiveness results. Differences between the results obtained on FR and WSJ by these approaches and those obtained by the classical document clustering D and the document clustering depending on passages similarities B and C have been shown to be statistically significant by a 99% Student t-test[6]. At last, the soft document clustering appears to obtain slightly better results than the hard one. This is due to the difficulty of hard assignment of the documents: the algorithm has to choose one aspect of each document among the clusters containing its passages. The selected topic is not necessarily the most connected to the user's subject. Note that, the hard method generally leads to a better precision (smaller clusters containing a better ratio of relevant documents) and the soft one leads to a better recall of the relevant documents.

4   Conclusion

This study proposes to consider topics of documents separately in order to produce clusters of documents better centered around a given subject than classical approaches. Even if intra and inter-cluster similarities may be not improved by the consideration of thematic segments, the set of resulting clusters appears actually to be more representative of the different topics addressed by the documents. Indeed, experiments, realized in the context of IR, have shown that considering the thematic passages of the documents allows to obtain clusters containing a better ratio of relevant documents.

This paper is part of an ongoing project which addresses the reorganization of responses furnished by an IR system. It has been shown here that considering passages may be useful to collect relevant information. Future works will concern the composition of response documents, by gathering parts of texts with regards to diverse criteria, whose aim is to furnish an overview of the different informations in connection with a user's subject that one could find in the corpus.

References

1. Voorhees, E.M., Harman, D.: Overview of the fifth text retrieval conference (trec-5). In: Proceedings of the Fifth Text Retrieval Conference, NIST Special Publication, pp. 1–28. NIST Special Publication 500-238 (1997)
2. Leuski, A.V.: Interactive information organization: techniques and evaluation. PhD thesis, University of Amhert, Massachussets, Director-James Allan (2001)

---

[6] The 99% Student t-test is based on the average, the standard deviation and the cardinality of a set of runs. It uses a p-value equal to 2.57 to insure with a confidence rate of 99% that the difference in means of two sets is significant.

3. Koenemann, J., Belkin, N.J.: A case for interaction: a study of interactive information retrieval behavior and effectiveness. In: CHI 1996: Proceedings of the SIGCHI conference on Human factors in computing systems, pp. 205–212. ACM Press, New York (1996)
4. Harper, D.J., Koychev, I., Yixing, S.: Query-based document skimming: A user-centred evaluation of relevance profiling. In: Sebastiani, F. (ed.) ECIR 2003. LNCS, vol. 2633, pp. 377–392. Springer, Heidelberg (2003)
5. Croft, W.B.: A model of cluster searching bases on classification. Information Systems 5(3), 189–195 (1980)
6. Willett, P.: Recent trends in hierarchic document clustering: a critical review. Information Processing & Management 24(5), 577–597 (1988)
7. Voorhees, E.M.: The effectiveness and efficiency of agglomerative hierarchic clustering in document retrieval. PhD thesis, Cornell University, Ithaca, NY, USA (1986)
8. Hearst, M.A., Pedersen, J.O.: Reexamining the cluster hypothesis: Scatter/gather on retrieval results. In: Proceedings of SIGIR-96, 19th ACM International Conference on Research and Development in Information Retrieval, Zürich, CH, pp. 76–84 (1996)
9. Leuski, A.: Evaluating document clustering for interactive information retrieval. In: CIKM 2001: Proceedings of the tenth international conference on Information and knowledge management, pp. 33–40. ACM, New York (2001)
10. Tombros, A., Villa, R., Rijsbergen, C.J.V.: The effectiveness of query-specific hierarchic clustering in information retrieval. Information Processing & Management 38(4), 559–582 (2002)
11. Jardine, N., van Rijsbergen, C.J.: The use of hierarchic clustering in information retrieval. Information Storage and Retrieval 7(5), 217–240 (1971)
12. Porter, M.: An algorithm for suffix stripping. Program 14(3), 130–137 (1980)
13. Callan, J.P.: Passage-level evidence in document retrieval. In: SIGIR 1994: Proceedings of the 17th annual international ACM SIGIR conference on Research and development in information retrieval, pp. 302–310. Springer, New York (1994)
14. Kaszkiel, M., Zobel, J.: Effective ranking with arbitrary passages. Journal of the American Society of Information Science 52(4), 344–364 (2001)
15. Salton, G., Buckley, C.: Term-weighting approaches in automatic text retrieval. Inf. Process. Manage. 24(5), 513–523 (1988)
16. Baeza-Yates, R., Ribeiro-Neto, B.: Modern Information Retrieval. ACM Press /Addison-Wesley, New York (1999)
17. Salton, G., Singhal, A., Buckley, C., Mitra, M.: Automatic text decomposition using text segments and text themes. In: Hypertext 1996, The Seventh ACM Conference on Hypertext, March 16-20, 1996, pp. 53–65. ACM, New York (1996)
18. Kozima, H.: Text segmentation based on similarity between words. In: Meeting of the Association for Computational Linguistics, pp. 286–288 (1993)
19. Hearst, M.: Texttiling: segmenting text into multi-paragraph subtopic passages. Computational Linguistics 23(1), 33–64 (1997)
20. Choi, F.: Advances in domain independent linear text segmentation. In: Proceedings of the first conference on North American chapter of the Association for Computational Linguistics, pp. 26–33. Morgan Kaufmann Publishers Inc., San Francisco (2000)
21. Lamprier, S., Amghar, T., Levrat, B., Saubion, F.: Seggen: A genetic algorithm for linear text segmentation. In: Veloso, M.M. (ed.) IJCAI, pp. 1647–1652 (2007)
22. Rasmussen, E.M.: Clustering algorithms. In: Information Retrieval: Data Structures & Algorithms, pp. 419–442. Prentice-Hall, Englewood Cliffs (1992)

23. Salton, G.: Automatic Information Organization and Retrieval. McGraw-Hill, New York (1968)
24. Mendes, M.E.S., Sacks, L.: Evaluating fuzzy clustering for relevance-based information access. In: The IEEE International Conference on Fuzzy Systems FUZZ-IEEE 2003, pp. 648–653 (2003)
25. Conrad, J.G., Al-Kofahi, K., Zhao, Y., Karypis, G.: Effective document clustering for large heterogeneous law firm collections. In: ICAIL 2005: Proceedings of the 10th international conference on Artificial intelligence and law, pp. 177–187. ACM Press, New York (2005)
26. Bezdek, J.C.: Pattern Recognition with Fuzzy Objective Function Algorithms. Kluwer Academic Publishers, Norwell (1981)
27. Kraft, D.H., Chen, J., Mikulcic, A.: Combining fuzzy clustering and fuzzy inference in information retrieval. In: The IEEE International Conference on Fuzzy Systems FUZZ-IEEE 2000, pp. 375–380 (2000)
28. Bradley, P.S., Reina, C., Fayyad, U.M.: Clustering very large databases using em mixture models. In: The International Conference on Pattern Recognition (ICPR 2000), vol. 2, pp. 2076–2080 (2000)
29. Chang, H.C., Hsu, C.C.: Using topic keyword clusters for automatic document clustering. In: ICITA 2005: Proceedings of the Third International Conference on Information Technology and Applications (ICITA 2005), vol. 2, pp. 419–424. IEEE Computer Society, Washington (2005)
30. Muscat, R.: Automatic document clustering using topic analysis. Master's thesis, University of Malta (2005)

Multilingual Plagiarism Detection

Zdenek Ceska, Michal Toman, and Karel Jezek

Department of Computer Science and Engineering, Faculty of Applied Sciences, University of
West Bohemia, Univerzitni 22, 306 14 Pilsen, Czech Republic
{zceska,mtoman,jezek_ka}@kiv.zcu.cz

Abstract. Multilingual text processing has been gaining more and more attention in recent years. This trend has been accentuated by the global integration of
European states and the vanishing cultural and social boundaries. Multilingual
text processing has become an important field bringing a lot of new and
interesting problems. This paper describes a novel approach to multilingual plagiarism detection. We propose a new method called MLPlag for plagiarism detection in multilingual environment. This method is based on analysis of word
positions. It utilizes the EuroWordNet thesaurus which transforms words into
language independent form. This allows to identify documents plagiarized from
sources written in other languages. Special techniques, such as semantic-based
word normalization, were incorporated to refine our method. It identifies the replacement of synonyms used by plagiarists to hide the document match. We
performed and evaluated our experiments on monolingual and multilingual corpora and results are presented in this paper.

Keywords: Plagiarism, Copy Detection, Nature Language Processing, EuroWordNet, Thesaurus, Lemmatization.

1 Introduction

A useful and continual progress showing sub area of document processing research is
the field of text document plagiarism detection. Particularly when documents are
written in various languages, the way of their comparison is not satisfactorily solved
so far. Simultaneously this implies a number of additional tasks.

Several systems have been developed for plagiarism detection; however, none of
them deals with such situation when documents are written in different languages.
Claugh published the fundamentals of plagiarism detection in [1]. Later, Maurer introduced an overview of current plagiarism detection systems in [7] and sketched some
future directions for the field of Natural Language Processing (NLP). The most popular is system SCAM [13] that employs single words contained in the examined documents. Another system that also uses words to measure the document similarity is
Detection of Duplicate Defect Reports, described in detail in [10]. The most recent
systems, such as Ferret [6] or KOPI Portal [9], utilize N-grams as features. Although
N-grams yield better results for plagiarism detection, they are inappropriate in multilingual environment. Therefore, we propose a new method called MLPlag that overcomes this issue.

D. Dochev, M. Pistore, and P. Traverso (Eds.): AIMSA 2008, LNAI 5253, pp. 83–92, 2008.

We can imagine several usage scenarios of our system. An example of successful usage of our system is a university environment where student works are often plagiarized. There are known cases where two thirds of the master theses were plagiarized from an internet source such as Wikipedia. Even if the work was translated from English to Czech language it cannot be considered as a novel work. The aim of our approach is to detect plagiarized documents and its sources even if they are written in foreign languages. Another problem we deal with is the replacement of words with synonymous expressions and slightly different word order in different languages. The resolution for these issues is a part of our proposal.

We started our experiments with only two languages – Czech and English. It should be noted that the principle remains the same for any number of processed languages. We designed a prototype of the system to detect Czech documents plagiarized from English documents and vice versa. It is able to deal with different word order in Czech and English sentences and synonymous word replacements.

The rest of this paper is organized as follows. Section 2 describes the EuroWordNet (EWN) thesaurus in detail. Section 3 proposes our method MLPlag for plagiarism detection in multilingual environment. Section 4 presents the results we achieved on two experimental multilingual corpora. And finally, Section 5 concludes our paper.

2 EuroWordNet Thesaurus

For language independent processing, we designed a technique which transforms multilingual texts into an easily processed form. The EWN thesaurus [4] is used for this task. It is a multilingual database of words and their relations for most European languages, i.e. English, Danish, Italian, Spanish, German, French, Czech, Bulgarian and Estonian. It contains sets of synonyms – synsets – and relations between them. A unique index is assigned to each synset; it interconnects the languages through an inter-lingual-index in such a way that the same synset in one language has the same index in another one. Thus, cross-language processing can easily be performed. We can, for example, detect a Czech article as a plagiarism of one or more English documents and vice versa. With EWN, completely language independent processing and storage can be carried out, and moreover, synonyms are identically indexed. Synonyms are often used to hide the plagiarism from readers. This fact plays an important role in plagiarism detection.

In order to use EWN, it is necessary to assign the EWN index to each term of a document. To accomplish that, the words must be firstly transformed into basic forms, i.e. words must be normalized. Lemmatization is the only possible way how to transform the word to obtain the basic form – lemma. The overview of the lemmatization system illustrates Fig. 1.

EWN-based lemmatization can be classified as a dictionary lemmatization method. A dictionary creation can be considered as the most difficult part of the EWN-based approach. We proposed a method for the lemmatization dictionary building based on the use of EWN thesaurus and Ispell dictionary [3]. The lemmatization dictionary was created by extraction of word forms using the Ispell utility. We are able to generate all existing word forms from the stem stored in the Ispell dictionary. It contains stems and attributes specifying the possible suffixes and prefixes, which are applied to stems in order to derive all possible word forms. We assume that one of the derived forms is a basic form (lemma). In order to recognize the basic form we look for the corresponding lemma in EWN. A fuzzy match routine based on [8] can be optionally used for searching lemmas in EWN thesaurus. This pre-processing helps especially in the case of highly inflected languages. Languages with a rich inflection are more difficult to process in general. A Czech morphological analyzer [5] was used to overcome this problem. Thanks to this module we achieved a further improvement of our method.

The lemmatization of English is relatively simple, thus it is possible to use the basic lemmatization algorithms with satisfying results. We implemented lemmatization modules for Czech and English languages, but the main principle remains the same for any number of languages. However, language specific pre-processing, such as morphological analysis and disambiguation, is needed in some cases to achieve better results.

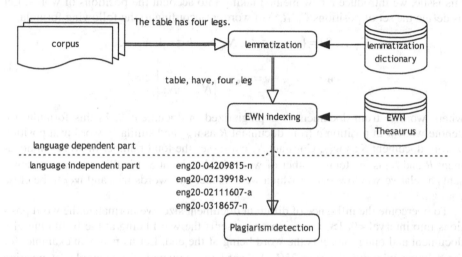

Fig. 1. Pre-processing of the multilingual corpus is split into two parts – lemmatization and EWN indexing. The independent language form is used as an input in the plagiarism detection module.

In case of monolingual pre-processing, when EWN is not used, we recommend using lemmatization followed by stop-word removal, which is the most fundamental pre-processing approach. Stop-words removal eliminates all common and inconvenient words from the text to reduce the amount of data.

When we process ambiguous word, each word sense is included into the processing. We consider two words from different languages as plagiarized if the one of senses matches with another one in foreign language.

3 MLPlag Method

In multilingual environment it is difficult to choose the right features that can unambiguously identify plagiarism among different languages. We need to choose such features that can be employed in most of languages regardless of the grammatical rules they apply. In monolingual environment, N-grams are mostly preferred since they contain meaningful phrases and achieve better results than single words. Generally, N-gram is a sequence of n successive words. In multilingual environment, it is impossible to employ N-grams due to the distinct word order in every language. Moreover, some words do not need to have any equivalent in other languages we are dealing with. This precludes the use of N-grams because they require all translated words to have the same order in various languages. Therefore, we recommend using single words as features.

The feature we can utilize is the word location in the examined documents. Let us imagine the situation when a word occurs in one document at the beginning and in the second one at the end. It is obvious that these two words cannot be plagiarized because they stand at absolutely different positions in both documents. To overcome this issue, we introduce a new method taking into account the positions of words. Let us define the set of positions $C_w(R,S)$ of word w according to the following formula

$$C_w(R,S) = \{a: \ a \in <1, N_R >, \ b \in <1, N_S >,$$

$$w = w_{R,a}, \quad w = w_{S,b}, \ \left| \frac{a}{N_R} - \frac{b}{N_S} \right| < \omega \}, \tag{1}$$

where word w from document R is plagiarized in document S. In this formula, we denote word w at position a from document R as $w_{R,a}$ and similarly word w at position b from document S as $w_{S,b}$. Constant N_R expresses the total number of words in document R and N_S is the total number of words in document S. Finally, constant ω represents a relative window size in which we consider two words $w_{R,a}$ and $w_{S,b}$ to be close enough.

To overcome the influence of different document size, we normalize the word positions into interval <0, 1>, where zero represents the word being at the beginning of a document and one represents the word being at the end. Let us make an example, for the relative window size $\omega = 25\%$ the word $w_{R,a}$ situated, for example, at position $a/N_R = 0.5$ is plagiarized if and only if the same word $w_{S,b}$ is situated at position $b/N_S \in <0.25, 0.75>$.

Now, let us define the occurrence frequency of plagiarized word w as $P_w(R,S)$ according to formula

$$P_w(R,S) = |C_w(R,S)| \cdot |C_w(S,R)|, \tag{2}$$

where $|C_w(R,S)|$ represents the number of positions of word w from document R such that are plagiarized in document S. Further, we define the occurrence frequency of word w in document R as $F_w(R)$ regardless if it is plagiarized or not. These both definitions are used in the following subsections where we introduce two measures of similarity.

3.1 Symmetric Similarity Measure (MLPlag$_{SYM}$)

The first measure of similarity is partially derived from the traditional VSM that has gained popularity in the Information Retrieval (IR) domain [12]. We define a measure of similarity between two documents R and S in multilingual environment as:

$$sim(R,S) = \frac{\sum_{w \in R \cap S} \alpha_w^2 \cdot P_w(R,S)}{\sqrt{\sum_{w \in R} \alpha_w^2 \cdot F_w^2(R) \cdot \sum_{w \in S} \alpha_w^2 \cdot F_w^2(S)}} \tag{3}$$

This expresses the symmetric measure for the document pair, where α_w is the weight associated with the occurrence of word w. Weight α_w in Equation (3) is composed of a local weight and a global weight, as depicted in [14]. The resulting similarity we obtain is in interval <0, 1>.

3.2 Asymmetric Similarity Measure (MLPlag$_{ASYM}$)

The second measure of similarity we introduce is an asymmetric one which is in contrast to the preceding measure. We define the subset measure of document R to be subset of document S according to formula

$$subset(R,S) = \frac{\sum_{w \in R \cap S} \alpha_w^2 \cdot P_w(R,S)}{\sum_{w \in R} \alpha_w^2 \cdot F_w^2(R)}, \tag{4}$$

where α_w is the weight associated with the occurrence of word w. Subsequently, we define the measure of similarity as the maximum of both asymmetric measures according to

$$sim(R,S) = \max\{subset(R,S),\ subset(S,R)\}. \tag{5}$$

Because we need to keep similarity in interval <0, 1>, we set $sim(R,S)$ to be 1 if it is greater than 1.

4   Experiments

To evaluate our proposed method MLPlag, we decided to use a standard mechanism well-known from IR. We define precision p and recall r according to Rijsbergen [10].

In the following figures, a more representative F_1-measure depending on threshold τ is used. F_1-measure combines precision and recall as follows

$$F_1 = \frac{2 \cdot p \cdot r}{p + r}. \tag{6}$$

Threshold τ determines a minimal level of similarity when two documents are considered plagiarized, see the following formula.

$$plagiarized(R, S) = \begin{cases} true & if\ sim(R, S) > \tau \\ false & if\ sim(R, S) \leq \tau \end{cases} \tag{7}$$

4.1 Pre-processing

In Section 2, we introduced the necessary pre-processing which is required for multi-lingual plagiarism detection of text documents. Fig. 2 presents the influence of multi-lingual pre-processing on the accuracy in comparison to monolingual pre-processing. In our case, monolingual pre-processing consists of lemmatization and stop-word removal. Multilingual pre-processing consists of lemmatization and the inter-lingual indexing process that partially substitutes the stop-words removal used for monolingual pre-processing.

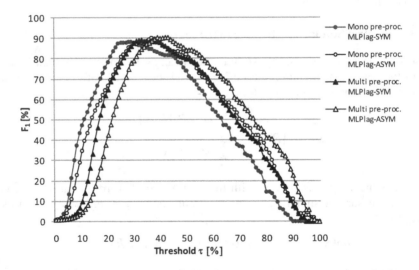

Fig. 2. The dependency of F_1-measure on the threshold τ. This experiment is performed for monolingual and multilingual pre-processing on the CTK (Czech Press Agency) corpus.

Although the dependencies of F_1-measure on the threshold τ are very similar for monolingual and multilingual pre-processing, we discovered that multilingual pre-processing slightly improves the accuracy of both $MLPlag_{SYM}$ and $MLPlag_{ASYM}$ methods, see Table 1. Another advantage is a certain separation of language dependent data processing, see Fig. 1. In higher layers of processing, i.e. plagiarism detection, we consider only inter-lingual indexes so it will be easier to include any other language in future. On the base of our expertise we strongly recommend employing multilingual pre-processing not only for multilingual environment.

4.2 MLPlag Evaluation

For our following experiments, we built up two two multilingual corpora, called JRC-EU and Fairy-tale. JRC-EU corpus is composed of 400 randomly selected European

Table 1. The results achieved on the CTK corpus

Pre-processing	Method	Threshold τ	F_1-measure
Monolingual	MLPlag$_{SYM}$	27%	88.38%
	MLPlag$_{ASYM}$	33%	90.24%
Multilingual	MLPlag$_{SYM}$	33%	89.01%
	MLPlag$_{ASYM}$	42%	90.35%

Union legislative texts [2]. As it includes a lot of topic-specific words not contained in the EWN thesaurus, we have to take into account that the indexation process some-times fails. This corpus contains 200 reports written in English and the same number of corresponding reports written in Czech. The second corpus, called Fairy-tale, represents a smaller set of text documents with a simplified vocabulary. This corpus consists of 54 documents, 27 English and 27 corresponding translations in Czech language.

Fig. 3 presents the dependency of F_1-measure on the threshold τ and the window size ω. The meaning of ω was given in Equation (1). As you can see, window size ω influences F_1-measure in multilingual environment. We examined both MLPlag$_{SYM}$ and MLPlag$_{ASYM}$ methods on JRC-EU and Fairy-tale corpus. During our examination, we tested window size ω in the interval from 1% to 100%. The case that no window is used corresponds with $\omega = 100\%$, i.e. the same words occurring at the beginning or at the end of documents are considered plagiarized. On the other hand, if $\omega = 1\%$ only such words having relative distance at most 1% are considered plagiarized. The same words having relative position greater than ω are regarded as absolutely distinct and are not included in computation of text similarity.

From our observation, too low values of ω significantly decrease F_1. On the other hand, high values of ω decrease F_1 too. Three of four experiments proved that the best results achieve ω between 8% and 10%, see Table 2. An exception is the MLPlag$_{ASYM}$ method on the JRC-EU corpus. Fig. 3 shows the best F_1-measure for $\omega = 55\%$. Some satisfactory results could be achieved with $\omega = 10\%$ as well. Both MLPlag$_{SYM}$ and MLPlag$_{ASYM}$ indicate a slightly better accuracy for ω specified in that interval compared with the case when no window size is used. For example, the MLPlag$_{SYM}$ method achieves 72.53% F_1-measure for $\omega = 8\%$ on the JRC-EU corpus, whereas only 65.91% F_1-measure if no window size is used. In Table 2, we denote the situation when no window is used as "---".

The second used parameter we use is threshold τ representing the minimal level of similarity when we consider two documents be plagiarized. This parameter signifi-cantly influences F_1-measure. The effective interval for MLPlag$_{SYM}$ is slightly nar-rower compared with the MLPlag$_{ASYM}$ method, see Fig. 3. In case of MLPlag$_{SYM}$, we talk about 5% to 12%, while MLPlag$_{ASYM}$ achieves good results in interval between 8% and 20%. Although MLPlag$_{ASYM}$ has a wider effective interval, the peak values occur about 16% for both corpora. Now, let us look at threshold τ in Table 1 and Table 2. You can see that for multilingual corpora the best results are achieved for τ being much lower than for monolingual corpus, i.e. CTK corpus. This is caused by the words that do not have any equivalent in other languages. Another issue can be

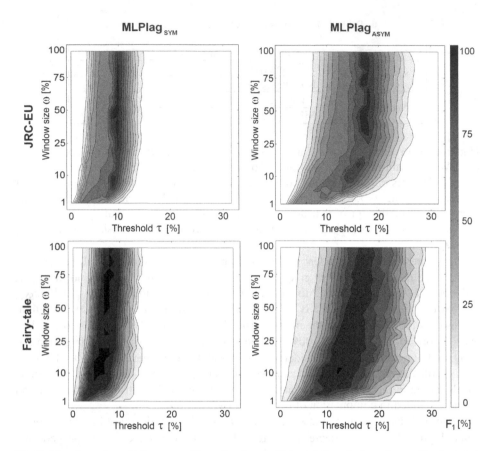

Fig. 3. The dependency of measure F_1 on the threshold τ and the window size ω. The top part of the figure presents the MLPlag$_{SYM}$ and MLPlag$_{ASYM}$ methods, from the left to right, on the JRC-EU corpus. The bottom part of the figure presents the same methods on the second corpus, called Fairy-tale.

Table 2. The results achieved on the multilingual JRC-EU and Fairy-tale corpora

Corpus	Method	Threshold τ	Window size ω	F_1-measure
JRC-EU	MLPlag$_{SYM}$	9%	---	65.91%
		9%	8%	72.53%
	MLPlag$_{ASYM}$	16%	---	70.52%
		17%	55%	71.47%
Fairy-tale	MLPlag$_{SYM}$	7%	---	100%
		5%	10%	100%
	MLPlag$_{ASYM}$	16%	---	94.73%
		11%	8%	100%

found in the EWN thesaurus because it is still under development. In any case, less word matches are found and therefore a much lower value of τ is required to achieve outstanding results in multilingual environment.

Generally, for both corpora, $MLPlag_{SYM}$ and $MLPlag_{ASYM}$ yield similar results. In case of monolingual environment, the decision is straightforward. We propose to employ $MLPlag_{ASYM}$ because it significantly overcomes $MLPlag_{SYM}$, see Table 1. In case of multilingual environment the decision is controversy because, from our observation, the differences between both methods are statistically insignificant, see Table 2. Nevertheless, we recommend employing $MLPlag_{ASYM}$ due to the easier determination of the right threshold τ. The reason is simple, an effective τ is spread over a wider interval so it is more likely to make an appropriate choice of threshold τ.

Fairy-tale corpus reaches outstanding results in comparison with the JRC-EU corpus. Both methods achieve 100% for F_1-measure because of the simplified vocabulary the corpus uses. On the other hand, JRC-EU corpus achieves at most 72.53% for the $MLPlag_{SYM}$ method. JRC-EU corpus is composed of a large amount of topic-specific words that do not occur in EWN. Therefore, the accuracy radically decreases.

During our experiments we found out that 42.0 % Czech words of JRC-EU corpus and 57.8 % words of Fairy-tale corpus were successfully transformed into EWN indexes. English EWN vocabulary is more developed and therefore the coverage is 51.4 % for JRC-EU and 66.4 % for Fairy-tale corpus.

5 Conclusion

As our experiments proved, the MLPlag method gives promising results. The method is able to process multilingual data without any significant negative impact on accuracy of plagiarism detection. It was not possible with other approaches. Actually, the experiments prove that the F_1-measure rises if multilingual preprocessing is applied.

Further improvement is obtained when we include relative window approach. The F_1-measure achieves 72.53% on JRC-EU corpus which means 7% increase compared with the approach where the window is not used. The second corpus called Fairy-tale consists of articles where a simplified language is used. Therefore, the F_1-measure is 100% for both setups – with and without the use of the window.

Our method is able to process any number of languages included in the EWN thesaurus. However, we should take into consideration that the EWN thesaurus is still under development. Its incompleteness can cause some difficulties in cross-language plagiarism detection. As EWN is gradually being completed, this problem will disappear and we expect even better results.

We are going to extend our work in several areas. We aim to replace the relative window with a more sophisticated approach based on structural features of languages. Further, we are working on an advanced word processing that includes word sense disambiguation. And finally, we plan to use the inter-word relationships stored in EWN.

Acknowledgments. This research was supported in part by National Research Programme II, project 2C06009 (COT-SEWing).

References

1. Clough, P.: Plagiarism in natural and programming languages: An overview of current tools and technologies. In: Internal Report CS-00-05, Department of Computer Science, University of Sheffield (2000)
2. European Commission - Joint Research Centre: The JRC-Acquis Multilingual Parallel Corpus, Version 3.0 (Last update 23/1/2008), http://langtech.jrc.it/JRC-Acquis.html
3. Gorin, R.: Ispell (Last update 5/6/1996),
   `http://fmgwww.cs.ucla.edu/fmgmembers/geoff/ispell.html`
4. Global WordNet Association: EuroWordNet (Last update 9/1/2001), `http://www.illc.uva.nl/EuroWordNet/`
5. Hajic, J.: Morphology analyzer (Last update 8/27/2001), `http://quest.ms.mff.cuni.cz/pdt/Morphology_and_Tagging/Morphology/index.html`
6. Lane, P., Lyon, C., Malcolm, J.: Demonstration of the Ferret Plagiarism Detector. In: Proceedings of the 2nd International Plagiarism Conference, Newcastle, UK (2006)
7. Maurer, H., Kappe, F., Zaka, B.: Plagiarism – A Survey. Journal of Universal Computer Science 12(8), 1050–1084 (2006)
8. Myers, E.: An O(ND) Difference Algorithm and Its Variations. Algorithmica 1, 251–266 (1986)
9. Pataki, M.: Distributed Similarity and Plagiarism Search. In: Proceedings of the Automation and Applied Computer Science Workshop, pp. 121-130, Budapest, Hungary (2006) ISBN 963-420-865-7
10. Rijsbergen, C.: Information Retrieval. Butterworth-Heinemann, 2nd rev. edn. (March 1979) ISBN 0-408-70929-4
11. Runeson, P., Alexanderson, M., Nyholm, O.: Detection of Duplicate Defect Reports Using Natural Language Processing. In: Proceedings of the IEEE 29th International Conference on Software Engineering, pp. 499-510 (2007)
12. Salton, G.: The state of retrieval system evaluation. International Journal of Information Processing & Management 24(4), 441–449 (1992)
13. Shivakumar, N., Garcia-Molina, H.: SCAM: A copy detection mechanism for digital documents. In: Proceedings of 2nd International Conference in Theory and Practice of Digital Libraries, Austin (1995)
14. Salton, G., Buckley, C.: Term-Weighting Approaches in Automatic Retrieval. Journal of Information Processing and Management 24(5), 513–523 (1988)

Dealing with Spoken Requests
in a Multimodal Question Answering System

Roberto Gretter, Milen Kouylekov, and Matteo Negri

Fondazione Bruno Kessler
Via Sommarive, 18 - Povo, Trento, Italy
{gretter,kouylekov,negri}@fbk.eu

Abstract. This paper reports on experiments performed in the development of the QALL-ME system, a multilingual QA infrastructure capable of handling input requests both in written and spoken form. Our objective is to estimate the impact of dealing with automatically transcribed (*i.e.* noisy) requests on a specific question interpretation task, namely the extraction of relations from natural language questions. A number of experiments are presented, featuring different combinations of manually and automatically transcribed questions datasets to train and evaluate the system. Results (ranging from 0.624 to 0.634 F-measure in the recogniton of the relations expressed by a question) demonstrate that the impact of noisy data on question interpretation is negligible with all the combinations of training/test data. This shows that the benefits of enabling speech access capabilities, allowing for a more natural human-machine interaction, outweight the minimal loss in terms of performance.

Keywords: Question Answering, Textual Entailment, Speech Recognition.

1 Introduction

Recent years have seen an increasing interest towards advanced information access applications, motivated by the huge market potential of systems providing natural human-machine interaction capabilities. Question Answering (QA) research plays an important role in this direction, focusing on the development of systems that return actual *answers* in response to *natural language questions*. In the same direction, enabling users to express their needs in the most natural way, the increased reliability of Automatic Speech Recognition (ASR) systems offers new opportunities for a simplified and more effective access to information.

The combination of QA technology and multimodal interaction capabilities is among the challenges addressed by the EU funded project QALL-ME[1]. The project aims at developing a distributed infrastructure for multilingual QA over structured data, in the domain of cultural events in a town. The foreseen multimodal capabilities of the QALL-ME system include the possibility of access by

---

[1] http://qallme.itc.it/

D. Dochev, M. Pistore, and P. Traverso (Eds.): AIMSA 2008, LNAI 5253, pp. 93–102, 2008.

means of mobile devices (*e.g.* mobile phones), to pose natural language questions either in *textual* form (*e.g.* sms), or in *speech* modality. From a research perspective, the speech access modality is particularly interesting, since it raises the need for robust methods capable of handling noisy and sub-optimal inputs. Often, in fact, spoken requests are more complex than written ones (*e.g.* they contain hesitations and repetitions), and their automatic transcription may contain errors. Investigating these aspects, which are currently out of the scope of traditional QA research, is the main purpose of this paper.

Our work builds on top of [1], which addresses QA over structured data reformulating the problem as a *Textual Entailment Recognition* (RTE) problem. Textual Entailment (TE) has been recently proposed as a unifying framework for applied semantics [2], where the need for an explicit representation of a mapping between linguistic objects and data objects can be, at least partially, bypassed through the definition of semantic inferences at the textual level. In this framework, a text (T) is said to entail a hypothesis (H) if the meaning of H can be derived from the meaning of T. According to the TE framework, [1] proposes that the interpretation of a given question can be addressed as a Relation Extraction task based on TE, where the text (T) is the question, and the hypothesis (H) is a relational pattern, which is associated to instructions for retrieving the answer to the question. Given a question q and a set of relational patterns $P=\{p_1, ..., p_n\}$, the basic operation is to select those patterns in P that are entailed by q. Instructions associated to patters may be viewed as high precision procedures for answer extraction, which are dependent on the specific data source accessed for answer extraction. For instance, in case of QA over structured data, instructions would be SQL queries to a database.

Building on the positive results reported in [1], we adopt a similar TE-based approach to question interpretation, to investigate the impact of handling noisy data obtained from automatic transcriptions. For this purpose, different experiments are carried out running the system under different training/test conditions. In the optimal situation, the system is trained and tested over datasets of manually transcribed Italian questions. The results achieved in this first setting are then compared with: *i)* those achieved by a system trained over manually transcripted questions, and tested over automatic transcriptions (to reproduce a situation that is closer to a real on-field evaluation), *ii)* those achieved by a system trained and tested over automatic transcriptions (to verify if the system can "learn" from systematic errors produced by the ASR), and *iii)* those achieved by training the system over both manual and automatic transcriptions, and testing it over automatic transcriptions (to check if the two sources used together give an added value at a training stage).

The paper is organized as follows. Section 2 introduces TE-based Relation Extraction as a question interpretation task. Section 3 describes the dataset used for experiments. Section 4 describes our automatic speech recognition system. Sections 5 and 6 respectively report experiment results, and concluding remarks.

2   TE-Based Relation Extraction

The TE-based approach to question interpretation has been defined in [1] as
a classification problem, where a question q has to be assigned to all the relations $R_1,...,R_n$ it expresses, selected from a predefined set R (in this work we
focus on binary relations, although extensions to n-ary relations are expected).
For instance, given the question *"What can I see today at cinema Astra?"*, the
following relations represent the expected system's output:

R1: HASMOVIESITE(MOVIE:?, SITE: "Astra")
R2: HASDATE(MOVIE:?, DATE: "today")

The classification (see Figure 1 for a schematic representaton of the overall
process) is carried out by means of a RTE engine, which is in charge of discovering
entailment relations between the input question q, and a set of textual patterns
stored in a *Pattern Repository (P)*. P contains n sets of textual patterns, each
set representing possible lexicalizations of one relation R_i in R.

Given a question q, the RTE engine attempts to verify whether an entailment
relation holds between q and each pattern in P. All the relations associated to
the patterns entailed by q are output by the system. If none of the patterns in P
is entailed by q, this is interpreted as evidence that the question is out of domain.
Sections 2.1 and 2.2 overview two crucial aspects of the proposed approach to
question interpretation, namely: *i)* the *type of the textual patterns stored in the
repository P*, and *ii)* our *TE recognition algorithm*.

2.1   Minimal Relational Patterns (MRPs)

According to our formulation of the task, we say that a relational pattern p
expresses a relation $R(arg1, arg2)$ in a certain language L if speakers of L agree
that the meaning of p expresses the relation R between $arg1$ and $arg2$, given

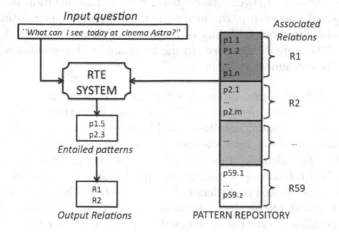

Fig. 1. Entailment-based Relation Extraction process

Table 1. Examples of relational patterns

(1)	<ARG2:MOVIE:X> *is shown at cinema* <ARG1:CINEMA:Y>
(2)	*What* <ARG2:*movie*> *is on at* <ARG1:CINEMA:Y>*?*
(3)	*Is there any* <ARG2:*movie*> *that I can see at* <ARG1:CINEMA:Y>*?*
(4)	*Can I see* <ARG2:MOVIE:X> *at cinema* <ARG1:CINEMA:Y> *on* <ARG?:DATE:Z>*?*

their knowledge about the entities. For instance, all the examples in Table 1 represent relational patterns for the relation HASMOVIESITE(MOVIE, SITE).

In order to be profitably used in the proposed entailment framework valid patterns have the additional property of representing only *one* relation. Pattern representing multiple relations, in fact, would be entailed only by questions containing all these relations, thus resulting limited in their usage. To describe one-relation patterns [1] introduces the notion of *Minimal Relational Pattern* (MRP), which can be formally defined in terms of TE. Given a set $P=\{p_1, p_n\}$ of relational patterns for a relation R, a pattern p_k belonging to P is a MRP for the relation R if condition (1) holds.

$$\forall p_i \in P, p_k \mapsto p_i = \emptyset \tag{1}$$

In other words, a pattern p_k is minimal if none of the other relational patterns contained in P can be derived from p_k (*i.e.* is logically entailed by p_k). According to such definition, patterns (1)-(3) in Table1 are MRPs for the relation HASMOVIESITE(MOVIE, SITE), while (4) is not, since it entails the others.

2.2   Distance Based Entailment Recognition

For each relation we train a RTE engine, which is an adaptation of our English system evaluated in the framework of the Pascal-RTE Challenge [3]. The system has been implemented within a distance-based framework, and is based on *Levenshtein Distance* (LD) or *Linear Distance* [4]. The intuition is that, given a question q and a pattern p, the probability of an entailment relation between q and p is related to the possibility of mapping the whole content of q into the content of p. The more straightforward the mapping can be established, the more probable is the entailment relation.

Algorithm. Edit distance approaches for RTE, such as the one proposed in [5], assume that the distance between T and H is a characteristic that separates the positive pairs, for which entailment holds, from the negative pairs, for which entailment does not hold. Such distance is computed as the cost of the editing operations (*i.e.* insertion, deletion and substitution) which are required to transform T into H. Each edit operation on two text fragments A and B (denoted as $A \rightarrow B$) has an associated cost (denoted as $\gamma(A \rightarrow B)$). The entailment score for a T-H pair is calculated on the minimal set of edit operations that transform T into H. An entailment relation is assigned to a T-H pair only if the overall cost of the transformation is below a certain threshold empirically estimated over training data. The entailment score function is defined in the following way:

$$score_{entailment}(T, H) = 1 - \frac{\gamma(T, H)}{\gamma_{nomap}(T, H)} \qquad (2)$$

where $\gamma(T, H)$ is the function that calculates the edit distance between T and H, and $\gamma_{nomap}(T, H)$ is the *no mapping* distance equivalent to the cost of inserting the entire text of H, and deleting the entire text of T. The entailment score function has a range from 0 (when T is identical to H), to 1 (when T is completely different from H).

LD is calculated by converting both the text T and the hypothesis H into sequences of words. Accordingly, edit operations have been defined as as follows:

- **Insertion** $(\Lambda \to A)$: insert a word A from H into T.
- **Deletion** $(A \to \Lambda)$: delete a word A from T.
- **Substitution** $(A \to B)$: substitute a word A in T with a word B from H.

Cost schemes for edit operations. The core of the edit distance approach is the mechanism for the definition of the cost of edit operations. This mechanism is defined apart from the distance algorithm, and should reflect the knowledge of the user about the processed data. The principle behind it is to capture certain phenomena that facilitate the algorithm to assign small distances to positive T-H pairs, and high distances to negative pairs. For instance, since our task consists in comparing questions (T) with MRPs (H) usually composed by few terms[2], for our experiments we adopted the following simple cost calculation scheme:

$$\gamma(\Lambda \to A) = length(T)$$
$$\gamma(B \to \Lambda) = length(H)$$
$$\gamma(A \to B) = \begin{cases} 0 & A = B \\ \gamma_{i+d}(A \to B) & otherwise \end{cases}$$

In this scheme the cost of inserting a text fragment from H in T is equal to the length of T, and the cost of deleting a text fragment from T is equal to the length of H. The cost of the substitution of two fragments is set to the sum of the insertion and the deletion of the text fragments, if they are not equal. This means that the algorithm would prefer to delete and insert text fragments rather than substituting them, in case they are not equal. During system development we discovered that this cost calculation scheme performs better than considering fixed costs for insertion and deletion operations.

3   The QALL-ME Benchmark

In order to experiment with the proposed TE-based approach to question interpretation, we used a dataset of 1487 Italian questions extracted from the Italian

---

[2] In the experiments reported in Section 5 we compare questions and MRPs of respective average lengths of around 12.5 and 4.5 words.

part of the QALL-ME benchmark[3] [6]. The benchmark contains several thousand questions, in the four languages involved in the project (English, German, Italian and Spanish), concerning cultural events in a town. Questions have been acquired at the telephone, then manually transcribed and annotated with all the relevant information necessary to train/test the core components of a QA system. The availability of both the original recorded questions and their manual transciptions, together with the annotation of part of the acquired data with the relations of interest in the selected domain[4], provides all the data necessary for our evaluation purposes. Sections 3.1 and 3.2 respectively report additional information about questions acquisition and annotation.

3.1 Data Acquisition

To obtain a reasonable linguistic variability in the acquired questions, more than 100 speakers for each language were involved in the data collection process. Each speaker was given a list of scenarios presented on a computer screen, and describing possible information needs in the selected domain. Scenarios were designed to allow the formulation of useful queries, without providing too many suggestions about "how" to formulate them. For this purpose, each scenario was presented as a request template, containing a limited amount of textual material.

Every speaker performed 30 spoken questions, based on 15 scenarios randomly chosen out of a set of 90. For each scenario speakers first generated a spontaneous request, and then read a written one previously prepared. As far as the Italian language is concerned, 161 speakers (93 females and 68 males), 12 of which non-native, were involved in data acquisition. The resulting database contains 4768 Italian utterances (2316 read + 2452 spontaneous), for a total speech duration of about 9 hours and 20 minutes. The average utterance duration is about 7 seconds. 104 utterances were marked as unusable, mainly due to technical problems experienced during the acquisition. As a result, the total number of valid utterances is 4664 (2290 read + 2374 spontaneous). The average word lengths of read and spontaneous utterances are respectively 11.2 and 14.1 words.

3.2 Data Annotation

Besides the original questions in the four languages, their orthographic transcription[5], and their translations into English, different annotation levels have been considered in the creation of the QALL-ME benchmark. These include pragmatic (speech acts) and semantic (Named Entities) annotations, the Expected Answer Type, the Expected Answer Quantifier, the Question Topical Target, and relations between entities appearing in the utterance.

---

[3] The QALL-ME benchmark is freely available on the project's website.

[4] Relation annotation is still work in progress: in the current version of the benchmark only 1487 questions out of 4664 are annotated at this level.

[5] Transcriptions were manually produced using Transcriber, a tool for assisting the manual annotation of speech signals freely downloadable from http://trans.sourceforge.net.

As far as relation annotation is concerned, in the current version of the benchmark questions have been manually marked as containing one or more relations chosen from a set of 59 binary relations defined in the QALL-ME ontology. As an example, the annotation of the question *"What is the name of the director of 007 Casino Royale, which is shown today at cinema Modena?"* contains three relations, namely:

HASDATE(MOVIE,DATE)
HASMOVIESITE(MOVIE,SITE)
HASDIRECTOR(MOVIE,DIRECTOR).

On average, spontaneous and read questions have been respectively annotated with 1.94 and 2.26 relations (ranging from 1 to 6 relations per question). A Kappa value of 0.94 (*almost perfect agreement*) was measured for the agreement between two annotators over part of the dataset (150 questions), demonstrating the reliability of relations annotation.

4   Automatic Speech Recognition (ASR)

ASR over the original recorded questions was carried out using the speech recognizer described in [7]. The system is based on a set of phonetic units represented by continuous density Hidden Markov Models (HMMSs). The acoustic features used are LPC Cepstral coefficients and log-energy, with the corresponding first and second order time derivatives. This feature vector is computed every 10 ms on overlapping windows of lenght 20 ms. HMMs were trained on a set of audio data completely disjoint from the QALL-ME benchmark.

Each word of the lexicon was transcribed and foreign words were hand-checked and possibly corrected. A class-based trigram language model was trained on the benchmark manual transcriptions, following the 10-fold cross validation paradigm (*i.e.* splitting the speech data into 10 blocks). All the utterances pronounced by the same speaker are contained in the same block, to avoid that utterances of the same person will appear both in train and in test data. In this way, each block contains utterances from 16 speakers, except one block that contains data from 17 speakers. After this division, for each block we defined two lists: a test list, which corresponds to the audio files of the given block, and a training list, which corresponds to the union of the other 9 blocks. For each test list, the manual transcriptions of the corresponding training list were used to train the class-based trigram language model. Hence, to evaluate speech recognition performance, 10 different recognizers were trained and 10 different tests were performed. After that, results were merged.

The main drawback of the resulting speech recognition system lies in the language model, which is trained on a very small amount of data. Anyway, despite this potential problem, the speech recognizer performed in a relatively satisfactory way. Following the cross validation paradigm, we evaluated the coverage of the test data in terms of Out-Of-Vocabulary (OOV) words:

Table 2. Sentence and Word Accuracy for two language models. These results merge the results obtained in the 10-fold cross validation paradigm.

	# sentence accuracy	# word accuracy
bigrams	24.1 %	74.3 %
trigrams	35.4 %	77.4 %

- *running words*: 98.9% - about 1 word every 100 words is an OOV word;
- *sentences*: 89.4 % - a sentence is covered if and only if all its words are known;
- *lexicon*: 92.7 % - most of the OOV occur only rarely.

It is worth noting that, since the scenarios considered a fixed time interval, with a fixed list of entities (*e.g.* movie titles, persons, theatre names), if pronounced correctly they will never appear in the OOV lists.

Speech recognition was computed following the cross validation paradigm. Results are reported in Table 2 in terms of Sentence and Word Accuracy. *Sentence Accuracy* measures the percentage of sentences recognized without any error, so that the interpretation will be exactly the same of the corresponding manual transcription. *Word Accuracy* gives the percentage of correctly recognized words and is a better indicator of the quality of ASR performance. These results, altough not very high if compared to state of the art speech recognizers for a small task like this, are quite interesting because, as we will see in the experiments, they do not affect very much the interpretation of the sentences. In fact, many of the errors concern either functional words or similar words carrying the same meaning.

5   Experiments and Results

A number of experiments have been carried out to evaluate our TE-based approach to Relation Extraction under different training/test conditions, depending on the use of *clean* data (*i.e* those resulting from manual transcriptions), or *noisy* data (*i.e.* question transciptions output by the ASR system).

Training/test sets. The benchmark questions annotated with the 59 selected relations have been used to create the *training* and *test* sets for our experiments. For this purpose, the question corpus was randomly split in two sets, respectively containing 999 and 448 questions. Such separation was carried out guaranteeing that, for each relation R, the questions marked with R are distributed in the two sets in proportion 2/3-1/3. In addition, the two sets contain a balanced random mixture of spontaneous and read questions, due to some differences noticed between the two types. On average, in fact, spontaneous questions are longer than the read ones, (the respective average lengths are 14.1, and 11.2 words), and involve more relations (2.26 vs. 1.94).

The larger set of 999 questions is used for the manual creation of MRPs and, together with the resulting Pattern Repository, is used to train our RTE system (*i.e.* to empirically estimate an entailment threshold for each relation, considering positive and negative examples). The smaller set of 448 questions (which remained "unseen" in the MRP acquisition phase) is used as test set to evaluate system's performance.

Pattern Repository. According to the definition given in Section 2.1, for each of the 59 relations R we manually[6] extracted a set of MRPs from the training questions annotated with R. The resulting Pattern Repository contains a total of 226 patterns, with at least 1 MRP per relation (4 on average).

Experiments. Evaluation has been carried out under the following conditions (*i.e.* using different combinations of *clean* and *noisy* training/test data):

- **Experiment 1: Clean/Clean (C/C).** In this configuration the system is trained and tested over manually transcribed questions. This is the same evaluation setting proposed in [1], and is used for comparison with the other configurations involving noisy data.
- **Experiment 2: Clean/Noisy (C/N).** In this configuration the system is trained over manually transcribed questions, and tested over automatic transcriptions. The idea is to check performance variations with non-homogeneous training/test data.
- **Experiment 3: Noisy/Noisy (N/N).** In this configuration both training and test are carried out over automatically transcribed questions. The idea is to verify if the errors produced by the ASR system can be "learned" by the system.
- **Experiment 4: Clean+Noisy/Noisy (C+N/N).** In this configuration the system is trained over the combination of manual and automatic transcriptions, and tested over automatic transcriptions. The idea is to check if the two sources used together give an added value at a training stage.

Results. For each configuration, system performance has been calculated comparing the relations recognized by the system, with those manually marked in the reference test questions. Precision, Recall, and F-measure scores are reported in Table 3.

Quite surprisingly, the impact of dealing with noisy data is negligible under all the training/test combinations. This can be explained by the fact that most

Table 3. Results obtained with different combinations of *clean* (C) and *noisy* (N) data

	C/C	C/N	N/N	C+N/N
Precision	0.543	0.546	0.534	0.55
Recall	0.76	0.763	0.75	0.744
F-measure	0.634	0.637	0.624	0.633

---

[6] Even though automatic pattern extraction (either from local corpora or from the Web) is a very active research area, this particular aspect falls beyond the scope of this work.

of the ASR errors typically concern functional words (articles and prepositions) which are not really important in determining the meaning of an input question. A crucial point could also be the way in which relations are learned, which demonstated to be quite robust. This also motivates the fact that little differences are observed when clean, noisy or clean + noisy data are used.

6   Conclusions

This paper addressed the problem of extracting relations from a natural language question, focusing on a comparative evaluation of system's performance under different training/test conditions, which depend on the type of data used (manual vs. automatic transcriptions of the same questions). A number of experiments have been described, showing that the use of noisy data has a negligible impact under all the training/test combinations. Such positive result demonstrates that, at least in the proposed task, the benefits of enabling speech access capabilities providing a more natural human-machine interaction outweight the minimal loss in terms of performance.

References

1. Negri, M., Kouylekov, M., Magnini, B.: Detecting Expected Answer Relations through Textual Entailment. In: Gelbukh, A. (ed.) CICLing 2008. LNCS, vol. 4919. Springer, Heidelberg (2008)
2. Dagan, I., Glickman, O.: Probabilistic Textual Entailment: Generic Applied Modeling of Language Variability. In: Proceedings of the PASCAL Workshop on Learning Methods for Text Understanding and Mining, Grenoble, France (2004)
3. Dagan, I., Glickman, O., Magnini, B.: The PASCAL Recognising Textual Entailment Challenge. In: Quiñonero-Candela, J., Dagan, I., Magnini, B., d'Alché-Buc, F. (eds.) MLCW 2005. LNCS (LNAI), vol. 3944, pp. 177–190. Springer, Heidelberg (2006)
4. Levenshtein, V.: Binary Codes Capable of Correcting Deletions, Insertions, and Reversals. Doklady Akademii Nauk SSSR 163 (1965)
5. Kouylekov, M., Magnini, B.: Combining Lexical Resources with Tree Edit Distance for Recognizing Textual Entailment. In: Quiñonero-Candela, J., Dagan, I., Magnini, B., d'Alché-Buc, F. (eds.) MLCW 2005. LNCS (LNAI), vol. 3944. Springer, Heidelberg (2006)
6. Cabrio, E., Coppola, B., Gretter, R., Kouylekov, M., Magnini, B., Negri, M.: Question Answering Based Annotation for a Corpus of Spoken Requests. In: Proceedings of the Workshop on Semantic Representation of Spoken Language (SRSL 2007), Salamanca, Spain, (2007)
7. Falavigna, D., Gretter, R.: Telephone Speech Recognition Applications at IRST. In: Proceedings of IVTTA, Turin, Italy (1998)

Noun Compound Interpretation Using Paraphrasing Verbs: Feasibility Study

Preslav Nakov*

Linguistic Modeling Department,
Institute for Parallel Processing,
Bulgarian Academy of Sciences
25A, Acad. G. Bonchev St., 1113 Sofia, Bulgaria
and
Department of Mathematics and Informatics,
Sofia University,
5, James Bourchier Blvd., 1164 Sofia, Bulgaria
nakov@lml.bas.bg
http://nakov.eu

Abstract. The paper addresses an important challenge for the automatic processing of English written text: understanding noun compounds' semantics. Following Downing (1977) [1], we define noun compounds as sequences of nouns acting as a single noun, e.g., *bee honey*, *apple cake*, *stem cell*, etc. In our view, they are best characterised by the set of all possible paraphrasing verbs that can connect the target nouns, with associated weights, e.g., *malaria mosquito* can be represented as follows: *carry* (23), *spread* (16), *cause* (12), *transmit* (9), etc. These verbs are directly usable as paraphrases, and using multiple of them simultaneously yields an appealing fine-grained semantic representation.

In the present paper, we describe the process of constructing such representations for 250 noun-noun compounds previously proposed in the linguistic literature by Levi (1978) [2]. In particular, using human subjects recruited through Amazon Mechanical Turk Web Service, we create a valuable manually-annotated resource for noun compound interpretation, which we make publicly available with the hope to inspire further research in paraphrase-based noun compound interpretation. We further perform a number of experiments, including a comparison to automatically generated weight vectors, in order to assess the dataset quality and the feasibility of the idea of using paraphrasing verbs to characterise noun compounds' semantics; the results are quite promising.

Keywords: Noun Compounds, Lexical Semantics, Paraphrasing.

---

* Part of this research was performed while the author was a PhD student at the EECS department, Computer Science division, University of California at Berkeley.

D. Dochev, M. Pistore, and P. Traverso (Eds.): AIMSA 2008, LNAI 5253, pp. 103–117, 2008.
© Springer-Verlag Berlin Heidelberg 2008

1 Introduction

An important challenge for the automatic analysis of English written text is posed by noun compounds – sequences of nouns acting as a single noun[1], e.g., *colon cancer tumor suppressor protein* – which are abundant in English: Baldwin&Tanaka'04 [3] calculated that noun compounds comprise 3.9% and 2.6% of all tokens in the *Reuters corpus* and the *British National Corpus*[2], respectively.

Understanding noun compounds' syntax and semantics is difficult but important for many natural language applications (NLP) including but not limited to question answering, machine translation, information retrieval, and information extraction. For example, a question answering system might need to determine whether '*protein acting as a tumor suppressor*' is a good paraphrase for *tumor suppressor protein*, and an information extraction system might need to decide whether *neck vein thrombosis* and *neck thrombosis* could possibly co-refer when used in the same document. Similarly, a machine translation system facing the unknown noun compound *WTO Geneva headquarters* might benefit from being able to paraphrase it as *Geneva headquarters of the WTO* or as *WTO headquarters located in Geneva*. Given a query like *migraine treatment*, an information retrieval system could use suitable paraphrasing verbs like *relieve* and *prevent* for page ranking and query refinement.

Throughout the rest of the paper, we hold the view that noun compounds' semantics is best characterised by the set of all possible paraphrasing verbs that can connect the target nouns, with associated weights, e.g., *malaria mosquito* can be represented as follows: *carry* (23), *spread* (16), *cause* (12), *transmit* (9), etc. Such verbs are directly usable as paraphrases, and using multiple of them simultaneously yields an appealing fine-grained semantic representation.

The remainder of the paper is organised as follows: Section 2 provides a short overview of the different representations of noun compounds' semantics previously proposed in the literature. Section 3 gives details on the process of creating a lexicon of human-proposed paraphrasing verbs for 250 noun-noun compounds. Section 4 describes the experiments we performed in order to assess the lexicon's quality and the feasibility of using paraphrasing verbs to characterise noun compounds' semantics. Section 5 contains a discussion on the applicability of the approach. Section 6 concludes and suggests possible directions for future work.

2 Related Work

The dominant view in theoretical linguistics is that noun compound semantics can be expressed by a small set of abstract relations. For example, in the theory of Levi [2], complex nominals – a general concept grouping together the partially overlapping classes of nominal compounds (e.g., *peanut butter*), nominalisations

---

[1] This is Downing's definition of noun compounds [1], which we adopt throughout the rest of the paper.

[2] There are 256K distinct noun compounds out of the 939K distinct wordforms in the 100M-word *British National Corpus*.

Table 1. Levi's recoverably deletable predicates (RDPs). Column 3 shows the modifier's function in the corresponding paraphrasing relative clause: when the modifier is the subject of that clause, the RDP is marked with the index 2.

RDP	Example	Subj/obj	Traditional Name
CAUSE$_1$	tear gas	object	causative
CAUSE$_2$	drug deaths	subject	causative
HAVE$_1$	apple cake	object	possessive/dative
HAVE$_2$	lemon peel	subject	possessive/dative
MAKE$_1$	silkworm	object	productive/composit.
MAKE$_2$	snowball	subject	productive/composit.
USE	steam iron	object	instrumental
BE	soldier ant	object	essive/appositional
IN	field mouse	object	locative
FOR	horse doctor	object	purposive/benefactive
FROM	olive oil	object	source/ablative
ABOUT	price war	object	topic

Table 2. Levi's nominalisation types with examples

	Subjective	Objective	Multi-modifier
Act	parental refusal	dream analysis	city land acquisition
Product	clerical errors	musical critique	student course ratings
Agent	—	city planner	—
Patient	student inventions	—	—

(e.g., *dream analysis*), and nonpredicate noun phrases (e.g., *electric shock*) – can be derived by the following two processes:

1. **Predicate Deletion.** It can delete the 12 abstract recoverably deletable predicates (RDPs) shown in Table 1, e.g., *pie made of apples* → *apple pie*. In the resulting nominals, the modifier is typically the object of the predicate; when it is the subject, the predicate is marked with the index 2;
2. **Predicate Nominalisation.** It produces nominals whose head is a nominalised verb, and whose modifier is derived from either the subject or the object of the underlying predicate, e.g., *the President refused general MacArthur's request* → *presidential refusal*. Multi-modifier nominalisations retaining both the subject and the object as modifiers are possible as well. Therefore, there are three types of nominalisations depending on the modifier, which are combined with the following four types of nominalisations the head can represent: *act, product, agent* and *patient*. See Table 2 for examples.

In the alternative linguistic theory of Warren [4], noun compounds are organised into a four-level hierarchy, where the top level is occupied by the following six major semantic relations: Possession, Location, Purpose, Activity-Actor, Resemblance, and Constitute. Constitute is further subdivided into finer-grained level-2 relations: Source-Result, Result-Source or `Copula`. Furthermore, `Copula` is sub-divided into the level-3 relations `Adjective-Like_Modifier`, `Subsumptive`, and `Attributive`. Finally, `Attributive` is divided into the level-4 relations `Animate_Head` (e.g., *girl friend*) and `Inanimate_Head` (e.g., *house boat*).

A similar view is dominant in computational linguistics. For example, Nastase&Szpakowicz [5] proposed a two-level hierarchy consisting of thirty fine-grained relations, grouped into the following five coarse-grained ones (the corresponding fine-grained relations are shown in parentheses): CAUSALITY (`cause, effect, detraction, purpose`), PARTICIPANT (`agent, beneficiary, instrument, object_property, object, part, possessor, property, product, source, whole, stative`), QUALITY (`container, content, equative, material, measure, topic, type`), SPATIAL (`direction, location_at, location_from, location`), and TEMPORALITY (`frequency, time_at, time_through`). For example, *exam anxiety* is classified as `effect` and therefore also as CAUSALITY.

Similarly, Girju&al. [6] propose a set of 21 abstract relations (POSSESSION, ATTRIBUTE-HOLDER, AGENT, TEMPORAL, PART-WHOLE, IS-A, CAUSE, MAKE/PRODUCE, INSTRUMENT, LOCATION/SPACE, PURPOSE, SOURCE, TOPIC, MANNER, MEANS, THEME, ACCOMPANIMENT, EXPERIENCER, RECIPIENT, MEASURE, and RESULT) and Rosario & Hearst [7] use 18 abstract domain-specific biomedical relations (e.g., `Defect, Material, Person_Afflicted`).

An alternative view is held by Lauer [8], who defines the problem of noun compound interpretation as predicting which among the following eight prepositions best paraphrases the target noun compound: `of, for, in, at, on, from, with`, and `about`. For example, *olive oil* is *oil from olives*.

Lauer's approach is attractive since it is simple and yields prepositions representing paraphrases directly usable in NLP applications. However, it is also problematic since mapping between prepositions and abstract relations is hard [6], e.g., `in, on`, and `at`, all can refer to both LOCATION and TIME.

Using abstract relations like CAUSE is problematic as well. First, it is unclear which relation inventory is the best one. Second, being both abstract and limited, such relations capture only part of the semantics, e.g., classifying *malaria mosquito* as CAUSE obscures the fact that mosquitos do not directly cause malaria, but just transmit it. Third, in many cases, multiple relations are possible, e.g., in Levi's theory, *sand dune* is interpretable as both HAVE and BE.

Some of these issues are addressed by Finin [9], who proposes to use a specific verb, e.g., *salt water* is interpreted as *dissolved in*. In a number of publications [10,11,12], we introduced and advocated an extension of this idea, where noun compounds are characterised by the set of all possible paraphrasing verbs, with associated weights, e.g., *malaria mosquito* can be *carry (23)*, *spread (16)*, *cause (12)*, *transmit (9)*, etc. These verbs are fine-grained, directly usable as paraphrases, and using multiple of them for a given noun compound approximates its semantics better.

Following this line of research, below we describe the process of building a lexicon of human-proposed paraphrasing verbs, and a number of experiments in assessing both the lexicon's quality and the feasibility of the idea of using paraphrasing verbs to characterise noun compounds' semantics.

3 Creating a Lexicon of Paraphrasing Verbs

Below we describe the process of creating a new lexicon for noun compound interpretation in terms of multi-sets of paraphrasing verbs. We used the Amazon Mechanical Turk Web Service[3] to recruit human subjects to annotate 250 noun-noun compounds previously proposed in the linguistic literature.

We defined a special noun-noun compound paraphrasing task, which, given a noun-noun compound, asks human subjects to propose verbs, possibly followed by prepositions, that could be used in a paraphrase involving *that*. For example, *nourish, run along* and *come from* are good paraphrasing verbs for *neck vein* since they can be used in paraphrases like '*a vein that <u>nourishes</u> the neck*', '*a vein that <u>runs along</u> the neck*' or '*a vein that <u>comes from</u> the neck*'. In an attempt to make the task as clear as possible and to ensure high quality of the results, we provided detailed instructions, we stated explicit restrictions, and we gave several example paraphrases. We instructed the participants to propose at least

---

Paraphrasing Noun-Noun Compounds

Introduction

Given a noun-noun compound like *malaria mosquito, olive oil, grain alcohol, canola leaves, fruit fly, evening ride, neck vein, disease victim, migraine drug, Google ads,* etc., you are asked to paraphrase it using verbs and prepositions.
For example, *neck vein* can be paraphrased as follows:

"*neck vein*" is a *vein* that **comes from** the *neck*
"*neck vein*" is a *vein* that **drains** the *neck*
"*neck vein*" is a *vein* that **descends in** the *neck*
"*neck vein*" is a *vein* that **emerges from** the *neck*
"*neck vein*" is a *vein* that **enters** the *neck*
"*neck vein*" is a *vein* that **feeds** the *neck*
"*neck vein*" is a *vein* that **flows in** the *neck*
"*neck vein*" is a *vein* that **is in** the *neck*
"*neck vein*" is a *vein* that **is located in** the *neck*
"*neck vein*" is a *vein* that **is found in** the *neck*
"*neck vein*" is a *vein* that **is terminated at** the *neck*
"*neck vein*" is a *vein* that **nourishes** the *neck*
"*neck vein*" is a *vein* that **passes through** the *neck*
"*neck vein*" is a *vein* that **runs through** the *neck*
"*neck vein*" is a *vein* that **runs from** the *neck*
"*neck vein*" is a *vein* that **runs along** the *neck*
"*neck vein*" is a *vein* that **goes into** the *neck*
"*neck vein*" is a *vein* that **supplies** the *neck*
"*neck vein*" is a *vein* that **terminates in** the *neck*
etc.

Fig. 1. Paraphrasing in Mechanical Turk: task introduction

---

[3] http://www.mturk.com

three paraphrasing verbs per noun-noun compound, if possible. The instructions we provided and the actual interface the human subjects were seeing are shown in Figures 1 and 2.

We used *Amazon Mechanical Turk* Web service, which represents a cheap and easy way to recruit subjects for various tasks that require human intelligence. The service provides an API allowing a computer programme to ask a human to perform a task and returns the results. *Amazon* calls the process *Artificial Artificial Intelligence*. The idea behind the latter term and behind the origin of the service's name come from the *Mechanical Turk*, a life-sized wooden chess-playing mannequin the Hungarian nobleman Wolfgang von Kempelen constructed in 1769, which was able to defeat skilled opponents including Benjamin Franklin and Napoleon Bonaparte. The audience believed the automaton was

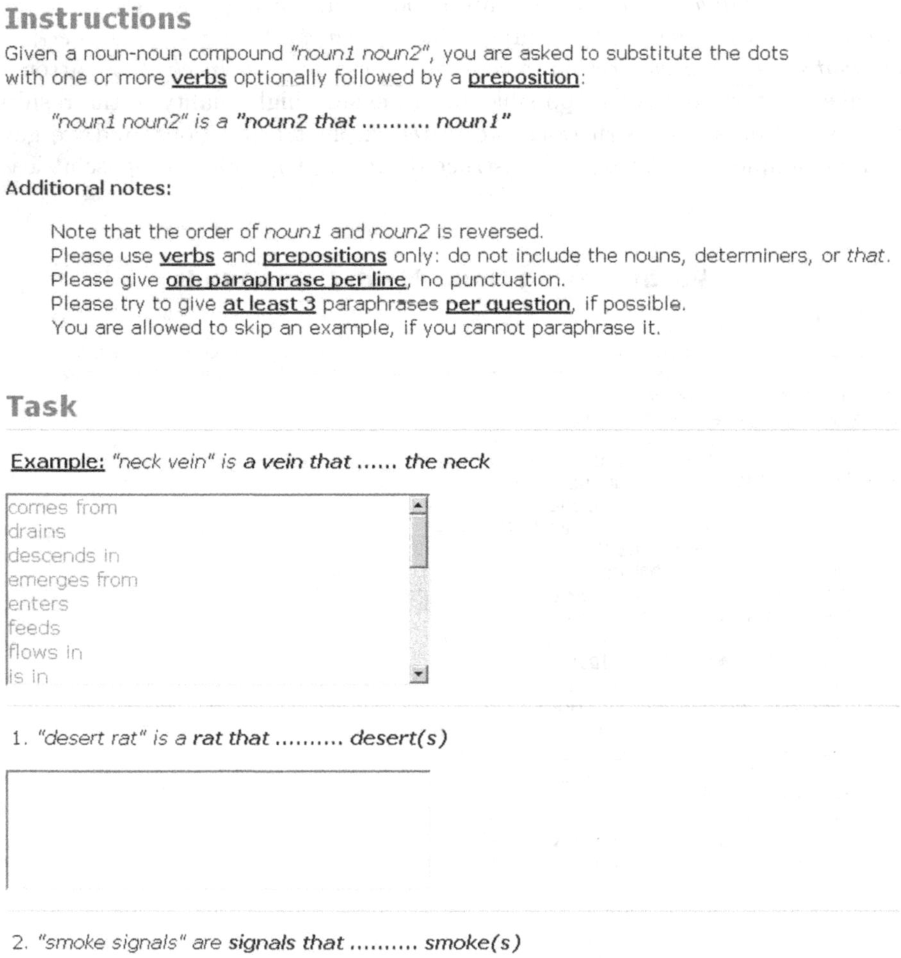

Instructions

Given a noun-noun compound "*noun1 noun2*", you are asked to substitute the dots with one or more **verbs** optionally followed by a **preposition**:

"*noun1 noun2*" is a "*noun2 that .... noun1*"

Additional notes:

Note that the order of *noun1* and *noun2* is reversed.
Please use **verbs** and **prepositions** only: do not include the nouns, determiners, or *that*.
Please give **one paraphrase per line**, no punctuation.
Please try to give **at least 3** paraphrases **per question**, if possible.
You are allowed to skip an example, if you cannot paraphrase it.

Task

Example: "*neck vein*" is a *vein that ......* the neck

```
comes from                    ▲
drains
descends in
emerges from
enters
feeds
flows in
is in                         ▼
```

1. "*desert rat*" is a *rat that .... desert(s)*

2. "*smoke signals*" are **signals that .... smoke(s)**

Fig. 2. Paraphrasing in Mechanical Turk: instructions, example, sample questions

making decisions using Artificial Intelligence, but the secret was a chess master hidden inside. Now *Amazon* provides a similar service to computer applications.

We used the 387 complex nominals Levi studied in her theory, listed in the appendix of [2]. We had to exclude the examples with an adjectival modifier, which are allowed in that theory, but do not represent noun compounds under our definition as was mentioned above. In addition, the following compounds were written concatenated and we decided to exclude them as well: *whistleberries, gunboat, silkworm, cellblock, snowball, meatballs, windmill, needlework, textbook, doghouse,* and *mothballs.* Some other examples contained a modifier that is a concatenation of two nouns, e.g., *wastebasket category, hairpin turn, headache pills, basketball season, testtube baby;* we decided to retain these examples. A similar example (which we chose to retain as well) is *beehive hairdo,* where both the modifier and the head are concatenations. As a result, we ended up with 250 good noun-noun compounds out of the original 387 complex nominals.

We randomly distributed these 250 noun-noun compounds (below, we will be referring to them as the *Levi-250 dataset*) into groups of 5, which yielded 50 Mechanical Turk tasks known as HITs (*Amazon* Human Intelligence Tasks), and we requested 25 different human subjects (*Amazon* workers) per HIT. We had to reject some of the submissions, which were empty or were not following the instructions, in which cases we requested additional workers in order to guarantee at least 25 good submissions per HIT. Each human subject was allowed to work on any number of HITs (between 1 and 50), but was not permitted to do the same HIT twice, which is controlled by the *Amazon Mechanical Turk* Web Service. A total of 174 different human subjects worked on the 50 HITs, producing 19,018 different verbs. After removing the empty and the bad submissions, and after normalising the verbs, we ended up with a total of 17,821 verbs, i.e., 71.28 verbs per noun-noun compound on average, not necessarily distinct.

Since many workers did not strictly follow the instructions, we performed some automatic cleaning of the results, followed by a manual check and correction, when it was necessary. First, some workers included the target nouns, the complementiser *that,* or determiners like *a* and *the,* in addition to the paraphrasing verb, in which cases we removed this extra material. For example, *star shape* was paraphrased as *shape that looks like a star* or as *looks like a* instead of just *looks like.* Second, the instructions required that a paraphrase be a sequence of one or more verb forms possibly followed by a preposition (complex prepositions like *because of* were allowed), but in many cases the proposed paraphrases contained words belonging to other parts of speech, e.g., nouns (*is in the _shape_ of, has _responsibilities_ of, has the _role_ of, makes _people_ have, is _part_ of, makes _use_ of*) or predicative adjectives (*are _local_ to, is _full_ of*); we filtered out all such paraphrases. In case a paraphrase contained an adverb, e.g., *occur only in, will _eventually_ bring,* we removed the adverb and kept the paraphrase. Third, we normalised the verbal paraphrases by removing the leading modals (e.g., *_can_ cause* becomes *cause*), perfect tense *have* and *had* (e.g., *_have_ joined* becomes *joined*), or continuous tense *be* (e.g., *_is_ donating* becomes *donates*). We converted complex verbal construction of the form '*<raising verb> to be*' (e.g., *appear to be, seems to be, turns to be, happens to be, is expected to be*) to just *be*. We further removed present participles introduced by *by*, e.g., *are caused by peeling* becomes *are caused*. Furthermore, we filtered out any paraphrase that involved *to* as part of the infinitive of a verb different from *be*, e.g., *is willing to donate* or *is painted to appear like* are not allowed. We also added *be* when it was missing in passive constructions, e.g., *made from* became *be made from*. Finally, we lemmatised the conjugated verb forms using *WordNet*, e.g., *comes from* becomes *come from*, and *is produced from* becomes *be produced from*. We also fixed some occasional spelling errors that we noticed, e.g., *b_olongs to, happens bec_asue of, is m_made from*.

The resulting lexicon of human-proposed paraphrasing verbs with corresponding frequencies, and some other lexicons, e.g., a lexicon of the first verbs proposed by each worker only, and a lexicon of paraphrasing verbs automatically extracted from the Web as described in [12], are released under the *Creative Commons License*[4], and can be downloaded from the *Multiword Expressions Website*: http://multiword.sf.net. See [13] for additional details.

4   Experiments and Evaluation

We performed a number of experiments in order to assess both the quality of the created lexicon and the feasibility of the idea of using paraphrasing verbs to characterise noun compounds' semantics.

For each noun-noun compound from the *Levi-250 dataset*, we constructed two frequency vectors $\overrightarrow{h}$ (human) and $\overrightarrow{p}$ (programme). The former is composed of the above-described human-proposed verbs (after lemmatisation) and their corresponding frequencies, and the latter contains verbs and frequencies that were automatically extracted from the Web, as described in [12]. We then calculated the cosine correlation coefficient between $\overrightarrow{h}$ and $\overrightarrow{p}$ as follows:

$$cos(\overrightarrow{h}, \overrightarrow{p}) = \frac{\sum_{i=1}^{n} h_i p_i}{\sqrt{\sum_{i=1}^{n} h_i^2} \sqrt{\sum_{i=1}^{n} p_i^2}} \qquad (1)$$

Table 3 shows human- and programme-proposed vectors for sample noun-noun compounds together with the corresponding cosine. The average cosine correlation (in %s) for all 250 noun-noun compounds is shown in Table 4. Since the workers were instructed to provide at least three paraphrasing verbs per noun-noun compound, and they tried to comply, some bad verbs were generated as a result. In such cases, the very first verb proposed by a worker for a given noun-noun compound is likely to be the best one. We tested this hypothesis by calculating the cosine using these first verbs only. As the last two columns of the table show, using all verbs produces consistently better cosine correlation, which suggests that there are many additional good human-generated verbs among those that follow the first one. However, the difference is 1-2% only and is not statistically significant.

---

[4] http://creativecommons.org

Table 3. Human- and programme-proposed vectors, and cosines for sample noun-noun compounds. The common verbs for each vector pair are underlined.

0.96 "blood donor" NOMINALIZATION:AGENT
Human: give(30), donate(16), supply(8), provide(6), share(2), contribute(1), volunteer(1), offer(1), choose(1), hand over(1), ...
Progr.: give(653), donate(395), receive(74), sell(41), provide(39), supply(17), be(13), match(11), contribute(10), offer(9), ...

0.93 "city wall" HAVE$_2$
Human: surround(24), protect(10), enclose(8), encircle(7), encompass(3), be in(3), contain(2), snake around(1), border(1), go around(1), ...
Progr.: surround(708), encircle(203), protect(191), divide(176), enclose(72), separate(49), ring(41), be(34), encompass(25), defend(25), ...

0.91 "disease germ" CAUSE$_1$
Human: cause(20), spread(5), carry(4), create(4), produce(3), generate(3), start(2), promote(2), lead to(2), result in(2), ...
Progr.: cause(919), produce(63), spread(37), carry(20), propagate(9), create(7), transmit(7), be(7), bring(5), give(4), ...

0.89 "flu virus" CAUSE$_1$
Human: cause(19), spread(4), give(4), result in(3), create(3), infect with(3), contain(3), be(2), carry(2), induce(1), ...
Progr.: cause(906), produce(21), give(20), differentiate(17), be(16), have(13), include(11), spread(7), mimic(7), trigger(6), ...

0.89 "gas stove" USE
Human: use(20), run on(9), burn(8), cook with(6), utilize(4), emit(3), be heated by(2), need(2), consume(2), work with(2), ...
Progr.: use(98), run on(36), burn(33), be(25), be heated by(10), work with(7), be used with(7), leak(6), need(6), consume(6), ...

0.89 "collie dog" BE
Human: be(12), look like(8), resemble(2), come from(2), belong to(2), be related to(2), be called(2), be classified as(2), be made from(1), be named(1), ...
Progr.: be(24), look like(14), resemble(8), be border(5), feature(3), come from(2), tend(2), be bearded(1), include(1), betoken(1), ...

0.87 "music box" MAKE$_1$
Human: play(19), make(12), produce(10), emit(5), create(4), contain(4), provide(2), generate(2), give off(2), include(1), ...
Progr.: play(104), make(34), produce(18), have(16), provide(14), be(13), contain(9), access(8), say(7), store(6), ...

0.87 "cooking utensils" FOR
Human: be used for(17), be used in(9), facilitate(4), help(3), aid(3), be required for(2), be used during(2), be found in(2), be utilized in(2), involve(2), ...
Progr.: be used for(43), be used in(11), make(6), be suited for(5), replace(3), be used during(2), facilitate(2), turn(2), keep(2), be for(1), ...

Table 4. Average cosine correlation (in %s) between human- and programme-generated verbs for the *Levi-250 dataset*. Shown are the results for different limits on the minimum number of programme-generated Web verbs. The last column shows the cosine when only the first verb proposed by each worker is used.

Min # of Web Verbs	Number of Compounds	Correlation with Humans	
		Using All Verbs	First Verb Only
0	250	31.8%	30.6%
1	236	33.7%	32.4%
3	216	35.4%	34.1%
5	203	36.9%	35.6%
10	175	37.3%	35.5%

A limitation of the Web-based verb-generating method is that it could not provide paraphrasing verbs for 14 of the noun-noun compounds, in which cases the cosine was zero. If the calculation was performed for the remaining 236 compounds only, the cosine increased by 2%. Table 4 shows the results when the cosine calculations are limited to compounds with at least 1, 3, 5 or 10 different verbs. We can see that the correlation increases with the minimum number of required verbs, which means that the extracted verbs are generally good, and part of the low cosines are due to an insufficient number of extracted verbs. Overall, all cosines in Table 4 are in the 30-37%, which corresponds to a medium correlation [14].

We further compared the human- and the programme-generated verbs aggregated by relation. Given a relation like $HAVE_1$, we collected all verbs belonging to noun-noun compounds from that relation together with their frequencies. From

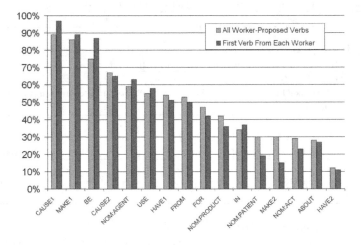

Fig. 3. Cosine correlation (in %s) between the human- and the programme-generated verbs from the *Levi-250 dataset* aggregated by relation: using all human-proposed verbs vs. only the first verb from each worker

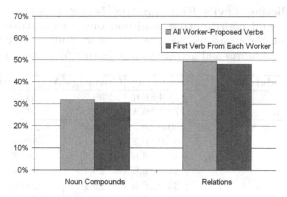

Fig. 4. Average cosine correlation (in %s) between the human- and the programme-generated verbs for the *Levi-250 dataset* calculated for each noun compound (left) and aggregated by relation (right): using all human-proposed verbs vs. only the first verb from each worker

a vector-space model point of view, we summed their corresponding frequency vectors. We did this separately for the human- and the programme-generated verbs, and we then compared the corresponding pairs of summed vectors separately for each relation.

Figure 3 shows the cosine correlations for each of the 16 relations using all human-proposed verbs and only the first verb from each worker. We can see a very-high correlation (mid-70% to mid-90%) for relations like $CAUSE_1$, $MAKE_1$, BE, but low correlation 11-30% for reverse relations like $HAVE_2$ and $MAKE_2$, and for most nominalisations (except for NOM:AGENT). Interestingly, using only the first verb improves the results for highly-correlated relations, but damages low-correlated ones. This suggests that when a relation is more homogeneous, the first verbs proposed by the workers are good enough, and the following verbs only introduce noise. However, when the relation is more heterogeneous, the extra verbs are more likely to be useful. As Figure 4 shows, overall the average cosine correlation is slightly higher when all worker-proposed verbs are used vs. the first verb from each worker only: this is true both when comparing the individual noun-noun compounds and when the comparison is performed for the 16 relations. The figure also shows that while the cosine correlation for individual noun-noun compounds is in the low-30%, for relations it is almost 50%.

Finally, we tested whether the paraphrasing verbs are good features to use in a nearest-neighbour classifier. Given a noun-noun compound, we used the human-proposed verbs as features to predict Levi's RDP for that compound. In this experiment, we only used those noun-noun compounds which are not nominalisations, i.e., for which Levi has an RDP provided; this left us with 214 examples (*Levi-214 dataset*) and 12 classes. We performed leave-one-out cross-validation experiments with a 1-nearest-neighbor classifier (using TF.IDF-weighting

Table 5. Predicting Levi's RDP on the *Levi-214 dataset* using verbs v, prepositions p, and coordinating conjunctions c as features: leave-one-out cross-validation. Shown are micro-averaged accuracy and coverage in %s, followed by average number of features and average sum of feature frequencies per example.

Model	Accuracy	Coverage	Avg #feats	Avg Σfeats
Human: all v	78.4±6.0	99.5	34.3	70.9
Human: first v from each worker	72.3±6.4	99.5	11.6	25.5
Web: $v + p + c$	50.0±6.7	99.1	216.6	1716.0
Web: $v + p$	50.0±6.7	99.1	208.9	1427.9
Web: $v + c$	46.7±6.6	99.1	187.8	1107.2
Web: v	45.8±6.6	99.1	180.0	819.1
Web: p	33.0±6.0	99.1	28.9	608.8
Web: $p + c$	32.1±5.9	99.1	36.6	896.9
Baseline (majority class)	19.6±4.8	100.0	–	–

and the Dice coefficient[5] as a similarity measure, as in [15]), trying to predict the correct RDP for the testing example. The results are shown in Table 5. We achieved 78.4% accuracy using all verbs, and 72.3% with the first verb from each worker. This result is very strong for a 12-way classification problem, and supports the hypothesis that the paraphrasing verbs are very important features for the task of noun-noun compound interpretation.

Table 5 also shows the results when verbs, prepositions and coordinating conjunctions automatically extracted from the Web are used as features. As we can see, using prepositions alone only yields about 33% accuracy, which is a statistically significant improvement over the majority-class baseline, but is well below the classifier performance when using verbs. Overall, the most important Web-derived features are the verbs: they yield 45.8% accuracy when used alone, and 50% when used together with prepositions. Adding coordinating conjunctions helps a bit with verbs, but not with prepositions. Note however that none of the differences between the different feature combinations involving verbs are statistically significant. However, the difference between using Web-derived verbs and using human-proposed verbs (78.4% vs. 50%) is very statistically significant, and suggests that the human-proposed verbs could be considered an upper bound on the accuracy that could be achieved with automatically extracted features.

Table 5 also shows the average number of distinct features and the sum of feature counts per example. As we can see, for Web-derived features, there is a strong positive correlation between number of extracted features and classification accuracy, the best result being achieved with more than 200 features per example. Note however, that using human-proposed verbs yields very high accuracy with seven times less features on average.

---

[5] Given two TF.IDF-weighted frequency vectors A and B, we compare them using the following generalised Dice coefficient: $Dice(A, B) = \frac{2 \times \sum_{i=1}^{n} \min(a_i, b_i)}{\sum_{i=1}^{n} a_i + \sum_{i=1}^{n} b_i}$.

5   Discussion

Interpreting noun compounds in terms of sets of fine-grained verbs that are directly usable in paraphrases of the target noun-noun compounds can be useful for a number of NLP tasks, e.g., for noun compound translation in isolation [3,16,17], for paraphrase-augmented machine translation [18,19,20,21], for machine translation evaluation [22,23], for summarisation evaluation [24], etc.

As we have shown above (see [11,12,15] for additional details and discussion on our experiments), assuming annotated training data, the paraphrasing verbs can be used as features to predict abstract relations like CAUSE, USE, MAKE, etc. Such coarse-grained relations can in turn be helpful for other applications, e.g., for recognising textual entailment as shown by Tatu&Moldovan [25]. Note however, that, for this task, it is possible to use our noun compound paraphrasing verbs directly as explained in Appendix B of [11].

In information retrieval, the paraphrasing verbs can be used for index normalisation [26], query expansion, query refinement, results re-ranking, etc. For example, when querying for *migraine treatment*, pages containing good paraphrasing verbs like *relieve* or *prevent* could be preferred.

In data mining, the paraphrasing verbs can be used to seed a Web search that looks for particular classes of NPs such as diseases, drugs, etc. For example, after having found that *prevent* is a good paraphrasing verb for *migraine treatment*, we can use the query[6] "* which prevents migraines" to obtain different treatments/drugs for migraine, e.g., *feverfew, Topamax, natural treatment, magnesium, Botox, Glucosamine*, etc. Using a different paraphrasing verb, e.g., using "* reduces migraine" can produce additional results: *lamotrigine, PFO closure, Butterbur Root, Clopidogrel, topamax, anticonvulsant, valproate, closure of patent foramen ovale, Fibromyalgia topamax, plant root extract, Petadolex, Antiepileptic Drug Keppra (Levetiracetam), feverfew, Propranolol*, etc. This is similar to the idea of a relational Web search of Cafarella&al. [27], whose system TEXTRUNNER serves four types of relational queries, among which there is one asking for all entities that are in a particular relation with a given target entity, e.g., *"find all X such that X prevents migraines"*.

6   Conclusion and Future Work

In this paper, we explored and experimentally tested the idea that, in general, the semantics of a given noun-noun compound can be characterised by the set of all possible paraphrasing verbs that can connect the target nouns, with associated weights. The verbs we used were fine-grained, directly usable in paraphrases, and using multiple of them for a given noun-noun compound allowed for better approximating its semantics.

Using Amazon's *Mechanical Turk*, we created a new resource for noun-noun compound interpretation based on paraphrasing verbs, and we demonstrated

---

[6] Here "*" is the Google star operator, which can substitute one or more words. In fact, it is not really needed in this particular case.

experimentally that verbs are especially useful features for predicting abstract relations like Levi's RDPs. We have already made the resource publicly available [13]; we hope that by doing so, we will inspire further research in the direction of paraphrase-based noun compound interpretation, which opens the door to practical applications in a number of NLP tasks including but not limited to machine translation, text summarisation, question answering, information retrieval, textual entailment, relational similarity, etc.

The present situation with noun compound interpretation is similar to that with word sense disambiguation: in both cases, there is a general agreement that the research is important and much needed, there is a growing interest in performing further research, and a number of competitions are being organised, e.g., as part of SemEval [28]. Still, there are very few applications of noun compound interpretation in real NLP tasks (e.g., [19] and [25]). We think that increasing this number is key for the advancement of the field, and we believe that turning to paraphrasing verbs could help bridge the gap between research interest and practical applicability for noun compound interpretation.

Acknowledgments. This research was supported in part by NSF DBI-0317510 and by FP7-REGPOT-2007-1 SISTER.

References

1. Downing, P.: On the creation and use of English compound nouns (53), 810–842 (1977)
2. Levi, J.: The Syntax and Semantics of Complex Nominals. Academic Press, New York (1978)
3. Baldwin, T., Tanaka, T.: Translation by machine of compound nominals: Getting it right. In: Proceedings of the ACL 2004 Workshop on Multiword Expressions: Integrating Processing, pp. 24–31 (2004)
4. Warren, B.: Semantic patterns of noun-noun compounds. In: Gothenburg Studies in English 41, Goteburg, Acta Universtatis Gothoburgensis (1978)
5. Nastase, V., Szpakowicz, S.: Exploring noun-modifier semantic relations. In: Fifth International Workshop on Computational Semantics (IWCS-5), Tilburg, The Netherlands, pp. 285–301 (2003)
6. Girju, R., Moldovan, D., Tatu, M., Antohe, D.: On the semantics of noun compounds. Journal of Computer Speech and Language - Special Issue on Multiword Expressions 4(19), 479–496 (2005)
7. Rosario, B., Hearst, M.: Classifying the semantic relations in noun compounds via a domain-specific lexical hierarchy. In: Proceedings of EMNLP, pp. 82–90 (2001)
8. Lauer, M.: Designing Statistical Language Learners: Experiments on Noun Compounds. PhD thesis, Dept. of Computing, Macquarie University, Australia (1995)
9. Finin, T.: The Semantic Interpretation of Compound Nominals. PhD thesis, University of Illinois, Urbana, Illinois (1980)
10. Nakov, P., Hearst, M.: Using verbs to characterize noun-noun relations. In: Euzenat, J., Domingue, J. (eds.) AIMSA 2006. LNCS (LNAI), vol. 4183, pp. 233–244. Springer, Heidelberg (2006)
11. Nakov, P.: Using the Web as an Implicit Training Set: Application to Noun Compound Syntax and Semantics. PhD thesis, EECS Department, University of California, Berkeley, UCB/EECS-2007-173 (2007)

12. Nakov, P., Hearst, M.: Solving relational similarity problems using the web as a corpus. In: Proceedings of ACL 2008: HLT, Columbus, OH (2008)
13. Nakov, P.: Paraphrasing verbs for noun compound interpretation. In: Proceedings of the LREC 2008 Workshop: Towards a Shared Task for Multiword Expressions (MWE 2008), Marrakech, Morocco (2008)
14. Cohen, J.: Statistical Power Analysis for the Behavioral Sciences, 2nd edn. Lawrence Erlbaum Associates, Inc., Hillsdale (1988)
15. Nakov, P., Hearst, M.: UCB: System description for SemEval Task #4. In: Proceedings of SemEval, Prague, Czech Republic, pp. 366–369 (2007)
16. Grefenstette, G.: The World Wide Web as a resource for example-based machine translation tasks. In: Translating and the Computer 21: Proceedings of the 21st International Conference on Translating and the Computer (1999)
17. Tanaka, T., Baldwin, T.: Noun-noun compound machine translation: a feasibility study on shallow processing. In: Proceedings of the ACL, workshop on Multiword expressions, pp. 17–24 (2003)
18. Callison-Burch, C., Koehn, P., Osborne, M.: Improved statistical machine translation using paraphrases. In: Proceedings of Human Language Technology Conference of the North American Chapter of the Association of Computational Linguistics, New York, pp. 17–24 (2006)
19. Nakov, P.: Improved statistical machine translation using monolingual paraphrases. In: Proceedings of the European Conference on Artificial Intelligence (ECAI 2008), Patras, Greece (2008)
20. Nakov, P., Hearst, M.: UCB system description for the WMT 2007 shared task. In: Proceedings of the Second Workshop on Statistical Machine Translation, Prague, Czech Republic, June 2007, pp. 212–215 (2007)
21. Nakov, P.: Improving English-Spanish statistical machine translation: Experiments in domain adaptation, sentence paraphrasing, tokenization, and recasing. In: Proceedings of the Third Workshop on Statistical Machine Translation (WMT), Columbus, OH (2008)
22. Russo-Lassner, G., Lin, J., Resnik, P.: A paraphrase-based approach to machine translation evaluation. Technical Report LAMP-TR-125/CS-TR-4754/UMIACS-TR-2005-57, University of Maryland (August 2005)
23. Kauchak, D., Barzilay, R.: Paraphrasing for automatic evaluation. In: Proceedings Human Language Technology Conference of the North American Chapter of the Association of Computational Linguistics, pp. 455–462 (2006)
24. Zhou, L., Lin, C.-Y., Hovy, E.: Re-evaluating machine translation results with paraphrase support. In: Proceedings of the 2006 Conference on Empirical Methods in Natural Language Processing, Sydney, Australia, July 2006, pp. 77–84 (2006)
25. Tatu, M., Moldovan, D.: A semantic approach to recognizing textual entailment. In: Proceedings of HLT, pp. 371–378 (2005)
26. Zhai, C.: Fast statistical parsing of noun phrases for document indexing. In: Proceedings of the fifth conference on Applied natural language processing, pp. 312–319. Morgan Kaufmann Publishers Inc., San Francisco (1997)
27. Cafarella, M., Banko, M., Etzioni, O.: Relational Web search. Technical Report 2006-04-02, University of Washington, Department of Computer Science and Engineering (2006)
28. Girju, R., Nakov, P., Nastase, V., Szpakowicz, S., Turney, P., Yuret, D.: Semeval-2007 task 04: Classification of semantic relations between nominals. In: Proceedings of SemEval, Prague, Czech Republic, pp. 13–18 (2007)

Optimising Predictive Control Based on Neural Models

Maciej Ławryńczuk

Institute of Control and Computation Engineering, Warsaw University of Technology
ul. Nowowiejska 15/19, 00-665 Warsaw, Poland, tel. +48 22 234-76-73
M.Lawrynczuk@ia.pw.edu.pl

Abstract. This paper presents a Model Predictive Control (MPC) algorithm for on-line economic optimisation of nonlinear technological processes. The economic profit is explicitly expressed in the minimised objective function. The algorithm uses only one dynamic neural model, which is linearised on-line. As a result, an easy to solve on-line quadratic programming problem is formulated. In contrast to the classical multilayer control system structure, the necessity of repeating two nonlinear optimisation problems at each sampling instant is avoided.

Keywords: Process control, Model Predictive Control, neural networks, optimisation, linearisation, quadratic programming.

1 Introduction

Model Predictive Control (MPC) algorithms based on linear models have been successfully used for years in numerous advanced industrial applications [11, 14, 16]. It is mainly because MPC algorithms have a unique ability to take into account constraints imposed on process inputs (manipulated variables) and outputs (controlled variables) which decide on quality, economic efficiency and safety. Moreover, MPC is very efficient in multivariable process control. To maximise economic gains MPC cooperates with economic optimisation [1, 2, 3, 4, 7, 8, 9, 15, 16].

In recent years neural models [5] have been frequently used for economic optimisation and control of nonlinear processes [6, 8, 9, 10, 13, 16, 17] rather than fundamental models. It is because fundamental models are usually not suitable for on-line control and optimisation as they are very complicated and may lead to numerical problems (ill-conditioning, stiffness, etc.). Conversely, neural models can be efficiently used on-line in MPC and economic optimisation because they have excellent approximation abilities, small numbers of parameters and simple structures. Furthermore, neural models directly describe input-output relations of process variables, complicated systems of algebraic and differential equations do not have to be solved on-line.

As far as the cooperation between MPC and economic optimisation is concerned, there are a few alternatives. In the classical multilayer (hierarchical) control system structure [4, 16] the control layer keeps the process at given operating points and the optimisation layer calculates these set-points. Typically,

D. Dochev, M. Pistore, and P. Traverso (Eds.): AIMSA 2008, LNAI 5253, pp. 118–129, 2008.

because of big computational burden, the nonlinear economic optimisation task is solved significantly less frequently than the MPC controller executes. Such an approach can result in low economic effectiveness [16]. In order to improve the economic efficiency, the MPC algorithm can be supplemented with an additional steady-state target optimisation task which recalculates the operating point as frequently as MPC executes [7,9,14,16]. Unfortunately, three optimisation problems have to be solved on-line. It is also possible to integrate economic optimisation and MPC optimisation into one task [8, 16]. Two models of the process are usually used on-line: a dynamic one for MPC and a steady-state one for economic optimisation. Alternatively, only one dynamic model can be used for both MPC and economic optimisation tasks [1].

This paper presents an efficient MPC algorithm for economic optimisation and control of nonlinear technological processes. The described approach is based on the MPC algorithm with Nonlinear Prediction and Linearisation (MPC-NPL) [10,16,17]. The economic profit is explicitly expressed in the minimised objective function. The algorithm uses only one dynamic neural model, which is linearised taking into account the current state of the process. As a result, an easy to solve on-line quadratic programming problem is formulated. Unlike the classical multilayer control system structure, the necessity of repeating two nonlinear optimisation problems at each sampling instant is avoided.

2   Classical Multilayer Control System Structure

The standard multilayer control system structure is depicted in Fig. 1. The objective of economic optimisation (named local steady-state optimisation [16]) is to maximise the production profit and to satisfy constraints, which determine safety and quality of production. Typically, the economic optimisation layer solves the following problem (for simplicity of presentation a Single-Input Single-Output (SISO) process is assumed)

$$\min_{u^s} \{J_E(k) = c_u u^s - c_y y^s\}$$

subject to

$$u^{\min} \leq u^s \leq u^{\max} \tag{1}$$
$$y^{\min} \leq y^s \leq y^{\max}$$
$$y^s = f^s(u^s, h^s)$$

where u is the input of the process (manipulated variable), y is the output (controlled variable) and h is the measured (or estimated) disturbance, the superscript 's' refers to the steady-state. The function $f^s : \Re^2 \longrightarrow \Re \in C^1$ denotes a steady-state model of the process. Quantities c_u, c_y represent economic prices, $u^{\min}$, $u^{\max}$, $y^{\min}$, $y^{\max}$ denote constraints.

In MPC at each sampling instant k future control increments are calculated

$$\Delta u(k) = [\Delta u(k|k)\ \Delta u(k+1|k) \ldots \Delta u(k+N_u-1|k)]^T \tag{2}$$

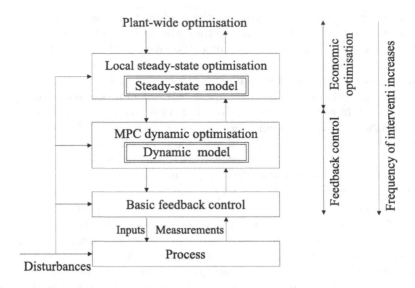

Fig. 1. The classical multilayer control system structure

Only the first element of the determined sequence (2) is applied to the process, the control law is $u(k) = \Delta u(k|k) + u(k-1)$. At the next sampling instant, $k+1$, the prediction is shifted one step forward and the whole procedure is repeated.

Let u_{eo}^s be the solution to the economic optimisation task (1). Using the nonlinear steady-state model, the value y_{eo}^s corresponding to u_{eo}^s is calculated. It is next passed as the desired set-point $(y^s(k) = y_{eo}^s)$ to the MPC optimisation problem

$$\min_{\Delta \mathbf{u}(k)} \left\{ J_{MPC}(k) = \sum_{p=1}^{N}(y^s(k) - \hat{y}(k+p|k))^2 + \sum_{p=0}^{N_u-1} \lambda_p(\Delta u(k+p|k))^2 \right\}$$

subject to

$$u^{\min} \leq u(k+p|k) \leq u^{\max}, \quad p = 0, \ldots, N_u - 1$$
$$-\Delta u^{\max} \leq \Delta u(k+p|k) \leq \Delta u^{\max}, \quad p = 0, \ldots, N_u - 1$$
$$y^{\min} \leq \hat{y}(k+p|k) \leq y^{\max}, \quad p = 1, \ldots, N$$

$$(3)$$

where N and N_u are prediction and control horizons, respectively, $\lambda_p > 0$.

In the standard multilayer control system structure two nonlinear optimisation problems have to be solved on-line. Typically, the economic optimisation problem (1) is solved less frequently than the MPC controller executes, while the MPC optimisation task (3) has to be solved at each sampling instant. The classical multilayer structure with low frequency of economic optimisation can be economically inefficient when disturbances (e.g. flow rates, properties of feed and energy streams) vary significantly and fast [16]. In the multilayer structure two nonlinear models are used: a steady-state one in economic optimisation, a dynamic one in MPC optimisation.

3   Optimising Predictive Control Based on Neural Models

The configuration of the discussed structure is shown in Fig. 2. It is based on the MPC algorithm with Nonlinear Prediction and Linearisation (MPC-NPL) [10, 16, 17]. Typically, in the MPC optimisation problem (3), differences between the desired set-point $y^s(k)$ and predictions $\hat{y}(k+p|k)$ as well as control increments $\Delta u(k+p|k)$ are minimised. In the discussed optimising MPC-NPL approach, the economic profit is explicitly expressed in the minimised cost function

$$J(k) = \sum_{p=0}^{N_u-1} \lambda_p (\Delta u(k+p|k))^2 + c_u \sum_{p=0}^{N_u-1} u(k+p|k) - c_y \sum_{p=1}^{N} \hat{y}(k+p|k) \quad (4)$$

The first part of the cost function is used to minimise future excessive control increments while the role of the second and the third part is to maximise economic profits. For economic optimisation the dynamic model of the process is used, i.e. future values of the process input and predictions of the output signal are taken into account. It is in contrast to the standard multilayer control system shown in Fig. 1, in which the optimisation problem (1) uses a steady-state model of the process. Hence, only one dynamic neural model is used on-line.

The neural model is linearised on-line taking into account the current state of the process. As a result, a quadratic programming problem is formulated, which can be easily solved on-line. Unlike the classical multilayer control system structure, the necessity of repeating two nonlinear optimisation problems at each sampling instant is avoided.

3.1   Optimising Predictive Control Optimisation Problem

As shown in Fig. 2, in the optimising MPC-NPL algorithm at each sampling instant k the nonlinear dynamic neural model is used on-line twice: to find a

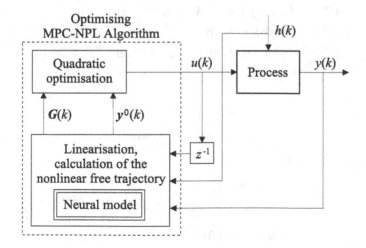

Fig. 2. The structure of the optimising MPC-NPL algorithm

local linearisation and a nonlinear free trajectory. Thanks to the linearisation, the output prediction can be expressed as the sum of a forced trajectory, which depends only on the future (on future input moves $\Delta \boldsymbol{u}(k)$) and a free trajectory $\boldsymbol{y}^0(k)$, which depends only on the past

$$\hat{\boldsymbol{y}}(k) = \boldsymbol{G}(k)\Delta \boldsymbol{u}(k) + \boldsymbol{y}^0(k) \tag{5}$$

where the matrix

$$\boldsymbol{G}(k) = \begin{bmatrix} s_1(k) & 0 & \dots & 0 \\ s_2(k) & s_1(k) & \dots & 0 \\ \vdots & \vdots & \ddots & \vdots \\ s_N(k) & s_{N-1}(k) & \dots & s_{N-N_u+1}(k) \end{bmatrix} \tag{6}$$

of dimensionality $N \times N_u$ contains step-response coefficients of the linearised model and

$$\hat{\boldsymbol{y}}(k) = [\hat{y}(k+1|k) \dots \hat{y}(k+N|k)]^T \tag{7}$$
$$\boldsymbol{y}^0(k) = [y^0(k+1|k) \dots y^0(k+N|k)]^T \tag{8}$$

Taking into account the minimised cost function (4), constraints from the rudimentary MPC optimisation task (3) and the suboptimal prediction (5), one obtains a quadratic programming problem which is solved at each sampling instant of the optimising MPC-NPL algorithm

$$\min_{\Delta \boldsymbol{u}(k),\ \varepsilon^{\min},\ \varepsilon^{\max}} \Big\{ (\Delta \boldsymbol{u}(k))^T \boldsymbol{\Lambda} \Delta \boldsymbol{u}(k)$$
$$+ c_u \left(\boldsymbol{J} \Delta \boldsymbol{u}(k) + \boldsymbol{u}^{k-1}(k) \right)$$
$$- c_y \left(\boldsymbol{G}(k) \Delta \boldsymbol{u}(k) + \boldsymbol{y}^0(k) \right)$$
$$+ \rho^{\min} \left(\varepsilon^{\min} \right)^T \varepsilon^{\min} + \rho^{\max} \left(\varepsilon^{\max} \right)^T \varepsilon^{\max} \Big\} \tag{9}$$

subject to
$$\boldsymbol{u}^{\min} \leq \boldsymbol{J} \Delta \boldsymbol{u}(k) + \boldsymbol{u}^{k-1}(k) \leq \boldsymbol{u}^{\max}$$
$$-\Delta \boldsymbol{u}^{\max} \leq \Delta \boldsymbol{u}(k) \leq \Delta \boldsymbol{u}^{\max}$$
$$\boldsymbol{y}^{\min} - \varepsilon^{\min} \leq \boldsymbol{G}(k)\Delta \boldsymbol{u}(k) + \boldsymbol{y}^0(k) \leq \boldsymbol{y}^{\max} + \varepsilon^{\max}$$
$$\varepsilon^{\min} \geq 0, \quad \varepsilon^{\max} \geq 0$$

where

$$\boldsymbol{y}^{\min} = \left[y^{\min} \dots y^{\min} \right]^T \tag{10}$$
$$\boldsymbol{y}^{\max} = \left[y^{\max} \dots y^{\max} \right]^T \tag{11}$$

are vectors of length N,

$$\boldsymbol{u}^{\min} = \left[u^{\min} \dots u^{\min} \right]^T \tag{12}$$
$$\boldsymbol{u}^{\max} = \left[u^{\max} \dots u^{\max} \right]^T \tag{13}$$
$$\Delta \boldsymbol{u}^{\max} = \left[\Delta u^{\max} \dots \Delta u^{\max} \right]^T \tag{14}$$
$$\boldsymbol{u}^{k-1}(k) = \left[u(k-1) \dots u(k-1) \right]^T \tag{15}$$

are vectors of length N_u, $\boldsymbol{J}$ is the all ones lower triangular matrix of dimensionality $N_u \times N_u$, $\boldsymbol{\Lambda} = diag(\lambda_0, \ldots, \lambda_{N_u-1})$. To cope with infeasibility problems, output constraints are softened by means of slack variables (vectors ε^{min}, ε^{max} of length N), ρ^{min}, $\rho^{max} > 0$ [11,16].

Considering the structure of the optimising MPC-NPL algorithm depicted in Fig. 2, at each sampling instant k the following steps are repeated:

1. Linearisation of the neural model: obtain the dynamic matrix $\boldsymbol{G}(k)$.
2. Find the nonlinear free trajectory $\boldsymbol{y}^0(k)$ using the neural model.
3. Solve the quadratic programming problem (9) to calculate $\Delta\boldsymbol{u}(k)$.
4. Apply the first element of the sequence $\Delta\boldsymbol{u}(k)$, i.e. $u(k) = \Delta\boldsymbol{u}(k|k)+u(k-1)$.
5. Set $k := k + 1$, go to step 1.

3.2   Neural Model of the Process

Let the dynamic model of the process under consideration be described by

$$y(k) = f(\boldsymbol{x}(k)) = f(u(k - \tau), \ldots, u(k - n_B), y(k - 1), \ldots, y(k - n_A), \quad (16)$$
$$h(k - \tau_h), \ldots, h(k - n_C))$$

where $f : \Re^{n_A+n_B+n_C-\tau-\tau_h+2} \longrightarrow \Re \in C^1$, $\tau \leq n_B$, $\tau_h \leq n_C$. A MultiLayer Perceptron (MLP) neural network with one hidden layer and a linear output [5] is used as the function f in (16). Structure of the model is shown in Fig. 3. Output of the model can be expressed as

$$y(k) = f(\boldsymbol{x}(k)) = w_0^2 + \sum_{i=1}^{K} w_i^2 v_i(k) = w_0^2 + \sum_{i=1}^{K} w_i^2 \varphi(z_i(k)) \quad (17)$$

where $z_i(k)$ are sums of inputs of the ith hidden node, $\varphi : \Re \longrightarrow \Re$ is the nonlinear transfer function, K is the number of hidden nodes. From (16) one has

$$z_i(k) = w_{i,0}^1 + \sum_{j=1}^{I_u} w_{i,j}^1 u(k - \tau + 1 - j) + \sum_{j=1}^{n_A} w_{i,I_u+j}^1 y(k - j) \quad (18)$$
$$+ \sum_{j=1}^{I_h} w_{i,I_u+n_A+j}^1 h(k - \tau_h + 1 - j)$$

Weights of the network are denoted by $w_{i,j}^1$, $i = 1, \ldots, K$, $j = 0, \ldots, n_A + n_B + n_C - \tau - \tau_h + 2$, and w_i^2, $i = 0, \ldots, K$, for the first and the second layer, respectively, $I_u = n_B - \tau + 1$, $I_h = n_C - \tau_h + 1$.

3.3   On-Line Linearisation and the Free Trajectory Calculation

The dynamic neural model is linearised on-line taking into account the current state $\bar{\boldsymbol{x}}(k)$ of the process determined by past input and output signal values

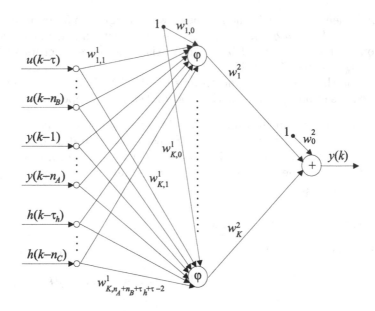

Fig. 3. The structure of the neural model

corresponding to the arguments of the nonlinear model (16). Using Taylor series expansion, the linear approximation of the model is

$$y(k) = f(\bar{\boldsymbol{x}}(k)) + \sum_{l=1}^{n_B} b_l(\bar{\boldsymbol{x}}(k))(u(k-l) - \bar{u}(k-l)) \tag{19}$$

$$- \sum_{l=1}^{n_A} a_l(\bar{\boldsymbol{x}}(k))(y(k-l) - \bar{y}(k-l))$$

where coefficients $a_l(\bar{\boldsymbol{x}}(k)) = -\frac{\partial f(\bar{\boldsymbol{x}}(k))}{\partial y(k-l)}$, $b_l(\bar{\boldsymbol{x}}(k)) = \frac{\partial f(\bar{\boldsymbol{x}}(k))}{\partial u(k-l)}$ of the linearised model are calculated on-line. Considering the structure of the neural model shown in Fig. 3 and described by (17), (18), these coefficients are calculated from

$$a_l(\bar{\boldsymbol{x}}(k)) = -\sum_{i=1}^{K} w_i^2 \frac{\mathrm{d}\varphi(z_i(\bar{\boldsymbol{x}}(k)))}{\mathrm{d}z_i(\bar{\boldsymbol{x}}(k))} w_{i,I_u+l}^1 \quad l = 1, \dots, n_A \tag{20}$$

$$b_l(\bar{\boldsymbol{x}}(k)) = \begin{cases} 0 & l = 1, \dots, \tau - 1 \\ \sum_{i=1}^{K} w_i^2 \frac{\mathrm{d}\varphi(z_i(\bar{\boldsymbol{x}}(k)))}{\mathrm{d}z_i(\bar{\boldsymbol{x}}(k))} w_{i,l-\tau+1}^1 & l = \tau, \dots, n_B \end{cases} \tag{21}$$

If hyperbolic tangent is used as the function φ, $\frac{\mathrm{d}\varphi(z_i(\bar{\boldsymbol{x}}(k)))}{\mathrm{d}z_i(\bar{\boldsymbol{x}}(k))} = 1 - \tanh^2(z_i(\bar{\boldsymbol{x}}(k)))$. Step-response coefficients $s_j(k)$ of the linearised model are

$$s_j(k) = \sum_{i=1}^{\min(j,n_B)} b_i(\bar{\boldsymbol{x}}(k)) - \sum_{i=1}^{\min(j-1,n_A)} a_i(\bar{\boldsymbol{x}}(k))s_{j-i}(k) \tag{22}$$

The nonlinear free trajectory $y^0(k + p|k)$, $p = 1, \ldots, N$, is calculated on-line

$$y^0(k + p|k) = w_0^2 + \sum_{i=1}^{K} w_i^2 \varphi(z_i^0(k + p|k)) + d(k) \tag{23}$$

The "DMC type" disturbance model is used [11, 16]. The unmeasured disturbance $d(k)$ is assumed to be constant over the prediction horizon

$$d(k) = y(k) - y(k|k - 1) = y(k) - \left(w_0^2 + \sum_{i=1}^{K} w_i^2 v_i(k) \right) \tag{24}$$

where $y(k)$ is measured while $y(k|k - 1)$ is calculated from the model (17). The quantities $z_i^0(k + p|k)$ are determined from (18) assuming no changes in the control signal from the sampling instant k onwards and replacing predicted output signals from $k + 1$ by corresponding values of the free trajectory

$$z_i^0(k + p|k) = w_{i,0}^1 + \sum_{j=1}^{I_{uf}(p)} w_{i,j}^1 u(k - 1) + \sum_{j=I_{uf}(p)+1}^{I_u} w_{i,j}^1 u(k - \tau + 1 - j + p)$$

$$+ \sum_{j=1}^{I_{yp}(p)} w_{i,I_u+j}^1 y^0(k - j + p|k) + \sum_{j=I_{yp}(p)+1}^{n_A} w_{i,I_u+j}^1 y(k - j + p)$$

$$+ \sum_{j=1}^{I_{hf}(p)} w_{i,I_u+n_A+j}^1 h(k - \tau_h + 1 - j + p|k)$$

$$+ \sum_{j=I_{hf}(p)+1}^{I_h} w_{i,I_u+n_A+j}^1 h(k - \tau_h + 1 - j + p) \tag{25}$$

where $I_{uf}(p) = \max(\min(p - \tau + 1, I_u), 0)$, $I_{yp}(p) = \min(p - 1, n_A)$, $I_{hf}(p) = \max(\min(p - \tau_h, I_h), 0)$. Typically, future values of the measured disturbance over the prediction horizon are not known at k, $h(k + p|k) = h(k)$ for $p \geq 1$.

4   Simulation Results

The considered process is a polymerisation reaction taking place in a jacketed continuous stirred tank reactor [12] depicted in Fig. 4. The reaction is the free-radical polymerisation of methyl methacrylate with azo-bis-isobutyronitrile as initiator and toluene as solvent. The output $NAMW$ (Number Average Molecular Weight) is controlled by manipulating the inlet initiator flow rate F_I, the flow rate F of the monomer is the measured disturbance ($u = F_I$, $y = NAMW$, $h = F$). Both steady-state and dynamic properties of the process are nonlinear. Hence, it is justified to use nonlinear neural models for economic optimisation and control.

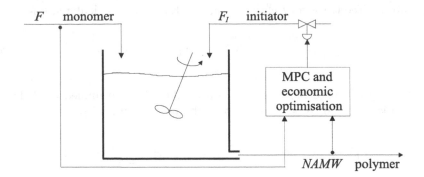

Fig. 4. Polymerisation reactor control system structure

Three models of the process are used. The fundamental model [12] is used as the real process during simulations. It is solved using the Runge-Kutta RK45 method. The model is simulated open-loop in order to generate data for identification. Two neural models are obtained, namely a dynamic one (it has 6 hidden nodes) and a steady-state one (it has 4 hidden nodes) which are next used in MPC and economic optimisation. Parameters of MPC are: $N = 10$, $Nu = 3$, $\lambda_p = 0.2$, $\rho^{\min} = \rho^{\max} = 10$.

To maximise the production rate the economic performance function (1) or (4) with $c_u = -1$, $c_y = 0$ is used. It means that economic optimisation maximises the amount of the substance flowing through the reactor. The following constraints are imposed on manipulated and controlled variables

$$F_I^{\min} \leq F_I, F_I^s \leq F_I^{\max}, \qquad NAMW^{\min} \leq NAMW, NAMW^s \qquad (26)$$

where $F_I^{\min} = 0.0035\ m^3/h$, $F_I^{\max} = 0.033566\ m^3/h$, $NAMW^{\min} = 20000$ $kg/kmol$. The same constraints are used in economic optimisation and MPC. The scenario of disturbance changes is

$$F(k) = 2 - 1.6(\sin(0.008k) - \sin(0.08)) \qquad (27)$$

Four different structures are compared:

a) The realistic multilayer structure with nonlinear economic optimisation repeated 50 times less frequently than the nonlinear MPC algorithm,
b) The realistic multilayer structure with nonlinear economic optimisation repeated 25 times less frequently than the nonlinear MPC algorithm,
c) The "ideal" multilayer structure with nonlinear economic optimisation repeated as frequently as the nonlinear MPC algorithm,
d) The optimising MPC-NPL algorithm with quadratic programming.

In the multilayer structure the MPC algorithm with on-line Nonlinear Optimisation (MPC-NO) is used [10, 16, 17]. In the first two cases nonlinear MPC optimisation is repeated at each sampling instant on-line whereas nonlinear economic optimisation every 50th or 25th sampling instant, respectively. In the third

Table 1. The economic performance index J_E in different structures

Structure	T_E	J_E
Realistic multilayer, low frequency of economic optimisation	50	−4.82543
Realistic multilayer, medium frequency of economic optimisation	25	−5.05209
"Ideal" multilayer, high frequency of economic optimisation	1	**−5.21406**
Optimising MPC-NPL algorithm	−	**−5.21409**

case two nonlinear optimisation problems are solved at each sampling instant. In the optimising MPC-NPL algorithm the neural model is linearised on-line and only one quadratic programming problem is solved.

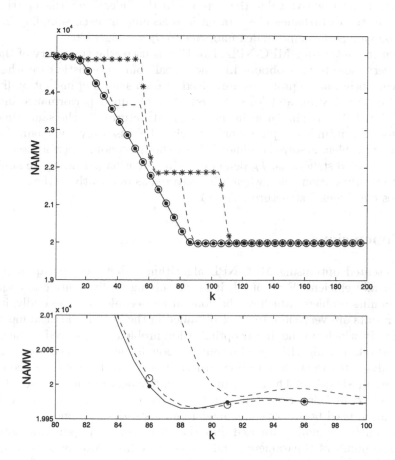

Fig. 5. Simulation results: the multilayer structure with nonlinear economic optimisation repeated 50 (*dashed line with asterisks*) and 25 (*dashed line*) times less frequently than the nonlinear MPC algorithm, the "ideal" multilayer structure with nonlinear economic optimisation repeated as frequently as the nonlinear MPC algorithm (*solid line with dots*) and the optimising MPC-NPL algorithm (*dashed line with circles*)

In the classical multilayer structure both steady-state and dynamic neural models are used in economic optimisation and MPC layers, respectively. In the optimising MPC-NPL algorithm only one dynamic neural model is used.

Simulation results are depicted in Fig. 5. In the first two cases the frequency of nonlinear economic optimisation is low. It means that changes in the disturbance F are taken into account infrequently. As a result the operating point is constant for long periods. For the whole simulation horizon the economic performance index

$$J_E = \sum_{k=1}^{200} J_E(k) = \sum_{k=1}^{200}(-F_I(k)) \tag{28}$$

is calculated. In the first case $J_E = -4.82543$, in the second case $J_E = -5.05209$ (the bigger the negative value the better). In the "ideal" multilayer structure changes in the disturbance F are taken into account at each sampling instant. The economic performance index improves to $J_E = -5.21406$.

When the optimising MPC-NPL algorithm is used, the trajectory of the system is very close to that obtained in the "ideal" but unrealistic case when two nonlinear optimisation problems are solved at each sampling instant on-line (at economic optimisation and MPC layers). The obtained performance index is $J_E = -5.21409$ (very similar as in the "ideal" structure). At the same time, the discussed algorithm is computationally efficient because only one quadratic programming problem is solved on-line. Table 1 shows economic performance index in all considered structures. T_E denotes the intervention period of the nonlinear economic optimisation layer, when it is activated as frequently as the MPC layer executes (the "ideal" structure), $T_E = 1$.

5 Conclusions

The presented optimising MPC-NPL algorithm is both computationally and economically efficient. First of all, it needs solving on-line only one quadratic programming problem, which can be done in foreseeable time. Secondly, its economic results are very close to those obtained in the "ideal" classical multilayer structure in which two nonlinear optimisation problems are solved at each sampling instant. The algorithm is also computationally efficient in comparison with the steady-state target optimisation structure in which three optimisation problems are solved on-line. The algorithm uses on-line only one neural model, while the standard multilayer structure and the structure with steady-state target optimisation need both a dynamic model and a steady-state one.

Neural network models are used since they have excellent approximation abilities, small numbers of parameters and simple structures. Moreover, they directly describe input-output relations of process variables while in the case of fundamental models complex systems of differential and algebraic equations have to be solved on-line. Although in this paper MLP neural networks are used, different types of networks can be considered (e.g. RBF). The presented approach can be relatively easy extended to deal with Multi-Input Multi-Output (MIMO) processes.

Acknowledgement. This work was supported by Polish national budget funds 2007-2009 for science as a research project.

References

1. Beccera, V.M., Roberts, P.D., Griffiths, G.W.: Novel Developments in process optimisation using predictive control. Journal of Process Control 8, 117–138 (1998)
2. Blevins, T.L., Mcmillan, G.K., Wojsznis, M.W.: Advanced control unleashed. ISA (2003)
3. Brdys, M., Tatjewski, P.: Iterative algorithms for multilayer optimizing control. Imperial College Press, London (2005)
4. Findeisen, W.M., Bailey, F.N., Brdyś, M., Malinowski, K., Tatjewski, P., Woźniak, A.: Control and coordination in hierarchical systems. J. Wiley and Sons, Chichester (1980)
5. Haykin, S.: Neural networks – a comprehensive foundation. Prentice-Hall, Englewood Cliffs (1999)
6. Hussain, M.A.: Review of the applications of neural networks in chemical process control – simulation and online implmementation. Artificial Intelligence in Engineering 13, 55–68 (1999)
7. Kassmann, D.E., Badgwell, T.A., Hawkins, R.B.: Robust steady-state target calculation for model predictive control. AIChE Journal 46, 1007–1024 (2000)
8. Ławryńczuk, M., Tatjewski, P.: Efficient predictive control integrated with economic optimisation based on neural models. In: Rutkowski, L., Tadeusiewicz, R., Zadeh, L.A., Zurada, J.M. (eds.) ICAISC 2008. LNCS (LNAI), vol. 5097, pp. 111–122. Springer, Heidelberg (2008)
9. Ławryńczuk, M.: Neural models in computationally efficient predictive control cooperating with economic optimisation. In: de Sá, J.M., Alexandre, L.A., Duch, W., Mandic, D.P. (eds.) ICANN 2007. LNCS, vol. 4669, pp. 650–659. Springer, Heidelberg (2007)
10. Ławryńczuk, M.: A family of model predictive control algorithms with artificial neural networks. Int. Journal of Applied Mathematics and Computer Science 17, 217–232 (2007)
11. Maciejowski, J.M.: Predictive control with constraints. Prentice-Hall, Englewood Cliffs (2002)
12. Maner, B.R., Doyle, F.J., Ogunnaike, B.A., Pearson, R.K.: Nonlinear model predictive control of a simulated multivariable polymerization reactor using second-order Volterra models. Automatica 32, 1285–1301 (1996)
13. Nørgaard, M., Ravn, O., Poulsen, N.K., Hansen, L.K.: Neural networks for modelling and control of dynamic systems. Springer, London (2000)
14. Qin, S.J., Badgwell, T.A.: A survey of industrial model predictive control technology. Control Engineering Practice 11, 733–764 (2003)
15. Saez, D., Cipriano, A., Ordys, A.W.: Optimisation of industrial processes at supervisory level: application to control of thermal power plants. Springer, Heidelberg (2002)
16. Tatjewski, P.: Advanced control of industrial processes, Structures and algorithms. Springer, London (2007)
17. Tatjewski, P., Ławryńczuk, M.: Soft computing in model-based predictive control. Int. Journal of Applied Mathematics and Computer Science. 16, 101–120 (2006)

Prototypes Based Relational Learning

Rocío García-Durán, Fernando Fernández, and Daniel Borrajo

Universidad Carlos III de Madrid
29011 Leganes, Madrid, Spain
rgduran@inf.uc3m.es, ffernand@inf.uc3m.es, dborrajo@ia.uc3m.es
http://www.plg.inf.uc3m.es

Abstract. Relational instance-based learning (RIBL) algorithms offer high prediction capabilities. However, they do not scale up well, specially in domains where there is a time bound for classification. Nearest prototype approaches can alleviate this problem, by summarizing the data set in a reduced set of prototypes. In this paper we present an algorithm to build Relational Nearest Prototype Classifiers (RNPC). When compared with RIBL approaches, the algorithm is able to dramatically reduce the number of instances by selecting the most relevant prototypes, maintaining similar accuracy. The number of prototypes is obtained automatically by the algorithm, although it can be also bounded by the user. Empirical results on benchmark data sets demonstrate the utility of this approach compared to other instance based approaches.

Keywords: Inductive Logic Programming, Instance Based learning, Nearest Prototype.

1 Introduction

The main activity in relational data mining consists on finding patterns in data, described by multiple relations. This difference on the representation of data with respect to the propositional case increases the complexity of finding these patterns. The different relational techniques generate models using diverse representation schemes, such as equations, classification/regression trees/rules, association rules, or instance-based. In the last type of representation, we need to define a relational distance between two data entries (similarity measure). Usually, this distance measure in the relational case must be adapted and extended from the ones used for propositional learning. This technique to learn patterns is also named Instance-Based Learning (IBL).

IBL methods rank among the best options for classification tasks when dealing both with propositional and relational representations. Relational Instance-Based Learning (RIBL) [1] techniques have been able to obtain good classifiers in challenging domains, such as those composed of biological data.

The accuracy of IBL methods is comparable (sometimes even better than) to other relational techniques like inductive classification logic (ICL), logical decision trees (TILDE), relational regression trees (S-CART), relational rules

D. Dochev, M. Pistore, and P. Traverso (Eds.): AIMSA 2008, LNAI 5253, pp. 130–143, 2008.

induction (CPROGOL4.4) or relational association rules (WARMR) [2]. However, IBL have problems scaling up when the training set grows.

In most domains in the machine learning literature, there is no restriction on the time to perform classification. Furthermore, given that for IBL techniques classification time is usually proportional to the size of the classifier, the number of instances in the learned "model" does not represent a hard constrain. But, there are application domains where classification time is an important constrain, as it the case of applying relational learning to AI planning tasks. For instance, if we want to learn a model that would predict the heuristic value of every node of a search process, it should return a prediction (heuristic value) quite fast, so that the use of the learned heuristic outperforms using other heuristics, or even performing an extended blind search.

In this paper we introduce the first implementation of a Relational Nearest Prototype Classifier (RNPC). The goal is to obtain a reduced set of prototypes that generalizes the data set such that it can predict the class of a new instance faster than using RIBL and with an equivalent accuracy. Specifically, this solution is based on an existing nearest protoype algorithm for propositional data, the Evolutionary Nearest Prototype Classifier (ENPC) [3]. There are two main differences with that work: RNPC uses a relational representation; and the prototypes are extracted by selection as in [4]. The algorithm is based on an evolutionary process, where different operators eliminate or select new protoypes from the original data set. RNPC has three main properties: (i) an accuracy performance similar to other IBL approaches; (ii) the automated selection of the number of prototypes, that can be bounded by the user, and (iii) a reduction in time and memory in future classifications required in some domains.

In the next section we explain the RNPC algorithm. The third section shows the experiments performed over four different data sets, and comparative results with IBL methods and other classical relational algorithms. Finally, we draw some conclusions and suggest future work.

2 The RNPC Technique

As the rest of RIBL techniques, the RNPC algorithm uses a relational representation of the data (we use the MTARFF format [5]) and a relational distance measure to compute the similarity between instances. The RNPC algorithm is independent of the distance measure, and we have experimented with different relational distances. We provide first a set of definitions, and later describe the algorithm.

2.1 Definitions

A classifier, C, is composed of a set of N relational prototypes $C = \{p_1, \ldots, p_N\}$. Each prototype p_i is a selected instance of the training set and represents a region of the full data set. One instance I_j belongs to a specific region, or to a prototype p_i, when p_i is the closest prototype to I_j.

Table 1. RNPC Algorithm

Input: TrainingSet, $\boldsymbol{X}$; MaxNumberIterations, it
Initialize C_0 =choose-randomly($\boldsymbol{X}$), $i = -1$

REPEAT
    $i = i + 1$
    $C_{i+1} = $ mutation($\boldsymbol{X}$, C_i)
    $C_{i+1} = $ reproduction($\boldsymbol{X}$, C_{i+1})
    $C_{i+1} = $ move($\boldsymbol{X}$, C_{i+1})
    $C_{i+1} = $ fight($\boldsymbol{X}$, C_{i+1})
    $C_{i+1} = $ die($\boldsymbol{X}$, C_{i+1})
UNTIL($C_i = C_{i+1}$) OR ($i = it$)

return C_{i+1}

Each prototype, p_i, is characterized by its quality, $quality(p_i)$, which is computed by equation 1.

$$quality(p_i) = min(1, accuracy(p_i) \times coverage(p_i)) \tag{1}$$

The quality of a prototype is high if it classifies correctly, *accuracy*, and if it represents a sufficient number of instances, *coverage*. The value of the quality is limited to the range $[0, 1]$.

2.2 The Algorithm

The RNPC algorithm is summarized in Algorithm 1. It receives as input a classifier of only one prototype randomly chosen from the set of instances. It also receives the maximum number of iterations, it, and, optionally (in case the user supplies this value), a bound on the number of prototypes, $MaxPrototypes$. It returns the last classifier (set of prototypes) obtained by iteratively applying some genetic-based operators described next. The loop stops when the maximum number of iterations has been completed.

Table 2. Algorithm of the Mutation operator

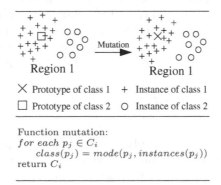

$\times$ Prototype of class 1   + Instance of class 1
$\square$ Prototype of class 2 $\circ$ Instance of class 2

Function mutation:
$for\ each\ p_j \in C_i$
    $class(p_j) = mode(p_j, instances(p_j))$
return C_i

Table 3. Algorithm of the Reproduction operator

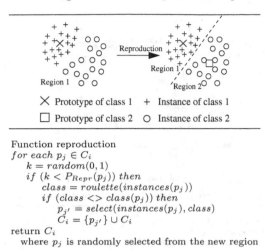

X Prototype of class 1   + Instance of class 1
□ Prototype of class 2   O Instance of class 2

Function reproduction
$for\ each\ p_j \in C_i$
   $k = random(0,1)$
   $if\ (k < P_{Repr}(p_j))\ then$
      $class = roulette(instances(p_j))$
      $if\ (class <> class(p_j))\ then$
         $p_{j'} = select(instances(p_j), class)$
         $C_i = \{p_{j'}\} \cup C_i$
  return C_i
    where p_j is randomly selected from the new region

The **mutation** operator (Table 2) labels each prototype in the current classifier C_i with the most popular class in each region. Following the nearest neighbour rule, each prototype knows the number of instances of each class located in its region. Then, the prototype class becomes the same class as the most abundant class of instances in its region (the mode).

The **reproduction** operator (Table 3) introduces new prototypes into the current classifier C_i. The insertion of new prototypes is a decision that is taken by each prototype in order to increase its own quality. Thus, regions with instances belonging to different classes can create new regions of different classes. Equation 2 defines the probability of reproduction of a propotype, where p_j is a prototype, N is the current number of prototypes. In case the user supplies it, $MaxPrototypes$ is the maximum number of prototypes. So, the probability is inversely proportional to the accuracy (first factor): if the classification success of the prototype is high there is no need to reproduce. Also, when the current number of prototypes tends to $MaxPrototypes$, the probability tends to zero (second factor). By default $MaxPrototypes$ is the number of training instances.

$$P_{Repr}(p_j) = (1 - accuracy(p_j)) \times (1 - \frac{N}{MaxPrototypes}) \qquad (2)$$

A roulette method decides the class of the new prototype. The roulette contains a slide per class with size proportional to the number of instances of each class in the region. If the selected class, j, is different from $class(p_i)$, the new region will have all instances belonging to class j and the new prototype will be chosen randomly among the instances of that class.

The **fight** operator (Table 4) allows the prototypes to capture instances from other regions. Each prototype chooses one rival prototype among its neighbours, adjacent regions, by a roulette mechanism. The probability of choosing one rival

Table 4. Algorithm of the Fight operator

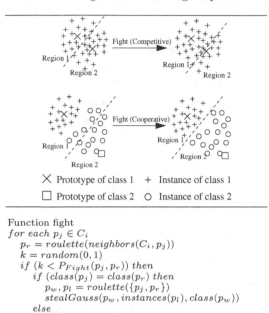

X Prototype of class 1   + Instance of class 1

□ Prototype of class 2   ○ Instance of class 2

Function fight
$for\ each\ p_j \in C_i$
$\quad p_r = roulette(neighbors(C_i, p_j))$
$\quad k = random(0, 1)$
$\quad if\ (k < P_{Fight}(p_j, p_r))\ then$
$\quad\quad if\ (class(p_j) = class(p_r))\ then$
$\quad\quad\quad p_w, p_l = roulette(\{p_j, p_r\})$
$\quad\quad\quad stealGauss(p_w, instances(p_l), class(p_w))$
$\quad\quad else$
$\quad\quad\quad stealClass(p_j, instances(p_r), class(p_j))$
$return\ C_i$
where p_w and p_l are the winner and loser prototypes

is proportional to the difference among the qualities of both prototypes. In a second step, the prototype decides whether to fight or not. This probability is also proportional to the difference of qualities between both rivals.

Finally, if they fight and both have the same class, the winner will be selected with a roulette mechanism using the quality measure. The winner prototype steals as many instances of its class as a Gaussian function dictates. However, if both rivals belong to different classes, they cooperate and the rival gives to the initial prototype all its instances belonging to the initial prototype class.

The **move** operator (Table 5) relocates the prototypes. It finds the medoid of a region (the relational instance with less distance to the rest) as was successfully applied in FORC [6]. This is the main difference with respect to ENPC, where the centroid (a new propositional instance placed in the middle of the region) was computed.

Finally, the **Die** operator (Table 6) eliminates weak prototypes. The probability of dying is proportional to the quality of the prototype and inversely proportional to the difference between the number of current prototypes and $MaxPrototypes$. Then, a prototype have more probability of dying if its quality is close to 0, and also if the current number of prototypes of the classifier is close to or greater than the expected one. The instances of a dead prototype are

Table 5. Algorithm of the Move operator

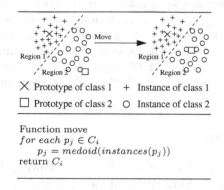

✕ Prototype of class 1	+ Instance of class 1
☐ Prototype of class 2	○ Instance of class 2

Function move
$for\ each\ p_j \in C_i$
$\quad p_j = medoid(instances(p_j))$
return C_i

Table 6. Algorithm of the Die operator

Function die
$for\ each\ p_j \in C_i$
$\quad k = random(0, 1)$
$\quad if\ (k < P_{Die}(p_j))\ then C_i = C_i - \{p_j\}$
return C_i

moved to the nearest prototype. The resulting probability is given by equation 3, with the same notation as in equation 2.

$$P_{Die}(p_j) = 1 - min(1, \frac{MaxPrototypes}{N} \times quality(p_j)) \tag{3}$$

The prototypes change in each iteration, so a different classifier is generated at the end of each iteration. The algorithm collects only the next classifier with best training accuracy. If the training accuracy is similar, it selects the one with less number of prototypes.

3   Experiments

In this section we study the RNPC algorithm for three data sets:[1]

1. **Protein Fingerprints:** A protein fingerprint consists of a list of motifs where one motif is a set of sequences of amino-acids [7]. The total number of instances is 1842 and there are three possible diagnostics [8].
2. **Diterpenes:** Diterpenes are organic compounds of low molecular weight that are based on a skeleton of 20 carbon atoms. The goal is to identify the type of diterpenoid compound skeletons given their 13C-NMR spectra. There are 1503 instances and 23 different classes [9].

---

[1] A description of the data sets can be found in:
$http : //cui.unige.ch/ \sim woznica/rel\_weka/$

3. **Mutagenesis:** The goal is to predict mutagenicity of a set of 188 aromatic and hetero-aromatic nitro-compounds. It has been described in two ways: in the first one, *Mutagenesis1*, the compounds are composed of atoms, which constitute bonds. In this case, the median of the number of atoms for each compound is 26 and the recursion is four. In the second one, *Mutagenesis2*, the compounds are composed of bonds, which consist of atoms. Now, the median of the number of bonds is 28 and the recursion level is limited to three.

3.1 Training in RNPC

First, we sample how RNPC behaves in training, showing the results of two executions with the Protein Fingerprints data set (Figure 1). They correspond to $MaxPrototypes = \#TrainingInstances$ and $MaxPrototypes = 92$ (5% of the training set) respectively. We refer as the *reduction factor* the percentage of reduction in the number of instances between the original data set and the one that RNPC returns. In both figures we can observe the accuracy over the training set (left Y axis) and the number of prototypes of the classifiers in each iteration (right Y axis).

In Figure 1(a) training accuracy rises to 100% in iteration 356, what means that there is overfitting in training. The algorithm selects the classifier with best training accuracy and, between these, the one with a lower number of prototypes: classifier resulting in iteration 356 with 617 prototypes. Then, the selected classifier is tested with the last fold of instances and its test accuracy decrease to 72.04%. However, in Figure 1 (b) where $MaxPrototypes = 92$, the best classifier is selected in iteration 372 with a training accuracy of 85.87% and 72 prototypes. This classifier is tested with the last fold of instances and obtains 84.41% test accuracy. We can see in this case how using a higher reduction in the number of prototypes, the test accuracy increases. Although we will show in the next section that this issue depends on the data set.

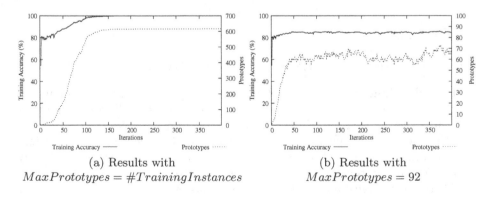

(a) Results with
$MaxPrototypes = \#TrainingInstances$

(b) Results with
$MaxPrototypes = 92$

Fig. 1. Training accuracy and number of prototypes in Fingerprints

3.2   Results in Biological Data Sets

The following experiment used the four data sets and the results are summarized from Table 7 to 10 respectively. The average number of prototypes are shown in Tables (a) and the accuracy results in Tables (b). We executed a 10-fold cross validation in each domain. For each fold, the experiments have been repeated 10 times for RNPC, since it is stochastic. In all cases, the algorithm was executed for a maximum of 400 iterations ($it = 400$). In all domains, the algorithm has been executed with different distance measures: the RIBL [6], AL (Average Linkage), CL (Maximum), Hausdorff (metric), SL (Minimum), SoMDM (Average sum of minimal), SoMDMNorm (Normalized SoMDM), and SymmetricDiff (Symmetric Difference) or Tanimoto (metric) distances [10,11,12].

In all cases we have run the RNPC algorithm with no reduction and with a 95% reduction (RNPC$_{95\%}$). To compare the generalization capability of the algorithm, we have also implemented the relational version of the IBL (RIBL) algorithm that processes instances incrementally [13]. It takes instance by instance selecting them as prototypes only if no prototype before covers it. The number of final prototypes tends to be similar to the number of prototypes in the RNPC algorithm. The RIBL

Table 7. Results in the Fingerprints data set, 1842 instances, 3 classes

(a) Average number of prototypes.

	RNPC	RNPC$_{95\%}$	RIBL
RIBL	620.38	69.09	507.1
AL	580.90	70.75	522.8
CL	412.98	64.35	581.7
Hausdorff	603.78	78.9	491.8
SL	649.45	76.37	514.9
SoMDM	642.64	76.94	513.6
SoMDMNo	2.85	2.97	955.4
SymDiff	705.97	71.62	556.2
Tanimoto	691.74	73.28	538.6

(b) Accuracy results.

	RNPC	RNPC$_{95\%}$	RIBL	IBk1	IBk3	IBk8
RIBL	73.87 ± 3	82.79 ± 2.1	69.71 ± 2.3	78.99 ± 3.7	83.66 ± 4	84.31 ± 3.4
AL	73.38 ± 9.4	78.70 ± 10.1	68.22 ± 10.5	79.04 ± 2.7	75.62 ± 3.3	85.02 ± 2.6
CL	73.59 ± 9.2	77.77 ± 10.1	65.12 ± 8.6	75.62 ± 3.3	81.11 ± 2.2	83.17 ± 2.4
Hausdorff	74.21 ± 2.8	82.07 ± 2.9	70.03 ± 2.4	79.48 ± 2.4	82.63 ± 2.5	84.04 ± 2.1
SL	71.66 ± 10.2	80.00 ± 10.2	67.94 ± 8.1	80.29 ± 2	84.09 ± 2.2	86.00 ± 2.5
SoMDM	72.55 ± 9.2	79.90 ± 10.6	67.62 ± 9.3	80.18 ± 2.4	83.82 ± 3.2	85.83 ± 2.4
SoMDMNo	54.17 ± 4.1	54.22 ± 4.1	43.49 ± 6.4	45.34 ± 6.5	43.87 ± 5.7	43.92 ± 5.9
SymDiff	70.97 ± 10.9	76.27 ± 9.5	68.49 ± 9.9	76.11 ± 1.8	76.87 ± 2.3	79.26 ± 1.6
Tanimoto	72.36 ± 9.4	77.32 ± 10.2	70.92 ± 7.6	78.61 ± 1.6	79.97 ± 2	82.30 ± 2.8

algorithm has two disadvantages: the resulting classifier depends on the order of the data; and it is not possible to reduce the number of final prototypes without deleting instances of the initial data set. For the RNPC and IBL algorithms we also show the average number of prototypes of the final classifiers.

Finally we compare against Relational IBk where k is 1, 3 and 8. This algorithm predicts the class of a new example, by selecting the majority class of the k nearest neighbours. The disadvantages in this algorithm are two: the need of trying with different values of k; and the test time cannot be reduced, because it must compute the distance with the complete data set.

Comparing distance metrics. First, we observe that not all distances are good for all the data sets. One example is the SoMDMNorm distance; its accuracies with all techniques is bad in the Fingerprints data set, while it behaves relatively well in the rest of data sets. In fact, it is the best in Diterpenes. However, the three algorithms behave similarly with the same distance.

Comparing each data set. Analyzing the results in the Fingerprints data set in Table 7 (b), we conclude that applying a reduction of 95% improves the

Table 8. Results of test accuracy in the Diterpenes data set, 1503 instances, 23 classes

(a) Average number of prototypes.

	RNPC	RNPC$_{95\%}$	**RIBL**
RIBL	311.52	64.77	327.5
AL	7.94	8.35	859.7
CL	9.38	11.02	1086.8
Hausdorff	478.28	54.18	442.9
SL	16.13	17.76	1012.7
SoMDM	238.81	67.27	256.8
SoMDMNo	213.41	69.17	227.9
SymDiff	3.75	4.17	996.8
Tanimoto	454.18	51.65	459.7

(b) Accuracy results.

	RNPC	RNPC$_{95\%}$	RIBL	IBk1	IBk3	IBk8
RIBL	86.03 ± 2.7	70.77 ± 4.2	87.87 ± 2.2	95.14 ± 2.9	92.68 ± 2	88.82 ± 2.5
AL	50.03 ± 11.5	49.47 ± 12	40.13 ± 6.9	39.39 ± 7.7	25.35 ± 3.7	26.01 ± 2.1
CL	30.45 ± 3.9	30.51 ± 4.2	19.87 ± 4.4	19.89 ± 4	19.23 ± 3.8	21.63 ± 4
Hausdorff	82.24 ± 3.1	61.38 ± 4.8	79.87 ± 3.2	84.90 ± 2.4	76.78 ± 2.3	68.12 ± 2.8
SL	36.48 ± 3.6	36.33 ± 3.6	24.53 ± 3.4	51.77 ± 2.9	51.70 ± 2.7	51.57 ± 2.8
SoMDM	91.32 ± 1.7	80.91 ± 2.8	91.2 ± 2.1	95.14 ± 1.5	92.22 ± 1.5	87.36 ± 2
SoMDMNo	92.93 ± 1.8	83.75 ± 3.8	91.73 ± 2.3	96.67 ± 1.1	94.15 ± 1.2	90.29 ± 1.8
SymDiff	35.21 ± 2.6	35.04 ± 2.4	26.4 ± 6.3	33.60 ± 2.2	33.60 ± 2.2	33.60 ± 2.2
Tanimoto	77.24 ± 4.3	57.02 ± 4.6	77.8 ± 2.3	88.02 ± 2.4	79.97 ± 2	80.11 ± 2.4

test accuracy in all cases with respect to not applying the reduction. These results are also better than the ones obtained in the relational IBL algorithm for all distances. However, the best classifier with respect to accuracy with all distances is IBk (k =8). The smallest difference between the test accuracies of the best two results for both algorithms is 1.5 points for the RIBL distance. The biggest difference is 6.5 points with the AL distance.

In the Diterpenes data set (Table 8), using the RNPC algorithm without reduction is better in most cases than with reduction (between 10 and 20 points). The explanation could be the high number of different classes in this data set (23). For the best distance, SoMDMNorm, it looks easier to predict instances of 23 different classes with 213 prototypes than with 69. The 95% reduction is too high for this domain. In fact, with a 90% reduction over the total set (or 88.9% over the training set) and with the best distance, we obtain a 87.58% in test accuracy. Globally, the IBk algorithm (k =1) is the best one, although its results are close to RNPC and relational IBL.

In both versions of Mutagenesis (Tables 9 and 10), the total number of instances is 188, which explains the high standard deviations. The best results in Mutagenesis1 are shared by RNPC algorithm in both versions and the IBk1. In

Table 9. Results of test accuracy in the Mutagenesis1 data set, 188 instances, 2 classes

(a) Average number of prototypes.

	RNPC	RNPC$_{95\%}$	**RIBL**
RIBL	70.41	7.39	57.9
AL	5.18	3.96	67.5
CL	8.77	5.07	67.3
Hausdorff	62.73	8.99	55.1
SL	25.92	8.51	65.4
SoMDM	50.32	9.4	43.5
SoMDMNo	47.71	9.61	46.9
SymDiff	4.58	3.18	66.2
Tanimoto	56.04	9.87	54.5

(b) Accuracy results.

	RNPC	RNPC$_{95\%}$	RIBL	IBk1	IBk3	IBk8
RIBL	73.16 ± 7.4	75.73 ± 8.4	65.82 ± 12.2	72.31 ± 9.4	67.95 ± 11.6	71.23 ± 7.3
AL	64.60 ± 9.4	67.21 ± 8.5	64.95 ± 9.7	55.29 ± 12.7	49.62 ± 14.5	68.10 ± 9.9
CL	63.31 ± 10.3	65.26 ± 10.6	63.31 ± 11.1	49.62 ± 14.5	65.41 ± 9.2	66.52 ± 6.4
Hausdorff	79.31 ± 11.5	70.67 ± 12.4	70.4 ± 14.1	76.52 ± 12	72.31 ± 11.6	76.61 ± 8.2
SL	75.30 ± 7.1	77.63 ± 7.6	60.62 ± 9	74.47 ± 9.2	76.08 ± 8.7	76.08 ± 8.3
SoMDM	81.76 ± 9.7	78.00 ± 10.5	80.93 ± 10.8	83.45 ± 9.4	78.22 ± 11.1	76.67 ± 9.9
SoMDMNo	83.85 ± 8.5	72.26 ± 10.5	81.33 ± 10.4	85.00 ± 9.8	82.40 ± 11.1	79.74 ± 9.4
SymDiff	64.93 ± 12.1	63.99 ± 11.2	62.26 ± 8.2	75.06 ± 9.9	75.06 ± 9.9	70.76 ± 7.1
Tanimoto	80.16 ± 7.3	71.89 ± 10.6	77.18 ± 8.5	78.25 ± 8.8	77.60 ± 9.9	69.15 ± 6.9

Table 10. Results of test accuracy in the Mutagenesis2 data set, 188 instances, 2 classes

(a) Average number of prototypes.

	RNPC	RNPC$_{95\%}$	**RIBL**
RIBL	53.99	8.24	75.5
AL	4.23	3.54	69.3
CL	3.93	3.48	72.5
Hausdorff	67.03	9.28	60.3
SL	25.44	9.03	65
SoMDM	44.75	9.61	47.6
SoMDMNo	25.13	5.94	42.9
SymDiff	3.36	2.56	63.1
Tanimoto	51.31	9.8	48.4

(b) Accuracy results.

	RNPC	RNPC$_{95\%}$	RIBL	IBk1	IBk3	IBk8
RIBL	76.21 ± 9.3	72.53 ± 10.3	59.91 ± 12.6	75.41 ± 11.1	67.98 ± 8.7	72.22 ± 10
AL	65.82 ± 12.5	66.13 ± 11.7	60.22 ± 11.9	55.29 ± 12.1	40.44 ± 10.2	74.39 ± 6.4
CL	57.29 ± 17.6	58.87 ± 16.3	50.87 ± 15.3	40.44 ± 10.2	62.81 ± 6.3	54.88 ± 15.5
Hausdorff	75.51 ± 11.6	63.06 ± 13.8	71.98 ± 12.2	76.55 ± 9.5	71.26 ± 11.8	66.02 ± 10.4
SL	77.16 ± 8.6	76.67 ± 8.6	57.59 ± 17.1	75.03 ± 9.6	76.08 ± 8.7	76.08 ± 9.3
SoMDM	85.03 ± 7.1	77.27 ± 8.7	78.76 ± 6.9	84.01 ± 8.8	78.27 ± 11.4	69.65 ± 8.5
SoMDMNo	71.75 ± 22.8	63.98 ± 21.6	81.46 ± 8.9	87.72 ± 8.7	81.84 ± 11.3	80.85 ± 11.2
SymDiff	66.54 ± 11	66.43 ± 11.3	60.09 ± 11.4	73.42 ± 10.7	73.42 ± 10.4	69.65 ± 8.5
Tanimoto	79.28 ± 7.8	76.46 ± 10.5	75.6 ± 6.4	82.46 ± 8.3	80.29 ± 9.4	73.45 ± 6.4

Mutagenesis2 all results are better using no reduction. Globally both algorithms, RNPC and IBk, obtain the best results.

In general terms, as almost always happens, the best algorithm depends on the data set. In the Fingerprints and Diterpenes data sets the IBk algorithm works better with k =8 and k =1. In Mutagenesis1 the best results are given by RNPC and in Mutagenesis2 the best results are good for both algortihms. RIBL is not so good as the others.

If we analyze the best results in the RNPC and IBk algorithms independently of the used distance and we compute the statistic significance tests, we can say that in Fingerprints and Diterpenes data sets, the differences are statistically significant with α =0.05. That is, IBk algorithm is the best one in Fingerprints and Diterpenes data sets. For both versions of Mutagenesis, it does not exist significant difference between both algorithms.

Comparing number of prototypes. In this subsection we study the reduction in the number of prototypes generated by RNPC and relational IBL with respect to the size of the training set. In general, and for all distances with

Table 11. Test accuracy results in Fingerprints data set using 10-fold cross validation

Algorithm	Test accuracy (%)
SVM-RBF algorithm	85.92
IBk 8 + RIBL distance	81.69
IBk 1 + RIBL distance	75.21
RNPC + RIBL distance	71.5 ± 1.41

Table 12. Test accuracy results in Mutagenesis1 data set using 10-fold cross validation

Algorithm	Test ac.(%)
CLAUDIEN [2]	90
RNPC + SoMDMNorm	82.1
ICL [2]	80.9
PROGOL [14]	79
TILDE [15]	75
MRDTL [16]	67
FOIL [15]	61

the best results, the number of protoypes is similar for the RNPC and IBL algorithms. However, observing the number of prototypes and the corresponding test accuracy, these two values are not directly proportional. This is the case of the Fingerprints domain (Table 7) where the test accuracy is better for all distances in the execution of RNPC$_{95\%}$, but the number of prototypes is much smaller than the resulting from the standard RNPC and the RIBL algorithms. That is, we can obtain better accuracy with a lower number of prototypes.

Comparing with other methods. In addition, we compared: i) the obtained results by RNPC; with ii) the ones reported in [7] using IBk ($k =1$ and $k =8$) and RIBL distance; and iii) the best result in [8] using the SVM-RBF algorithm. All these results refer only to the Fingerprints data set, where the instances belonging to the training and test sets of the 10-fold cross validation are the first 1487 instances used in the experiments explained before, and the final test was performed with the last 355 instances. The results are shown in Figure 11, where the test result of RNPC is 3.71 points worse than the results of IBk for $k =1$, 10.19 points worse than the IBk algorithm for $k =8$ with the RIBL distance, and 14.42 points worse than the SVM-RBF algorithm. The results of RNPC and IBk follow the same behavior as in Table 7 (b) for the RIBL distance.

More experiments with the Mutagenesis1 data set and other algorithms can be seen in Figure 12. We can see how RNPC with the SoMDMNorm distance (the best one selected in Table 9 (b)) improves the results of ICL, PROGOL, TILDE, MRDTL and FOIL. However, it does not improve CLAUDIEN.

4 Conclusions and Future Work

In this paper, we have presented RNPC as a first approach for nearest prototype classification of relational data. It is an efficient distance-based method that can use any relational distance, reducing dramatically the size of the database to compare new instances with. Furthermore, the number of prototypes can be either automatically generated by the algorithm, or bounded by the user. The reduction in the size of the data set permits to classify new instances faster than with the original data set.

RNPC obtains a kind of generalization of the training instances by generating prototypes. The algorithm has, as a secondary effect, the automatic clustering of the set of training instances such that the instances belonging to the same cluster (or prototype) are similar to each other. Thus, we can use the final prototypes as a generalized description of the clusters. This opens the possibility of human understanding of the resulting classifier and of using those reduced sets of prototypes.

We would like to extend this work in three ways. First, to decide the best distance metric for each domain automatically. Second, to improve the distance metrics used by RNPC, introducing feature selection or weighting methods following previous attribute-value approaches. Third, to extend the application to learning in planning, where the learning protoypes can be seen as policies to guide the search of a solution. This learning can substitute the planner or even it can be combined with the base planner to act as a heuristic.

Acknowledgments

This work has been partially supported by the Spanish MEC project TIN2005-08945-C06-05, a grant from the Spanish MEC, and regional CAM-UC3M project CCG06-UC3M/TIC-0831.

References

1. Emde, W., Wettschereck, D.: Relational instance-based learning. In: Proceedings of the Thirteen International Conference on Machine Learning, pp. 122–130. Morgan Kaufmann, San Francisco (1996)
2. Dzeroski, S., Lavrac, N.: Relational Data Mining. Springer, Heidelberg (2001)
3. Fernández, F., Isasi, P.: Evolutionary design of nearest prototype classifiers. Journal of Heuristics 10(4), 431–454 (2004)
4. Kuncheva, L., Bezdek, J.: Nearest prototype classfication: Clustering, genetic algorithms, or random search? IEEE Transactions on Systems, Man, and Cybernetics (1998)
5. Witten, I., Frank, E.: Data Mining. Practical Machine Learning Tools and Techniques. Elsevier and Morgan Kaufmann (2005)
6. Kirsten, M., Wrobel, S., Horváth, T.: Distance Based Approaches to Relational Learning and Clustering. In: Relational Data Mining, pp. 213–232. Springer, Heidelberg (2001)

7. Woznica, A., Kalousis, A., Hilario, M.: Distance-based learning over extended relational algebra structures. In: Proceedings of the 15th International Conference of Inductive Logic Programming (2005)
8. Hilario, M., Kim, J., Bradley, P., Attwood, T.: Classifying protein fingerprints. In: Proceedings of the 8th Conference on Principles and Practice of Knowledge Discovery in Databases (2004)
9. Dzeroski, S., Schulze-Kreme, S., Heidtke, K., Siems, K., Wettschereck, D.: Intelligent Data Analysis in Medicine and Pharmacology. In: Diterpenes structure elucidation from 13C NMR spectra with machine learning, pp. 207–225 (1997)
10. Kalousis, A., Hilario, M.: Representational issues in meta-learning. In: Proceedings of the 20th International Conference on Machine Learning (2003)
11. Ramón, J., Bruynooghe, M.: A polinomial time computable metric between point sets. In: Acta Informática (2001)
12. Duda, P.H., Stork, D.: Nonparametric Techniques. In: Pattern Classification, 2nd edn. John Wiley & Sons, Chichester (2001)
13. Aha, D.W., Kibler, D., Albert, M.K.: Instance-based learning algorithms. Machine Learning 6, 37–66 (1991)
14. Srinivasan, A., King, R., Muggleton, S.: The role of background knowledge: using a problem from chemistry to examine the performance of an ILP program, Under review for Intelligent Data Analysis in Medicine and Pharmacology (1996)
15. Bolckeel, H.: Top-down induction of first order logical decision trees. PhD thesis, Departament of Computer Science, Katholieke Universiteit Leuven (1998)
16. Leiva, H., Atramentov, A., Honavar, V.: Experiments with MRDTL – a multirelational decision tree learning algorithm. In: Proceedings of the Workshop on Multi-Relational Decision Tree Learning (2002)

Robustness Analysis of SARSA(λ): Different Models of Reward and Initialisation

Marek Grześ and Daniel Kudenko

Department of Computer Science
University of York
York, YO10 5DD, UK
{grzes,kudenko}@cs.york.ac.uk

Abstract. In the paper the robustness of SARSA(λ), the reinforcement learning algorithm with eligibility traces, is confronted with different models of reward and initialisation of the Q-table. Most of the empirical analyses of eligibility traces in the literature have focused mainly on the step-penalty reward. We analyse two general types of rewards (final goal and step-penalty rewards) and show that learning with long traces, i.e., with high values of λ, can lead to suboptimal solutions in some situations. Problems are identified and discussed. Specifically, obtained results show that SARSA(λ) is sensitive to different models of reward and initialisation. In some cases the asymptotic performance can be significantly reduced.

Keywords: reinforcement learning, temporal-difference learning, SARSA, Q-learning, eligibility traces.

1 Introduction

Reinforcement learning (RL) agents learn how to act given observations of the world. They execute actions which have some impact on the environment and the environment subsequently provides numerical feedback which can be used to guide the learning process. Agents use this information to find a policy which maximises the accumulated reward when acting in the environment. A policy determines which action should be taken in a given state and is usually represented as a value function $Q(s, a)$ which estimates 'how good' it is to execute action a in state s. The value function is usually expressed in terms of the discounted or expected total reward which can be obtained when action a will be chosen in state s and the given policy followed thereafter [1].

The same behaviour can be learned from different rewards, e.g., to solve a stochastic shortest path problem the penalty for each step can be given or the final reward can be back-propagated in a discounted way. In this paper we present an empirical analysis of SARSA(λ) with different models of reward and initialisation of the Q-table. Specifically, obtained results show that SARSA(λ) is sensitive to these factors. In some cases the asymptotic performance can be significantly reduced.

D. Dochev, M. Pistore, and P. Traverso (Eds.): AIMSA 2008, LNAI 5253, pp. 144–156, 2008.

In contrast to supervised learning, RL agents are not given instructive feedback on what the best decision in a particular situation is. This leads to the *temporal credit assignment* problem, that is, the problem of determining which part of the behaviour deserves the reward [2]. To tackle this problem, the iterative approach to RL applies the back-propagation of the value function in the state space. Because this is a delayed, iterative approach, it usually leads to a slow convergence especially when the state space is huge. In fact, the state space grows exponentially with each variable added to the encoding of the environment when the Markov property needs to be preserved.

The concept of eligibility traces in reinforcement learning represents one of the methods of dealing more efficiently with the temporal credit assignment problem and with non-Markovian state spaces [3,4]. In basic temporal difference learning the back-propagation is performed only on one preceding state, that is, the temporal difference:

$$\delta = r + \gamma Q(s', a') - Q(s, a) \tag{1}$$

is used to update only state s:

$$Q(s, a) = Q(s, a) + \alpha \delta, \tag{2}$$

where α is the learning rate, γ the discount factor, r an immediate reward, s' is the current state, a' an action to be taken in state s', s the previous state and a the action taken in state s (the computation of the temporal difference in Equation 1 is according to the SARSA algorithm [5]). The idea of eligibility traces is to propagate temporal difference δ not only to state s but also to states which were recently visited (trace) and the measure of this recency is named eligibility. If we assume $e(s, a)$ to be the eligibility of pair (s, a) the SARSA update takes the form:

$$Q(s, a) = Q(s, a) + \alpha \delta e(s, a). \tag{3}$$

When state s is the most recent state to be updated, eligibility for this state is set to 1 and after each time step it is reduced by the multiplicative factor $\lambda \gamma$ where λ controls how eligibility decays in time. SARSA with updates of this type is named SARSA(λ). For the clarity of presentation the standard version of this algorithm with aforementioned type of eligibility traces is shown in Algorithm 1. In this type of traces the eligibility of the most recent state to be updated is reset to the value of one (line 9 in Algorithm 1). These traces are named *replacing eligibility traces* and are used in this paper since they were shown in the literature to work better than accumulating traces [6,5]. The analysis of eligibility traces in this paper is based on SARSA because (in contrast to Q-learning) it is an on-policy method and the back-propagation can be performed along the entire trace of preceding states. In this way we try to avoid obscuration in our analysis and results which may be caused by problems of Q-learning when eligibility traces are used. A detailed discussion of this issue can be found in Sutton [5] and Peng [7].

Algorithm 1. The SARSA(λ) algorithm with replacing eligibility traces [5,6].

1: Initialise $Q(s,a)$ arbitrarily and $e(s,a) = 0$ for all s, a
2: **repeat** {for each episode}
3:     Initialise s
4:     Choose a from s using policy derived form Q
5:     **repeat** {for each step of episode}
6:         Take action a, observe reward r, s'
7:         Choose a' from s' using policy derived form Q
8:         $\delta \leftarrow r + \gamma Q(s',a') - Q(s,a)$
9:         $e(s,a) = 1$ {replacing traces}
10:         **for all** s,a **do**
11:             $Q(s,a) = Q(s,a) + \alpha \delta e(s,a)$
12:             $e(s,a) \leftarrow \gamma \lambda e(s,a)$
13:         **end for**
14:         $s \leftarrow s'$, $a \leftarrow a'$
15:     **until** state s is terminal
16: **until**

The motivation for this paper can be summarised as follows. Firstly, in the existing literature most evaluations of RL algorithms with different values of λ present results for a given, small number of learning episodes, e.g., 10 [6,5]. Secondly, sensitivity of RL with $\lambda > 0$ has never been confronted with different models of reward, i.e, whether the positive reward is given only upon reaching the goal state or negative penalty is given after each step. If for all pairs (s,a) zero is initially assigned to $Q(s,a)$, the modification of the reward model leads to a straightforward change of the character of the initialisation (i.e., optimistic or pessimistic). This leads to an additional dimension of the analysis, i.e., different initialisations for each reward type can be considered.

2   Experimental Design

In this section, the evaluated algorithm with its parameters and the problem domain are discussed. Some characteristics of the experimental domain are identified. They allow for generalisation of our findings to other domains which have these characteristics.

2.1   Parameters

The SARSA(λ) algorithm is used [5]. The eligibility traces are implemented in an efficient way [8,5]. They are truncated when the eligibility becomes negligible. Specifically, the trace of N most recently visited state-action pairs is stored where $(\lambda\gamma)^N \geq 10^{-9}$. The value of eligibility is evaluated as: $e(s,a) = (\lambda\gamma)^{t(s,a)}$ where $t(s,a)$ is the number of time steps since pair (s,a) has been added to the trace. In this way, for the most recent pair $t(s,a) = 0$ and it makes $e(s,a) = 1$. It means that in accordance with the previous section the replacing eligibility traces are used (line 9 in Algorithm 1).

The ϵ-greedy exploration is used. It is an instance of undirected exploration strategies which are based on the current content of the Q-table. This allows us to gain better insight into the influence of the investigated characteristics of RL. However, the evaluated range of configurations allows drawing conclusions about potential positive effects of directed exploration in some situations.

The analysis of eligibility traces is in our case equivalent to the analysis of the impact of λ. Experiments with a range of values of this parameter are reported. A high value of learning rate α can have a negative impact on asymptotic performance with higher values of λ, because the value function can converge too fast [8]. For this reason α starts with a relatively small 0.1 value in the first episode and is further decreased linearly to 0.01 in the last episode (value 0.1 was also used in the famous practical application of temporal difference learning: TD-gammon [9] and as it is a relatively small value it should not obscure our results). A lower value of the learning rate is also advised [5] when the environment is stochastic like in our case because the agent needs more time to adjust to stochastic effects of actions. The ϵ-greedy exploration starts with $\epsilon = 0.3$ and this value is also decreased linearly to 0.01 in the last episode. The discount factor is $\gamma = 0.99$ for the goal-based reward and $\gamma = 1.0$ when step penalty is given. Goal-based reward is 100, i.e., all rewards are 0 and only upon entering the goal state 100 is given[1]. In the second configuration with a step penalty the reward -1 is given after each step.

All runs of experimental configurations were repeated 10 times and average results are presented in graphs where the reward for getting to the goal state from the start state is used. For the goal-based reward this is the discounted reward and for the step penalty reward the sum of penalties to reach the goal state from the start state. In each graph a number of curves for different values of λ is shown. To make the interpretation of these graphs easier, the curve for $\lambda = 0$ is plotted with a thicker line style in each case.

2.2   Experimental Domain

In this experimental analysis we are interested in a class of reinforcement learning problems in which certain properties can be identified. The problem of learning a policy to reach goal state g from any starting state s_0 in a stochastic environment is considered. The trajectory to reach a goal state when following policy π can be $\omega_\pi = (s_0, s_1, ..., g)$. We point out here that in some domains, like for example navigation problems, for many pairs of states (s_i, s_j) where $s_i \in \omega_\pi$ and $s_j \in \omega_\pi$ such that $i < j$, the path between these states has many opportunities to be suboptimal (see Figure 1). It means that the trajectory ω_π may be not connecting states s_i and s_j via the shortest path. This raises the possibility of suboptimal solutions in, e.g., navigation problems. As an example of such tasks a maze-based domain is investigated. An extreme example of a maze task in which this

---

[1] For this reason, even though the task is episodic the discount factor γ has to be less than 1. Otherwise, all Q-values would converge to the value 100. With $\gamma < 1$ they are appropriately discounted allowing expressing a policy for a stochastic shortest path problem.

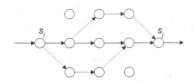

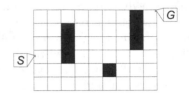

Fig. 1. Domain properties

Fig. 2. The navigation maze task used for experimental analysis

problem does not exist is a one dimensional maze (1D) which is also used for comparison in the final part of the paper.

A policy which represents the solution to the RL task can be obtained from different definitions of the reward. Specifically, the policy for the stochastic shortest path problem can lead to the same optimal behaviour when learning takes place either according to the discounted goal reward or penalties given for each step. Even thought these models of rewards are much different they determine the same optimal behaviour: navigating to the goal state under the constraints of transition probabilities of the underlying MDP. The particular feature of this task, is that regardless which reward is used, the reward sets only one objective, that is, the behaviour of reaching the goal state as fast as possible. Thus, the following categories of RL tasks can be distinguished:

- The reward sets one goal regardless which type of the reward is used (e.g., the goal-based or penalty-based reward in stochastic shortest path problems),
- The reward sets more than one objective; one reward can encourage the agent to reach a particular goal state with a positive reward (winning a game) subject to constraints imposed by another reward which penalises for some other behaviour (an objective of avoiding a particular situation, for example, loosing a game).

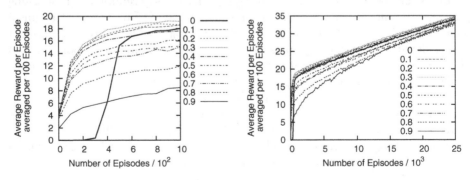

Fig. 3. Results for the goal-based reward model with a pessimistic initialisation. Each curve corresponds to a different value of λ.

In this paper a thorough analysis of the instance of the first category is carried out. The categorisation presented here may help RL practitioners to determine to which kind of domains our findings can be generalised.

Experiments are performed on the domain which has been commonly used in the RL literature [10]. This navigation maze task is shown in Figure 2. In our case, a scaled version is used. Each grid position from the base configuration is uniformly divided into 64 squares yielding 72×48 states. There are eight actions which lead to an adjacent cell if it is not the border nor an obstacle. In such situations actions do not have any effect. Actions are stochastic. With probability 0.1 an action can fail in which case one of the remaining actions is chosen with a uniform probability.

3   Standard Initialisation

In the first experiment both reward models (i.e., goal-based and penalty-based) are used with the default initialisation of the Q-table:

$$\forall_s \forall_a Q(s,a) = 0.$$

This yields a pessimistic initialisation for the goal-based reward and an optimistic initialisation for the penalty-based reward. Results with a range of λ values are shown in Figures 3 and 4 (left-hand parts of these figures show first 10^2 episodes). In the case of the goal-based reward it can be observed in Figure 3 that learning with $\lambda = 0$ has the largest delay in reaching a cumulative reward which is much earlier obtained with higher values of λ. However an interesting insight into this learning process is shown by the asymptotic convergence when learning takes place up to 2.5×10^4 episodes. Higher values of λ, especially $\lambda > 0.5$, show worse asymptotic performance when compared with $\lambda = 0$. Even though the learning rate is relatively low ($\alpha = 0.1$ chosen not to obscure our results), an undirected exploration which is based on the current content of the Q-table is significantly dominated by the suboptimal paths propagated by longer eligibility traces. This

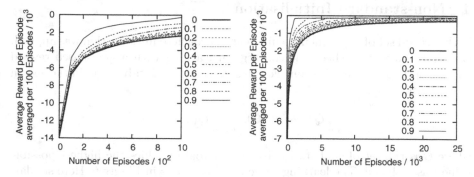

Fig. 4. Results for the penalty-based reward model with an optimistic initialisation. Each curve corresponds to a different value of λ.

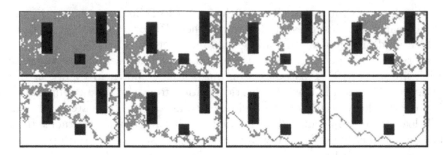

Fig. 5. Trajectories during the learning process with goal-based reward, pessimistic initialisation and $\lambda = 0.9$. When counting figures from the left to the right the following numbers of learning episodes correspond to these figures: 1, 3, 5, 7 in the first row and 13, 15, 1.2×10^3, 2.5×10^4 in the second row.

can be well observed in the trajectories for $\lambda = 0.9$ which are shown in Figure 5. A general shape of the trajectory which is followed after few dozens of episodes is reinforced and may stay up to the end of learning yielding a final solution which is suboptimal.

Learning with a penalty-based reward and zero-initialised Q-table shown in Figure 4 is accelerated during the entire period of learning when $\lambda > 0$ and the convergence improves monotonically when λ increases. The best result is for $\lambda = 0.9$. This can be justified in a relatively straightforward way. The Q-table is initialised optimistically and these values correspond to a trivial heuristic, although this heuristic is admissible [11]. The agent is forced to do exhaustive exploration. The propagation of temporal differences when $\lambda > 0$ allows for a faster 'propagation of pessimism' (the negative reward) when enough optimism is still preserved to have broad exploration. Long eligibility traces do not accelerate the convergence to the suboptimal solution. Rather, they accelerate the propagation of the negative reward which needs to be propagated anyway due to an optimistic and trivial (from heuristics point of view [11]) initialisation.

4   Non-standard Initialisation

In the second set of experiments a different approach to initialisation was applied. Our first aim was to check the converge of learning with a goal-based reward and an optimistic initialisation. For this setting the Q-table was initialised in the following way:

$$\mathop{\forall}_{s}\mathop{\forall}_{a} Q(s,a) = 100, \quad \mathop{\forall}_{g \in G}\mathop{\forall}_{a} Q(g,a) = 0,$$

where G is the set of goal states. To all non-goal Q-values the highest possible value was assigned. The learning curve for this case is in Figure 6. Here similarity to the penalty-based reward with an optimistic initialisation can be found. Specifically, higher values of λ lead also to a better convergence. But in this case

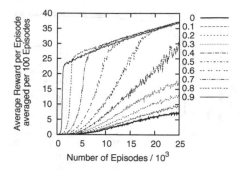

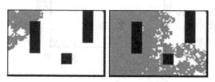

Fig. 6. Results for the goal-based reward model with an optimistic initialisation. Each curve corresponds to a different value of λ.

Fig. 7. Explored area when penalty-based reward is used with the negative initialisation of the Q-table and $\gamma = 0.99$: after 10^4 iterations in the left part and after 10^6 iterations in the right part of the figure

difference is noted for $\lambda = 0.9$. Even though the initial convergence is the most rapid, the asymptotic performance is decreased. We conjecture that this difference was caused by the different reward type. Even though the optimism under uncertainty is used in both cases, the different character of rewards (0 or -1) can influence the propagation of the value function differently.

To make the comparison complete the last configuration to be checked is a penalty-based reward with a pessimistic initialisation. But, this is more tricky than it might look at first glance. If the Q-table is simply initialised to a negative value (e.g., -200) for all state-action pairs, then under the condition $\gamma = 1$ this initialisation does not change anything when compared to initialisation with zero (all Q-values are simply decreased not from 0 but from another value, e.g., -200). The first thing which can be done is the different initialisation of goal and non-goal states:

$$\forall_{s \notin G} \forall_a Q(s, a) = -200, \ \forall_{g \in G} \forall_a Q(g, a) = 0. \tag{4}$$

This yields a configuration which we call a semi-optimistic initialisation, because it leads to unchanged learning when the value function from only non-goal states is being propagated. The propagation of the value function from the goal state gives properties of the pessimistic initialisation. Results with this initialisation are shown in Figure 8. Learning with a penalty-based reward is in this case accelerated by increasing values of λ during the entire period of learning. Even though there was an element of the pessimism it was not spoiling the asymptotic performance.

There is, however, one more factor which can make learning with the penalty-based reward more difficult. A short analysis of how the temporal difference is computed in the previous experiment shows that if instead of $\gamma = 1$ one uses $\gamma < 1$ this can ensure pessimism even before the propagation of the value

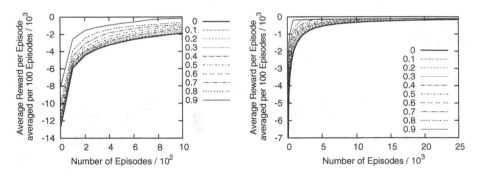

Fig. 8. Results for the penalty-based reward model with a semi-optimistic initialisation. Each curve corresponds to a different value of λ.

function from the goal state starts. But, a short analysis also shows that such learning will be very destructive for the exploration based on the Q-table. With initialisation according to Equation 4 and $\gamma = 0.99$, the temporal difference is positive since the first update and in this way an updated entry Q(s,a) for a given state gains the highest value when compared to remaining actions of the same state. Thus if any loop appears in the current trajectory, the agent will stay following this loop with only small divergence. To support this theoretical discussion we present empirical results of this configuration. In the left-hand side of Figure 7 the area of the state space in which the agent was at least once during the first 10^4 steps of the first episode is shown. The right-hand side shows first 10^6 steps. It is worth noting that in both cases it was the first episode of learning and during given numbers of steps the agent was not able to reach the goal even once. This experiment is of minor significance for this paper but makes our analysis of the pessimistic initialisation with the penalty-based reward complete.

5   Further Analysis

In the last analysis we use a popular 1D maze also known as a random walk [4]. This domain does not have problems which are depicted in Figure 1. The importance of the choice of this task is that it has been used in the literature with eligibility traces as well. Particularly we are interested in the performance of the goal-based reward with a pessimistic Q-table in this task. There are 128 states. The agent starts in the most left position, and has to reach the most right position where reward 100 is given upon entering the final state. All other rewards are equal to zero. There are two stochastic actions: left and right which have their intended effect with probability 0.9. All learning parameters are the same as in the previous experiments. Results of this analysis are in Figure 9. The convergence in the initial phase of learning (the left-hand side of Figure 9) shows that $\lambda > 0$, in contrast to $\lambda = 0$, accelerates learning at the beginning. The expected observation is that in the latter period of learning (up to 10^4

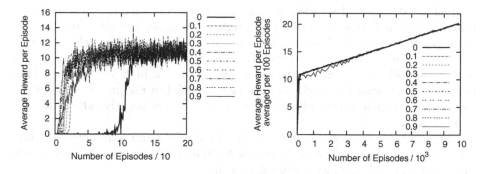

Fig. 9. Results for the goal-based reward model with a pessimistic initialisation on the 1D maze. Each curve corresponds to a different value of λ.

learning episodes in the right-hand side of Figure 9) no suboptimal convergence is observed for all values of λ. Only for λ = 0.9 initially lower performance can be observed. This domain does not yield suboptimal solutions which exist in more general navigation problems (Figure 1) and learning with even high values of λ can be successful in a given configuration of reward and initialisation. This task is more specific and this fact should be taken into account when designing empirical evaluations of RL algorithms. It is worth noting that this domain has been often used in analyses of RL with eligibility traces.

6 Conclusion

In this paper the robustness of SARSA(λ) the RL algorithm with eligibility traces is confronted with different models of reward and initialisation. The analysis is focused on both the initial learning rate and an asymptotic convergence which is usually omitted in empirical evaluations when eligibility traces are used. Most of the empirical analyses of eligibility traces existing in the literature have focused also mainly on the step-penalty reward [6].

The experimental setup and findings reported in the paper are novel as according to our best knowledge the problem has never been discussed and analysed in the literature from this perspective. Results are important not only from context-dependent but also from a more general methodological point of view.

Obtained results show that the performance of SARSA(λ) can be influenced by the reward type and significantly influenced by the initialisation of the Q-table when an undirected exploration (i.e, based only on the current content of the Q-table) is used.

Learning with λ > 0 when a pessimistic initialisation is used and reward is given only upon entering the goal state can lead to a final solution which is suboptimal. An undirected exploration strategy which according to its principles uses only the current content of the Q-table is not able to deal with this problem when the value of λ is increased. Even though the initial convergence is accelerated, the final solution can be suboptimal. In learning with the goal-based reward and with an optimistic initialisation longer eligibility traces have more positive influence, however, highest values of λ can negatively influence the final result. Experiments on the 1D maze show that the goal-based reward with pessimistic initialisation can work well even with high values of λ. This task does not belong to the class of problems defined in Figure 1 and the most radical propagation of the value function does not have negative effects on the final solution.

The penalty-based reward when tested on optimistic and semi-optimistic initialisation converges well at all stages of learning and the improvement grows monotonically with the increasing value of λ.

Results of similar experiments with three versions of $Q(\lambda)$ (i.e, Watkins', Peng's and naive [7,5]) may be different particularly because of the off-policy character of Q-learning and its implications on eligibility traces.

Convergence has not been proven so far for both SARSA(λ) and $Q(\lambda)$ for $0 < \lambda < 1$ [5, pp. 192]. But, to make our discussion more illuminating we assume hypothetically that the convergence of SARSA(λ) has been proven. Such a proof would require certain conditions on the learning process. For example the proof that Q-learning will converge to the optimal value function requires that all actions in all states are performed infinitely often [5]. In case of proofs for SARSA(λ) or $Q(\lambda)$ we can expect similar if not more restricted conditions. It means that at least each action would need to be attempted infinitely often. In practice, such exploration is infeasible when huge state spaces are considered and thus most model-free RL algorithms explore by deviating randomly form the learned policy. In theory every sequence of actions is possible under this assumption, but a sequence becomes exponentially unlikely the more it deviates from the current policy [12]. According to the proof for the Q-learning algorithm this approach does not guarantee the convergence to the optimal solution, but it allowed for the most significant applications of RL to real-life problems with enormously huge state spaces: TD-gammon [9] and elevator dispatching [13]. The results presented in this paper do not contradict proofs because we invalidate their assumptions by using exploration based on the learned policy (a common approach to reinforcement learning). When learning with a pessimistic initialisation, the suboptimal policy is more rapidly reinforced by long eligibility traces, and, as it is difficult to deviate from it, the final solution is more likely to be suboptimal. Even if we continue learning performing a huge number of episodes the improvement is exponentially unlikely (see Figure 5), albeit possible.

It was highlighted in Section 2.2 that the empirical evaluation in this paper was carried out on the instance of the class of RL problems in which regardless of its type (i.e., goal-based reward or step penalty) the reward determines one objective. In this case it is to navigate as fast as possible to the gaol state, subject to constraints imposed by transition probabilities of the underlying MDP. We conjecture that our findings in this paper can be generalised to at least similar problems. The evaluations with rewards which provide more than one objective would be enlightening in the context of our investigation and are planned for the future.

One more important factor which should be stressed here is that our results are based on the tabular representation of the state space with a distinct value for each state-action pair. We conjecture that function approximation would mitigate the convergence to a suboptimal solution with the goal-based reward and a pessimistic initialisation. In function approximation the value function is shared between at least neighbouring states and such a knowledge transfer may help in avoiding convergence to suboptimal solutions. But, we leave the analysis of eligibility traces with function approximation for future work, particulary as the convergence mechanisms of learning with function approximation have not been well understood in the field [14] and function approximation in itself can pose problems.

In this paper our analysis was focused on the robustness of eligibility traces when different models of reward and initialisation are used. The insight into the problem which we gained from this work shows that interesting conclusions can be drawn about the benefits and disadvantages of different reward configurations. We are currently working on such evaluations.

Acknowledgment

This research was sponsored by the United Kingdom Ministry of Defence Research Programme.

References

1. Puterman, M.L.: Markov Decision Processes: Discrete Stochastic Dynamic Programming. John Wiley & Sons, Inc., New York (1994)
2. Sutton, R.S.: Temporal credit assignment in reinforcement learning. PhD thesis, Department of Computer Science, University of Massachusetts, Amherst (1984)
3. Loch, J., Singh, S.: Using eligibility traces to find the best memoryless policy in partially observable Markov Decision Processes. In: Proceedings of the 15th International Conferrence on Machine Learning, pp. 323–331 (1998)
4. Sutton, R.: Learning to predict by the methods of temporal differences. Machine Learning 3, 9–44 (1988)
5. Sutton, R.S., Barto, A.G.: Reinforcement Learning: An Introduction. MIT Press, Cambridge (1998)
6. Singh, S.P., Sutton, R.S.: Reinforcement learning with replacing eligibility traces. Machine Learning 22(1–3), 123–158 (1996)
7. Peng, J., Williams, R.J.: Incremental multi-step Q-learning. Machine Learning 22, 283–290 (1996)
8. Cichosz, P.: Truncating temporal differences: On the efficient implementation of TD(λ) for reinforcement learning. Journal of Artificial Intelligence Research 2, 287–318 (1995)
9. Tesauro, G.: Practical issues in temporal difference learning. Machine Learning 8, 257–277 (1992)

10. Sutton, R.S.: Integrated architectures for learning, planning, and reacting based on approximating dynamic programming. In: Proceedings of the 7th International Conference on Machine Learning, pp. 216–224 (1990)
11. Russell, S.J., Norvig, P.: Artificial Intelligence: A Modern Approach, 2nd edn. Prentice-Hall, Englewood Cliffs (2002)
12. Jong, N.K., Stone, P.: Model-based exploration in continuous state spaces. In: The 7th Symposium on Abstraction, Reformulation, and Approximation (2007)
13. Crites, R.H., Barto, A.G.: Improving elevator performance using reinforcement learning. Advances in Neural Information Processing Systems 8, 1017–1023 (1996)
14. Stone, P., Sutton, R.S., Kuhlmann, G.: Reinforcement learning for RoboCup-soccer keepaway. Adaptive Behavior 13(3), 165–188 (2005)

Thematic Segment Retrieval Revisited

Sylvain Lamprier*, Tassadit Amghar, Bernard Levrat, and Frédéric Saubion

LERIA - University of Angers, 2 Bd Lavoisier 49000 Angers, France
{lamprier,amghar,levrat,saubion}@info.univ-angers.fr

Abstract. Documents, especially long ones, may contain very diverse passages related to different topics. Passages Retrieval approaches have shown that, in most cases, there is a great potential benefit in considering these passages independently when computing the similarity of a document with a user's query. Experiments have been realized in order to identify the kinds of passage which are the best suited for such a process. Contrarily to what could have been expected, working with thematic segments, which are likely to represent only one topic each, has led to greatly lower effectiveness results than the use of arbitrary sequences of words. In this paper, we show that this paradoxical observation is mainly due to biases induced by the great length diversity of the thematic passages. Therefore, we propose here to cope with these biases by using a more powerful text length normalization technique. Experiments show that, when length biases are laid aside, the use of thematic passages is better suited than arbitrary sequences of words to retrieve relevant informations as response to a user's query.

Keywords: Passage Retrieval, Thematic Segmentation, Similarity Measures.

1 Introduction

Browsing document collections in order to identify interesting information is one of the main task in Information Retrieval (IR). Most of IR systems provide a ranked list of documents as response to an information need [1]. Two main families of approaches allow to build such list by estimating the relevance likelihood of each document: while the keyword based methods rely on the distribution of query terms occurrences in the documents, the linked based approaches analyse the relations between documents to determine their connection to the concerned subject. We focus here on keyword based methods. In that field, several measures have been proposed to compute the similarity between each document and the query [2]. Most of them rely on the vectorial model [3], on the probabilistic model [4] or on simple co-occurrence of terms[1]. Usually, the relevance of a document is estimated w.r.t. its whole set of terms. Nevertheless, despite the fact

---

* With the support of Angers Loire Metropole.
[1] Terms are indexing units of the texts. Here, units used are meaningful words, obtained after having removed the stop words and applied a stemming process [5].

D. Dochev, M. Pistore, and P. Traverso (Eds.): AIMSA 2008, LNAI 5253, pp. 157–166, 2008.
© Springer-Verlag Berlin Heidelberg 2008

they own some interesting passages (contiguous blocks of text), some documents, especially long ones, may be ranked far from the top of the list because of their large amount of irrelevant information. Alternative approaches, called Passage Retrieval approaches [6,7,8,9], address this problem by ranking documents w.r.t. the similarity (with the query) of the passages they contain rather than considering their whole set of terms. With such approaches, the fact that relevant information is surrounded by parts of text deviating from the interesting topic does not penalize the document. Moreover, most of IR systems do not consider relative word positions in documents [8]. Considering passages reintroduces a notion of locality of terms in the text since terms have to belong to the same part of text to be taken into account together.

In [8], authors have shown that, in most cases, retrieving fragments of texts or using them to compute the proximity of a document to a query may turn out to be more effective than considering documents in their whole. Experiments have been realized to determine what kind of fragment of text is the best suited for such a process. Contrarily to what could have been expected, working with thematic segments resulting from a text segmentation process [10], which are likely to better represent topics of the texts, has led to greatly lower effectiveness results than the use of fixed length words sequences. Authors claimed that these surprising results are due to biases induced by the great length diversity of the thematic segments. By introducing even more homogeneously sized passages than words sequences, we confirm that the length diversity of the passages has actually a great impact on the effectiveness. So, we propose to reproduce the experiments of [8] by using a length normalization technique which relies on a statistical study of text similarities [11] in order to judge fragments more fairly w.r.t. their size. First, in section 2, the paper presents the similarity measures and the normalization techniques used in the experiments. The different investigated fragments of text are then described in section 3. Finally, the experimental process and results are detailed in section 4.

2 Similarity Computation

Three main factors affect the weight of a term in a text: the term frequency factor (tf), the inverse document frequency factor (idf) and the text length normalization [12]. Whereas the tf factor corresponds to the number of occurrences of a term within the text, the idf factor renders its ability of discrimination. Since longer texts contain more terms, their probability to contain query's terms is higher. Moreover, their terms may more probably be repeated several times, query's terms are thus likely to have a better frequency in the text. Length normalization aims at reducing the advantage of that long texts. This section describes the similarity computation techniques used in our experiments. First, the classical Cosine measure [2] and its term weighting scheme are described. Then, two variants of this measure are presented, the Pivoted Cosine [13] used in [8] to attempt to improve the thematic approach results and the Cosine normalized by statistical regression [11] which is expected to better cope with the length biases whose the thematic approach suffers.

2.1 Cosine Measure

The similarity computation between the documents/passages and the user's query is based, as in [8] for instance, on the well-known vectorial model [3]. Queries and documents/passages are encoded in vectors of weights w.r.t. their set of meaningful terms. The proximity of a query Q with a text T can be computed as a Cosine measure [2]:

$$Sim(T, Q) = \frac{\sum_{i=1}^{t} w_{T,i} \times w_{Q,i}}{\sqrt{\sum_{i=1}^{t} w_{T,i}^2 \times \sum_{i=1}^{t} w_{Q,i}^2}} \tag{1}$$

where t is the number of unique meaningful terms[2] contained in $T \cap Q$ and $w_{T,i}$ and $w_{Q,i}$ the weights of the term i in the text and the query respectively:

$$w_{T,i} = 1 + \ln(tf_{T,i})$$
$$w_{Q,i} = (1 + \ln(tf_{Q,i})) \times \ln \frac{N}{n_i} \tag{2}$$

where $tf_{T,i}$ and $tf_{Q,i}$ are the numbers of occurrences of the term i respectively in the text T and the query Q, N the total number of documents/passages in the collection and n_i the number of documents/passages in which the term i is present. In formula 1, the denominator corresponds to the length normalization factor ($\sqrt{\sum_{i=1}^{t} w_{T,i}^2}$ for text term weights and $\sqrt{\sum_{i=1}^{t} w_{Q,i}^2}$ for query ones).

2.2 Pivoted Cosine Measure

In [13], authors have shown that long documents, being related to more topics, have a greater probability to be relevant for the user's request than shorter ones. They thus proposed to favor these documents thanks to a measure called Pivoted Normalized Cosine. The basic idea is to pivot and tilt the Cosine normalization factor of the text term weights. In that way, a pivot point and a slope degree are chosen to compute the new term weights:

$$PivotSim(T, Q) = \frac{\sum_{i=1}^{t} w_{T,i} \times w_{Q,i}}{((1 - slope) \times pivot + slope \times \sqrt{\sum_{i=1}^{t} w_{T,i}^2}) \times \sqrt{\sum_{i=1}^{t} w_{Q,i}^2}} \tag{3}$$

In [13], authors suggest to set the pivot equal to the average of the Cosine text weights normalization factor ($\sqrt{\sum_{j=1}^{t} w_{T,j}^2}$) computed over every documents of the corpus[3]. Experiments led them to fix 0.2 as the optimal value for the

---

[2] The number of unique terms is obtained by considering one occurrence of each term.
[3] In Passage Retrieval, this average is realized over every passages of the corpus.

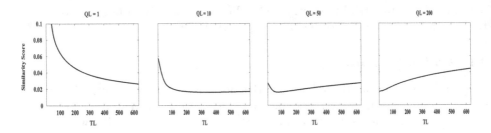

Fig. 1. Cosine tendencies w.r.t. query and text lengths

slope. Nevertheless, in [14], experiments have shown that the size of the query influences the optimal slope. The evolution of the optimal slope w.r.t. the query size has been studied in order to determine an analytic form for its estimation: $slope = 0.0921 \times \log(QL) + 0.0658$, with QL the number of unique query terms. Our experiments use this slope value.

2.3   Cosine Normalization by Statistical Regression

Contrarily to the Pivoted Cosine which aims at favoring long documents over short ones, the normalization by statistical regression we have introduced in [11] adjusts similarities of documents w.r.t. their expectation in order to judge their relevance probability more fairly. In order to study the similarity expectation, the 75000 shortest documents of the TREC's ZIFF corpus [1] have been divided into 75 bins of 1000 documents according to their size. Then, after having produced artificial queries of different lengths, the average Cosine similarity score between documents of a same bin and queries of a given size has been computed. Results, reported in figure 1, have shown that the Cosine similarity expectation of a text with a query is really influenced by the length, in term of unique terms, of the text (TL) and of the query (QL). The idea is to model these score tendencies by statistical regression (using the least squares method) in order to determine a normalization of the similarity scores:

$$NormSim(T,Q) = \frac{1 + Sim(T,Q) - ExpSim(T_L,Q_L)}{2} \qquad (4)$$

with $ExpSim(T_L,Q_L)$ the function of similarity tendency of a text with a query according to their sizes. The statistical regression realized in [11] on the ZIFF corpus led to the following equation:

$$
\begin{aligned}
ExpSim(TL,QL) = &\ (1.00586 + 0.186851 \times \ln(QL) - 1.02757 \times \ln(\ln(QL)+1)) \\
&+ \ln(TL) \times \\
&\quad (0.090365 + 0.02671 \times \ln(QL) - 0.10388 \times \ln(\ln(QL)+1)) \\
&+ \ln(\ln(TL)+1) \times \\
&\quad (-0.77143 - 0.16194 \times \ln(QL) + 0.8030 \times \ln(\ln(QL)+1))
\end{aligned}
\qquad (5)
$$

This estimation of the Cosine similarity score expectation has been shown in [11] to be strongly correlated with observations realized on other corpora. In our experiments, the formula 5 has thus been used as it is on each corpus, without the need of a specific training.

3   Passages

This section focuses on the two main fragments types experimented in the field of passage retrieval [8]: the thematic/semantical passages and the fixed length arbitrary passages (or window-based passages). Other types of passages such as paragraphs or sections, i.e., discourse passages, are not considered hereafter since they rely on the assumption that the documents present an explicit structure of their contents. Moreover, even if documents of a corpus are supplied with their structure, it may be loosely related or even unrelated with the topics of the text. It's the case, for example, when it is only a matter of presentation [7].

3.1   Window-Based Passages

An alternative to discourse or thematic passages is to segment documents into passages of fixed length. Usually, windows are defined as contiguous sequences of words [7]. In several studies [15,16], documents are partitioned into non-overlapping windows of a given fixed size. Nevertheless, the use of overlapping windows may limit the probability of a small block of relevant information to be split into different passages [7]. Therefore, as in [8], we also experiment here this other kind of window-based passages. In that way, while passages are sequences of N words taken every N consecutive words in the W_N approach (non-overlapped windows), sequences of N words are taken every 25 words[4] in the OW_N one (overlapping windows). In order to consider words located at the end of a document, additional passages containing the last N words of each document are added.

According to [8], the main advantage of arbitrary passages lies on the fact that every passages (except those incoming from documents containing less than N words) contain N words and thus that length normalization problems are laid aside. Nevertheless, due to the $tf\Theta idf$ weighting scheme, a lot of words, since being very generic and thus appearing in a lot of documents, may have an insignificant weight in the computation. The final score is then the same whether or not such words are removed by using a stop-word list. Consequently, it is likely to obtain some passages containing few terms and others containing a lot. Length normalization problems are therefore still present. This observation led us to introduce two other kinds of window-based passages which consider only meaningful terms rather than all the words of the documents. In T_N, passages of N consecutive terms are taken every N terms and in OT_N, passages of N consecutive terms are taken every 25 words (the step size is still a number of words in order to obtain a comparable number of windows).

---

[4] [17] showed this interval as effective as starting at each possible position.

In [11], it has been shown that the similarity score expectation is really dependent on the number of unique terms of the text in concern. We then investigate two last kinds of window-based passages which depend on the number of unique terms: In U_N, N unique terms are taken every N consecutive unique terms and in OU_N, N unique terms are taken every 25 consecutive words.

3.2 Thematic Passages

The purpose of thematic text segmentation is to identify the most important thematic breaks in a document in order to cut it into homogeneous units, disconnected from other adjacent parts [18]. More precisely, the segmentation process partitions a text by determining boundaries between contiguous segments related to different topics, defining so semantically coherent parts of text that are sufficiently large to expose some aspect of a given subject. Thematic segmentation of texts can also be seen as a grouping process of basic units (words, sentences, paragraphs...) in order to highlight local semantical coherences. Several segmentation methods have been proposed and most of them rely on statistical approaches [10]. SegGen, which uses a genetic algorithm to maximize the internal cohesion of the segments and to minimize the similarity between adjacent ones, appears to create the most accurate thematic segmentation of texts [10]. This method is used here to produce the thematic passages.

4   Experiments

4.1   Passage Retrieval Evaluation

The Passage Retrieval effectiveness evaluation could have been achieved by assessing the relevance of each retrieved passage manually. Nevertheless, this task may turn out to be really costly because of the multiplicity of the passage types, the number of units to assess and the difficulty in evaluating passages extracted from their context, especially when considering window-based passages. As in [8], we have chosen to assess the effectiveness of the approaches same manner as in the case of a classical document retrieval process, by evaluating ranked lists of the documents (for which relevance judgments are available). In that way, each passage of a ranked list, that has been built in a first step w.r.t. the similarity scores they obtained, is replaced by the document containing it. Double occurrences are then removed to get the final list of documents (the highest ranked occurrence is kept). In the following, Doc stands for the classical ranked list of documents and W_N, T_N, U_N, OW_N, OT_N, OU_N, and SG (SegGen) stand for ranked lists of documents built w.r.t. the best passage of the given type each document owns.

4.2   Material Details

Four document collections from the TREC corpus are used: the ZIFF corpus includes computer science articles, FR whole issues of the Federal Register, and

Table 1. Statistics of the collections

Statistics	ZIFF	AP	WSJ	FR
Documents	6939	6740	7898	4151
Words per document	445	479	557	2879
Terms per document	213	236	269	1495
Unique terms per document	119	153	165	350
Words of the 1000 shortest documents	89	154	67	238
Words of the 1000 longest documents	1822	894	1513	9521
Terms of the 1000 shortest documents	46	80	36	137
Terms of the 1000 longest documents	840	434	717	4834
Uniques of the 1000 shortest documents	36	59	28	88
Uniques of the 1000 longest documents	387	275	411	838
Thematic passages	12977	12806	17288	21997
Queries with relevant documents	39	43	47	27
Relevant documents per query	25.89	20.22	29.21	12.22

AP and WSJ full articles from the Associated Press and the Wall Street Journal respectively. The same set of queries, the topics 1-50 of TREC, have been used for each corpus. Each collection contains thousands of documents that have been evaluated by experts in TREC as relevant or not as response to each query. Nevertheless, given the number of window-based passages, the use of such large sets of documents is difficult. Therefore, we chose to constitute smaller sets of documents by gathering the two hundred first retrieved documents as response to each query[5] by a classical IR system using the cosine measure. Each corpus presented here is a subset of the TREC's corresponding collection (Double occurrences are removed from these resulting sets). Whereas the FR corpus owns a lot of very long documents, AP contains rather short ones. The ZIFF and FR corpora appear to present the greatest document length heterogeneity.

4.3   Effectiveness Measures

The IR effectiveness of the different approaches is evaluated w.r.t. their ability to rank the relevant documents at the top of the list of documents retrieved. The most usually used criterion to assess this ability, based on the classical precision measure *Prec* which computes the ratio of relevant documents within a set of first documents retrieved, is the mean average precision (*MAP*) which corresponds to the average of the precisions computed after each relevant document of the list. Table 2 gives the average of the MAP results over the four corpora, using the classical cosine measure (Cos), the pivoted cosine measure (Piv) and the normalized cosine by statistical regression (Reg). In this table, Title refers to experiments realized by only using title keywords of queries (2.95 terms per query, 2.95 unique) and Narra to results obtained with their full narrative description (88.08 terms per query, 52.6 unique)[6]. For each window-based

---

[5] In all experiments, only queries having at least one relevant document are used.

[6] Title and Narrative correspond to the topics fields names.

Table 2. Mean Average Precision of different approaches on four corpora

FR								ZIFF							
Title	Cos	Piv	Reg	Narra	Cos	Piv	Reg	Title	Cos	Piv	Reg	Narra	Cos	Piv	Reg
Doc	0.131	0.229	0.159	Doc	0.327	0.329	0.318	Doc	0.205	0.218	0.236	Doc	0.354	0.253	0.355
W-500	0.230	0.222	0.248	W-400	0.332	0.327	0.346	W-400	0.223	0.210	0.236	W-300	0.358	0.298	0.363
T-400	0.232	0.223	0.247	T-350	0.341	0.332	0.346	T-250	0.226	0.208	0.237	T-200	0.361	0.301	0.365
U-350	0.234	0.223	0.250	U-200	0.344	0.332	0.348	U-150	0.227	0.209	0.237	U-100	0.361	0.300	0.364
OW-450	0.243	0.226	0.268	OW-350	0.343	0.339	0.357	OW-400	0.230	0.216	0.242	OW-300	0.377	0.310	0.379
OT-350	0.262	0.238	0.269	OT-200	0.359	0.347	0.359	OT-200	0.243	0.214	0.244	OT-150	0.379	0.310	0.380
OU-250	0.264	0.239	0.272	OU-150	0.361	0.342	0.360	OU-150	0.244	0.214	0.244	OU-100	0.380	0.306	0.379
SG	0.187	0.207	0.270	SG	0.322	0.326	0.354	SG	0.213	0.214	0.243	SG	0.340	0.294	0.377

WSJ								AP							
Title	Cos	Piv	Reg	Narra	Cos	Piv	Reg	Title	Cos	Piv	Reg	Narra	Cos	Piv	Reg
Doc	0.206	0.279	0.245	Doc	0.416	0.385	0.400	Doc	0.172	0.198	0.191	Doc	0.326	0.318	0.330
W-300	0.228	0.243	0.252	W-250	0.409	0.390	0.411	W-300	0.185	0.186	0.186	W-300	0.327	0.318	0.328
T-250	0.235	0.246	0.255	T-150	0.411	0.390	0.411	T-200	0.184	0.184	0.190	T-150	0.329	0.321	0.331
U-100	0.240	0.247	0.255	U-100	0.413	0.393	0.412	U-100	0.186	0.187	0.190	U-150	0.330	0.321	0.331
OW-300	0.249	0.265	0.274	OW-200	0.417	0.403	0.416	OW-300	0.193	0.194	0.197	OW-250	0.334	0.329	0.336
OT-200	0.260	0.267	0.276	OT-100	0.419	0.402	0.418	OT-200	0.193	0.193	0.196	OT-150	0.335	0.330	0.336
OU-100	0.262	0.265	0.278	OU-100	0.420	0.405	0.418	OU-100	0.194	0.194	0.197	OU-100	0.338	0.332	0.338
SG	0.219	0.263	0.273	SG	0.410	0.390	0.416	SG	0.172	0.187	0.198	SG	0.307	0.302	0.335

approach, experiments are realized with N in $[50, 600]$, N varying by a step of 50. Only the results of the best approaches are reported in table 2.

4.4 Results Analysis

From the results, one first can note that document retrieval based on passage ranking may actually turn out to be more effective than classical document retrieval, especially when considering corpora owning a great diversity of document lengths (ZIFF) or containing a lot of very long documents (FR). With the Cosine measure, short texts are favored when using short queries and long texts are favored when using long ones (see figure 1). Considering text sets owning a great diversity of lengths, this observation may turn out to be really problematic since some texts are more likely to be retrieved thanks to their fitted size. The great dominance of the term windows approaches (T_N and OT_N) over the word ones (W_N and OW_N) highlights this bias since the only difference between both approaches lies on the length diversity level of their passages (the term windows approaches work with more homogeneously sized passages). As claimed in [11], the fact that the similarity expectation is more dependent on the number of unique terms allows the unique term approaches (U_N and OU_N) to obtain the best results. From these observations, we may conclude that the passages length diversity level has a great impact on the IR effectiveness.

In [8], authors have tried to solve length normalization problems by using the Pivoted Cosine measure. The use of this measure appears to improve the classical document retrieval effectiveness results in most cases. Some other results have been slightly improved, notably those of the thematic approach (SG) which presents the greatest passages length diversity level. Nevertheless, thematic approach results remain lower than window-based approaches ones. The Pivoted Cosine measure has been initially designed in [13] in order to adapt the retrieval probability of documents to their probability of relevance. Long documents being more often relevant than short ones, this measure tends to favor long texts. Nevertheless, while this last remark can be applied when considering whole documents as units, it does not hold when considering thematic passages. Indeed, the high relevance probability of the long documents is mainly due to the great number of topics they may address [11]. Now, from their definition, thematic passages are supposed to address only one topic each. Short thematic passages are thus as likely to be relevant as long ones and then, there is no reason to prefer long thematic passages to short ones. Therefore, the Pivoted Cosine does not appear to be suitable for the problem of Passage Retrieval. Consequently, we experiment here the cosine measure normalized by statistical regression [11] which aims at giving the same similarity score expectation to every documents/passages.

First, the fact that results obtained by the word windows approaches W_N and OW_N catch up with those obtained by other window based ones shows that length biases are mostly released by this measure. The thematic approach appears to greatly benefit from this normalization: contrarily to what has been observed in [8], the thematic approach obtains really better results than the non-overlapped window-based approaches when using this normalized cosine. Moreover, a Student t-test has shown that differences between the thematic approach results and those obtained by the overlapped window-based approaches are not significant. Overlapped windows approaches, due to their huge number of considered passages (an average around 115 passages per document on the FR corpus for the overlapped windows approaches against 2.5 for the thematic ones), require a greatly higher processing time to furnish the answers list than the thematic approaches. Thematic passages appear then to be better suited to retrieve relevant information to a user's query.

5   Conclusion

In [8], experiments have shown that window-based passages are the best suited to retrieve relevant information to a user's query. Nevertheless, we were convinced that thematic approaches, which use passages resulting from a segmentation process and thus being likely to better represent the topics of the documents, have a better potential of effectiveness. Indeed, in the case of overlapping windows approaches, the great number of windows allows to consider at least one passage starting (or ending) at an optimal place of the document but the length of such a passage, contrarily to the one of a thematic passage, is arbitrarily predefined. Experiments reported in this paper highlight the great impact of the passages length diversity level on the IR effectiveness. It has been shown here that the dominance of the window-based approaches observed in [8] was only due to length biases. From its greatly lower processing cost and the expected further segmentation accuracy improvements, the use of thematic passages appears as a very promising approach to identify the relevant information. Furthermore, the use of a thematic approach is more appropriate to retrieve relevant information since a window-based passage, presented to the user without any additional information describing its context, may be difficult to understand.

References

1. Voorhees, E.M., Harman, D.: Overview of the fifth text retrieval conference (trec-5). In: TREC 1996 (1996)
2. Zobel, J., Moffat, A.: Exploring the similarity space. SIGIR Forum 32(1), 18–34 (1998)
3. Baeza-Yates, R., Ribeiro-Neto, B.: Modern Information Retrieval. ACM Press /Addison-Wesley (1999)
4. Robertson, S.E., Walker, S., Hancock-Beaulieu, M., Gull, A., Lau, M.: Okapi at TREC. In: TREC 1992, pp. 21–30 (1992)
5. Porter, M.: An algorithm for suffix stripping. Program 14(3), 130–137 (1980)
6. Salton, G., Allan, J., Buckley, C.: Approaches to passage retrieval in full text information systems. In: SIGIR 1993, pp. 49–58. ACM, New York (1993)
7. Callan, J.P.: Passage-level evidence in document retrieval. In: SIGIR 1994, pp. 302–310. Springer, Heidelberg (1994)
8. Kaszkiel, M., Zobel, J.: Effective ranking with arbitrary passages. Journal of the American Society of Information Science 52(4), 344–364 (2001)
9. Liu, X., Croft, W.B.: Passage retrieval based on language models. In: CIKM 2002, pp. 375–382. ACM, New York (2002)
10. Lamprier, S., Amghar, T., Levrat, B., Saubion, F.: Seggen: A genetic algorithm for linear text segmentation. In: Veloso, M.M. (ed.) IJCAI 2007, pp. 1647–1652 (2007)
11. Lamprier, S., Amghar, T., Levrat, B., Saubion, F.: Document length normalization by statistical regression. In: ICTAI 2007, vol. (2), pp. 19–26. IEEE, Los Alamitos (2007)
12. Salton, G., Buckley, C.: Term-weighting approaches in automatic text retrieval. Information Processing and Management 24(5), 513–523 (1988)
13. Singhal, A., Salton, G., Mitra, M., Buckley, C.: Document length normalization. Information Processing and Management 32(5), 619–633 (1996)
14. Chung, T.L., Luk, R.W.P., Wong, K.F., Kwok, K.L., Lee, D.L.: Adapting pivoted document-length normalization for query size: Experiments in chinese and english. In: TALIP 2006, vol. 5(3), pp. 245–263 (2006)
15. Zobel, J., Moffat, A., Wilkinson, R., Sacks-Davis, R.: Efficient retrieval of partial documents. Information Processing and Management 31(3), 361–377 (1995)
16. Stanfill, C., Waltz, D.L.: Statistical methods, artificial intelligence, and information retrieval, pp. 215–225 (1992)
17. Kaszkiel, M., Zobel, J.: Passage retrieval revisited. In: SIGIR 1997, pp. 178–185. ACM Press, New York (1997)
18. Salton, G., Singhal, A., Buckley, C., Mitra, M.: Automatic text decomposition using text segments and text themes. In: Hypertext 1996, pp. 53–65. ACM Press, New York (1996)

Implementing Prioritized Circumscription by Computing Disjunctive Stable Models*

Emilia Oikarinen** and Tomi Janhunen

Helsinki University of Technology TKK
Department of Information and Computer Science
P.O. Box 5400, FI-02015 TKK, Finland
{Emilia.Oikarinen,Tomi.Janhunen}@tkk.fi

Abstract. The stable model semantics of disjunctive logic programs is based on minimal models which assign atoms false by default. While this feature is highly useful and leads to concise problem encodings, it occasionally makes knowledge representation with disjunctive rules difficult. Lifschitz' parallel circumscription provides a remedy by introducing atoms that are allowed to vary or to have fixed values while others are falsified. Prioritized circumscription further refines this setting in terms of priority classes for atoms being falsified. In this paper, we present a linear and faithful transformation to embed prioritized circumscription into disjunctive logic programming in a systematic fashion. The implementation of the method enables the use of disjunctive solvers for computing prioritized circumscription. The results of an experimental evaluation indicate that the method proposed herein compares favorably with other existing implementations.

Keywords: Prioritized circumscription, disjunctive stable models, linear transformation, answer set programming.

1 Introduction

In *answer set programming* (ASP) a problem at hand is formalized as a logic program so that its answer sets, or more formally *(disjunctive) stable models* [2,3], correspond to the solutions of the problem—to be computed using a solver for disjunctive logic programs. Numerous applications of disjunctive logic programs have emerged since efficient solvers such as, for example, DLV [4], and GNT [5], for computing answer sets became available. The stable model semantics of disjunctive logic programs is based on *minimal models* which makes every atom appearing in a disjunctive logic program false by default. While this feature is highly useful—leading to concise encodings of problems as disjunctive programs—it occasionally makes knowledge representation with disjunctive rules difficult. For instance, the representation of Reiter-style minimal diagnoses [6] is complicated by the fact that in addition to abnormality atoms, also atoms describing the state of the system being diagnosed become subject to falsification.

---

* This research has been partially funded by the Academy of Finland under project #122399. A preliminary version of this work appeared in [1].

** The financial support from Helsinki Graduate School in Computer Science and Engineering, Nokia Foundation, Finnish Foundation for Technology Promotion TES, Emil Aaltonen foundation, and Finnish Cultural Foundation is gratefully acknowledged.

D. Dochev, M. Pistore, and P. Traverso (Eds.): AIMSA 2008, LNAI 5253, pp. 167–180, 2008.
© Springer-Verlag Berlin Heidelberg 2008

This problem can be alleviated by a more refined control of minimization provided by *parallel circumscription* [7] which allows certain atoms to *vary* or to have *fixed* truth values. In particular, varying atoms enhance the knowledge representation capability over ordinary circumscription [8]. The scheme of *prioritized circumscription* [7] generalizes this setting with priority classes for atoms being minimized. Together, these principles provide an elegant solution to the problem raised in the first paragraph: atoms describing state should vary and priorities can be specified among those representing abnormality. We want to bring these enhanced notions of minimality to the realm of disjunctive logic programming so that they can be advisedly exploited in problem encodings. To this end, we have already addressed parallel circumscription and provided a *linear* and *faithful* translation into disjunctive logic programming [9].

The goal of this paper is to cover the prioritized case accordingly, extending the preliminary version of this work presented in [1]. Our translation-based approach effectively removes varying and fixed atoms as well as priority classes in terms of a transformation. We have also implemented the method, and give an experimental evaluation contrasting our tool with others developed for the same purpose [10,11,12].

The rest of this paper is organized as follows. In Section 2, we review the syntax and semantics of disjunctive logic programs. Section 3 is devoted to an introduction of parallel and prioritized circumscription [7] in the propositional case. A linear transformation from prioritized circumscription to disjunctive logic programs is presented in Section 4. The results from an experimental evaluation of the tool that implements the linear transformation are reported in Section 5. We end this paper with a brief discussion about related work in Section 6.

2   Disjunctive Logic Programs

We review in this section the basic concepts of disjunctive logic programs (DLPs) in the propositional case. A *disjunctive logic program* is a finite set of *disjunctive rules* of the form

$$a_1 \vee \cdots \vee a_n \leftarrow b_1, \ldots, b_m, \sim c_1, \ldots, \sim c_k \tag{1}$$

where $n, m, k \geq 0$, and $a_1, \ldots, a_n, b_1, \ldots, b_m$, and $c_1, \ldots, c_k$ are propositional atoms. Since the order of atoms is considered insignificant, we use $A \leftarrow B, \sim C$ as a shorthand for rules of form (1), where A, B, and C denote the sets of atoms $\{a_1, \ldots, a_n\}$, $\{b_1, \ldots, b_m\}$, and $\{c_1, \ldots, c_k\}$, respectively, and $\sim C = \{\sim c \mid c \in C\}$ for any set of atoms C.

The basic intuition behind (1) is that if each atom in the positive body B and none of the atoms in the negative body C can be inferred, then some atom in the head A can be inferred. If $C = \emptyset$, we have a *positive* rule written $A \leftarrow B$. When both B and C are empty, we have a *disjunctive fact*, written $A \leftarrow$. If A is empty, then we have a *constraint*, written $\perp \leftarrow B, \sim C$. Given a DLP Π, we write $\mathrm{At}(\Pi)$ for its *signature*, that is, the set of atoms appearing in the rules of Π.

Given any DLP Π, we define an *interpretation* M for Π as a subset of $\mathrm{At}(\Pi)$. An atom $a \in \mathrm{At}(\Pi)$ is *true* under M (symbolically $M \models a$) if and only if $a \in M$, otherwise a is *false* under M. For a negative literal $\sim a$, we define $M \models \sim a$ if and only if $M \not\models a$. A set L of literals is satisfied by M (denoted by $M \models L$) if and only if

$M \models l$, for every $l \in L$. We also define $M \models \bigvee L$, providing $M \models l$ for some $l \in L$. An interpretation $M \subseteq \text{At}(\Pi)$ is a (*classical*) *model* of a DLP Π, denoted $M \models \Pi$, if and only if for every rule $A \leftarrow B, \sim C \in \Pi$, it holds that $M \models B \cup \sim C$ implies $M \models \bigvee A$. It is typical in logic programming that atoms are assumed false by default. In case of a *positive* DLP (PDLP) Π this is formalized in terms of *minimal models* $M \models \Pi$ for which there is no $N \subset M$ such that $N \models \Pi$. We write $\text{MM}(\Pi)$ for the set of minimal models associated with a PDLP Π. To extend the semantics for DLPs involving negation a *partial evaluation* technique proposed by Gelfond and Lifschitz [2] is applied.

Definition 1. *Given a DLP Π, an interpretation $M \subseteq \text{At}(\Pi)$ is a* stable model *of Π if and only if $M \in \text{MM}(\Pi^M)$, where*

$$\Pi^M = \{A \leftarrow B \mid A \leftarrow B, \sim C \in \Pi \text{ and } M \cap C = \emptyset\}$$

is the reduct of Π with respect to M.

In the sequel, the set of stable models of a DLP Π is denoted by $\text{SM}(\Pi)$.

3   Parallel and Prioritized Circumscription

In this section we introduce *parallel* and *prioritized circumscription* [7] in the propositional case. Parallel circumscription is based on a notion of minimality which partitions atoms in three disjoint categories.

Definition 2. *Let Π be a PDLP and let $P, V, F \subseteq \text{At}(\Pi)$ be disjoint sets of atoms such that $\text{At}(\Pi) = P \cup V \cup F$. A model $M \models \Pi$ is $\langle P, V, F \rangle$-minimal if and only if there is no $N \models \Pi$ such that (i) $N \cap P \subset M \cap P$ and (ii) $N \cap F = M \cap F$.*

By this definition, atoms in P are subject to minimization, that is, falsified as far as possible, while the truth values of atoms in V may vary freely and the truth values of atoms in F are kept fixed. Note that in the notation of $\langle P, V, F \rangle$-minimality one of the sets P, V, and F is actually redundant, as given any two of the sets, the third one is implicitly clear from the context. Example 1 illustrates the use of varying atoms in a concise representation for the model-based diagnosis [6] of digital circuits.

Example 1. Consider a positive DLP[1]

$$\Pi_{\text{diag}} = \{a \vee b \vee ab_1. \quad ab_1 \leftarrow a, b. \quad b \vee c \vee ab_2. \quad ab_2 \leftarrow b, c.$$
$$c \vee d \vee ab_3. \quad ab_3 \leftarrow c, d. \quad \bot \leftarrow a. \quad \bot \leftarrow d\}$$

representing a sequence of three *inverters* with observations $\neg a$ and $\neg d$ indicating a fault. By setting $P = \{ab_1, ab_2, ab_3\}$ and $V = \{a, b, c, d\}$ we obtain minimal diagnoses $\{ab_1, c\}$, $\{ab_2, b, c\}$, and $\{ab_3, b\}$ as $\langle P, V, \emptyset \rangle$-minimal models. In contrast, the $\langle P \cup V, \emptyset, \emptyset \rangle$-minimal models of Π_{diag}, that is, the stable models of Π_{diag}, include a spurious non-minimal diagnosis $\{ab_1, ab_2, ab_3\}$.   ∎

---

[1] A full stop is used to separate rules in a program and the symbol "$\leftarrow$" is omitted in the case of disjunctive facts, that is, when the body of a rule is empty.

The set of all $\langle P, V, F \rangle$-minimal models of Π is denoted by $\mathrm{MM}_{P,V,F}(\Pi)$. The conventional case that all atoms are subject to minimization is covered by $\mathrm{MM}(\Pi) = \mathrm{MM}_{\mathrm{At}(\Pi),\emptyset,\emptyset}(\Pi) = \mathrm{SM}(\Pi)$ for a PDLP Π. We write $\mathrm{Circ}(\Pi, P, V, F)$ to denote the parallel circumscription of a PDLP Π.

An extended notation $\mathrm{Circ}(\Pi, P_1 > \cdots > P_k, V, F)$ is introduced to represent the prioritized circumscription of Π which includes the parallel circumscription of Π as its special case, that is, when $k = 1$. The idea is that atoms in P_1 are falsified with the highest priority, those in P_2 with the next highest priority, and so on. Lifschitz [7] shows that $\mathrm{Circ}(\Pi, P_1 > \cdots > P_k, V, F)$ corresponds to the conjunction

$$\bigwedge_{i=1}^{k} \mathrm{Circ}(\Pi, P_i, P_{i+1} \cup \ldots \cup P_k \cup V, P_1 \cup \ldots \cup P_{i-1} \cup F). \tag{2}$$

The formula (2) does not have a direct interpretation as a DLP but such a representation can be obtained using the translation from [9]. A drawback is that $2k$ copies of Π must be created which gives a quadratic nature for the overall transformation [10] because $k \leq |\mathrm{At}(\Pi)| \leq \|\Pi\|$, that is, the length of Π in symbols. Therefore, we get better premises for the development of a linear representation if Definition 2 is generalized to the case of prioritized circumscription.

Definition 3 ([11]). *A model $M \models \Pi$ for a positive DLP Π is $\langle P_1 > \cdots > P_k, V, F \rangle$-minimal if and only if there is no $N \models \Pi$ such that*

(i) $N \cap (P_1 \cup \ldots \cup P_{i-1}) = M \cap (P_1 \cup \ldots \cup P_{i-1})$ and $N \cap P_i \subset M \cap P_i$ for some $1 \leq i \leq k$; and
(ii) $N \cap F = M \cap F$.

Example 2. Recalling Π_{diag} from Example 1, we note that it has a unique $\langle \{ab_1\} > \{ab_2\} > \{ab_3\}, V, \emptyset \rangle$-minimal model $M = \{ab_3, b\}$, that is, ab_1 is falsified first, then ab_2, and finally ab_3, but the last minimization fails as it holds $\Pi_{\mathrm{diag}} \cup \{\neg ab_1, \neg ab_2\} \models ab_3$. ∎

Our next objective is to characterize $\langle P_1 > \cdots > P_k, V, F \rangle$-minimality for a model M of a PDLP Π in terms of propositional satisfiability. The idea is to check whether the set of disjunctive rules $\mathrm{Tr}_{\mathrm{U}}(\Pi, P_1 > \cdots > P_k, F, M)$ defined as

$$\begin{aligned}
&\{(A \setminus F) \leftarrow (B \setminus F) \mid A \leftarrow B \in \Pi, M \not\models \bigvee(A \cap F), \text{ and } M \models B \cap F\} \cup \\
&\{e_0 \leftarrow\} \cup \{\bot \leftarrow e_k\} \cup \\
&\bigcup_{i=1}^{k} \{e_i \leftarrow (P_i \cap M) \cup \{e_{i-1}\}\} \cup \bigcup_{i=1}^{k} \{\bot \leftarrow a, e_{i-1} \mid a \in P_i \setminus M\},
\end{aligned} \tag{3}$$

is *unsatisfiable in the classical sense.* The new atoms $e_1, \ldots, e_k$ in $\mathrm{Tr}_{\mathrm{U}}(\Pi, P_1 > \cdots > P_k, F, M)$ correspond to the strata in $P_1 > \cdots > P_k$. The intuitive reading of e_i is that the truth values of all atoms in $P_1 \cup \ldots \cup P_i$ coincide with those assigned by M.

Lemma 1. *Given a positive DLP Π, a model $M \subseteq \mathrm{At}(\Pi)$ is $\langle P_1 > \cdots > P_k, V, F \rangle$-minimal if and only if $\mathrm{Tr}_{\mathrm{U}}(\Pi, P_1 > \cdots > P_k, F, M)$ is unsatisfiable.*

Proof sketch. ($\Longrightarrow$) Assume that there is $N \subseteq (\mathrm{At}(\Pi) \setminus F) \cup \{e_i \mid 0 \leq i \leq k\}$ such that $N \models \mathrm{Tr_U}(\Pi, P_1 > \cdots > P_k, F, M)$, that is, $N \models (A \setminus F) \leftarrow (B \setminus F)$ for each $A \leftarrow B \in \Pi$ such that $M \not\models \bigvee(A \cap F)$ and $M \models B \cap F$; $N \models e_0$; and $N \not\models e_k$. Furthermore, there is $1 \leq i \leq k$ such that $N \models e_j$ and $N \models P_j \cap M$ for each $j < i$, $N \not\models a$ for each $a \in P_j \setminus M$ for $j \leq i$, and $N \not\models P_i \cap M$. It is easy to see that $M' = (N \cap \mathrm{At}(\Pi)) \cup (M \cap F)$ is a counter-example for the $\langle P_1 > \cdots > P_k, V, F \rangle$-minimality of M, as $M' \models \Pi$, $M' \cap (P_1 \cup \cdots \cup P_{i-1}) = M \cap (P_1 \cup \cdots \cup P_{i-1})$ and $M' \cap P_i \subset M \cap P_i$.

($\Longleftarrow$) Assume that $M \models \Pi$ and M is not $\langle P_1 > \cdots > P_k, V, F \rangle$-minimal, that is, there is $N \models \Pi$ such that for some $1 \leq i \leq k$, $N \cap (P_1 \cup \cdots \cup P_{i-1}) = M \cap (P_1 \cup \cdots \cup P_{i-1})$ and $N \cap P_i \subset M \cap P_i$. We define $N' = (N \setminus F) \cup \{e_j \mid 0 \leq j \leq i - 1\}$. It is straightforward to verify that $N' \models \mathrm{Tr_U}(\Pi, P_1 > \cdots > P_k, F, M)$. $\qquad\square$

4   Translation-Based Approach to Prioritized Circumscription

In [9] we present a translation function which enables the removal of varying atoms from a PDLP Π in a *faithful* way, that is , the $\langle P, V, F \rangle$-minimal models M of Π and the stable models N of its translation are in a bijective relationship such that $M = N \cap \mathrm{At}(\Pi)$ holds for each pair of models. In [1] we propose a way to generalize this method to the case of prioritized circumscription. In this section, we present the details of the translation and justify the correctness of the method in general.

The translation $\mathrm{Tr_{circ2dlp}}(\Pi, P_1 > \cdots > P_k, V, F)$ consists of two parts. Instead of presenting the parts as regular DLPs, we exploit the theory from [13] and describe the parts as *DLP-modules* or *DLP-functions* with input/output interface. Syntactically, a DLP-module is a triple $\langle \Pi, I, O \rangle^2$ where Π is a set of disjunctive rules and I and O are disjoint sets of atoms such that $\mathrm{At}(\Pi) = I \cup O$, and all occurrences of *input atoms* $a \in I$ are in the bodies of rules in Π. The idea is to keep the interpretation of input atoms fixed, that is, an interpretation $M \subseteq \mathrm{At}(\Pi)$ is a stable model of a DLP-module $\langle \Pi, I, O \rangle$ if and only if $M \in \mathrm{MM}_{O,\emptyset,I}(\Pi^M)$. One may notice that for DLP-modules with $I = \emptyset$, this definition results in the standard stable model semantics of DLPs. The *composition* or *join* of two DLP-modules $\langle \Pi_1, I_1, O_1 \rangle$ and $\langle \Pi_2, I_2, O_2 \rangle$ is defined syntactically as $\langle \Pi_1, I_1, O_1 \rangle \sqcup \langle \Pi_2, I_2, O_2 \rangle = \langle \Pi_1 \cup \Pi_2, (I_1 \cup I_2) \setminus (O_1 \cup O_2), O_1 \cup O_2 \rangle$. The overall translation $\mathrm{Tr_{circ2dlp}}(\Pi, P_1 > \cdots > P_k, V, F)$ is a join of two DLP-modules:

$$\mathrm{Tr_{gen}}(\Pi, P_1 > \cdots > P_k, V, F) \sqcup \mathrm{Tr_{min}}(\Pi, P_1 > \cdots > P_k, V, F). \qquad (4)$$

For the sake of brevity, we omit the sets $P_1, \ldots, P_k$, V, and F when they are clear from the context.

Next, we describe the structure of modules $\mathrm{Tr_{gen}}(\Pi)$ and $\mathrm{Tr_{min}}(\Pi)$ in more detail. They involve a number of atoms which are new to Π:

---

[2] In [13] a more general setting is introduced in which a DLP-function may contain *hidden atoms*, which are local to the module in question. See [13] for more details, for example, of the conditions under which the composition of two DLP-modules is defined.

Module : $\mathrm{Tr_{gen}}(\Pi, P_1 > \cdots > P_k, V, F)$
Input : $\emptyset$
Output : $\mathrm{At}(\Pi) \cup \overline{\mathrm{At}(\Pi)}$
1 : $a \leftarrow \sim\overline{a}$ for each $a \in V \cup F$
2 : $\overline{a} \leftarrow \sim a$ for each $a \in \mathrm{At}(\Pi)$
3 : $(A \cap P) \leftarrow (B \cap P), \sim(A \setminus P), \sim\overline{(B \setminus P)}$ for each $A \leftarrow B \in \Pi$

Fig. 1. Module $\mathrm{Tr_{min}}(\Pi, P_1 > \cdots > P_k, V, F)$ from Definition 4 with $P = P_1 \cup \cdots \cup P_k = \mathrm{At}(\Pi) \setminus (V \cup F)$

- An atom $\overline{a}$ denoting that a is false is introduced for each $a \in \mathrm{At}(\Pi)$.
- For each atom $a \in \mathrm{At}(\Pi)$, a renamed copy a^* of a is created in order to formulate the test for $\langle P_1 > \cdots > P_k, V, F \rangle$-minimality. Thus, the meaning of a^* is that a is true in a potential counter-model for $\langle P_1 > \cdots > P_k, V, F \rangle$-minimality.
- An atom e_i is introduced for each priority class P_i. The intuitive reading of e_i is the same as in (3), that is, the truth values of all atoms in $P_1 \cup \ldots \cup P_i$ in the potential counter-example coincide with those assigned by the model candidate.
- A renamed copy a^{d} is introduced for each atom $a \in P_1 \cup \cdots \cup P_k$. The meaning of a^{d} is that the model candidate and the potential counter-example for its $\langle P_1 > \cdots > P_k, V, F \rangle$-minimality assign *different* truth values to atom $a \in P_i$.
- Finally, an atom u is introduced to denote the unsatisfiability of (3).

We use shorthands $\overline{A} = \{\overline{a} \mid a \in A\}$ and $A^* = \{a^* \mid a \in A\}$ for any $A \subseteq \mathrm{At}(\Pi)$. Likewise A^{d} denotes $\{a^{\mathrm{d}} \mid a \in A\}$ for any $A \subseteq P_1 \cup \cdots \cup P_k$. The modules $\mathrm{Tr_{gen}}(\Pi)$ and $\mathrm{Tr_{min}}(\Pi)$ forming the join (4) are as follows.

Definition 4. *Let Π be a PDLP subject to a prioritized circumscription* $\mathrm{Circ}(\Pi, P_1 > \cdots > P_k, V, F)$. *The translation* $\mathrm{Tr_{circ2dlp}}(\Pi) = \langle \Pi_{\mathrm{g}} \cup \Pi_{\mathrm{m}}, \emptyset, O_{\mathrm{g}} \cup O_{\mathrm{m}} \rangle$ *is the join of the DLP-modules* $\mathrm{Tr_{gen}}(\Pi) = \langle \Pi_{\mathrm{g}}, \emptyset, O_{\mathrm{g}} \rangle$ *presented in Fig. 1 and* $\mathrm{Tr_{min}}(\Pi) = \langle \Pi_{\mathrm{m}}, O_{\mathrm{g}}, O_{\mathrm{m}} \rangle$ *presented in Fig. 2.*

The module $\mathrm{Tr_{gen}}(\Pi)$ takes no input but it produces a model candidate for circumscription as its output. The rules in lines 1–2 choose truth values for atoms in $V \cup F$ and define the complementary atoms $\overline{a}$ for all atoms $a \in \mathrm{At}(\Pi)$. The rules in the third line make sure that Π is satisfied. As a matter of optimization, the satisfaction of rules is focussed on the atoms subject to minimization, that is, those in $P = P_1 \cup \ldots \cup P_k$.

An actual input for $\mathrm{Tr_{min}}(\Pi)$ is a candidate for a $\langle P_1 > \cdots > P_k, V, F \rangle$-minimal model of Π but represented as a set $M \cup \overline{(\mathrm{At}(\Pi) \setminus M)}$ instead of $M \subseteq \mathrm{At}(\Pi)$. Rules in line 4 create a renamed copy of Π to check the $\langle P_1 > \cdots > P_k, V, F \rangle$-minimality of M. The atoms in F are not renamed to maintain the semantics of fixed atoms. The rules in lines 5–7 are activated for each set P_i only if e_j is true for all $j < i$. Each rule in line 5 captures a rule $e_i \leftarrow (P_i \cap M) \cup \{e_{i-1}\}$ from (3). This rule *depends dynamically* on M and effectively states, using the disjunction $\bigvee P_i^{\mathrm{d}}$, the *falsity* of at least one atom a that is both subject to minimization with priority i ($a \in P_i$) and true in M ($a \in M$).

Module	:	$\mathrm{Tr}_{\min}(\Pi, P_1 > \cdots > P_k, V, F)$
Input	:	$\mathrm{At}(\Pi) \cup \overline{\mathrm{At}(\Pi)}$
Output	:	$\mathrm{At}(\Pi)^* \cup P_1^{\mathrm{d}} \cup \cdots \cup P_k^{\mathrm{d}} \cup \{e_i \mid 0 \leq i \leq k\} \cup \{u\}$

4 : $(A \setminus F)^* \cup \{u\} \leftarrow (B \setminus F)^*, \sim(A \cap F), \sim\overline{(B \cap F)}$ for each $A \leftarrow B \in \Pi$

5 : $P_i^{\mathrm{d}} \cup \{e_i, u\} \leftarrow e_{i-1}$ for each P_i

6 : $u \leftarrow a^{\mathrm{d}}, \sim a, e_{i-1}$ and $u \leftarrow a^*, \sim a, e_{i-1}$ for each P_i and $a \in P_i$

7 : $u \leftarrow a^{\mathrm{d}}, a^*, \sim\overline{a}, e_{i-1}$ and $u \vee a^{\mathrm{d}} \vee a^* \leftarrow \sim\overline{a}, e_{i-1}$ for each P_i and $a \in P_i$

8 : $a^* \leftarrow u$ for each $a \in \mathrm{At}(\Pi)$

9 : $a^{\mathrm{d}} \leftarrow u$ for each $a \in P_1 \cup \cdots \cup P_k$

10 : $e_i \leftarrow u$ for $0 \leq i \leq k$

11 : $u \leftarrow e_k$; $e_0 \vee u \leftarrow$; and $\perp \leftarrow \sim u$

Fig. 2. Module $\mathrm{Tr}_{\min}(\Pi, P_1 > \cdots > P_k, V, F)$ from Definition 4

The rules in line 6 cover the case that a is false in M ($a \in P_i \setminus M$). Conforming to (3), both a^{d} and a^* are implicitly assigned to false, as they imply u. Otherwise, a is true in M which activates the rules in line 7, enforcing a^{d} equivalent to the negation of a^*. The net effect of the rules in lines 5–7 is that any *potential counter-model* $N \models \Pi$ for the $\langle P_1 > \cdots > P_k, V, F \rangle$-minimality of M, expressed in $(\mathrm{At}(\Pi) \setminus F)^* \cup (\mathrm{At}(\Pi) \cap F)$ instead of $\mathrm{At}(\Pi)$, must satisfy conditions (i) and (ii) from Definition 3. The rules in lines 8–11 are directly related to the unsatisfiability check which effectively proves that counter-models like N above do not exist.

Finally, we need to justify the *faithfulness* of the translation $\mathrm{Tr}_{\mathrm{circ2dlp}}(\Pi)$, that is, to show that the $\langle P_1 > \cdots > P_k, V, F \rangle$-minimal models of a PDLP Π are in a bijective relationship with the stable models of $\mathrm{Tr}_{\mathrm{circ2dlp}}(\Pi)$. A key observation is that the stable model semantics is *compositional* for the join of DLP-modules, that is, stable models for $\mathrm{Tr}_{\mathrm{circ2dlp}}(\Pi)$ can be computed for one submodule at a time. The following lemma is a direct consequence of the *module theorem* [13, Theorem 1].

Lemma 2. *Let* $M \subseteq \mathrm{At}(\mathrm{Tr}_{\mathrm{gen}}(\Pi))$ *and* $N \subseteq \mathrm{At}(\mathrm{Tr}_{\min}(\Pi)) \setminus (\mathrm{At}(\Pi) \cup \overline{\mathrm{At}(\Pi)})$. *Then* $M \in \mathrm{SM}(\mathrm{Tr}_{\mathrm{gen}}(\Pi))$ *and* $M \cup N \in \mathrm{SM}(\mathrm{Tr}_{\min}(\Pi))$ *if and only if* $M \cup N \in \mathrm{SM}(\mathrm{Tr}_{\mathrm{circ2dlp}}(\Pi))$.

The classical models of Π and the stable models of $\mathrm{Tr}_{\mathrm{gen}}(\Pi)$ are in bijective correspondence.

Lemma 3. *For a PDLP* Π *subject to a prioritized circumscription* $\mathrm{Circ}(\Pi, P_1 > \cdots > P_k, V, F)$ *and* $M \subseteq \mathrm{At}(\Pi)$, $M \models \Pi$ *if and only if* $M' = M \cup \overline{(\mathrm{At}(\Pi) \setminus M)} \in \mathrm{SM}(\mathrm{Tr}_{\mathrm{gen}}(\Pi))$.

Next, we characterize the connection between the existence of stable models for $\mathrm{Tr}_{\min}(\Pi)$ and the unsatisfiability of the translation $\mathrm{Tr}_{\mathrm{U}}(\Pi, P_1 > \cdots > P_k, F, M)$ in (3).

Lemma 4. *Let Π be a PDLP subject to a prioritized circumscription* $\mathrm{Circ}(\Pi, P_1 > \cdots > P_k, V, F)$, $M \subseteq \mathrm{At}(\Pi)$ *a model of Π, and $M' = M \cup \overline{(\mathrm{At}(\Pi) \setminus M)}$.*

(i) *If $\mathrm{Tr}_{\min}(\Pi)$ has a stable model N such that $N \cap (\mathrm{At}(\Pi) \cup \overline{\mathrm{At}(\Pi)}) = M'$, then $N = M' \cup P^{\mathrm{d}} \cup E \cup \{u\} \cup \mathrm{At}(\Pi)^*$ where $P = P_1 \cup \ldots \cup P_k$ and $E = \{e_i \mid 0 \leq i \leq k\}$.*

(ii) *The set of disjunctive rules $\mathrm{Tr}_{\mathrm{U}}(\Pi, P_1 > \cdots > P_k, F, M)$ is unsatisfiable if and only if there is $N \in \mathrm{SM}(\mathrm{Tr}_{\min}(\Pi))$ such that $N \cap (\mathrm{At}(\Pi) \cup \overline{\mathrm{At}(\Pi)}) = M'$.*

The proof of Lemma 4 is similar to the proof of [9, Proposition 2] as the same technique is used to encode propositional unsatisfiability check with the primitives of DLPs [14].

The correctness of the translation $\mathrm{Tr}_{\mathrm{circ2dlp}}(\Pi)$ now follows from Lemmas 1–4.

Theorem 1. *Given a PDLP Π, M is a $\langle P_1 > \cdots > P_k, V, F \rangle$-minimal model of Π if and only if $N \in \mathrm{SM}(\mathrm{Tr}_{\mathrm{circ2dlp}}(\Pi))$ where $N \cap (\mathrm{At}(\Pi) \cup \overline{\mathrm{At}(\Pi)}) = M \cup \overline{(\mathrm{At}(\Pi) \setminus M)}$.*

Proof sketch. ($\Longrightarrow$) Assume that M is a $\langle P_1 > \cdots > P_k, V, F \rangle$-minimal model of Π. Since $M \models \Pi$, we have $N = M \cup \overline{(\mathrm{At}(\Pi) \setminus M)} \in \mathrm{SM}(\mathrm{Tr}_{\mathrm{gen}}(\Pi))$ by Lemma 3. Since M is $\langle P_1 > \cdots > P_k, V, F \rangle$-minimal, $\mathrm{Tr}_{\mathrm{U}}(\Pi, P_1 > \cdots > P_k, F, M)$ is unsatisfiable by Lemma 1. By Lemma 4 (ii), there is $N' \in \mathrm{SM}(\mathrm{Tr}_{\min}(\Pi))$ such that $N' \cap (\mathrm{At}(\Pi) \cup \overline{\mathrm{At}(\Pi)}) = N$, and by Lemma 4 (i) $N' = N \cup P_1^{\mathrm{d}} \cup \cdots \cup P_k^{\mathrm{d}} \cup \{e_i \mid 0 \leq i \leq k\} \cup \{u\} \cup \mathrm{At}(\Pi)^*$. Furthermore, by Lemma 2, $N' \in \mathrm{SM}(\mathrm{Tr}_{\mathrm{circ2dlp}}(\Pi))$. ($\Longleftarrow$) Assume $M \in \mathrm{SM}(\mathrm{Tr}_{\mathrm{circ2dlp}}(\Pi))$. By Lemma 2, $M' = M \cap \mathrm{At}(\mathrm{Tr}_{\mathrm{gen}}(\Pi)) \in \mathrm{SM}(\mathrm{Tr}_{\mathrm{gen}}(\Pi))$ and $M \in \mathrm{SM}(\mathrm{Tr}_{\min}(\Pi))$. By Lemma 3, we have $N = M' \cap \mathrm{At}(\Pi) \models \Pi$. By Lemma 4 (ii), $\mathrm{Tr}_{\mathrm{U}}(\Pi, P_1 > \cdots > P_k, F, N)$ is unsatisfiable, and finally, by Lemma 1, N is a $\langle P_1 > \cdots > P_k, V, F \rangle$-minimal model of Π.   □

The following example illustrates the use of the translation for computing prioritized circumscription.

Example 3. Recall the positive DLP Π_{diag} in Example 1. In order to compute the $\langle \{ab_1\} > \{ab_2\} > \{ab_3\}, \{a, b, c, d\}, \emptyset \rangle$-minimal models of Π_{diag}, we consider the translation $\mathrm{Tr}_{\mathrm{circ2dlp}}(\Pi_{\mathrm{diag}}, \{ab_1\} > \{ab_2\} > \{ab_3\}, \{a, b, c, d\}, \emptyset)$ which is the join of modules $\mathrm{Tr}_{\mathrm{gen}}(\Pi_{\mathrm{diag}})$ and $\mathrm{Tr}_{\min}(\Pi_{\mathrm{diag}})$ presented in Figure 3. The translation $\mathrm{Tr}_{\mathrm{circ2dlp}}(\Pi_{\mathrm{diag}})$ has a unique stable model

$$N = \{\bar{a}, b, \bar{c}, \bar{d}, \overline{ab_1}, \overline{ab_2}, ab_3\} \cup \mathrm{At}(\Pi_{\mathrm{diag}})^* \cup \{ab_1^{\mathrm{d}}, ab_2^{\mathrm{d}}, ab_3^{\mathrm{d}}, e_0, e_1, e_2, e_3, u\}.$$

By Theorem 1, $N \cap \mathrm{At}(\Pi_{\mathrm{diag}}) = \{ab_3, b\} = M$ is the unique $\langle \{ab_1\} > \{ab_2\} > \{ab_3\}, \{a, b, c, d\}, \emptyset \rangle$-minimal model of Π_{diag} as already discussed in Example 2.   ∎

5   Experiments

We use the problem of finding Reiter-style minimal diagnoses [6] for digital circuits encoded as parallel/prioritized circumscription as a benchmark. The circuits are generated as follows. First a random tree is generated. The leaves of the tree, that is, the inputs of the circuit, are assigned random Boolean values. The intermediate nodes are assigned random logical operations which correspond to the gates of the circuit. The gate at the

Module :	$\mathrm{Tr}_{\mathrm{gen}}(\Pi_{\mathrm{diag}}, \{ab_1\} > \{ab_2\} > \{ab_3\}, \{a,b,c,d\}, \emptyset)$
Input :	$\emptyset$
Output :	$\mathrm{At}(\Pi_{\mathrm{diag}}) \cup \overline{\mathrm{At}(\Pi_{\mathrm{diag}})}$

1 : $a \leftarrow \sim\overline{a}.\ \ b \leftarrow \sim\overline{b}.\ \ c \leftarrow \sim\overline{c}.\ \ d \leftarrow \sim\overline{d}$

2 : $\overline{a} \leftarrow \sim a.\ \ \overline{b} \leftarrow \sim b.\ \ \overline{c} \leftarrow \sim c.\ \ \overline{d} \leftarrow \sim d.\ \ \overline{ab_1} \leftarrow \sim ab_1.\ \ \overline{ab_2} \leftarrow \sim ab_2.$

$\overline{ab_3} \leftarrow \sim ab_3$

3 : $ab_1 \leftarrow \sim a, \sim b.\ \ ab_1 \leftarrow \sim\overline{a}, \sim\overline{b}.\ \ ab_2 \leftarrow \sim b, \sim c.\ \ ab_2 \leftarrow \sim\overline{b}, \sim\overline{c}.$

$ab_3 \leftarrow \sim c, \sim d.\ \ ab_3 \leftarrow \sim\overline{c}, \sim\overline{d}.\ \ \bot \leftarrow \sim\overline{a}.\ \ \bot \leftarrow \sim\overline{d}$

Module :	$\mathrm{Tr}_{\mathrm{min}}(\Pi_{\mathrm{diag}}, \{ab_1\} > \{ab_2\} > \{ab_3\}, \{a,b,c,d\}, \emptyset)$
Input :	$\mathrm{At}(\Pi_{\mathrm{diag}}) \cup \overline{\mathrm{At}(\Pi_{\mathrm{diag}})}$
Output :	$\mathrm{At}(\Pi_{\mathrm{diag}})^* \cup \{ab_1^{\mathrm{d}}, ab_2^{\mathrm{d}}, ab_3^{\mathrm{d}}, e_0, e_1, e_2, e_3, u\}$

4 : $a^* \vee b^* \vee ab_1^* \vee u.\ \ ab_1^* \vee u \leftarrow a^*, b^*.\ \ b^* \vee c^* \vee ab_2^* \vee u.\ \ ab_2^* \vee u \leftarrow b^*, c^*.$

$c^* \vee d^* \vee ab_3^* \vee u.\ \ ab_3^* \vee u \leftarrow c^*, d^*.\ \ u \leftarrow a^*.\ \ u \leftarrow d^*$

5 : $ab_1^{\mathrm{d}} \vee e_1 \vee u \leftarrow e_0.\ \ ab_2^{\mathrm{d}} \vee e_2 \vee u \leftarrow e_1.\ \ ab_3^{\mathrm{d}} \vee e_3 \vee u \leftarrow e_2$

6 : $u \leftarrow ab_1^{\mathrm{d}}, \sim ab_1, e_0.\ \ u \leftarrow ab_1^*, \sim ab_1, e_0.\ \ u \leftarrow ab_2^{\mathrm{d}}, \sim ab_2, e_1.$

$u \leftarrow ab_2^*, \sim ab_2, e_1.\ \ u \leftarrow ab_3^{\mathrm{d}}, \sim ab_3, e_2.\ \ u \leftarrow ab_3^*, \sim ab_3, e_2$

7 : $u \leftarrow ab_1^{\mathrm{d}}, ab_1^*, \sim\overline{ab_1}, e_0.\ \ u \vee ab_1^{\mathrm{d}} \vee ab_1^* \leftarrow \sim\overline{ab_1}, e_0.$

$u \leftarrow ab_2^{\mathrm{d}}, ab_2^*, \sim\overline{ab_2}, e_1.\ \ u \vee ab_2^{\mathrm{d}} \vee ab_2^* \leftarrow \sim\overline{ab_2}, e_1.$

$u \leftarrow ab_3^{\mathrm{d}}, ab_3^*, \sim\overline{ab_3}, e_2.\ \ u \vee ab_3^{\mathrm{d}} \vee ab_3^* \leftarrow \sim\overline{ab_3}, e_2$

8 : $a^* \leftarrow u.\ \ b^* \leftarrow u.\ \ c^* \leftarrow u.\ \ d^* \leftarrow u.\ \ ab_1^* \leftarrow u.\ \ ab_2^* \leftarrow u.\ \ ab_3^* \leftarrow u$

9 : $ab_1^{\mathrm{d}} \leftarrow u.\ \ ab_2^{\mathrm{d}} \leftarrow u.\ \ ab_3^{\mathrm{d}} \leftarrow u$

10 : $e_0 \leftarrow u.\ \ e_1 \leftarrow u.\ \ e_2 \leftarrow u.\ \ e_3 \leftarrow u$

11 : $u \leftarrow e_3.\ \ e_0 \vee u.\ \ \bot \leftarrow \sim u$

Fig. 3. The modules for translation $\mathrm{Tr}_{\mathrm{circ2dlp}}(\Pi_{\mathrm{diag}})$ in Example 3

root node produces the output for the entire circuit. Its value is calculated and flipped to obtain faulty behavior for the circuit. The $\langle\{ab_i \mid i \leq N\}, \{high_i \mid i \leq N\}, \emptyset\rangle$-minimal models of the resulting program correspond to minimal diagnoses, where N is the number of nodes in the tree forming the circuit. For each value of N we select priorities for the atoms ab_j according to the following scheme. An atom ab_j is given priority i, if $(i-1) \cdot \lfloor N/k \rfloor + 1 \leq j \leq i \cdot \lfloor N/k \rfloor$, where k is the number of priority classes.

The tool CIRC2DLP (v. 2.1)[3] implements the translation $\mathrm{Tr}_{\mathrm{circ2dlp}}$ described in Section 4. We compare the performance of CIRC2DLP with our previous translator PRIO_CIRC2DLP [10] which implements Lifschitz' scheme (2). We use GNT (v. 2.1) and DLV (2006-07-14) for the computation of stable models. We also compare the performance of our translation-based approach with that of CIRCUM2 system [12]. The

---

[3] See http://www.tcs.hut.fi/Software/circ2dlp/ for binaries and benchmarks.

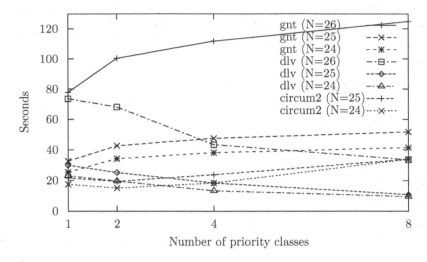

Fig. 4. The averages of running times for computing *all* minimal diagnoses for faulty digital circuits for $N = 24, 25, 26$ nodes with the numbers of priority classes $k = 1, 2, 4, 8$ for minimization

measured running time for CIRC2DLP and PRIO_CIRC2DLP is the sum of the translation time and the time of finding the stable models using GNT/DLV. The translation times are negligible, however. For CIRCUM2 the measured running time is the duration of the search for models of circumscription using the tool with DLV as its back-end. We report the sum of user and system time of /usr/bin/time. All the tests were run under Linux on a 1.7GHz AMD Athlon XP 2000+ with a timeout 1800 seconds and a memory limit 512MB.

First, we compare the performances of CIRC2DLP and PRIO_CIRC2DLP with an instance with $N = 15$ nodes in the circuit and use GNT for the computation of stable models. The running times for $k = 1, 2, 3, 4$ priorities were 0.29, 0.38, 0.39 and 0.43 seconds for CIRC2DLP, and respectively 0.30, 44.3, 88.8 and 127.25 seconds for PRIO_CIRC2DLP. The performance of PRIO_CIRC2DLP is poor for $k > 1$ even with a very small circuit with $N = 15$ and CIRC2DLP shows a promising improvement in the performance.

Next, we use CIRC2DLP with GNT and DLV as back-ends to see whether the choice of a solver has an effect on the running times. We also compare the performance of CIRC2DLP to that of CIRCUM2. To see how different approaches scale on the number of priority classes $k = 1, 2, 4, 8$ we generate randomly 20 circuits for each of the values $N = 24, 25, 26$. The average running times from this experiment are shown in Figure 4.[4] First, observe that in contrast to the experimental results in [10] computing models with DLV is faster than with GNT. This illustrates the flexibility of the translation-based method compared to developing a specialized solver as one can directly benefit from solver development. For values $k = 1, 2$ CIRCUM2 is slightly

---

[4] The average running times for $N = 26$ cannot be reported for CIRCUM2, since it was not able to solve all instances within the memory limit.

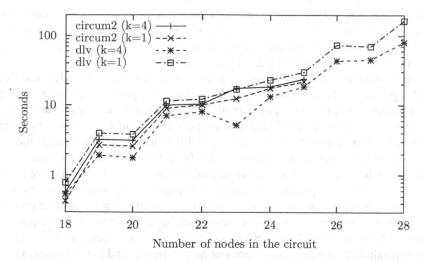

Fig. 5. The averages of running times for computing *all* minimal diagnoses for faulty digital circuits for $N = 18, \ldots, 28$ nodes with $k = 1, 4$ priorities for minimization

faster than CIRC2DLP, but CIRCUM2 consumed over 512MB of memory with some rather moderately sized instances with $N = 26$. For $k = 4, 8$, CIRC2DLP+DLV becomes faster than CIRCUM2. Also, we note that the average running times of CIRC2DLP+DLV decrease when the number of priorities k increases. This reflects the fact that eventually, that is, when $k = |\text{At}(\Pi) \setminus (V \cup F)|$, it becomes easier to decide $\langle \{a_1\} > \cdots > \{a_k\}, V, F \rangle$-minimality using a total order of atoms $a_1, \ldots, a_k$ under minimization [15].

Finally, we study how the performances of CIRC2DLP+DLV[5] and CIRCUM2 scale up when the number of nodes N grows. We consider $k = 1$ and $k = 4$ priorities for minimization, and generate randomly 20 circuits with $N = 18, \ldots, 28$ nodes. The average running times are presented in Figure 5. Again, CIRCUM2 could not solve all instances for $N \geq 26$ without exceeding the memory limit. Both systems performed similarly for values $N = 18, \ldots, 25$, but using CIRC2DLP we could easily compute models for the respective circumscriptions up to $N = 28$.

6 Discussion

We present a transformation from prioritized circumscription to DLPs. The translation function $\text{Tr}_{\text{circ2dlp}}$ generalizes the one designed for parallel circumscription [9] and it has a distinctive combination of features: (i) arbitrary propositional theories Π subject to prioritized circumscription are covered, (ii) the translation $\text{Tr}_{\text{circ2dlp}}(\Pi, P_1 > \cdots >$

---

[5] As computing stable models with DLV is faster than with GNT, the choice of DLV as the back-end reflects the best performance of our approach. Notice, however, that the results of the previous experiment show that even with GNT as back-end, CIRC2DLP can handle circuits with at least $N = 26$ nodes.

$P_k, V, F)$ can be produced in linear time and space before computing any models for it, (iii) the models of $\text{Circ}(\Pi, P_1 > \cdots > P_k, V, F)$ and the stable models of its translation are in a bijective relationship, (iv) the signature $\text{At}(\Pi)$ is preserved under $\text{Tr}_{\text{circ2dlp}}$, and (v) there is no need for incremental updating.

In contrast, all previous approaches lack some of these features. De Kleer and Konolige [16] present the basic technique for eliminating fixed predicates. The case of varying predicates is addressed by Cadoli et al. [17] but a query-based equivalence rather than an exact correspondence of models is of their interest. In addition to these general results, a number of attempts to reduce parallel/prioritized circumscription into logic programming have been made. Gelfond and Lifschitz [18] address prioritized circumscription but their translation scheme is amenable to *stratified* circumscriptive theories only. The translation of parallel circumscription presented by Sakama and Inoue [19] is based on *characteristic clauses* resulting in exponential space and time complexity in the worst case. In [20], the same authors embed prioritized circumscription into *prioritized logic programming* based on a different semantics. Lee and Lin [21] characterize parallel circumscription in terms of *loop formulas* and exploit them to obtain an embedding in disjunctive logic programming. However, the number of loops can be exponential in the worst case. Thus, it remains open whether an efficient translation is feasible in general using their approach. Wakaki and Inoue [11] concentrate on prioritized circumscription and design a two-phase procedure for the computation of minimal models. The first phase generates model candidates which are then tested for minimality in the sense of prioritized circumscription. Both the model generator and the tester are represented as *separate* disjunctive logic programs. There is an implementation of the procedure, named CIRCUM1, but it is rather inefficient since all model candidates are computed first. Wakaki and Tomita [12] improve the procedure by Wakaki and Inoue [11] and integrate the generating and testing programs into one in analogy to our approach [9]. However, this is not a one-shot transformation because the answer sets of the generating program have to be computed and counted before the testing part can be created. The resulting implementation, CIRCUM2 was used as one of the reference systems in our experiments.

The experiments reported in Section 5 suggest that our translation-based approach compares favorably with CIRCUM2. Due to linearity of the transformation, CIRC2DLP with a disjunctive solver as its back-end needs far less memory than CIRCUM2 and thus it eventually scales up better as demonstrated in our experiments. The performance of CIRC2DLP with disjunctive solvers is encouraging—suggesting that there is no need to develop dedicated solvers for prioritized circumscription. Furthermore, we can take full advantage of the ongoing development of disjunctive solvers.

Our results enable the use of prioritized circumscription as a primitive in disjunctive logic programming. Consequently, we expect that more concise encodings can be devised in applications like model-based diagnosis [6] formalized in the experiments of Section 5. In fact, we can now view prioritized circumscription as syntactic sugar as it can be translated away using $\text{Tr}_{\text{circ2dlp}}$. However, it may be wise to store the original representation, rather than the translation, for easier maintainability. A further goal is the generalization of stable models with prioritized minimization of models. In fact, the design of CIRC2DLP already includes support for negative body literals in rules.

This readily enables the computation of $\langle P_1 > \cdots > P_k, V, F\rangle$-*stable models* M of an arbitrary (not just positive) DLP Π based on the reduct Π^M.

References

1. Oikarinen, E., Janhunen, T.: A linear transformation from prioritized circumscription to disjunctive logic programming. In: Dahl, V., Niemelä, I. (eds.) ICLP 2007. LNCS, vol. 4670, pp. 440–441. Springer, Heidelberg (2007)
2. Gelfond, M., Lifschitz, V.: Classical negation in logic programs and disjunctive databases. New Generation Computing 9(3/4), 365–385 (1991)
3. Przymusinski, T.: Stable semantics for disjunctive programs. New Generation Computing 9(3/4), 401–424 (1991)
4. Leone, N., Pfeifer, G., Faber, W., Eiter, T., Gottlob, G., Perri, S., Scarcello, F.: The DLV system for knowledge representation and reasoning. ACM Transactions on Computational Logic 7(3), 499–562 (2006)
5. Janhunen, T., Niemelä, I., Seipel, D., Simons, P., You, J.H.: Unfolding partiality and disjunctions in stable model semantics. ACM Transactions on Computational Logic 7(1), 1–37 (2006)
6. Reiter, R.: A theory of diagnosis from first principles. Artificial Intelligence 32(1), 57–95 (1987)
7. Lifschitz, V.: Computing circumscription. In: Joshi, A.K. (ed.) Proceedings of the 9th International Joint Conference on Artificial Intelligence, Los Angeles, California, August 18–23, 1985, pp. 121–127. Morgan Kaufmann, San Francisco (1985)
8. McCarthy, J.: Circumscription — A form of non-monotonic reasoning. Artificial Intelligence 13(1–2), 27–39 (1980)
9. Janhunen, T., Oikarinen, E.: Capturing parallel circumscription with disjunctive logic programs. In: Alferes, J.J., Leite, J.A. (eds.) JELIA 2004. LNCS (LNAI), vol. 3229, pp. 134–146. Springer, Heidelberg (2004)
10. Oikarinen, E., Janhunen, T.: CIRC2DLP — translating circumscription into disjunctive logic programming. In: Baral, C., Greco, G., Leone, N., Terracina, G. (eds.) LPNMR 2005. LNCS (LNAI), vol. 3662, pp. 405–409. Springer, Heidelberg (2005)
11. Wakaki, T., Inoue, K.: Compiling prioritized circumscription into answer set programming. In: Demoen, B., Lifschitz, V. (eds.) ICLP 2004. LNCS, vol. 3132, pp. 356–370. Springer, Heidelberg (2004)
12. Wakaki, T., Tomita, K.: Compiling prioritized circumscription into general disjunctive programs. In: Provetti, A., Son, T.C. (eds.) Proceedings of PREFS 2006: Preferences and their Applications in Logic Programming Systems, August 16, 2006, pp. 1–15 (2006)
13. Janhunen, T., Oikarinen, E., Tompits, H., Woltran, S.: Modularity aspects of disjunctive stable models. In: Baral, C., Brewka, G., Schlipf, J. (eds.) LPNMR 2007. LNCS (LNAI), vol. 4483, pp. 175–187. Springer, Heidelberg (2007)
14. Eiter, T., Gottlob, G.: On the computational cost of disjunctive logic programming: Propositional case. Annals of Mathematics and Artificial Intelligence 15(3–4), 289–323 (1995)
15. Gottlob, G.: The complexity of default reasoning under the stationary fixed point semantics. Information and Computation 121(1), 81–92 (1995)
16. de Kleer, J., Konolige, K.: Eliminating the fixed predicates from a circumscription. Artificial Intelligence 39(3), 391–398 (1989)
17. Cadoli, M., Eiter, T., Gottlob, G.: An efficient method for eliminating varying predicates from a circumscription. Artificial Intelligence 54(2), 397–410 (1992)

18. Gelfond, M., Lifschitz, V.: Compiling circumscriptive theories into logic programs. In: Reinfrank, M., Ginsberg, M.L., de Kleer, J., Sandewall, E. (eds.) Non-Monotonic Reasoning 1988. LNCS, vol. 346, pp. 74–99. Springer, Heidelberg (1988)
19. Sakama, C., Inoue, K.: Embedding circumscriptive theories in general disjunctive programs. In: Marek, V.W., Truszczyński, M., Nerode, A. (eds.) LPNMR 1995. LNCS, vol. 928, pp. 344–357. Springer, Heidelberg (1995)
20. Sakama, C., Inoue, K.: Prioritized logic programming and its application to commonsense reasoning. Artificial Intelligence 123(1–2), 185–222 (2000)
21. Lee, J., Lin, F.: Loop formulas for circumscription. Artificial Intelligence 170(2), 160–185 (2006)

Mapping Properties of Heterogeneous Ontologies

Chiara Ghidini and Luciano Serafini

FBK-IRST, Via Sommarive 18, I-38040 Trento, Italy
surname@fbk.eu

Abstract. In recent papers, we have proposed an extension of the formalism of Distributed Description Logic (DDL) to represent a wide set of homogeneous and heterogeneous mappings involving relations and concepts. Here we investigate in depth how to provide an effective reasoning algorithm for reasoning within this logic in the case of homogeneous mappings.

1   Introduction

In [5] we have provided a Distributed Description Logic (DDL) [11] able to represent a wide set of mappings for ontology alignment, namely concept-mappings, role-mappings and concept/role-mappings. We have done this in order to meet the need for expressive mapping languages, underlined in reports of the Ontology Management Working Group[1] and in several works such as [3,10,12].

In that work the descriptive language $\mathcal{ALCQI}_b$ [13] was chosen to formalise the single ontologies. This choice originated from the fact that $\mathcal{ALCQI}_b$ provided the right expressivity to investigate the theoretical properties of the general framework. One of the limitations of this choice is that it makes the definition of effective reasoning procedures for DDL extremely difficult because $\mathcal{ALCQI}_b$ lacks efficient reasoning algorithms to be used as a starting point in the definition of a distributed reasoning algorithm for DDL, following the approach shown in [11][2]. As a consequence the only effective reasoning procedure for DDL available so far is the one for the original logic which only includes concept-mappings (see [11]).

In this paper, we aim at extending effective reasoning procedure for DDL. To do this we focus on a DDL containing homogeneous concept-mappings and role-mappings between ontologies described using languages weaker or at most equivalent to $\mathcal{SHIQ}$. We focus on $\mathcal{SHIQ}$ because it is one of the relevant DL languages used in the Semantic Web, and because it provides effective tableau based reasoning algorithms to be used in the definition of reasoning procedures for DDL. We call this logic $DDL_o^{\mathcal{SHIQ}}$. In the paper we show that the straightforward $DDL_o^{\mathcal{SHIQ}}$ obtained by imposing homogeneous mappings between $\mathcal{SHIQ}$ ontologies leads to an *undecidable* logical system. Moreover, we show that some

---

[1] (www.omwg.org)

[2] The tableau for $\mathcal{ALCQI}_b$ provided in [13] only deals with concept satisfiability. The reasoning algorithms for concept satisfiability with general TBoxes are instead based on looping tree automata[14].

D. Dochev, M. Pistore, and P. Traverso (Eds.): AIMSA 2008, LNAI 5253, pp. 181–193, 2008.

logical consequences of role-mappings between $\mathcal{SHIQ}$ ontologies are of the form $P \sqsubseteq Q \sqcup R$, with P, Q, and R roles, and cannot be expressed in $\mathcal{SHIQ}$. We show that this is mainly due to number restriction, and we focus our attention to the $\mathcal{SHI}$ language ($\mathcal{SHIQ}$ without number restriction), for which we define an effective decision procedure for the corresponding $DDL_o^{\mathcal{SHI}}$ logic.

2   The Distributed Description Logic $DDL_o^{\mathcal{SHIQ}}$

Here we introduce the syntax and semantics of $DDL_o^{\mathcal{SHIQ}}$. For lack of space we assume familiarity with Description Logic [1] and DDL [5,11].

The syntax. Given a non empty set I of indexes, used to identify ontologies, let $\{\mathcal{DL}_i\}_{i\in I}$ be a collection of description logics. For each $i \in I$ we denote a T-box of $\mathcal{DL}_i$ with $\mathcal{T}_i$. In this paper, we assume that each $\mathcal{DL}_i$ is weaker or at most equivalent to $\mathcal{SHIQ}$. We call $\mathbf{T} = \{\mathcal{T}_i\}_{i\in I}$ a family of T-Boxes indexed by I. Intuitively, $\mathcal{T}_i$ is the description logic formalization of the i-th ontology. To make every description distinct, we will prefix it with the index of the ontology it belongs to. For instance, the concept C that occurs in the i-th ontology is denoted as $i : C$.

Semantic mappings between different ontologies are expressed via *bridge rules*. The bridge rules from i to j we consider in this paper are expressions of the form:

- $i : X \xrightarrow{\sqsupseteq} j : Y$ *(onto-bridge rule)*
- $i : X \xrightarrow{\sqsubseteq} j : Y$ *(into-bridge rule)*

where X and Y are both concepts or both atomic roles. These expressions are called homogeneous bridge rules in [5]. As the semantics will make clear, into and onto bridge rules are used to express that, from the j-th point of view the element X in i is less general (into rule) or more general (onto rule) than its local element Y. As an example, $i : \mathit{marriedTo} \xrightarrow{\sqsubseteq} j : \mathit{partnerOf}$ says that according to ontology j, the role $\mathit{marriedTo}$ in ontology i is less general than its own role $\mathit{partnerOf}$.

A *distributed T-box* (DTB) $\mathfrak{T} = \langle \mathcal{T}_i, \mathfrak{B} \rangle$ consists of a collection of T-boxes $\mathcal{T}_i$, and a collection of bridge rules $\mathfrak{B} = \{\mathfrak{B}_{ij}\}_{i \neq j \in I}$ between them.

The semantics. The first component of an interpretation of a DTB is a family of interpretations $\{\mathcal{I}_i\}_{i\in I}$, one for each T-box $\mathcal{T}_i$. Each $\mathcal{I}_i$ is called a *local interpretation* and consists of a, possibly empty, local domain $\Delta^{\mathcal{I}_i}$ and a valuation function $\cdot^{\mathcal{I}_i}$, which maps every concept to a subset of $\Delta^{\mathcal{I}_i}$, and every role to a subset of $\Delta^{\mathcal{I}_i} \times \Delta^{\mathcal{I}_i}$.

The second component of the semantics is a family of domain relations. Formally, a *domain relation* r_{ij} from $\Delta^{\mathcal{I}_i}$ to $\Delta^{\mathcal{I}_j}$ is a subset of $\Delta^{\mathcal{I}_i} \times \Delta^{\mathcal{I}_j}$. Domain relations define how the different T-boxes interact: a domain relation r_{ij} represents a possible way of mapping the elements of $\Delta^{\mathcal{I}_i}$ into its domain $\Delta^{\mathcal{I}_j}$, seen from j's perspective. Notationally, we use $r_{ij}(d)$ to denote $\{d' \in \Delta^{\mathcal{I}_j} \mid \langle d, d' \rangle \in r_{ij}\}$;

for a subset D of $\Delta^{\mathcal{I}_i}$, we use $r_{ij}(D)$ to denote $\bigcup_{d \in D} r_{ij}(d)$; for $R \subseteq \Delta^{\mathcal{I}_i} \times \Delta^{\mathcal{I}_i}$ we use $r_{ij}(R)$ to denote $\bigcup_{\langle d,d' \rangle \in R} r_{ij}(d) \times r_{ij}(d')$.

Local interpretations and domain relations are used to define an interpretation for a DTB. Formally, a *distributed interpretation* $\mathfrak{I}$ of a distributed T-box $\mathfrak{T}$ consists of the pair $\mathfrak{I} = \langle \{\mathcal{I}_i\}_{i \in I}, \{r_{ij}\}_{i \neq j \in I} \rangle$. A distributed interpretation $\mathfrak{I}$ *satisfies* a bridge rule $\mathfrak{b}$, in symbols $\mathfrak{I} \models \mathfrak{b}$, according with the following:

1. $\mathfrak{I} \models i : X \xrightarrow{\sqsubseteq} j : Y$, if $r_{ij}(X^{\mathcal{I}_i}) \subseteq Y^{\mathcal{I}_j}$;
2. $\mathfrak{I} \models i : X \xrightarrow{\sqsupseteq} j : Y$, if $r_{ij}(X^{\mathcal{I}_i}) \supseteq Y^{\mathcal{I}_j}$.

A distributed interpretation $\mathfrak{I}$ *satisfies* a DTB $\mathfrak{T}$ if (i) all the T-boxes $\mathcal{T}_i$ are satisfied by their local interpretation $\mathcal{I}_i$, and (ii) $\mathfrak{I}$ satisfies all the bridge rules of $\mathfrak{T}$. Entailment and satisfiability are defined in the usual way. See [5].

3   The Effects of Mappings

DDL mappings can be thought of as inter-theory axioms that allow to derive new knowledge via logical consequence. In this section we characterise the effects of mappings in the simple case of a DTB $\mathfrak{T}_{ij}$ composed of two T-boxes, $\mathcal{T}_i$ and $\mathcal{T}_j$, connected with bridge rules $\mathfrak{B}_{ij}$ from i to j. Proofs of the properties and theorems stated here can be found in [4].[3] A first important feature of the DDL mappings is that they are *directional*, i.e., they allow to transfer knowledge from a source ontology i to a target ontology j without any back-flow effect. This property holds also for $DDL_o^{\mathcal{SHIQ}}$. A consequence of directionality is that we can characterise the effects of bridge rules using *propagation rules* of the form:

axioms in i	*Reading*: if $\mathcal{T}_i$ entails all the axioms in i, and
bridge rules from i to j	$\mathfrak{B}_{ij}$ contains the bridge rules from i to j, then
axiom in j	$\langle \mathcal{T}_i, \mathcal{T}_j, \mathfrak{B}_{ij} \rangle$ satisfies axioms in j.

From the characterisation of the effects of homogeneous mappings in DDL provided in [5], we know that homogeneous bridge rules allow to obtain two specific types of effects: (i) propagation of hierarchies across ontologies; and (ii) propagation of role domain and range restriction. Simple rules showing these effects are:

$$
\begin{array}{ccc}
\begin{array}{c}
i : A \xrightarrow{\sqsupseteq} j : C \\
i : B \xrightarrow{\sqsubseteq} j : D \\
\hline
i : A \sqsubseteq B \\
\hline
j : C \sqsubseteq D
\end{array}
&
\begin{array}{c}
i : P \xrightarrow{\sqsupseteq} j : R \\
i : Q \xrightarrow{\sqsubseteq} j : S \\
\hline
i : P \sqsubseteq Q \\
\hline
j : R \sqsubseteq S
\end{array}
&
\begin{array}{c}
i : S \xrightarrow{\sqsupseteq} j : R \\
i : A_1 \xrightarrow{\sqsubseteq} j : B_1 \\
i : A_2 \xrightarrow{\sqsubseteq} j : B_2 \\
i : \exists S. \neg A_2 \sqsubseteq A_1 \\
\hline
j : \exists R. \neg B_2 \sqsubseteq B_1
\end{array} \\[4mm]
(3.1) & (3.2) & (3.3)
\end{array}
$$

Into and onto bridge rules are combined in rules (3.1) and (3.2) to propagate hierarchies from $\mathcal{T}_i$ to $\mathcal{T}_j$: (3.1) propagate the concept hierarchy of $\mathcal{T}_i$ into

---

[3] For the sake of notation, we assume that for each bridge rule between roles $i : P \longrightarrow j : R$ in $\mathfrak{B}_{ij}$, also $i : inv(P) \longrightarrow j : inv(R)$ is in $\mathfrak{B}_{ij}$ (where $inv(X)$ is the inverse of X).

the analogous hierarchy of $\mathcal{T}_j$; (3.2) does the same with role hierarchies. The combination of homogeneous bridge rules on concepts and roles propagates information about the domain and range of roles as expressed by rule (3.3). This rule propagates the restriction between the domain and range of S to an analogous restriction on R. Note that if A_2 and B_2 are the empty concept $\bot$, (3.3) allows one to propagate knowledge on the domain of S to the domain of R. In an analogous manner we can obtain the propagation of range.

Other effects of role-mappings concern the propagation of role properties such as symmetry, transitivity, functionality, and cardinality restriction. If P is a symmetric relation and $\mathsf{Symm}(P) \in \mathcal{T}_i$, then the mapping $i : P \xrightarrow{\equiv} j : R$ (and $i : P^- \xrightarrow{\sqsupseteq} j : R^-$ which is contained in $\mathfrak{B}_{ij}$ because of the closure condition on $\mathfrak{B}_{ij}$) force R also to be a symmetric relation in $\mathcal{T}_j$. This effect is captured by rule (3.2). In fact $\mathsf{Symm}(P)$ is equivalent to $P \sqsubseteq P^-$, and using rule (3.2) with the two bridge rules between P and R above, one can infer $R \sqsubseteq R^-$ which states the symmetry of R in $\mathcal{T}_j$. On the contrary, if P is a transitive relation in $\mathcal{T}_i$, then the bridge rule $1 : P \xrightarrow{\equiv} 2 : R$ and $1 : P \xrightarrow{\sqsupseteq} 2 : R$ do not force R to be transitive. An example of P transitive and a R non transitive, connected by the above bridge rules is $P^{\mathcal{I}_1} = \{(a,b),(c,d)\}$, $Q^{\mathcal{I}_2} = \{(1,2),(2,3)\}$ and $r_{ij} = \{(a,1),(b,2),(c,2),(d,3)\}$. Finally, functionality (and more in general of (qualified) general number restriction) does not propagate. The distributed interpretation depicted below satisfies the functionality of P, the most stringent bridge rules we can impose between P and Q, that is $1 : P \xrightarrow{\equiv} 2 : Q$, but it does not interpret Q in a function.

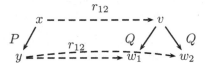

The propagation of transitivity and (qualified) number restriction can be guaranteed by imposing suitable cardinality restrictions on the domain relation. For instance, the inverse-functionality of r_{ij} guarantees the propagation of transitivity, and the functionality of r_{ij} guarantees the propagation of $(\leq n)$ number restriction. This restrictive semantics for DDL, though interesting, is out of the scope of this paper and will be the subject of further research.

3.1   Undecidability and Incompleteness of $DDL_o^{\mathcal{SHIQ}}$

Rules (3.1)–(3.3) describe sound effects of mappings in $DDL_o^{\mathcal{SHIQ}}$. Thus we tried to generalise them in order to encode the sound and complete propagation of knowledge from $\mathcal{T}_i$ to $\mathcal{T}_j$. Nevertheless, finding the correct generalisation of (3.1)–(3.3) involves solving two different problems. The first problem concerns decidability: decidability of $\mathcal{SHIQ}$ is guaranteed under the syntactic restriction that number restrictions are restricted to simple roles, that is roles having no transitive sub-roles according to the role hierarchy. Propagation of the role hierarchy, modelled by rule (3.2), could allow to infer a subsumption between roles which breaks this syntactic restriction. Thus, appropriate restrictions need to be

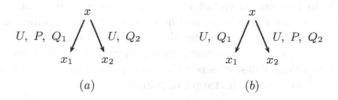

Fig. 1. (a) and (b) satisfy $\mathcal{T}$, but (a) does not satisfy $P \sqsubseteq Q_2$ and (b) does not satisfy $P \sqsubseteq Q_1$

enforced to ensure that no subsumption is inferred of the form $j : R \sqsubseteq S$ where S is a role affected by a number restriction and R is transitive.

The second, and more complex problem concerns the possibility of generalising rule (3.2). To illustrate this problem let us start noticing that if the role union primitive were available in $\mathcal{SHIQ}$, then a general form of rule (3.2)

$$\frac{i : P \sqsubseteq \bigsqcup_{k=1}^{n} Q_k \qquad i : P \xrightarrow{\sqsupseteq} j : R \qquad i : Q_k \xrightarrow{\sqsubseteq} j : S_k, \text{ for } 1 \leq k \leq n}{j : R \sqsubseteq \bigsqcup_{k=1}^{n} S_k} \tag{3.4}$$

would be also correct. This rule is primitive, that is, cannot be reconstructed from rules (3.1)–(3.3). Despite the fact that the formula $P \sqsubseteq \bigsqcup_{k=1}^{n} Q_k$ is not directly expressible in a $\mathcal{SHIQ}$ T-box, there are special cases in which the inclusion between a role and the union of other n roles can be entailed by a set of $\mathcal{SHIQ}$ axioms. Consider the following example, where we restrict to $n = 2$ for the sake of simplicity.

Example 1. Consider the $\mathcal{SHIQ}$ T-box $\mathcal{T}$ containing the following axioms:

$$P \sqsubseteq U \qquad\qquad \exists P.\top \sqsubseteq \exists Q_1.\top \qquad\qquad \top \sqsubseteq (\leq 2)U$$
$$Q_1 \sqsubseteq U \qquad\qquad \exists P.\top \sqsubseteq \exists Q_2.\top$$
$$Q_2 \sqsubseteq U \qquad\qquad \exists Q_1^-.\top \sqsubseteq \forall Q_2^-.\bot$$

All models $\mathcal{I}$ of $\mathcal{T}$ are such that $P^{\mathcal{I}} \subseteq Q_1^{\mathcal{I}} \cup Q_2^{\mathcal{I}}$. Indeed for every $(x, y) \in P^{\mathcal{I}}$ we have that there is an x_1 such that $(x, x_1) \in Q_1^{\mathcal{I}}$, and there is an x_2 such that $(x, x_2) \in Q_2^{\mathcal{I}}$. From the inclusion of P, Q_1 and Q_2 in U, we have that y, x_1 and x_2, are U-successors of x. From the fact that x has at most two U-successors and that x_1 and x_2 are different, (the axiom $\exists Q_1^-.\top \sqsubseteq \forall Q_2^-.\bot$ guarantees this) we conclude that either $y = x_1$ or $y = x_2$, which implies that $(x, y) \in Q_1^{\mathcal{I}}$ or $(x, y) \in Q_2^{\mathcal{I}}$. Finally, notice that, $\mathcal{T} \not\models P \sqsubseteq Q_1$ and $\mathcal{T} \not\models P \sqsubseteq Q_2$ as it is shown by the two models of $\mathcal{T}$ depicted in Figure 1. Summarizing, if we add the construct of role union to our DL language, we have that $\mathcal{T} \models U \sqsubseteq Q_1 \sqcup Q_2$, but $\mathcal{T} \not\models U \sqsubseteq Q_1$ and $\mathcal{T} \not\models U \sqsubseteq Q_2$.

The previous example describes a T-box in which it is possible to entail an inclusion between a role and a union of roles in $\mathcal{SHIQ}$. However, the formula

$P \sqsubseteq Q_1 \sqcup Q_2$ in this case does not create problems as it is a consequence of a very specific set of $\mathcal{SHIQ}$ axioms, and holds only in very specific sets of $\mathcal{SHIQ}$ models. Nevertheless if we insert the T-box $\mathcal{T}$ described above in a DTB and connect P, Q_1 and Q_2 to other relations R, S_1 and S_2 in a target ontology $\mathcal{T}_t$ via appropriate bridge rules, it is possible to express a generic inclusions axiom of the form $R \sqsubseteq S_1 \sqcup S_2$ in the target ontology.

Theorem 1. *Let* $\mathcal{T}_\sqcup = \{R \sqsubseteq S_1 \sqcup S_2\}$, *and* $\mathfrak{I}_{\mathcal{SHIQ}} = \langle \mathcal{T}, \mathcal{T}_t, \mathfrak{B} \rangle$, *where* $\mathcal{T}$ *is the Tbox defined above,* $\mathcal{T}_t = \emptyset$ *and* $\mathfrak{B}$ *contains the following bridge rules*

$$1 : P \xrightarrow{\sqsupseteq} 2 : R \qquad 1 : Q_1 \xrightarrow{\sqsupseteq} 2 : S_1 \qquad 1 : Q_2 \xrightarrow{\sqsupseteq} 2 : S_2 \qquad (3.5)$$

then $\mathcal{T}_\sqcup \models \phi$ *if and only if* $\mathfrak{I}_{\mathcal{SHIQ}} \models t : \phi$, *for any concept and role inclusion statement* ϕ *of* $\mathcal{SHIQ}(\sqcup)$, *where* $\mathcal{SHIQ}(\sqcup)$ *is* $\mathcal{SHIQ}$ *extended with the role union operator.*

Proof. We show that or any interpretation $\mathcal{I}$ of $\mathcal{T}_\sqcup$, (such that $R^{\mathcal{I}_\sqcup} \subseteq S_1^{\mathcal{I}_\sqcup} \cup S_2^{\mathcal{I}_\sqcup}$) there is an interpretation $\mathcal{I}_1$ of $\mathcal{T}_1$ and a domain relation r_{12} such that $\langle \mathcal{I}_1, \mathcal{I}_\sqcup, r_{12} \rangle \models \mathfrak{I}_{\mathcal{SHIQ}}$. For any $(v, w) \in R^{\mathcal{I}}$ we consider the following two cases

(i) $(v, w) \in S_1^{\mathcal{I}}$. Let $\mathcal{I}_1^{R(v,w)}$ be the interpretation (a) of Figure 1, and let
$r_{12}^{R(v,w)} = \{(x, v), (x_1, w)\}$

(ii) $(v, w) \in S_2^{\mathcal{I}}$. Let $\mathcal{I}_1^{R(v,w)}$ be the interpretation (b) of Figure 1, and let
$r_{12}^{R(v,w)} = \{(x, v), (x_1, w)\}$

With no loss of generality, we can assume that the domains of all $\mathcal{I}_1^{R(v,w)}$'s are disjoint. If we define $\mathcal{I}_1$ as the union of each $\mathcal{I}_1^{R(v,w)}$, and r_{12} and the union of $r_{12}^{R(v,w)}$, which formally corresponds to

$$\mathcal{I}_1 = \bigcup_{(v,w) \in R^{\mathcal{I}}} \mathcal{I}_1^{R(v,w)} \qquad r_{12} = \bigcup_{(v,w) \in R^{\mathcal{I}}} r_{12}^{R(v,w)}$$

it is easy to show that $\langle \mathcal{I}_1, \mathcal{I}, r_{12} \rangle \models \mathfrak{I}_{\mathcal{SHIQ}}$.

Theorem 1 says that if we connect two $\mathcal{SHIQ}$ T-boxes with appropriate bridge rules we generate the exact semantics of a generic role union axiom of the form $R \sqsubseteq S_1 \sqcup S_2$, which does not depend upon specific $\mathcal{SHIQ}$ axioms of $\mathcal{T}_t$. This is out of the expressivity of $\mathcal{SHIQ}$ and prevents us from the possibility of finding a complete generalisation of rule (3.2) inside the expressivity of $\mathcal{SHIQ}$.

If we look carefully to the example, we can see that part of the problem is due to the presence of a number restriction axiom in $\mathcal{T}$ and to the fact that number restriction does not propagate across ontologies. To illustrate this problem consider two T-boxes connected with the bridge rule $1 : P \xrightarrow{\equiv} 2 : Q$ and where all objects have at most one $1 : P$ successor. The following distributed interpretation:

satisfies the bride rule and does not propagate the number restriction over $1 : P$ to $2 : Q$. To fix this problem we have several options:

1. Extend the source and target logic to a more expressive DL. This was the choice we made in [5], where the effects of bridge rules are studied in the context of $\mathcal{ALCQI}_b$. As we already said in the introduction the problem of this approach consists in finding effective reasoning procedures.
2. Modify the semantics of DDL to force the domain relation to propagate number restriction, e.g., by imposing cardinality restrictions over r_{ij}. Note that one of the main reasons to define r_{ij} as a relation is the assumption that ontologies can have different domains of interpretation, possibly at different levels of granularity, and that different objects in one ontology can correspond (be abstracted in or collapsed in) a single object in another ontology. On the one hand, a restriction of the domain relation to preserve the granularity of the domains could help us to control granularity. on the other hand, this will also limit the capability to map ontologies with specific domain mismatches due to granularity.
3. Restrict the source and target language to a less expressive DL. In particular remove number restriction and check whether the propagation rules (3.1)–(3.3) can be generalised for some logic weaker than $\mathcal{SHIQ}$. This is the approach we follow in the remaining of the paper.

3.2 Soundness and Completeness of $DDL_o^{\mathcal{SHI}}$

We first generalise rules (3.1)–(3.3) to characterise the effects of the bridge rules in $\mathcal{SHI}$ in a sound and complete manner, then in the next section we use these general rules to define a distributed reasoning tableau for $\mathcal{SHI}$.

$$
\frac{\begin{array}{l} i : A \sqsubseteq \bigsqcup_{k=1}^{n} B_k \\ i : A \xrightarrow{\sqsupseteq} j : C \\ i : B_k \xrightarrow{\sqsubseteq} j : D_k, \text{ for } 1 \leq k \leq n \end{array}}{j : C \sqsubseteq \bigsqcup_{k=1}^{n} D_k} \tag{R1}
$$

$$
\frac{\begin{array}{l} i : P \xrightarrow{\sqsupseteq} j : R \\ i : Q \xrightarrow{\sqsubseteq} j : S \\ i : P \sqsubseteq Q \end{array}}{j : R \sqsubseteq S} \tag{R2}
$$

$$
\frac{\begin{array}{l} i : \exists P.(\neg \bigsqcup_{k=1}^{p} A_k) \sqsubseteq (\bigsqcup_{k=1}^{m} B_k) \\ i : P \xrightarrow{\sqsupseteq} j : R \\ i : A_k \xrightarrow{\sqsubseteq} j : C_k, \text{ for } 1 \leq k \leq p \\ i : B_k \xrightarrow{\sqsubseteq} j : D_k, \text{ for } 1 \leq k \leq m \end{array}}{j : \exists R.(\neg \bigsqcup_{k=1}^{p} C_k) \sqsubseteq (\bigsqcup_{k=1}^{m} D_k)} \tag{R3}
$$

Following the approach illustrated in [11] we define an operator $\mathfrak{B}_{12}(\cdot)$, taking as input a T-box in $\mathcal{DL}_1$ and producing a T-box in $\mathcal{DL}_2$, enriched with the conclusions of rules (R1)–(R3).

Theorem 2. *Let $\mathfrak{T}_{12} = \langle \mathcal{T}_1, \mathcal{T}_2, \mathfrak{B}_{12} \rangle$ be a distributed T-box, where $\mathcal{T}_1$ and $\mathcal{T}_2$ are expressed in the $\mathcal{SHI}$ descriptive language. Then $\mathfrak{T}_{12} \models 2 : X \sqsubseteq Y$ iff $\mathcal{T}_2 \cup \mathfrak{B}_{12}(\mathcal{T}_1) \models X \sqsubseteq Y$.*

The proof can be found in [4].

4   A Distributed Tableaux Algorithm for $\mathcal{SHI}$

In this section we use the previous theoretical results to define a tableaux-based decision procedure for $\mathfrak{T} \models i : X \sqsubseteq Y$, for the case of simple distributed T-boxes of the form $\langle \mathcal{T}_1, \mathcal{T}_2, \mathfrak{B}_{12} \rangle$. The generalization to the case of general acyclic DTBs is similar to the one presented in [11]. With no loss of generality we can assume that the consequences of bridge rules between concepts are atomic[4]. In addition we use the usual notion of axiom internalization, as in [7]: given a T-box $\mathcal{T}_i$, the concept $C_{\mathcal{T}_i}$ is defined as $C_{\mathcal{T}_i} = \bigsqcap_{E \sqsubseteq D \in \mathcal{T}_i} \neg E \sqcup D$; also, the role hierarchy $R_{\mathcal{T}_i}$ contains the role axioms of $\mathcal{T}_i$, plus additional axioms $P \sqsubseteq U$, for each role P of $\mathcal{T}_i$, with U some fresh role.

We follow the approach proposed in [11]: we build a decision procedure which is distributed among the different ontologies, and combines local decision procedures, that is, procedures for testing the satisfiability of concept expressions, in specific ontologies. More in detail, the algorithm for testing j-satisfiability of a concept expression X (i.e., checking $\mathfrak{T} \not\models_{\epsilon} j : X \sqsubseteq \bot$) builds, as usual, a finite representation of a distributed interpretation $\mathfrak{J}$, by running local *autonomous* $\mathcal{SHI}$ tableaux procedures (as decribed in [7]) to find each local interpretation $\mathcal{I}_i$ of $\mathfrak{J}$.

Definition 1. *The function $\mathbf{DTab}_2$ takes as input a concept X and tries to build a representation of $\mathcal{I}_2$ with $X^{\mathcal{I}_2} \neq \emptyset$ (called a completion tree [7]) for the concept $X \sqcap C_{\mathcal{T}_2} \sqcap \forall U.C_{\mathcal{T}_2}$, using the $\mathcal{SHI}$ expansion rules, w.r.t. the role hierarchy $R_{\mathcal{T}_2}$, plus the "bridge" expansion rules described in Figure 2.*

The idea of these rules is inspired by the three propagation rules (R1)–(R3) upon which the operator $\mathfrak{B}_{ij}(.)$ is defined in Section 3.2. Rule Unsat-$\mathfrak{C}_{12}$ corresponds to the usage of rule (R1), and was first introduced in [11]. The idea behind this rule is that whenever $\mathbf{DTab}_2$ encounters a node x that contains a label C which is a consequence of a concept-onto-concept bridge rule, then if $C \sqsubseteq \sqcup\mathbf{D}$ is entailed by the bridge rules, the label $\bigsqcup \mathbf{D}$, is added to x. To determine if $C \sqsubseteq \sqcup\mathbf{D}$ is entailed by the bridge rules in $\mathfrak{B}_{12}$, $\mathbf{DTab}_2$ invokes $\mathbf{DTab}_1$ on the satisfiability of the concept $A \sqcap \neg(\sqcup\mathbf{B})$. $\mathbf{DTab}_1$ will build (independently from $\mathbf{DTab}_2$) an interpretation $\mathcal{I}_1$, as we do in Figure 3.

---

[4] Non atomic mappings can easily be modeled by introducing names for the complex concepts.

Unsat-$\mathfrak{C}_{12}$-rule

> if 1. $C \in \mathcal{L}(x)$, $1 : A \xrightarrow{\exists} 2 : C \in \mathfrak{B}_{12}$,
> $\mathbf{B} \subseteq \{B | 1 : B \xrightarrow{\sqsubseteq} 2 : D \in \mathfrak{B}_{12}\}$,
> $\mathbf{D} = \{D | 1 : B \xrightarrow{\sqsubseteq} 2 : D \in \mathfrak{B}_{12} \text{ and } B \in \mathbf{B}\}$, and
> 2. $IsSat_1(A \sqcap \neg \bigsqcup \mathbf{B}) = \textit{False}$ and $\mathbf{D} \not\subseteq \mathcal{L}(x)$,
> then $\mathcal{L}(x) \longrightarrow \mathcal{L}(x) \cup \{\bigsqcup \mathbf{D}\}$

New-$\mathfrak{C}_{12}$-rule

> if 1. $C \in \mathcal{L}(x)$, $1 : A \xrightarrow{\exists} 2 : C \in \mathfrak{B}_{12}$,
> $\mathbf{B} \subseteq \{B | 1 : B \xrightarrow{\sqsubseteq} 2 : D \in \mathfrak{B}_{12}\}$,
> 2. for no $\mathbf{B}' \subseteq \mathbf{B}$ is $IsSat_1(A \sqcap \neg \bigsqcup \mathbf{B}') = \textit{False}$, and
> 3. for no $\mathbf{B}' \supseteq \mathbf{B}$ is $IsSat_1(A \sqcap \neg \bigsqcup \mathbf{B}') = \textit{True}$,
> then $IsSat_1(A \sqcap \neg \bigsqcup \mathbf{B}) = \begin{cases} \textit{True}, & \text{if } \mathbf{DTab}_1(A \sqcap \neg \bigsqcup \mathbf{B}) = \textit{Satisfiable} \\ \textit{False}, & \text{if } \mathbf{DTab}_1(A \sqcap \neg \bigsqcup \mathbf{B}) = \textit{Unsatisfiable} \end{cases}$

Unsat-$\mathfrak{R}_{12}$-rule

> if 1. y is a R-neighbour of x, $1 : P \xrightarrow{\exists} 2 : R \in \mathfrak{B}_{12}$, and
> 2. $P \sqsubseteq Q \in R_{\mathcal{T}_1}$, with $1 : Q \xrightarrow{\sqsubseteq} 2 : S \in \mathfrak{B}_{12}$
> then y is a S-neighbour of x

Unsat-$\mathfrak{CR}_{12}$-rule

> if 1. y is a R-neighbour of x, $1 : P \xrightarrow{\exists} 2 : R \in \mathfrak{B}_{12}$,
> $\mathbf{B} \subseteq \{B | 1 : B \xrightarrow{\sqsubseteq} 2 : D \in \mathfrak{B}_{12}\}$,
> $\mathbf{D} = \{D | 1 : B \xrightarrow{\sqsubseteq} 2 : D \in \mathfrak{B}_{12} \text{ and } B \in \mathbf{B}\}$, and
> $\mathbf{A} \subseteq \{A | 1 : A \xrightarrow{\sqsubseteq} 2 : C \in \mathfrak{B}_{12}\}$,
> $\mathbf{C} = \{C | 1 : A \xrightarrow{\sqsubseteq} 2 : C \in \mathfrak{B}_{12} \text{ and } A \in \mathbf{A}\}$, and
> 2. $\neg \bigsqcup \mathbf{C} \subseteq \mathcal{L}(y)$, and
> 3. $IsSat_1(\exists R. \neg \bigsqcup \mathbf{A} \sqcap \neg \bigsqcup \mathbf{B}) = \textit{False}$ and $\mathbf{D} \not\subseteq \mathcal{L}(x)$,
> then $\mathcal{L}(x) \longrightarrow \mathcal{L}(x) \cup \{\bigsqcup \mathbf{D}\}$

New-$\mathfrak{CR}_{12}$-rule

> if 1. y is a R-neighbour of x, $1 : P \xrightarrow{\exists} 2 : R \in \mathfrak{B}_{12}$,
> $\mathbf{A} \subseteq \{A | 1 : A \xrightarrow{\sqsubseteq} 2 : C \in \mathfrak{B}_{12} \text{ and } \neg C \in \mathcal{L}(y)\}$, and
> 3. $\mathbf{B} \subseteq \{B | 1 : B \xrightarrow{\sqsubseteq} 2 : D \in \mathfrak{B}_{12}\}$, and
> 3. for no $\mathbf{B}' \subseteq \mathbf{B}$ is $IsSat_1(\exists R. \neg \bigcup \mathbf{A} \sqcap \neg \bigsqcup \mathbf{B}') = \textit{False}$, and
> 4. for no $\mathbf{B}' \supseteq \mathbf{B}$ is $IsSat_1(\exists R. \neg \bigcup \mathbf{A} \sqcap \neg \bigsqcup \mathbf{B}') = \textit{True}$,
> then $IsSat_1(\exists R. \neg \bigcup \mathbf{A} \sqcap \neg \bigsqcup \mathbf{B}) = \begin{cases} \textit{True}, & \text{if } \mathbf{DTab}_1(\exists R. \neg \bigcup \mathbf{A} \sqcap \neg \bigsqcup \mathbf{B}) = \textit{Satisfiable} \\ \textit{False}, & \text{if } \mathbf{DTab}_1(\exists R. \neg \bigcup \mathbf{A} \sqcap \neg \bigsqcup \mathbf{B}) = \textit{Unsatisfiable} \end{cases}$

Fig. 2. Additional expansion rules for $\mathbf{DTab}_2$

Rule Unsat-$\mathfrak{R}_{12}$ corresponds to the rule (R2). The idea behind this rule is that whenever $\mathbf{DTab}_2$ encounters a node x with a R-neighbour y, and R is a consequence of a role-onto-role bridge rule $1 : P \xrightarrow{\exists} 2 : R$, then if $R \sqsubseteq S$ can be entailed with the help of role-into-role bridge rules of the form $1 : Q \xrightarrow{\sqsubseteq} 2 : S$, and therefore y is also a S-neighbour of x. To determine if $R \sqsubseteq S$ is entailed by the bridge rules $\mathfrak{B}_{12}$, we simply check if $P \sqsubseteq Q$ belongs to the role hierarchy of $\mathcal{T}_1$ for all role-into-role bridge rules $1 : Q \xrightarrow{\sqsubseteq} 2 : S \in \mathfrak{B}_{12}$. This because $\mathcal{SHI}$ does not allow any reasoning on roles.

Rule Unsat-$\mathfrak{CR}_{12}$ corresponds to the rule (R3). The idea behind this rule is that whenever $\mathbf{DTab}_2$ encounters a node x with a R-neighbour y, and R is a consequence of a role-onto-role bridge rule $1 : P \xrightarrow{\exists} 2 : R$, then if y contains a label $\neg C_i$ which is a consequence of a concept-into-concept bridge rule $1 : A \xrightarrow{\sqsubseteq} 2 : C$, and $\exists R.\neg C \sqsubseteq D$ is entailed by the bridge rules, then the label D, is added to x. To determine if $\exists R.\neg C \sqsubseteq D$ is entailed by the bridge rules $\mathfrak{B}_{12}$, $\mathbf{DTab}_2$ invokes $\mathbf{DTab}_1$ on the satisfiability of the concept $\exists P.\neg A \sqcap \neg B$. $\mathbf{DTab}_1$ will build (independently from $\mathbf{DTab}_2$) an interpretation $\mathcal{I}_1$, in a manner similar to the construction illustrated in Figure 3.

To avoid redundant calls, $\mathbf{DTab}_2$ caches the calls to $\mathbf{DTab}_1$ in a data structure $IsSat_1$, which caches the subsumption propagations that have been computed so far.

Theorem 3 (Termination,Soundness, and completeness). $\mathbf{DTab}_2(X)$ *terminates, and* $2{:}X$ *is satisfiable in* $\mathfrak{T}_{12}$ *if and only if* $\mathbf{DTab}_2(X)$ *can generate a complete and clash-free completion tree.*

The proof is contained in [4]. To clarify how the distributed tableaux works we present here a specific example.

Example 2. To clarify how the distributed tableaux works let us consider the DTBox $\mathfrak{T}_{12} = \langle \mathcal{T}_1, \mathcal{T}_2, \mathfrak{B}_{12} \rangle$ where $\mathcal{T}_1$ contains the axioms $\exists R.\neg A_1 \sqsubseteq B_1$ and $\exists R.\neg A_2 \sqsubseteq B_2$, $\mathcal{T}_2$ does not contain any axiom, and $\mathfrak{B}_{12}$ contains :

$$1 : R \xrightarrow{\exists} 2 : S \qquad (4.1)$$

$$1 : A_1 \xrightarrow{\sqsubseteq} 2 : G_1 \qquad (4.2) \qquad\qquad 1 : B_1 \xrightarrow{\sqsubseteq} 2 : H_1 \qquad (4.4)$$

$$1 : A_2 \xrightarrow{\sqsubseteq} 2 : G_2 \qquad (4.3) \qquad\qquad 1 : B_2 \xrightarrow{\sqsubseteq} 2 : H_2 \qquad (4.5)$$

Let us show that $\mathfrak{T}_{12} \models_d 2 : \exists S.\neg(G_1 \sqcup G_2) \sqsubseteq H_1 \sqcap H_2$, i.e. that for any distributed interpretation $\mathfrak{I} = \langle \mathcal{I}_1, \mathcal{I}_2, r_{12} \rangle$, $(\exists S.\neg(G_1 \sqcup G_2))^{\mathcal{I}_2} \subseteq (H_1 \sqcap H_2)^{\mathcal{I}_2}$.

1. Suppose that by contradiction there is an $x \in \Delta_2$ such that $x \in (\exists S.\neg(G_1 \sqcup G_2))^{\mathcal{I}_2}$ and $x \notin (H_1 \sqcap H_2)^{\mathcal{I}_2}$.
2. Then there exists a y such that $\langle x, y \rangle \in S^{\mathcal{I}_2}$ and $y \notin G_1^{\mathcal{I}_2}$, $y \notin G_2^{\mathcal{I}_2}$, and either $x \notin H_1^{\mathcal{I}_2}$ or $x \notin H_2^{\mathcal{I}_2}$.
3. Because of the bridge rule (4.1) there is a pair $\langle x', y' \rangle \in \Delta_1$ such that $\langle r_{12}(x'), r_{12}(y') \rangle = \langle x, y \rangle$ and $\langle x', y' \rangle \in R^{\mathcal{I}_1}$.
4. Let us consider the case where $x \notin H_1^{\mathcal{I}_2}$. From the fact that $y \notin G_1^{\mathcal{I}_2}$, then by the bridge rule (4.2), $y' \notin A_1^{\mathcal{I}_1}$.
5. Since $\exists R.\neg A_1 \sqsubseteq B_1$ is an axiom of $\mathcal{T}_1$, then $y' \in B_1^{\mathcal{I}_1}$, and by bridge rule (4.4) $y \in H_1^{\mathcal{I}_1}$. But this is a contradiction.
6. The case where $x \notin H_2^{\mathcal{I}_2}$ is analogous and we can conclude that $\mathfrak{T}_{12} \models_d 2 : \exists S.\neg(G_1 \sqcup G_2) \sqsubseteq H_1 \sqcap H_2$.

The above reasoning can be seen as a combination of a tableau in $\mathcal{T}_2$ with a tableau in $\mathcal{T}_1$. In Figure 3 we depict the construction of a tableau for $\mathcal{T}_2$ where

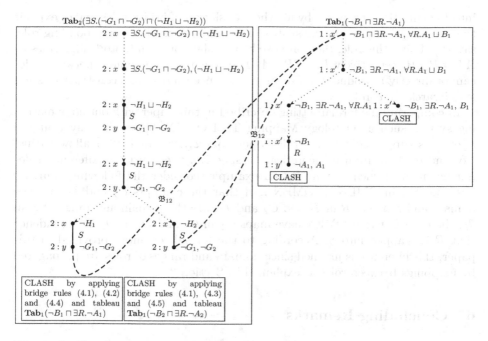

Fig. 3. An illustration of distributed tableaux for computation of DDL subsumption

dotted arrow lines represent the evolution of the tree while solid lines represent connections between neighbouring nodes. For the sake of space we depict only the tableau $\mathbf{Tab}_1(\neg B_1 \sqcap \exists R.\neg A_1)$. The tableau $\mathbf{Tab}_1(\neg B_2 \sqcap \exists R.\neg A_2)$ is analogous.

5   Related Work

The semantics of mappings proposed in this paper is in line with the semantic of mappings between relations proposed by Calvanese et al. in [2]. The only difference between our approach is that DDL admits multiple domains connected via domain relations rather than a single domain.

Role mappings can be encoded also in a unique top global ontology. To encode the semantics of role-mappings with heterogeneous domain, one needs to index each concept and role with the T-box it comes from, add concepts for local domains, and introduce a role R_{12} representing the domain relation from T-box 1 to T-box 2. The into-mapping $1 : R \xrightarrow{\sqsubseteq} 2 : S$ can be encoded with the role-axioms $R_{12}^- \circ R_1 \circ R_{12} \sqsubseteq S_2$. Similarly the onto mapping $1 : R \xrightarrow{\sqsupseteq} 2 : S$ is encoded with $S_2 \sqsubseteq R_{12}^- \circ R_1 \circ R_{12}$. The main drawback of this translation is that it is based on role composition, which in the general case is undecidable.

The formalism of $\mathcal{E}$-connections, proposed by Wolter et. al in, e.g., [8] allows to express mappings between concepts in different DL-based ontologies. To express mappings between roles in the formalism of $\mathcal{E}$-connections we need to consider the operator of role composition. Indeed, if we consider the translation of DDL into $\mathcal{E}$-connections proposed by [6], the translation of bridge rules makes explicit the domain relation r_{ij} in the syntax of $\mathcal{E}$-connections via a corresponding role, say R_{ij}. Using this role, they propose to translate onto and into bridge rules as $A \sqsubseteq \forall R_{ij}.B$ respectively $B \sqsubseteq \exists R_{ij}^-.A$. The semantic of mappings between roles cannot directly reproduced in $\mathcal{E}$-connection since there are no construct to state isa-hierarchy of $\mathcal{E}$-roles.

In comparing the formal setting described in this paper with ontology matching system such as Ontology Mapping Tool OMEN [9], we can say that the inferences supported by the bridge operators $\mathfrak{C}_{ij}, \mathfrak{R}_{ij}$ and $\mathfrak{C}\mathfrak{R}_{ij}$ allow to justify and explain, from a theoretical perspective, the heuristics inference rules implemented in such system. As an example, consider the following heuristic implemented in OMEN: If OMEN find that the concepts C_1 and D_1 are the domain and range of R in $\mathcal{T}_1$ and C_2 and D_2 are the domain and range of S in $\mathcal{T}_2$, then the fact that OMEN have maps C_1 in C_2 and D_1 in D_2 is an evidence that R is mapped into S. According to the formal semantics provided in this paper, this inference is justified since domain and ranges of roles are propagated by mappings between roles as explained in Section 3.

6   Concluding Remarks

The language, the semantics and the decision procedure presented in this paper constitute a genuine contribution in the direction of the axiomatization and reasoning support for ontology mapping. The language proposed in this paper makes it possible to directly bind relations in different ontologies as well as concepts. We have shown that unrestricted mapping of $\mathcal{SHIQ}$ roles leads to undecidability and generates inclusion axioms between unions of roles. To solve these problems we have restricted ourselves to the case of $\mathcal{SHI}$. The complete characterization for $\mathcal{SHI}$, together with the decision procedure presented in the paper, make the logic ready to be implemented in a reasoning system for distributed ontologies, implementation that we leave for future work.

Acknowledgments

Thanks to Sergio Tessaris for many useful insights on Description Logics. Thanks to Ulrike Sattler for comments about decidability in $\mathcal{SHIQ}$.

References

1. Baader, F., Calvanese, D., McGuinness, D.L., Nardi, D., Patel-Schneider, P.F. (eds.): The Description Logic Handbook: Theory, Implementation, and Applications. Cambridge University Press, Cambridge (2003)
2. Calvanese, D., Giacomo, G.D., Lenzerini, M., Rosati, R.: Logical foundations of peer-to-peer data integration. In: Proc. of the 23rd ACM SIGACT SIGMOD SIGART Sym. on Principles of Database Systems (PODS 2004), pp. 241–251 (2004)

3. Euzenat, J., Scharffe, F., Zimmermann, A.: Expressive alignment language and implementation. deliverable D2.2.10, Knowledge Web NoE (2007)
4. Ghidini, C., Serafini, L.: Mapping properties of heterogeneous ontologies. Technical Report 200804002, FBK-irst (2008), http://dkm.fbk.eu
5. Ghidini, C., Serafini, L., Tessaris, S.: On relating heterogeneous elements from different ontologies. In: Kokinov, B., Richardson, D.C., Roth-Berghofer, T.R., Vieu, L. (eds.) CONTEXT 2007. LNCS (LNAI), vol. 4635, pp. 234–247. Springer, Heidelberg (2007)
6. Grau, B.C., Parsia, B., Sirin, E.: Combining owl ontologies using e-connections. Journal of Web Semantics 4(1) (2005)
7. Horrocks, I., Sattler, U.: A description logic with transitive and inverse roles and role hierarchy. Journal of Logic and Computation 9(3), 385–410 (1999)
8. Kutz, O., Lutz, C., Wolter, F., Zakharyaschev, M.: E-connections of abstract description systems. Artificial Intelligence 156(1), 1–73 (2004)
9. Mitra, P., Noy, N.F., Jaiswal, A.R.: Omen: A probabilistic ontology mapping tool, Hisroshima, Japan (2004)
10. Scharffe, F., de Bruijn, J.: A language to specify mappings between ontologies. In: Proceedings of the Internet Based Systems IEEE Conference (SITIS 2005) (2005)
11. Serafini, L., Borgida, A., Tamilin, A.: Aspects of distributed and modular ontology reasoning. In: 19th Joint Conference on Artificial Intelligence (IJCAI 2005), August 2005, pp. 570–575 (2005)
12. Stuckenschmidt, H., Uschold, M.: Representation of semantic mappings. In: Semantic Interoperability and Integration, Schloss Dagstuhl, Germany. Dagstuhl Seminar Proceedings. Internationales Begegnungs- und Forschungszentrum (IBFI), vol. 04391 (2005)
13. Tobies, S.: Complexity Results and Practical Algorithms for Logics in Knowledge Representation. PhD thesis, RWTH Aachen, Germany (2001)
14. Vardi, M., Wolper, P.: Reasoning about infinite computations. Information and Computation 115, 1–37 (1994)

A Logical Approach to
Dynamic Role-Based Access Control

Philippe Balbiani, Yannick Chevalier, and Marwa El Houri

Institut de recherche en informatique de Toulouse, Toulouse University, France

Abstract. Since its formalization RBAC has become the yardstick for the evaluation of access control formalisms. In order to meet organizational needs, it has been extended along several directions: delegation, separation of duty, history-based access control, etc. We propose in this paper an access control language in which RBAC and all the above-listed extensions can be encoded. In contrast with Cassandra, we have not promoted role management mechanism to first-class citizenship, and have based our model on the assumption that access control systems could be separated into a dynamic part that evolves according to actions performed by users and a static part. We solve in this paper decision problems related to access control for policies expressed in this language.

Keywords: Access control language, RBAC, privacy policy.

1 Introduction

The academic foundations of the access control problems have been formalized in [4,8], where an information system is defined by a set of *objects*, a set of *subjects* performing *operations* on objects, and a finite number of relations called *rights* between subjects and objects. This system has later been refined with *Role-based access control* (RBAC) [6], where subjects are organized in groups called roles, and these groups are *hierarchically* structured. This structure permits one to define an inheritance of rights from one role to another, and thus to express complex policies in a less error-prone way. Several extensions of this "core RBAC" policy expression language have been developed over the years to address needs arising from real-case problems such as role hierarchy, separation of duty and delegation.

RBAC, along with several of its extensions, was a core component in many highly expressive policy languages. The notion of delegation was explored in the works of [1,11], in which delegation was seen as a subject who grants some access control rights to another one. In [9] one distinguished between delegation from a role to another and delegation from a subject active in a role to another subject active in another role to avoid potential inconsistencies. Actually, a study on a large information system [3] aiming at implementing a security policy in Cassandra [2] has shown that different flavors of delegation had to be employed in different parts of the system. Role hierarchy and separation of duty in its static

D. Dochev, M. Pistore, and P. Traverso (Eds.): AIMSA 2008, LNAI 5253, pp. 194–208, 2008.

and dynamic aspects were also subject of study, wether explicitly as extensions of the *RT* language [9] or within the Cassandra policy specification.

However we believe that such policy languages although highly expressive in terms of access control management, lack a dynamic aspect. In fact in Cassandra one can see a dynamic aspect in terms of activation and deactivation of a role as a mean to step in and out of a role and thus acquire the adequate rights, but the effect of performing an action other that activating a role cannot be specified within the policy language. We propose in this paper a language that takes into account RBAC extensions but also introduces a dynamic aspect to the expression of access control rules.

The concept underlying this language is that an access control system is characterised by decision contexts and an invariant datalog program. A decision context is defined by a set of permissions. Within each decision context, an access control decision is based on the computation of whether the requested permission is obtainable using the datalog program from the set of permissions defining the current decision context. The access control system evolves from one decision context to another according to actions performed by a user. A drawback of this simplicity is the declaration of every action that can alter the decision context, including *e.g.* the action stating that a client is active in a given role.

In spite of its simplicity, this model permits us to express seamlessly core RBAC policies as well as their different extensions presented above. It is similar in spirit to Cassandra, though we have not built-in role management: being active in a role is an action similar to the invocation of a service. Finally, in contrast with Transaction Logic [5] the side effects are not attached to a rule but to an effective action of the client. The consequence of this choice is the determinism of the system w.r.t. client's actions (instead of client's choice of rules to apply).

In Sect. 2 we present a novel language for expressing access control policies. Then, in Sect. 3 we present how RBAC can be encoded into this language. Section 4 is devoted to the definition and complexity analysis of decision problems related to access control in our language.

2   Access Control Policies

In this section, we present our framework for expressing role-based access control policies and their extensions.

2.1   Domains

Active processes acting on behalf of users are referred to as subjects whereas passive resources accessible on a computer system are referred to as objects. A key feature of our access control policies is that all actions are done through roles, i.e. subjects receive permissions to execute actions on objects only through the roles to which they are assigned. In our policies, following the notions considered in RBAC, subjects are organized in groups called roles, and these groups are hierarchically structured. In our setting, a domain is a tuple

$$\mathcal{D} = \langle S, A, O, R \rangle$$

such that S is a set of subjects, A is a set of actions, O is a set of objects and R is a set of roles. We will assume that:

- $S \subseteq O$ with S and R pairwise disjoint,
- A and O are pairwise disjoint.

2.2   Security States

Consider a domain $\mathcal{D} = \langle S, A, O, R \rangle$. A security state based on $\mathcal{D}$ is a subset Π of $S \times A \times O \times R$. For all s in S, for all a in A, for all o in O and for all r in R, we will write $\Pi(s, a, o, r)$ instead of $(s, a, o, r) \in \Pi$. $\Pi(s, a, o, r)$ means that "subject s has in Π the permission to execute action a on object o through role r". A primary relation of interest between security states is that of set inclusion, under which the set of security states based on $\mathcal{D}$ forms a complete lattice.

2.3   Atomic Formulae and Conditions

Consider a domain $\mathcal{D} = \langle S, A, O, R \rangle$ and a security state Π based on $\mathcal{D}$. We assume an alphabet of variable symbols: X, Y, etc, possibly with subscripts. A term based on $\mathcal{D}$ is either an element of $S \cup A \cup O \cup R$ or a variable symbol. An interpretation function for $\mathcal{D}$ is a function I mapping the variable symbols to elements of $S \cup A \cup O \cup R$. The value $\tilde{I}(t)$ of a term t is defined as follows:

- if t is an element of $S \cup A \cup O \cup R$ then $\tilde{I}(t) = t$,
- if t is a variable symbol then $\tilde{I}(t) = I(t)$.

A 4-tuple (t_1, t_2, t_3, t_4) of terms based on $\mathcal{D}$ is said to be correct iff the following conditions are satisfied:

- t_1 is either a variable symbol or an element of S,
- t_2 is either a variable symbol or an element of A,
- t_3 is either a variable symbol or an element of O,
- t_4 is either a variable symbol or an element of R.

An atomic formula based on $\mathcal{D}$ is an expression of the form $P(t_1, t_2, t_3, t_4)$ were (t_1, t_2, t_3, t_4) is a correct 4-tuple of terms based on $\mathcal{D}$. We define the well-formed conditions (ϕ, ψ, etc, possibly with subscripts) based on $\mathcal{D}$ by the rule

$$\phi ::= P(t_1, t_2, t_3, t_4) \mid \bot \mid \top \mid (\phi_1 \vee \phi_2) \mid (\phi_1 \wedge \phi_2).$$

The satisfiability relation $\Pi, I \models \phi$ between a security state Π, an interpretation function I and a condition ϕ is defined as follows:

- $\Pi, I \models P(t_1, t_2, t_3, t_4)$ iff $\Pi(\tilde{I}(t_1), \tilde{I}(t_2), \tilde{I}(t_3), \tilde{I}(t_4))$,
- $\Pi, I \not\models \bot$,
- $\Pi, I \models \top$,
- $\Pi, I \models \phi_1 \vee \phi_2$ iff $\Pi, I \models \phi_1$ or $\Pi, I \models \phi_2$,
- $\Pi, I \models \phi_1 \wedge \phi_2$ iff $\Pi, I \models \phi_1$ and $\Pi, I \models \phi_2$.

2.4   Static Clauses and Static Policies

Security states are dynamic in nature, i.e. they are likely to change over time in reflection of ever evolving environmental conditions. This observation leads us to the central concept of the paper: rule-based access control policies. Rule-based access control policies will be built up using static clauses and dynamic clauses. We define in this section the concept of static clauses whereas we define in the next section the concept of dynamic clauses. Consider a domain $\mathcal{D} = \langle S, A, O, R \rangle$ and a security state Π based on $\mathcal{D}$. A static clause based on $\mathcal{D}$ is an expression of the form

$$P(t_1, t_2, t_3, t_4) \leftarrow \phi.$$

For example, the expression $P(X, a, Y, r) \leftarrow P(X, a, o, r)$ is a static clause. It says that "if X has the permission to execute a on o through r then X has the permission to execute a on Y through r". A static policy based on $\mathcal{D}$ is a finite set SP of static clauses based on $\mathcal{D}$. We shall say that Π is a model of SP, in symbols $\Pi \models SP$, iff

- for all interpretation functions I for $\mathcal{D}$ and for all static clauses $P(t_1, t_2, t_3, t_4)$ $\leftarrow \phi$ in SP, if $\Pi, I \models \phi$ then $\Pi, I \models P(t_1, t_2, t_3, t_4)$.

The reader may easily verify that the set $\{\Pi : \Pi \models SP\}$ has a least element under set inclusion. Let $l(SP)$ be this least element.

2.5   Dynamic Clauses and Dynamic Policies

So far, there is nothing really new. But a simple idea is going to ensure that security states change over time: dynamic policies. If an action is permitted, it is not necessarily executed. As such the dynamic policy manages the consequence of executing (or not) a permitted action.

Consider a domain $\mathcal{D} = \langle S, A, O, R \rangle$ and security states Π, Π' based on $\mathcal{D}$. A dynamic clause based on $\mathcal{D}$ is an expression of the form

$$\phi \rightarrow (\psi_1, \psi_2)$$

where neither ψ_1 nor ψ_2 contain occurrences of the Boolean connective $\vee$. The dynamic clause $\phi \rightarrow (\psi_1, \psi_2)$ is said to be consistent iff neither ψ_1 nor ψ_2 contain occurrences of the Boolean connective $\bot$. For example, the dynamic clause $P(X, a, Y, r) \rightarrow (P(X, a, o, r), \top)$ is consistent. It says that "if X has the permission to execute a on Y through r then either X executes a on Y through r and X next obtain the permission to execute a on o through r or X does not execute a on Y through r". Informally, each dynamic clause $\phi \rightarrow (\psi_1, \psi_2)$ defines a transition relation from a security state Π to a security state Π' as follows:

- If ϕ is satisfied in Π then
    - if every action in ϕ is executed then ψ_1 will be true in the next state Π'
    - if some action in ϕ is not executed then ψ_2 will be true in Π'
- else the rule is not applied.

As such, let $\mathcal{A} \subseteq \Pi$ denote the set of the permitted actions that are actually executed. A (consistent) dynamic policy based on $\mathcal{D}$ is a finite set DP of (consistent) dynamic clauses based on $\mathcal{D}$. We shall say that the pair (Π, Π') is a transition of DP through $\mathcal{A}$, in symbols $\Pi \Longrightarrow^{\mathcal{A}}_{DP} \Pi'$, iff

- for all interpretation functions I for $\mathcal{D}$ and for all dynamic clauses $\phi \rightarrow (\psi_1, \psi_2)$ in DP, if $\Pi, I \models \phi$ then either $\mathcal{A}, I \models \phi$ and $\Pi', I \models \psi_1$ or $\mathcal{A}, I \not\models \phi$ and $\Pi', I \models \psi_2$.

The reader may easily verify that if the set $\{\Pi' : \Pi \Longrightarrow^{\mathcal{A}}_{DP} \Pi'\}$ is not empty, then it has a least element under set inclusion. Let $L(\Pi, DP, \mathcal{A})$ be this least element.

2.6   Rule-Based Policies

Consider a domain $\mathcal{D} = \langle S, A, O, R \rangle$ and security states Π, Π' based on $\mathcal{D}$. A (consistent) rule-based access control policy based on $\mathcal{D}$ is a tuple

$$\mathcal{P} = \langle SP, DP \rangle$$

whose first component is a static policy based on $\mathcal{D}$ and second component is a (consistent) dynamic policy based on $\mathcal{D}$. For all subsets $\mathcal{A}$ of Π, we shall say that the pair (Π, Π') is a transition of $\mathcal{P}$ through $\mathcal{A}$, in symbols $\Pi \Longrightarrow^{\mathcal{A}}_{\mathcal{P}} \Pi'$, iff

- for all interpretation functions I for $\mathcal{D}$ and for all dynamic clauses $\phi \rightarrow (\psi_1, \psi_2)$ in DP, if $\Pi, I \models \phi$ then either $\mathcal{A}, I \models \phi$ and $\Pi', I \models \psi_1$ or $\mathcal{A}, I \not\models \phi$ and $\Pi', I \models \psi_2$,
- $\Pi' \models SP$.

The reader may easily verify that the if the set $\{\Pi' : \Pi \Longrightarrow^{\mathcal{A}}_{\mathcal{P}} \Pi'\}$ is not empty, then it has a least element under set inclusion. Let $L(\Pi, \mathcal{P}, \mathcal{A})$ be this least element.

3   RBAC Features

In this section, we present an encoding of RBAC and some of its extensions into our access control language.

3.1   Terminology

In the rest of this section, for the purpose of characterizing RBAC features, we consider special actions. Namely, the special actions can-play and is-active express role membership and role activation respectively. The special action Acquire denotes the acquiring of permissions relative to a role. The special actions delegate and d-play express delegation of a role and playing the role by delegation respectively.

3.2   Role Activation

The essential notion in RBAC is that permissions are associated with roles and users are assigned to appropriate roles. To this end we define role membership by the predicate $P(X, \text{can-play}, X, r)$ saying that subject X has the permission to play role r. On the other hand $P(X, \text{is-active}, X, r)$ expresses that X is currently active in role r. The assignment of permissions to roles is expressed with the special action Acquire in the body of rules as follows:

$$P(X, a, o, r) \leftarrow P(X, \text{Acquire}, X, r)$$

That is, if user X has acquired the permissions associated with role r then X will have the permission to do action a on the object o. There is one rule for each triple (a, o, r) in the permission-role assignment relation.

One way for user X to acquire the permissions associated with role r is to activate the role r. This is done by executing the action can-play, and induces the addition of $P(X, \text{is-active}, X, r)$ at the next state. This is expressed by the two rules:

$$\begin{cases} P(X, \text{can-play}, X, r) \rightarrow (P(X, \text{is-active}, X, r), \top) \\ P(X, \text{Acquire}, X, r) \leftarrow P(X, \text{is-active}, X, r) \end{cases}$$

The user X can step out of a role r by simply choosing not to activate $P(X, \text{can-play}, X, r)$. In this case she automatically looses all privileges associated with role r in the next state.

Finally we impose that when user X becomes active in the role r, she must acquire the associated permissions, and this acquisition is modeled by explicit actions. This is guaranteed by the following two dynamic rules:

$$\begin{cases} P(X, \text{is-active}, X, r) \rightarrow (\top, \bot) \\ P(X, \text{Acquire}, X, r) \rightarrow (\top, \bot) \end{cases}$$

If the actions is-active and Acquire are not executed the system enters in an inconsistent state. We note that in a real system, these mandatory actions can be performed on a server as a consequence of the explicit actions of a client.

3.3   Role Hierarchy

Role hierarchy is very useful in structuring roles within a certain organization. As such role, r_1 will be junior to role r_2 if the permissions associated with r_1 are inherited by members of r_2. We express this by the rule

$$P(X, \text{Acquire}, X, r_1) \leftarrow P(X, \text{Acquire}, X, r_2)$$

For example, a cardiologist can inherit the permissions associated with role intern. This can be expressed by

$$P(X, \text{Acquire}, X, \text{intern}) \leftarrow P(X, \text{Acquire}, X, \text{cardiologist})$$

If $P(\text{Mary}, \text{can-play}, \text{Mary}, \text{cardiologist})$ is true in SP and if Mary activates the role cardiologist, then $P(\text{Mary}, \text{Acquire}, \text{Mary}, \text{cardiologist})$ will be true in the next state Π'. In this case $P(\text{Mary}, \text{AcquireMary}, \text{intern})$ will also be true in Π'. That is Mary automatically acquires the permissions for role intern.

3.4   Role Delegation

Delegation is the act of authorizing or requesting someone to act on one's behalf. In order to be able to delegate a role r, an entity should be active in some role r_1 or allowed to the set of permissions associated with that role:

$$P(X, \text{delegate}, Y, r) \leftarrow P(X, \text{Acquire}, X, r_1) \wedge P(Y, \text{can-play}, Y, r_2)$$

The dynamic rules

$$\begin{cases} P(X, \text{delegate}, Y, r) \rightarrow (P(Y, \text{d-play}, Y, r), \top) \\ P(X, \text{d-play}, X, r) \rightarrow (P(X, \text{Acquire}, X, r), \top) \end{cases}$$

grant the permission for Y to play the role by delegation. If Y chooses to activate this delegation, Y will be active in the role r at the next state. Note that if X chooses not to activate the action delegate, Y looses his privileges at the next state.

3.5   Separation of Duties

The separation of duty principle can be seen in both its static and dynamic aspects. In the dynamic separation of duty, a subject may have the permission to play two mutually exclusive roles, but can become active in only one of them. For example, a subject X can be both a doctor and a patient at a hospital, however X will not have the right to activate the role doctor if X is currently playing the role patient in the same hospital. We express this constraint as follows:

$$P(X, \text{Acquire}, X, \text{doctor}) \wedge P(X, \text{Acquire}, X, \text{patient}) \rightarrow (\bot, \top)$$

Note that $P(X, \text{Acquire}, X, r)$ is true only if a role r is activated, inherited or delegated, that is if X is active in the role. In that case no matter what X does, the system enters into an inconsistent state.

In the static separation of duty, a subject having the right to play the role teller in a bank for example will not be allowed to be a member of the role auditor of the same bank. A subject is considered as member of a role if she has the permission to play the role, was delegated the role or inherited the permissions associated with the role. Taking that into consideration, we define a special action $\widehat{\text{play}}$ such that

$$P(X, \widehat{\text{play}}, X, r) \leftarrow P(X, a, X, r) \text{ for } a \in \{\text{play}, \text{d-play}, \text{Acquire}\}$$

Then the static separation of duty can be expressed by the rule:

$$P(X, \widehat{\text{play}}, X, \text{teller}) \wedge P(X, \widehat{\text{play}}, X, \text{auditor}) \rightarrow (\bot, \bot)$$

If the subject X acquires both permissions at the same state of the dynamic model, then the system will enter into an inconsistent state, no matter what X decides to do.

3.6   Synchronization of Actions

By synchronization, we mean the necessity to execute two actions at the same time. For example, to open the safe in a bank both an agent and a manager should enter a password otherwise the system would block:

$$P(X, \text{enter}, \text{pswrd}, \text{agent}) \wedge P(Y, \text{enter}, \text{pswrd}, \text{manager}) \rightarrow (\top, \bot)$$

In this case we do not take into consideration the difference between not executing any action or executing only one action. However, in real life, executing only one may be considered as an intrusion in the system and should not be accepted. To express this case, the above rule must be replaced by the following set of rules:

$$P(X, \text{enter}, \text{pswrd}, \text{agent}) \wedge P(Y, \text{enter}, \text{pswrd}, \text{manager}) \rightarrow (\top, p),$$
$$P(X, \text{enter}, \text{pswrd}, \text{agent}) \rightarrow (p_{\text{agent}}, \top),$$
$$P(Y, \text{enter}, \text{pswrd}, \text{manager}) \rightarrow (p_{\text{manager}}, \top),$$
$$p \wedge (p_{\text{agent}} \vee p_{\text{manager}}) \rightarrow (\bot, \bot).$$

where p, p_{agent} and p_{manager} are ground predicates that will only be used in the last rule above. This will guarantee that if the agent or the manager execute the action alone, then the system will enter into an inconsistent state.

4   Assigning Permissions

This section is an encounter with the computational complexity of assigning permissions. The proofs of our results are given in the appendix. See [10] for details concerning computational complexity.

4.1   Static Assignments

The $STATIC(\exists)$ problem is the following decision problem:

- $STATIC(\exists)$: given a domain $\mathcal{D}$, a static policy SP based on $\mathcal{D}$ and a condition ϕ based on $\mathcal{D}$, determine whether there exists an interpretation function I for $\mathcal{D}$ such that $l(SP), I \models \phi$.

Proposition 1. $STATIC(\exists)$ is NP-complete.

4.2   Dynamic Assignments

The $DYNAMIC(\exists, \exists)$ problem and the $DYNAMIC_{con}(\exists, \exists)$ problem are the following decision problems:

- $DYNAMIC(\exists, \exists)$ $(DYNAMIC_{con}(\exists, \exists))$: given a domain $\mathcal{D}$, a security state Π based on $\mathcal{D}$, a (consistent) dynamic policy DP based on $\mathcal{D}$ and a condition ϕ based on $\mathcal{D}$, determine whether there exists a subset $\mathcal{A}$ of Π, there exists an interpretation function I for $\mathcal{D}$ such that $L(\Pi, DP, \mathcal{A}), I \models \phi$.

Proposition 2. $DYNAMIC_{con}(\exists, \exists)$ is NP-complete.

Proposition 3. $DYNAMIC(\exists, \exists)$ *is in* $\Sigma_2 P$.

The $DYNAMIC^{path}(\exists, \exists)$ problem and the $DYNAMIC^{path}_{con}(\exists, \exists)$ problem are the following decision problems:

- $DYNAMIC^{path}(\exists, \exists)$ $(DYNAMIC^{path}_{con}(\exists, \exists))$: given a domain $\mathcal{D}$, a security state Π based on $\mathcal{D}$, a (consistent) dynamic policy DP based on $\mathcal{D}$ and a condition ϕ based on $\mathcal{D}$, determine whether there exists an integer $n \geq 0$ and there exists security states $\Pi_0, \ldots, \Pi_n$ based on $\mathcal{D}$ such that
  - $\Pi_0 = \Pi$,
  - for all integers $i \geq 0$, if $1 \leq i \leq n$ then there exists a subset $\mathcal{A}$ of Π_{i-1} such that $\Pi_i = L(\Pi_{i-1}, DP, \mathcal{A})$,
  - there exists an interpretation function I for $\mathcal{D}$ such that $\Pi_n, I \models \phi$.

Proposition 4. $DYNAMIC^{path}_{con}(\exists, \exists)$ *is PSPACE-complete.*

Proposition 5. $DYNAMIC^{path}(\exists, \exists)$ *is PSPACE-complete.*

4.3 Rule-Based Assignments

The $RULEBASED(\exists, \exists)$ problem and the $RULEBASED_{con}(\exists, \exists)$ problem are the following decision problems:

- $RULEBASED(\exists, \exists)$ $(RULEBASED_{con}(\exists, \exists))$: given a domain $\mathcal{D}$, a security state Π based on $\mathcal{D}$, a (consistent) rule-based policy $\mathcal{P}$ based on $\mathcal{D}$ and a condition ϕ based on $\mathcal{D}$, determine whether there exists a subset $\mathcal{A}$ of Π, there exists an interpretation function I for $\mathcal{D}$ such that $L(\Pi, \mathcal{P}, \mathcal{A}), I \models \phi$.

Proposition 6. $RULEBASED_{con}(\exists, \exists)$ *is NP-complete.*

Proposition 7. $RULEBASED(\exists, \exists)$ *is in* $\Sigma_2 P$.

The $RULEBASED^{path}(\exists, \exists)$ problem and the $RULEBASED^{path}_{con}(\exists, \exists)$ problem are the following decision problems:

- $RULEBASED^{path}(\exists, \exists)$ $(RULEBASED^{path}_{con}(\exists, \exists))$: given a domain $\mathcal{D}$, a security state Π based on $\mathcal{D}$, a (consistent) rule-based policy $\mathcal{P}$ based on $\mathcal{D}$ and a condition ϕ based on $\mathcal{D}$, determine whether there exists an integer $n \geq 0$ and there exists security states $\Pi_0, \ldots, \Pi_n$ based on $\mathcal{D}$ such that
  - $\Pi_0 = \Pi$,
  - for all integers $i \geq 0$, if $1 \leq i \leq n$ then there exists a subset $\mathcal{A}$ of Π_{i-1} such that $\Pi_i = L(\Pi_{i-1}, \mathcal{P}, \mathcal{A})$,
  - there exists an interpretation function I for $\mathcal{D}$ such that $\Pi_n, I \models \phi$.

Proposition 8. $RULEBASED^{path}_{con}(\exists, \exists)$ *is PSPACE-complete.*

Proposition 9. $RULEBASED^{path}(\exists, \exists)$ *is PSPACE-complete.*

5   Conclusion

In our framework, a role-based access control policy consists in a set of static clauses and a set of dynamic clauses defined in terms of permissions. Static clauses characterize what remains true during the life of a system whereas dynamic clauses characterize the different ways according to which the system can change. We have provided examples on how to express RBAC features using static clauses and dynamic clauses and we have addressed the complexity issue of some decision problems related to the assignment of permissions with respect to such-or-such policy.

In other respect, we have seen how the Boolean connective $\perp$ in the right-hand side of well-formed conditions can be used in order to express RBAC features such as separation of duties and synchronisation of actions. Seeing that it is essential for the system to never enter into an inconsistent state, we plan to study the computational complexity of the decision problem consisting, given a domain $\mathcal{D}$, a security state Π based on $\mathcal{D}$ and a rule-based policy $\mathcal{P}$ based on $\mathcal{D}$, to determine whether there exists an infinite sequence of transitions of $\mathcal{P}$ beginning with Π. Finally, permissions are often associated to temporal constraints: permissions are given for such-or-such periods of time. Extending our current language with such temporal constraints would permit the expression of more useful policies.

Acknowledgements

The work presented in this paper was partially supported by the FP7-ICT-2007-1 Project no. 216471, "AVANTSSAR: Automated Validation of Trust and Security of Service-oriented Architectures" (www.avantssar.eu). We are also grateful for the support of the project ANR-05-SSIA-0007-01 *Cops* financed by the "Agence nationale de la recherche" (www.irit.fr/COPS/Accueil.htm).

References

1. Abadi, M.: Logic in access control. In: 18th IEEE Symposium on logic in Computer Science (LICS 2003), pp. 228–233. IEEE Computer Society, Los Alamitos (2003)
2. Becker, M., Sewell, P.: Cassandra: distributed access control policies with tunable expressiveness. In: 5th IEEE International Workshop on Policies for Distributed Systems and Networks (POLICY 2004), pp. 159–168. IEEE Computer Society, Los Alamitos (2004)
3. Becker, M., Sewell, P.: Cassandra: flexible trust management, applied to electronic health records. In: Proceedings of the 17th IEEE Computer Science Foundations Workshop (CSFW 2004), pp. 139–154. IEEE Computer Society Press, Los Alamitos (2004)
4. Bell, D., LaPadula, L.: Secure Computer Systems: Mathematical Foundations. MITRE Corporation (1973)
5. Bonner, A.J., Kifer, M.: An overview of transaction logic. Theoretical Computer Science 133(2), 205–265 (1994)

6. Ferraiolo, D., Kuhn, D.: Role-based access controls. In: 15th NIST-NCSC National Computer Security Conference, pp. 554–563 (1992)
7. Garey, M., Johnson, D.: Computers and Intractability. A Guide to the Theory of NP-Completeness. W.H. Freeman, New York (1979)
8. Harrison, M., Ruzzo, W., Ullman, J.: On protection in operating systems. Communications of the ACM 19, 461–471 (1976)
9. Li, N., Mitchell, J., Winsborough, W.: Design of a role-based trust-management framework. In: Symposium on Security and Privacy, pp. 114–130. IEEE Computer Society Press, Los Alamitos (2002)
10. Papadimitriou, C.: Computational Complexity. Addison-Wesley, Reading (1994)
11. Zhang, X., Oh, S., Sandhu, R.: PDBM: A flexible delegation model in RBAC. In: Proceedings of the 8th ACM Symposium on Access Control Models and Technologies (SACMAT 2003), pp. 149–157. ACM, New York (2003)

Appendix

In this appendix, we provide the proofs of propositions 1 to 9.

Proof of proposition 1. $STATIC(\exists)$ is in NP. It suffices to prove the existence of an algorithm in NP that solves $STATIC(\exists)$. Let us consider the following algorithm:

1. First, choose an enumeration A_1, ..., A_n of some subset of the set of all ground atomic formulas based on $\mathcal{D}$.
2. Second, choose a ground instance ϕ' of ϕ based on $\mathcal{D}$.
3. Third, for all integers $i \geq 0$, if $1 \leq i \leq n$ then check whether A_i can be inferred in one step from A_1, ..., A_{i-1} and some static clause in SP.
4. Fourth, check whether $\{A_1, \ldots, A_n\} \models \phi'$.

The reader may easily verify that this algorithm can be executed in nondeterministic polynomial time.

$STATIC(\exists)$ is NP-hard. It suffices to prove the existence of a reduction from an NP-hard problem to $STATIC(\exists)$. We shall say that a connected graph $G = (V, E)$ is 3-colorable iff there exists a function f in $\{0, 1, 2\}^V$ such that for all u, v in V, if (u, v) is in E then $f(u) \neq f(v)$. Let us consider the following decision problem:

– 3-*COLORABILITY*: given a connected graph G, determine whether G is 3-colorable.

It is well-known that 3-*COLORABILITY* is NP-hard [7]. Given a connected graph $G = (V, E)$, the instance $\rho(G)$ of the $STATIC(\exists)$ problem that we construct is given by the domain $\mathcal{D}_G$, the static policy SP_G based on $\mathcal{D}_G$ and the condition ϕ_G based on $\mathcal{D}_G$ defined by

– $\mathcal{D}_G = \langle \{0, 1, 2\}, \{a\}, \{0, 1, 2, \}, \{r\} \rangle$,
– $SP_G = \{P(i, a, j, r) : 0 \leq i, j \leq 2 \text{ and } i \neq j\}$,
– $\phi_G = \bigwedge \{P(X_u, a, X_v, r) : (u, v) \text{ is in } E\}$.

This completes the construction. Obviously, ρ can be computed in logarithmic space. Moreover, the reader may easily verify that G is 3-colorable iff there exists an interpretation function I for $\mathcal{D}_G$ such that $l(SP_G), I \models \phi_G$. Hence, ρ is a reduction from 3-$COLORABILITY$ to $STATIC(\exists)$.

Proof of proposition 2. $DYNAMIC_{con}(\exists, \exists)$ is in NP. It suffices to prove the existence of an algorithm in NP that solves $DYNAMIC_{con}(\exists, \exists)$. Let us consider the following algorithm:

1. First, choose a subset $\mathcal{A}$ of Π.
2. Second, choose an enumeration $A_1, \ldots, A_n$ of some subset of the set of all ground atomic formulas based on $\mathcal{D}$.
3. Third, choose a ground instance ϕ' of ϕ based on $\mathcal{D}$.
4. Fourth, for all integers $i \geq 0$, if $1 \leq i \leq n$ then check whether A_i can be inferred in one step from Π, $\mathcal{A}$ and some dynamic clause in DP.
5. Fifth, check whether $\{A_1, \ldots, A_n\} \models \phi'$.

The reader may easily verify that this algorithm can be executed in nondeterministic polynomial time.

$DYNAMIC_{con}(\exists, \exists)$ is NP-hard. It suffices to prove the existence of a reduction from an NP-hard problem to $DYNAMIC_{con}(\exists, \exists)$. Let us consider the 3-$COLORABILITY$ problem. Given a connected graph $G = (V, E)$, the instance $\rho(G)$ of the $DYNAMIC_{con}(\exists, \exists)$ problem that we construct is given by the domain $\mathcal{D}_G$, the security state Π_G based on $\mathcal{D}_G$, the consistent dynamic policy DP_G based on $\mathcal{D}_G$ and the condition ϕ_G based on $\mathcal{D}_G$ defined by

- $\mathcal{D}_G = \langle \{0, 1, 2\}, \{a\}, \{0, 1, 2, r\}, \{r\} \rangle$,
- $\Pi_G = \emptyset$,
- $DP_G = \{\top \to (\bigwedge\{P(i, a, j, r) : 0 \leq i, j \leq 2 \text{ and } i \neq j\}, \top)\}$,
- $\phi_G = \bigwedge\{P(X_u, a, X_v, r) : (u, v) \text{ is in } E\}$.

This completes the construction. Obviously, ρ can be computed in logarithmic space. Moreover, the reader may easily verify that G is 3-colorable iff there exists a subset $\mathcal{A}$ of Π_G, there exists an interpretation function I for $\mathcal{D}_G$ such that $L(\Pi_G, DP_G, \mathcal{A}), I \models \phi_G$. Hence, ρ is a reduction from 3-$COLORABILITY$ to $DYNAMIC_{con}(\exists, \exists)$.

The $INCONSISTENT$ problem is the following decision problem:

- $INCONSISTENT$: given a domain $\mathcal{D}$, a security state Π based on $\mathcal{D}$, a subset $\mathcal{A}$ of Π and a dynamic policy DP based on $\mathcal{D}$, determine whether there exists a dynamic clause $\phi \to (\psi_1, \psi_2)$ in DP and there exists an interpretation function I for $\mathcal{D}$ such that $\Pi, I \models \phi$ and either $\mathcal{A}, I \models \phi$ and $\psi_1 = \bot$ or $\mathcal{A}, I \not\models \phi$ and $\psi_2 = \bot$.

We will use the following result in the proof of proposition 3.

Proposition 10. $INCONSISTENT$ *is NP-complete.*

Proof of proposition 10. INCONSISTENT is in NP. It suffices to prove the existence of an algorithm in NP that solves $INCONSISTENT$. Let us consider the following algorithm:

1. First, choose a dynamic clause $\phi \rightarrow (\psi_1, \psi_2)$ in DP.
2. Second, choose an interpretation function I for $\mathcal{D}$.
3. Third, check whether $\Pi, I \models \phi$ and either $\mathcal{A}, I \models \phi$ and $\psi_1 = \bot$ or $\mathcal{A}, I \not\models \phi$ and $\psi_2 = \bot$.

The reader may easily verify that this algorithm can be executed in nondeterministic polynomial time.

$INCONSISTENT$ is NP-hard. It suffices to prove the existence of a reduction from an NP-hard problem to $INCONSISTENT$. Let us consider the 3-$COLORABILITY$ problem. Given a connected graph $G = (V, E)$, the instance $\rho(G)$ of the $INCONSISTENT$ problem that we construct is given by the domain $\mathcal{D}_G$, the security state Π_G based on $\mathcal{D}_G$, the subset $\mathcal{A}_G$ of Π_G and the dynamic policy DP_G based on $\mathcal{D}_G$ defined by

- $\mathcal{D}_G = \langle \{0, 1, 2\}, \{a\}, \{0, 1, 2\}, \{r\} \rangle$,
- $\Pi_G = \{(i, a, j, r) : 0 \leq i, j \leq 2 \text{ and } i \neq j\}$,
- $\mathcal{A}_G = \{0, 1, 2\} \times \{a\} \times \{0, 1, 2\} \times \{r\}$,
- $DP_G = \{\bigwedge\{P(X_u, a, X_v, r) : (u, v) \text{ is in } E\} \rightarrow (\bot, \top)\}$.

This completes the construction. Obviously, ρ can be computed in logarithmic space. Moreover, the reader may easily verify that G is 3-colorable iff there exists a dynamic clause $\phi \rightarrow (\psi_1, \psi_2)$ in DP_G and there exists an interpretation function I for $\mathcal{D}_G$ such that $\Pi_G, I \models \phi$ and either $\mathcal{A}, I \models \phi$ and $\psi_1 = \bot$ or $\mathcal{A}, I \not\models \phi$ and $\psi_2 = \bot$. Hence, ρ is a reduction from 3-$COLORABILITY$ to $INCONSISTENT$.

Proof of proposition 3. It suffices to prove the existence of an algorithm in NP^{NP} that solves $DYNAMIC(\exists, \exists)$. Let us consider the following algorithm:

1. First, choose a subset $\mathcal{A}$ of Π.
2. Second, choose an enumeration $A_1, \ldots, A_n$ of some subset of the set of all ground atomic formulas based on $\mathcal{D}$.
3. Third, choose a ground instance ϕ' of ϕ based on $\mathcal{D}$.
4. Fourth, check whether $INCONSISTENT(\Pi, \mathcal{A}, DP)$ returns "no".
5. Fifth, for all integers $i \geq 0$, if $1 \leq i \leq n$ then check whether A_i can be inferred in one step from Π, $\mathcal{A}$ and some dynamic clause in DP.
6. Sixth, check whether $\{A_1, \ldots, A_n\} \models \phi'$.

The reader may easily verify that this algorithm with oracle $INCONSISTENT$ in NP can be executed in nondeterministic polynomial time. In contrast to the aforementioned decision problems, we still do not know if $DYNAMIC(\exists, \exists)$ is complete with respect to the class $\Sigma_2 P$.

Proof of proposition 4. $DYNAMIC_{con}^{path}(\exists, \exists)$ is in $PSPACE$. It suffices to prove the existence of an algorithm in $NPSPACE$ that solves $DYNAMIC_{con}^{path}(\exists, \exists)$. Let us informally describe such an algorithm. First, choose an interpretation function I for $\mathcal{D}$. Second, check whether $\Pi, I \models \phi$. Third, if a negative answer is returned then we choose a subset $\mathcal{A}$ of Π, we let $\Pi := L(\Pi, DP, \mathcal{A})$ and we move to the first step. The reader may easily verify that this algorithm can be executed in nondeterministic polynomial space.

$DYNAMIC_{con}^{path}(\exists, \exists)$ is $PSPACE$-hard. It suffices to prove that any language in $PSPACE$ can be reduced to $DYNAMIC_{con}^{path}(\exists, \exists)$. Let $L \subseteq \{0, 1\}^\star$ be a language in PSPACE. Hence, there is a polynomial-space-bounded Turing machine M such that for all x in $\{0, 1\}^\star$, x is in L iff M accepts x. The machine M has three components (Q, Σ, δ) where Q is the set of states of M, Σ is the set of symbols of M, and δ is the transition function of M. We assume that there is q_0 in Q, the start state of M, and there is q_1 in Q, the final state of M, such that $q_0 \neq q_1$. We suppose that 0 is in Σ and 1 is in Σ. We also assume that there is B in Σ, the blank symbol of M, such that $0 \neq B$ and $1 \neq B$. The transition function δ assigns $\delta(q, A)$ in $Q \times \Sigma \times \{L, R\}$ to each q in Q and each A in Σ. If $\delta(q, A) = (q', A', L)$ or $\delta(q, A) = (q', A', R)$ then it means that whenever the machine is in state q and scans an A on its tape, it changes its state to q', replaces the A by a A' and moves its tape head leftward or rightward. We suppose that q_0 is not in the range of δ and q_1 is not in the domain of δ. We also assume that M never falls off the left end of its input string $x \in \{0, 1\}^\star$. The machine accepts x in $\{0, 1\}^\star$ iff starting in state q_0 and scanning the left end of x on its tape, preceded and followed by an infinity of blanks, M finally enters in state q_1. The machine M is allowed to use an amount of space that is polynomial in the size of its input, no matter how much time it uses. Consequently, there is a polynomial $p(n)$ such that when given input x in $\{0, 1\}^\star$ of length n, M never visits more than $p(n)$ cells of its tape. Hence, there is a positive integer k such that for all x in $\{0, 1\}^\star$, M uses at most 2^{N^k} positions on its tape when given input x in $\{0, 1\}^\star$ of length N. We are now ready to reduce the following decision problem to $DYNAMIC_{con}^{path}(\exists, \exists)$:

- $P(L)$: given x in $\{0, 1\}^\star$, determine whether M accept x.

Given x in $\{0, 1\}^\star$, it is convenient to write $x = x_1 \ldots x_N$ where N is the length of x. For all q in Q and for all A in Σ, if $\delta(q, A) = (q', A', L)$ then for all A'' in Σ and for all i in $\{1, \ldots, N^k\}$, we let $dc_L(q, A, A'', i)$ be the following dynamic clause:

$$\phi_L(q, A, A'', i) \to (\psi_L(q, A, A'', i), \psi_L(q, A, A'', i))$$

where

$$\phi_L(q, A, A'', i) = P(X_1, 1, X_1, r) \wedge \ldots \wedge P(X_{i-2}, i - 2, X_{i-2}, r)$$
$$\wedge P(A'', i - 1, A'', r) \wedge P((q, A), i, (q, A), r)$$
$$\wedge P(X_{i+1}, i + 1, X_{i+1}, r) \wedge \ldots \wedge P(X_{N^k}, N^k, X_{N^k}, r)$$

and

$$\psi_L(q, A, A'', i) = P(X_1, 1, X_1, r) \wedge \ldots \wedge P(X_{i-2}, i-2, X_{i-2}, r)$$
$$\wedge P((q', A''), i-1, (q', A''), r) \wedge P(A', i, A', r)$$
$$\wedge P(X_{i+1}, i+1, X_{i+1}, r) \wedge \ldots \wedge P(X_{N^k}, N^k, X_{N^k}, r).$$

For all q in Q and for all A in Σ, if $\delta(q, A) = (q', A', R)$ then for all A'' in Σ and for all i in $\{1, \ldots, N^k\}$, we let $dc_R(q, A, A'', i)$ be the following dynamic clause:

$$\phi_R(q, A, A'', i) \rightarrow (\psi_R(q, A, A'', i), \psi_R(q, A, A'', i))$$

where

$$\phi_R(q, A, A'', i) = P(X_1, 1, X_1, r) \wedge \ldots \wedge P(X_{i-1}, i-1, X_{i-1}, r)$$
$$\wedge P((q, A), i, (q, A), r) \wedge P(A'', i+1, A'', r)$$
$$\wedge P(X_{i+2}, i+2, X_{i+2}, r) \wedge \ldots \wedge P(X_{N^k}, N^k, X_{N^k}, r)$$

and

$$\psi_R(q, A, A'', i) = P(X_1, 1, X_1, r) \wedge \ldots \wedge P(X_{i-1}, i-1, X_{i-1}, r)$$
$$\wedge P(A', i, A', r) \wedge P((q', A''), i+1, (q', A''), r)$$
$$\wedge P(X_{i+2}, i+2, X_{i+2}, r) \wedge \ldots \wedge P(X_{N^k}, N^k, X_{N^k}, r).$$

The instance $\rho(x)$ of the $DYNAMIC_{con}^{path}(\exists, \exists)$ problem that we construct is given by the domain $\mathcal{D}_x$, the security state Π_x based on $\mathcal{D}_x$, the consistent dynamic policy DP_x based on $\mathcal{D}_x$ and the condition ϕ_x based on $\mathcal{D}_x$ defined by

- $\mathcal{D}_x = \langle (Q \times \Sigma) \cup \Sigma, \{1, \ldots, N^k\}, (Q \times \Sigma) \cup \Sigma, \{r\} \rangle$,
- $\Pi_x = \{((q_0, x_1), 1, (q_0, x_1), r), (x_2, 2, x_2, r), \ldots, (x_{N^k}, N^k, x_{N^k}, r)\}$,
- $DP_x = \{dc_L(q, A, A'', i) : \ q \text{ is in } Q, A \text{ is in } \Sigma, \delta(q, A) = (q', A', L), A''$ is in Σ and i in $\{1, \ldots, N^k\}\} \cup \{dc_R(q, A, A'', i) : \ q$ is in Q, A is in Σ, $\delta(q, A) = (q', A', R)$, A'' is in Σ and i in $\{1, \ldots, N^k\}\}$,
- $\phi_x = \bigvee\{P((q_1, A), i, (q_1, A), r) : \ A \text{ is in } \Sigma \text{ and } i \text{ is in } \{1, \ldots, N^k\}\}$.

This completes the construction. Obviously, ρ can be computed in logarithmic space. Moreover, the reader may easily verify that M accepts x iff there exists an integer $n \geq 0$ and there exists security states $\Pi_0, \ldots, \Pi_n$ based on $\mathcal{D}_x$ such that

- $\Pi_0 = \Pi_x$,
- for all integers $i \geq 0$, if $1 \leq i \leq n$ then there exists a subset $\mathcal{A}$ of Π_{i-1} such that $\Pi_i = L(\Pi_{i-1}, DP_x, \mathcal{A})$,
- there exists an interpretation function I for $\mathcal{D}_x$ such that $\Pi_n, I \models \phi_x$.

Hence, ρ is a reduction from $P(L)$ to $DYNAMIC_{con}^{path}(\exists, \exists)$.

Proof of proposition 5. Similar to the proof of proposition 4.

Proof of proposition 6. Similar to the proof of proposition 2.

Proof of proposition 7. Similar to the proof of proposition 3.

Proof of proposition 8. Similar to the proof of proposition 4.

Proof of proposition 9. Similar to the proof of proposition 5.

Interpolative Boolean Logic

Dragan Radojević[1], Aleksandar Perović[2], Zoran Ognjanović[3],
and Miodrag Rašković[2]

[1] Mihailo Pupin Institute, Volgina 15, 11000 Belgrade, Serbia
Dragan.Radojevic@automatika.imp.bg.ac.yu
[2] Faculty of Transportation and Traffic Engineering, Vojvode Stepe 305, 11000
Belgrade, Serbia
pera@sf.bg.ac.yu
[3] Mathematical Institute SANU, Kneza Mihaila 36, 11000 Belgrade, Serbia
{zorano,miodragr}@mi.sanu.ac.yu

Abstract. A polyvalent propositional logic $\mathcal{L}$ is in Boolean frame if the
set of all $\mathcal{L}$-valid formulas coincides with the set of all tautologies. It is
well known that the polyvalent logics based on the truth functionality
principle are not in the Boolean frame. Interpolative Boolean logic (IBL)
is a real-valued propositional logic that is in Boolean frame. The term
"interpolative" cames from the fact that semantics of IBL is based on
the notion of a generalized Boolean polynomial, where multiplication can
be substituted by any continuous t-norm $* : [0,1]^2 \longrightarrow [0,1]$ such that
$xy \leqslant x * y$. Possible applications are illustrated with several examples.

1 Introduction

Classical propositional logic is inadequate in many problems with inherited uncertainty. In order to overcome this difficulty, L. Zadeh introduced fuzzy sets in
his famous paper [26]. Since then, the mathematics of "fuzzy" have had a rapid
development, particulary in the last two decades.

A fuzzy logic is a polyvalent logic based on the truth functionality principle,
where conjunction is evaluated by some t-norm, disjunction by the corresponding
s-norm, implication by the residua of the t-norm, and negation is defined by
means of implication and the truth constant $\underline{0}$ ($\underline{0}$ is evaluated as 0). There are
many relevant papers and books related to fuzzy logics. Here we will just mention
[1,2,6,10].

As it is well known, fuzzy logics do not preserve tautologies: for each fuzzy
logic $\mathcal{L}$ there is a tautology ϕ which is not $\mathcal{L}$-valid. In order to overcome this, we
introduce the Interpolative Boolean logic (IBL) and show that it is a real-valued
propositional logic that is in Boolean frame. The term "interpolative" comesl
from the fact that semantics of IBL is based on the notion of a generalized
Boolean polynomial, where multiplication can be substituted by any continuous
t-norm $* : [0,1]^2 \longrightarrow [0,1]$ such that $xy \leqslant x * y$. IBL is a special kind of probabilistic logic. As it is well known, probabilistic logics are another prominent

D. Dochev, M. Pistore, and P. Traverso (Eds.): AIMSA 2008, LNAI 5253, pp. 209–219, 2008.

kind of polyvalent logics, that have found applications in many problems involving uncertainty. There are many relevant papers and books about probabilistic logics. Here we will just mention [3,7,8,9,11,12,13,14,15,16,17,22,23,24].

This paper continues work initiated in [18,19,20,21] on the real-valued realization of Boolean laws. We give a strict definition of the interpolative Boolean logic in the general case (the set Var of propositional letters may be infinite) and prove that IBL is in Boolean frame. Consequently, any complete axiomatization of the classical propositional logic is a complete axiomatization of IBL. To illustrate possible applications, we have listed several examples.

The rest of the paper is organized as follows: in Section 2 we introduce the interpolative Boolean logic; in Section 3 we give some examples of application of IBL; concluding remarks are in Section 4; the Appendix contains some definitions omitted from the main text.

2   Interpolative Boolean Logic

Let Var be a nonempty set of propositional letters. By For_C we will denote the corresponding set of propositional formulas. For $\phi \in For_C$, let $Var(\phi)$ be the set of all propositional letters appearing in ϕ.

Let X be a nonempty finite subset of Var. For all $Y \subseteq X$ we define the formula $\alpha_X(Y)$ in the following way:

$$\alpha_X(Y) = \begin{cases} \bigwedge_{p \in X} \neg p, & Y = \emptyset \\ \bigwedge_{p \in X} p, & Y = X \\ \bigwedge_{p \in Y} p \wedge \bigwedge_{q \in X \setminus Y} \neg q, & \text{otherwise.} \end{cases}$$

By $M(\phi)$ we will denote the set of all $a : Var(\phi) \longrightarrow \{0,1\}$ such that $a \models \phi$. The normal form of $\phi \in For_C$ is the formula

$$\bigvee_{a \in M(\phi)} \alpha_{Var(\phi)}(a^{-1}(1)),$$

where $a^{-1}(1) = \{p \in Var(\phi) \mid a(p) = 1\}$. Clearly, ϕ and its normal form are equivalent.

Let $\mathcal{L} = \mathcal{L}_{of} \cup \{\otimes\} \cup Var$, where each $p \in Var$ is a symbol of constant, $\otimes$ is a binary function symbol, and $\mathcal{L}_{of} = \{+, \cdot, -, ^{-1}, \leqslant, 0, 1\}$ is the language of the ordered fields. By $RCF_\otimes$ we will denote the $\mathcal{L}$-theory

$RCF +$ "$\otimes$ is a continuous t-norm" $+ (\forall x, y \in [0,1])(xy \leqslant x \otimes y) +$ "$Var \subseteq [0,1]$". Recall that RCF is the theory of real closed fields (axioms of ordered fields + axioms that provide that every polynomial of the odd degree has a root).

Definition 1. *For all $\phi \in For_C$ such that $M(\phi) \neq \emptyset$, we will define the $\mathcal{L}$-term $\phi^\otimes$ in the following way:*

1. $\phi^{\otimes} = \sum\limits_{a \in M(\phi)} (\alpha_{Var(\phi)}(a^{-1}(1)))^{\otimes}$

2. $(\alpha_{Var(\phi)}(a^{-1}(1)))^{\otimes} = \bigotimes\limits_{p \in a^{-1}(1)} p, \ a^{-1}(1) = Var(\phi)$

3. $(\alpha_{Var(\phi)}(a^{-1}(1)))^{\otimes} = 1 + \sum\limits_{X \subseteq a^{-1}(0), X \neq \emptyset} (-1)^{|X|} \bigotimes\limits_{q \in X} q, \ a^{-1}(1) = \emptyset$

4. $(\alpha_{Var(\phi)}(a^{-1}(1)))^{\otimes} = \bigotimes\limits_{p \in a^{-1}(1)} p + \sum\limits_{X \subseteq a^{-1}(0), X \neq \emptyset} (-1)^{|X|} \bigotimes\limits_{p \in a^{-1}(1)} p \otimes \bigotimes\limits_{q \in X} q,$ other-

wise.

$\phi^{\otimes}$ will be called the generalized Boolean polynomial of ϕ.   □

Strictly speaking, $\phi^{\otimes}$ is a finite set of terms, not a single term. However, if f and g are any such terms, then clearly $RCF_{\otimes} \vdash f = g$: it is enough to apply commutativity and associativity of $\otimes$. Furthermore, we will only consider expansions of the ordered field of reals that are models of $RCF_{\otimes}$, so we will treat $\phi^{\otimes}$ as a single term.

The reader familiar with Boolean polynomials will recognize that generalized Boolean polynomials are obtained by the following substitution: each appearance of $\cdot$ in a Boolean polynomial f is replaced by $\otimes$. Boolean polynomials are used for the standard propositional semantics. Namely, if $f_{\phi}(p_1, \ldots, p_n)$ is the Boolean polynomial of $\phi \in For_C$, then $a : \{p_1, \ldots, p_n\} \longrightarrow 2$ satisfies ϕ iff $f_{\phi}[a(p_1), \ldots, a(p_n)] = 1$.

If ϕ is a contradiction, then we adopt the convention $\phi^{\otimes} = 0$.

Example 1. To make Definition 1 more clear, we will list all $\alpha^{\otimes}$'s for the set of variables $\{p, q, r\}$.

α	$\alpha^{\otimes}$
$p \wedge q \wedge r$	$p \otimes q \otimes r$
$p \wedge q \wedge \neg r$	$p \otimes q - p \otimes q \otimes r$
$p \wedge \neg q \wedge r$	$p \otimes r - p \otimes q \otimes r$
$p \wedge \neg q \wedge \neg r$	$p - p \otimes q - p \otimes r + p \otimes q \otimes r$
$\neg p \wedge q \wedge r$	$q \otimes r - p \otimes q \otimes r$
$\neg p \wedge q \wedge \neg r$	$q - p \otimes q - q \otimes r + p \otimes q \otimes r$
$\neg p \wedge \neg q \wedge r$	$r - p \otimes r - q \otimes r + p \otimes q \otimes r$
$\neg p \wedge \neg q \wedge \neg r$	$1 - p - q - r + p \otimes q + p \otimes r + q \otimes r - p \otimes q \otimes r$

□

Lemma 1. *Let* $X = \{p_1, \ldots, p_n\}$ *be a set of propositional letters such that* $p_i \neq p_j$ *for* $i \neq j$. *Then,*

$$RCF_{\otimes} \vdash \sum\limits_{a : X \longrightarrow 2} (\alpha_X(a^{-1}(1)))^{\otimes} = 1.$$

Proof. Terms of the form $p_{i_1} \otimes \cdots \otimes p_{i_k}$, $1 \leqslant k \leqslant n$, appear in each $(\alpha_X(a^{-1}(1)))^{\otimes}$ such that $|a^{-1}(0)| \geqslant n - k$. The number of appearances is equal to

$$\binom{k}{0} + \binom{k}{1} + \cdots + \binom{k}{k} = 2^k.$$

Here $\binom{k}{l}$ represent the number of cases where the exactly l of p_{i_j}'s are negated in $\alpha(a^{-1}(1))$. Furthermore, the number of positive appearances is equal to 2^{k-1}, so terms $p_{i_1} \otimes \cdots \otimes p_{i_k}$ vanish in the $\sum_{a:X \longrightarrow 2} (\alpha_X(a^{-1}(1)))^{\otimes}$ (by means of the axioms of commutative rings, which are part of $RCF_{\otimes}$). The only remaining term is 1, so we have our claim.   ⊣

Theorem 1. *Suppose that* $\phi, \psi \in For_C$ *are equivalent. Then,*

$$RCF_{\otimes} \vdash \phi^{\otimes} = \psi^{\otimes}.$$

Proof. If $Var(\phi) = Var(\psi)$, then $\phi^{\otimes} = \psi^{\otimes}$ by Definition 1. Thus, let $Var(\phi) = \{p_1, \ldots, p_n\}$ and $Var(\psi) = \{p_1, \ldots, p_n, q_1, \ldots, q_m\}$. Since ϕ and ψ are equivalent, we obtain the normal form of ψ by the following substitution:

$$\alpha_{Var(\phi)}(a^{-1}(1)) \mapsto \bigvee_{b:\{q_1,\ldots,q_m\} \longrightarrow 2} \alpha_{Var(\phi)}(a^{-1}(1)) \wedge \alpha_{Var(\psi)\setminus Var(\phi)}(b^{-1}(1)),$$

$a \in M(\phi)$. Now, using the same argument as in Lemma 1, we obtain equality (by $RCF_{\otimes}$ means) of

$$\sum_{b:\{q_1,\ldots,q_m\} \longrightarrow 2} (\alpha_{Var(\phi)}(a^{-1}(1)) \wedge \alpha_{Var(\psi)\setminus Var(\phi)}(b^{-1}(1)))^{\otimes}$$

and $(\alpha_{Var(\phi)}(a^{-1}(1)))^{\otimes}$. Hence, $RCF_{\otimes} \vdash \phi^{\otimes} = \psi^{\otimes}$.   ⊣

The immediate consequence of Theorem 1 is the fact that generalized Boolean polynomials can be used for the definition of the real-valued evaluation of propositional formulas. In the next two definitions we will introduce the notion of a basic model and use it to define satisfiability and validity of $\phi \in For_C$.

Definition 2. *A* basic model *is any* $\mathcal{L}$-structure $\mathcal{M}$ *with the following two properties:*

1. *$\mathcal{M} \models RCF_{\otimes}$*
2. *$\mathcal{M}$ is an expansion of the ordered field of reals.*   □

Definition 3. *Let $\phi \in For_C$. We say that ϕ is* satisfiable *if there exists a basic model $\mathcal{M}$ such that $\mathcal{M} \models \phi^{\otimes} = 1$; ϕ is* valid *if $\neg\phi$ is not satisfiable.*   □

The logic semantically introduced by Definition 3 will be called the *interpolative Boolean logic* (IBL).

Theorem 2. *A propositional formula $\phi \in For_C$ is valid (in the sense of Definition 3) if and only if it is a tautology.*

Proof. If ϕ is a tautology, then $RCF_{\otimes} \vdash \phi^{\otimes} = 1$, so ϕ is valid. Conversely, suppose that ϕ is valid. Then, ϕ is satisfied in all basic models $\mathcal{M}$ such that $p^{\mathcal{M}} \in 2$ for all $p \in Var$. Since $\otimes^{\mathcal{M}}$ is a t-norm, we obtain that all $a : Var(\phi) \longrightarrow 2$ are models of ϕ. Thus, ϕ is a tautology.   ⊣

As immediate consequence of the previous theorem we have the following corollary:

Corollary 1. *Any complete axiomatization of the classical propositional logic is also a complete axiomatization of IBL.*

We will conclude this section with the motivation for the introduction of the "probabilistic consistency" axiom $(\forall x, y \in [0,1])(xy \leqslant x \otimes y)$. Consider the formula $\neg p \wedge \neg q \wedge \neg r$. Let T be the theory

$$RCF_\otimes - \text{"probabilistic consistency"}$$

and let $\mathcal{M}$ be a model of T such that $\otimes^{\mathcal{M}}$ is the Lukasiewicz t-norm $L(x, y) = \min(0, x + y - 1)$ and that $p^{\mathcal{M}} = q^{\mathcal{M}} = r^{\mathcal{M}} = \frac{1}{2}$. Then,

$$\mathcal{M} \models (\neg p \wedge \neg q \wedge \neg r)^\otimes = -\frac{1}{2}.$$

In particular, t-norms lesser or equal than the product norm will give negative truth values for some formulas, so they cannot be used for the real-valued truth evaluation. This is the reason why we have introduced the probabilistic consistency axiom. If $|Var| \leqslant 2$, then we can use any continuous t-norm that is greater or equal to the Lukasiewicz t-norm.

3   Some Applications of IBL

As we have seen, IBL is a real-valued logic, so it can be applied in all problems that require graded truth evaluations. In addition, IBL is in Boolean frame (all Boolean laws are preserved), so we have a natural generalization of the classical propositional reasoning to the real valued case.

Concrete applications are based on the logical representation of a problem and on the choice of the adequate t-norm. Here we will give several examples of the application of IBL. More examples and general motivation and insights can be found in [18,19,20,21].

3.1   IBA Based Decision Making - Monotone Case

Let us consider the following example. We need to classify compounds C_1, C_2 and C_2 of the substances p and q according to the criteria of minimal harmfulness of a compound. It is known that both p and q are harmful, but they neutralize each other. The ratio of the given compounds with respect to p and q is given in the following table:

compound	p	q
C_1	.95	.05
C_2	.15	.85
C_3	.65	.35

According to the available information, the adequate logical representation of the problem is the formula $p \leftrightarrow q$. The normal form of $p \leftrightarrow q$ is the formula $(p \wedge q) \vee (\neg p \wedge \neg q)$, so

$$(p \leftrightarrow q)^{\otimes} = (p \wedge q)^{\otimes} + (\neg p \wedge \neg q)^{\otimes}$$
$$= 1 - p - q + 2(p \otimes q).$$

The truth evaluations for $\otimes = \min$ and $\otimes = \cdot$ are given in the following table:

compound	p	q	$(p \leftrightarrow q)^{\min}$	$(p \leftrightarrow q)^{\cdot}$
C_1	.95	.05	.1	.095
C_2	.15	.85	.3	.255
C_3	.65	.35	.7	.455

Now we have the following classification: the least harmful compound is C_3, then follows C_2, and the most harmful compound is C_1.

Let us consider another example. Again, we need to classify compounds C_1, C_2 and C_3 of the substances p and q, but now according to the maximal usefulness criteria. It is known that both p and q are useful, but they neutralize each other. The ratio table for compounds is the same as in the previous example. Here we have a different logical representation: it is the exclusive disjunction $(p \wedge \neg q) \vee (\neg p \wedge q)$. The corresponding generalized Boolean polynomial is equal to

$$p + q - 2(p \otimes q).$$

The truth evaluations for $\otimes = \min$ and $\otimes = \cdot$ are given in the following table:

compound	p	q	$((p \wedge \neg q) \vee (\neg p \wedge q))^{\min}$	$((p \wedge \neg q) \vee (\neg p \wedge q))^{\cdot}$
C_1	.95	.05	.9	.905
C_2	.15	.85	.7	.745
C_3	.65	.35	.3	.545

Now we have the following classification: the most useful compound is C_2, then follows C_2, and the least useful compound is C_3.

3.2   IBL Based Decision Making - Non-monotone Case

The following example (see [4,21]) will illustrate the IBL based logical aggregation (see [21]). We define IBL based logical aggregation as a linear convex combination of generalized Boolean polynomials. That is, for the given formulas $\phi_1, \ldots, \phi_n \in For_C$ and $\xi_1, \ldots, \xi_n \in [0,1]$ such that $\sum_{i=1}^{n} \xi_i = 1$, the aggregation of $\phi_1, \ldots, \phi_n$ by $\xi_1, \ldots, \xi_n$ is defined by

$$\mathrm{Agg}_{\xi_1, \ldots, \xi_n}(\phi_1, \ldots, \phi_n) = \sum_{i=1}^{n} \xi_i \cdot \phi_i^{\otimes}.$$

Now we are ready to proceed with the example. Objects A, B, C and D are described by quality attributes, whose values are given in the following table:

Object	a	b	c
A	.75	.9	.3
B	.75	.8	.4
C	.3	.65	.1
D	.3	.55	.2

We classify objects according to the global quality, which is specified by the following conditions:

ϕ_1: The average value of quality attributes;
ϕ_2: If the analyzed object is good with respect to a, then c is more important than b. Otherwise, b is more important than c.

Obviously, ϕ_1 should be evaluated by the arithmetic mean of the values of a, b and c. Concerning ϕ_2, it can be logically expressed by the formula $(a \wedge c) \vee (\neg a \wedge b)$, so it should be evaluated by

$$\phi_2^{\otimes} = b + a \otimes c - a \otimes b.$$

Since there are no other demands, we will assume that ϕ_1 and ϕ_2 are equally important. Thus, the aggregation of ϕ_1 and ϕ_2 is given by

$$\text{Agg}_{\frac{1}{2},\frac{1}{2}}(\phi_1, \phi_2) = \frac{1}{2} \cdot \frac{a+b+c}{3} + \frac{1}{2}(b + a \otimes c - a \otimes b).$$

Notice that the standard aggregation by the discrete Choquet integral cannot be applied since the corresponding aggregation measure μ, given in the table below, is not monotone.

S	$\mu(S)$
$\emptyset$	0
$\{a\}$	1/6
$\{b\}$	2/3
$\{c\}$	1/6
$\{a,b\}$	5/6
$\{a,c\}$	5/6
$\{b,c\}$	1/3
$\{a,b,c\}$	1

If we interpret $\otimes$ as the Gödel t-norm $G(x,y) = \min(x,y)$, then the objects A, B, C and D are indiscernible:

Object	$\phi^{\min}$
A	.45
B	.45
C	.45
D	.45

However, if we interpret $\otimes$ as the product norm, then we obtain the classification:

Object	$\text{Agg}_{\frac{1}{2},\frac{1}{2}}(\phi_1,\phi_2)$
A	.5500
B	.5750
C	.4175
D	.3725

These results completely reflect all specified demands.

4   Conclusion

Interpolative Boolean logic is a real-valued propositional logic that is in Boolean frame. The real-valued truth allows application of IBL in all problems previously treated by fuzzy logics. On the other hand, the preservation of the Boolean laws implies that IBL may be more adequate than fuzzy logics in some particular problems. An example of this kind is the logical aggregation and the generalized discrete Choquet integral (see [21]).

IBL may be seen as a special kind of probabilistic logic. Namely, if $\mathcal{M}$ is any basic model of $RCF_{\otimes}$, than we can define a probabilistic Kripke model $\mathcal{K_M} = \langle W, H, \mu, v \rangle$ in the following way:

- W is the set of all classical propositional models $w : Var \longrightarrow 2$.
- $v : For_C \times W \longrightarrow 2$ is defined by $v(\phi, w) = 1$ iff $w \models \phi$.
- $H = \{[\phi] \mid \phi \in For_C\}$, where $[\phi] = \{w \in W \mid v(\phi, w) = 1\}$.
- $\mu : H \longrightarrow [0,1]$ is defined by $\mu[\phi] = (\phi^{\otimes})^{\mathcal{M}}$.

It is easy to see that H is an algebra of sets on W, and that μ is a finitely additive probability measure.

A fragment of IBL that involves reasoning with RCF-definable t-norms can be axiomatized within the logic of polynomial weight formulas (see [3,17]). The future work will include the complete finitary axiomatization of the definable fragment of IBL, which is based on the compactness theorem for the hyper-real valued propositional probabilistic logics, as well as the study of the model theoretical properties of $RCF_{\otimes}$.

References

1. Cintula, P., Hájek, P., Horčik, R.: Formal systems of fuzzy logic and their fragments. Annals of Pure and Applied Logic 150, 40–65 (2007)
2. Esteva, F., Godo, L., Hájek, P., Navara, M.: Reziduated fuzzy logics with an involutive negation. Arch. Math. Logic. 39, 103–124 (2000)
3. Fagin, R., Halpern, J., Megiddo, N.: A logic for reasoning about probabilities. Information and Computation 87(1–2), 78–128 (1990)
4. Grabisch, M., Labreuche, C.: Bi-capacities for decision making on bipolar scales. In: EUROFUSE Workshop in Information Systems, Varena, Italy, September 2002, pp. 185–190 (2002)

5. Grabisch, M., Labrenche, C.: Bi-capacities - II: the Choquet integral. Fuzzy sets and systems 151(2), 237–259 (2005)
6. Hájek, P.: Metamathematics of fuzzy logic. Kluwer academic publishers, Dordrecht (1998)
7. Lehmann, D.: Generalized qualitative probability: Savage revisited. In: Horvitz, E., Jensen, F. (eds.) Procs. of 12 th Conference on Uncertainty in Artificial Intelligence (UAI 1996), pp. 381–388 (1996)
8. Lukasiewicz, T.: Probabilistic Default Reasoning with Conditional Constraints. Annals of Mathematics and Artificial Intelligence 34, 35–88 (2002)
9. Lukasiewicz, T.: Nonmonotonic probabilistic logics under variable-strength inheritance with overriding: Complexity, algorithms, and implementation. International Journal of Approximate Reasoning 44(3), 301–321 (2007)
10. Marchioni, E., Godo, L.: A Logic for Reasoning about Coherent Conditional Probability: A Modal Fuzzy Logic Approach. In: Leite, J., Alferes, J. (eds.) JELIA 2004. LNCS (LNAI), vol. 3229, pp. 213–225. Springer, Heidelberg (2004)
11. Nilsson, N.: Probabilistic logic. Artificial intelligence 28, 71–87 (1986)
12. Ognjanović, Z., Rašković, M.: A logic with higher order probabilities. In: Publications de l'institut mathematique, Nouvelle série, tome, vol. 60(74), pp. 1–4 (1996)
13. Ognjanović, Z., Rašković, M.: Some probability logics with new types of probability operators. J. Logic Computat. 9(2), 181–195 (1999)
14. Ognjanović, Z., Rašković, M.: Some first-order probability logics. Theoretical Computer Science 247(1–2), 191–212 (2000)
15. Ognjanović, Z., Marković, Z., Rašković, M.: Completeness Theorem for a Logic with imprecise and conditional probabilities. Publications de L'Institute Matematique (Beograd) 78(92), 35–49 (2005)
16. Ognjanović, Z., Perović, A., Rašković, M.: Logic with the qualitative probability operator. Logic journal of IGPL 16(2), 105–120 (2008)
17. Perović, A., Ognjanović, Z., Rašković, M., Marković, Z.: A probabilistic logic with polynomial weight formulas. In: Hartmann, S., Kern-Isberner, G. (eds.) FoIKS 2008. LNCS, vol. 4932, pp. 239–252. Springer, Heidelberg (2008)
18. Radojević, D.: [0, 1]-valued logic: a natural generalization of Boolean logic. Yugoslav Journal on Operations Research 10(2), 185–216 (2000)
19. Radojević, D.: Interpolative realization of Boolean algebra. NEUREL 2006, Eight Seminar on Neural Network Applications in Electrical Engineering, 201–206 (2006)
20. Radojević, D.: Interpolative realization of Boolean algebra as a consistent frame for gradation and/or fuzziness. In: Nikravesh, M., Kacprzyk, J., Zadeh, L.A. (eds.) Forging New Frontiers: Fuzzy Pioneers II. Studies in Fuzziness and Soft Computing, pp. 326–351. Springer, Heidelberg (2007)
21. Radojević, D.: Logical aggregation based on Interpolative realization of Boolean algebra. Mathware and soft computing (accepted for publication)
22. Rašković, M.: Classical logic with some probability operators. In: Publications de l'institut mathematique, Nouvelle série, tome, vol. 53(67), pp. 1–3 (1993)
23. Rašković, M., Ognjanović, Z.: A first order probability logic LP_Q, Publications de l'institut mathematique. In: Nouvelle série, tome, vol. 65(79), pp. 1–7 (1999)
24. Rašković, M., Ognjanović, Z., Marković, Z.: A logic with Conditional Probabilities. In: Alferes, J.J., Leite, J.A. (eds.) JELIA 2004. LNCS (LNAI), vol. 3229, pp. 226–238. Springer, Heidelberg (2004)
25. Sugeno, M.: Theory of fuzzy integrals and its applications. PhD Thesis, Tokyo Institute of Technology, Japan (1974)
26. Zadeh, L.: Fuzzy sets. Information and Control 8(3), 856–865 (1965)

Appendix

This appendix contains only some necessary definitions of certain notions (t-norms, for instance) that have been omitted from the main text in order to make it more readable.

A t-norm is any binary operation $*$ on the real unit interval $[0,1]$ that is commutative ($x * y = y * x$), associative ($x * (y * z) = (x * y) * z$), 1 is the neutral element for $*$ ($x * 1 = x$) and $*$ is compatible with the standard ordering of the reals, i.e.

$$x \leqslant x_1 \text{ and } x \leqslant x_1 \; y \leqslant y_1 \text{ implies } x * y \leqslant x_1 * y_1,$$

for all $x, y, x_1, y_1 \in [0, 1]$.

The prime examples of t-norms are:

- Lukasiewicz t-norm: $x * y = \max(0, x + y - 1)$
- Product t-norm: $x * y = xy$
- Gödel t-norm: $x * y = \min(x, y)$.

A t-norm $*$ is said to be continuous if it is continuous in the usual Euclidean topology on $\mathbb{R}$, i.e. for all $x, y \in [0, 1]$ and all $\varepsilon > 0$, there exists $\delta > 0$ such that for all $x_1, y_1 \in [0, 1]$,

$$(x - x_1)^2 + (y - y_1)^2 \leqslant \delta \text{ implies } |x * y - x_1 * y_1| \leqslant \varepsilon.$$

The theory of the real closed fields RCF is a first order theory of the language $\mathcal{L}_{of} = \{+, \cdot, \leqslant, 0, 1\}$ with the following axioms:

1. $\forall x \forall y (x + y = y + x)$
2. $\forall x \forall y \forall z (x + (y + z) = (x + y) + z)$
3. $\forall x (x + 0 = x)$
4. $\forall x \exists y (x + y = 0)$
5. $\forall x \forall y (x \cdot y = y \cdot x)$
6. $\forall x \forall y \forall z (x \cdot (y \cdot z) = (x \cdot y) \cdot z)$
7. $\forall x (x \cdot 1 = x)$
8. $\forall x (x \neq 0 \rightarrow \exists y (x \cdot y = 1))$
9. $\forall x \forall y \forall z (x \cdot (y + z) = (x \cdot y) + (x \cdot z))$
10. $\forall x \forall y ((x \leqslant y \wedge y \leqslant x) \rightarrow x = y)$
11. $\forall x \forall y \forall z ((x \leqslant y \wedge y \leqslant z) \rightarrow x \leqslant z)$
12. $\forall x \forall y (x \leqslant y \vee y \leqslant x)$
13. $\forall x \forall y \forall z (x \leqslant y \rightarrow x + z \leqslant y + z)$
14. $\forall x \forall y (x \leqslant y \rightarrow \forall z (z > 0 \rightarrow x \cdot z \leqslant y \cdot z))$,
    where $z > 0$ is the usual abbreviation of the formula $0 \leqslant z \wedge z \neq 0$
15. $\forall y_{2n+1} \forall y_{2n} \ldots \forall y_0 (y_{2n+1} \neq 0 \rightarrow \exists x (\sum_{k=0}^{2n+1} y_k x^k = 0))$.

The first fourteen axioms are axioms of the ordered fields. The last axiom is actually an axiom scheme with countably many instances, and it provides that each polynomial of odd degree has a root.

We will conclude this appendix with some basic definitions from model theory. A first order language $\mathcal{L}$ is a disjoint union of the sets $Const\mathcal{L}$, $Fun\mathcal{L}$ and $Rel\mathcal{L}$. Each $C \in Const\mathcal{L}$ is called a symbol of constant, each $F \in Fun\mathcal{L}$ is called a symbol of function and each $R \in Rel\mathcal{L}$ is called a symbol of relation. In addition, for each $s \in Fun\mathcal{L} \cup Rel\mathcal{L}$ is given its arity, i.e. the number of argument places.

Let $\mathcal{L}$ be a first order language. A model of $\mathcal{L}$, or $\mathcal{L}$-structure, is a pair (M, I), where M is a nonempty set and I (interpretation) is a function which domain is $\mathcal{L}$, and I satisfies the following conditions:

- $I(C) \in M$, $C \in Const\mathcal{L}$
- $I(F) : M^n \longrightarrow M$, where n is the arity of $F \in Fun\mathcal{L}$
- $I(R) \subseteq M^n$, where n is the arity of $R \in Rel\mathcal{L}$.

Suppose that $\mathcal{L}$ and $\mathcal{L}_1$ are first order languages such that $\mathcal{L} \subseteq \mathcal{L}_1$, and that (M, I) is an $\mathcal{L}$-structure. An expansion of the model (M, I) into language $\mathcal{L}_1$ is any $\mathcal{L}_1$ structure (M, J) such that, for all $s \in \mathcal{L}$,

$$J(s) = I(s).$$

Abstract Argumentation Scheme Frameworks

Katie Atkinson and Trevor Bench-Capon

Department of Computer Science
University of Liverpool
Liverpool L69 3BX UK
{K.M.Atkinson,tbc}@liverpool.ac.uk

Abstract. This paper presents an approach to modelling and reasoning about arguments that exploits and combines two of the most popular mechanisms used within computational modelling of argumentation: argumentation schemes and abstract argumentation frameworks. Our proposal combines the desirable properties of each by representing the components of argumentation schemes as argumentation frameworks. This allows us to make use of the structure provided by the schemes to guide dialogues and provide contextual elements of evaluation, whilst retaining the desirable properties of abstract frameworks to enable evaluation with respect to the logical relations between arguments. Our proposal takes account of dialogical aspects within a debate, such as burden of proof, and we illustrate our approach through a particular argumentation scheme, namely argument from expert opinion.

1 Introduction

Two of the most significant developments in the computational modelling of argumentation in recent years have been *abstract argumentation frameworks*, introduced by Dung in [9], which emerged from logic programming, and *argumentation schemes*, e.g. [18] which emerged from informal logic. Interestingly, these seem to pull in opposite directions: while abstract argumentation considers arguments as structureless atomic entities related only by a binary attack relation, argumentation schemes articulate the varied structures that can constitute arguments, thus adding enriching detail, rather than abstracting from this detail to give the clean semantic properties offered by argumentation frameworks. Both approaches have clear attractions: in this paper we will attempt to provide a means of capturing the variety of structures offered by argumentation schemes in a way in which the properties of abstract frameworks can still be exploited.

Abstract argumentation frameworks have been widely studied as a means of exploring issues relating to defeasible reasoning and non-monotonic logics. There have been various proposals for different semantics for these frameworks, and these have been investigated and compared (e.g. [6]). Complexity questions relating to decision problems regarding such frameworks have been resolved (e.g. [8]), and particular constrained frameworks explored (e.g. [10]). All in all, abstract argumentation frameworks give a clean and well understood basis for considering the status of a related collection of arguments.

Turning to argumentation schemes, these have been exploited in argument diagramming tools such as Arucaria [16]. There, however, they are no more than annotations on

D. Dochev, M. Pistore, and P. Traverso (Eds.): AIMSA 2008, LNAI 5253, pp. 220–234, 2008.

the diagram serving to group premises and conclusions, and require the user to employ them in a principled fashion. Argumentation schemes are also used in the Carneades framework [11]. There the nature of the scheme is used to distinguish between ordinary premises, which must be shown, assumptions, which must be shown on demand, and exceptions, which vitiate the argument if shown. These distinctions are made on the basis of the *critical questions* characteristic of the argumentation scheme being represented.

Central to the notion of argumentation schemes, as described in [18], is that the claim they support is merely presumptive. Associated with each scheme is a set of critical questions which if posed must be answered successfully, or else the claim withdrawn. As well as their use in Carneades noted above, critical questions have been used in [3] to identify the various ways in which arguments made by instantiating a particular scheme can be attacked. A formal characterisation of one particular argumentation scheme, for reasoning about action, and its associated critical questions allowing the identification of attacking arguments, is given in [2].

Argumentation frameworks and argumentation schemes have different strengths. Argumentation frameworks are at their best when we have completely identified the set of relevant arguments and the attack relations between them. Very often, however, this complete set is not available: in many applications the argumentation framework is created, often through a dialogue between participants advocating different points of view, e.g. [13]. Here argumentation schemes come to the fore: these schemes help to identify the ways in which arguments can be attacked and defended, and the dialogical burdens of production and proof [11] [14] which are proper to the various participants, and which may impact on the evaluation of the arguments. If a point cannot be decisively established, it is important to know which participant has the burden of proof with respect to that point. Essentially, argumentation schemes guide the generation of arguments, and supply contextual aspects for their evaluation. Once the contextual issues have been resolved, the arguments can be abstracted to an argumentation framework, and evaluated with respect to their logical relations.

In this paper we will draw upon several of the above approaches to provide a means of firmly integrating argumentation schemes with abstract argumentation frameworks. We first recapitulate the notions of abstract argumentation frameworks introduced in [9]. We then discuss the notions of argumentation schemes and critical questions, and the interpretation of them given in [11], illustrated by a particular example scheme, *Argument from Expert Opinion*, as formulated in that paper. We will pay particular attention to the way in which argumentation schemes identify the responsibilities of participants in a dialogue, and how this affects the status of the arguments within the scheme. We then provide definitions to represent argumentation schemes in a form reducible to argumentation frameworks, and to link them together to form *abstract argumentation scheme frameworks*. We again illustrate our approach by applying it to the scheme *Argument from Expert Opinion*. We conclude by identifying directions in which this work can be built upon: most especially how argumentation schemes can be used to drive dialogue and to confer properties on arguments that are required to differentiate arguments acceptable to different audiences.

2 Argumentation Frameworks and Argumentation Schemes

In this section we provide a brief overview of the two approaches to argumentation that we later combine within a single framework.

2.1 Abstract Argumentation Frameworks

Abstract argumentation frameworks (AFs) have proven to be an influential approach to non-monotonic reasoning over the past decade. The underlying idea of AFs is to model and evaluate arguments by considering how well they can be defended against other arguments that can attack and defeat them. The relationships between arguments can be modelled as directed graphs showing which arguments attack one another. No concern is given to the internal structure of the arguments, so the status of an argument can be evaluated by considering whether or not it is able to be defended from attack from other arguments with respect to a set of arguments. Essentially, an argument can be justified with respect to a set of arguments if it is not attacked by a member of that set, and all its attackers are attacked by a member of that set. AFs were first introduced by Dung in [9] but numerous subsequent works have extended the basic frameworks to incorporate properties such as preferences [1] and values [7], as well as extending the semantics associated with the frameworks, e.g. [6].

Here we recall the following basic concepts that were introduced by Dung in [9][1].

Definition 1. *An argumentation framework (AF) is a pair $\mathcal{H} = \langle \mathcal{X}, \mathcal{A} \rangle$, in which $\mathcal{X}$ is a finite set of arguments and $\mathcal{A} \subset \mathcal{X} \times \mathcal{X}$ is the attack relationship for $\mathcal{H}$. A pair $\langle x, y \rangle \in \mathcal{A}$ is referred to as 'y is attacked by x' or 'x attacks y'. For R, S subsets of arguments in the system $\mathcal{H}(\langle \mathcal{X}, \mathcal{A} \rangle)$, we say that*

a. $s \in S$ is attacked by R if there is some $r \in R$ such that $\langle r, s \rangle \in \mathcal{A}$.
b. $x \in \mathcal{X}$ is acceptable with respect to S if for every $y \in \mathcal{X}$ that attacks x there is some $z \in S$ that attacks y.
c. S is conflict-free if no argument in S is attacked by any other argument in S.
d. A conflict-free set S is admissible if every argument in S is acceptable with respect to S.
e. S is a preferred extension if it is a maximal (with respect to $\subseteq$) admissible set.

2.2 Argumentation Schemes and Critical Questions

Having recapitulated abstract argumentation frameworks, we now provide an overview of argumentation schemes.

In [18] Walton has provided a number of different argumentation schemes that capture particular patterns of reasoning. Although we note that different schemes have been proposed by others, we base our view of argumentation schemes on Walton. Argumentation schemes are stereotypical patterns of reasoning. Like deductive arguments they have premises and a conclusion, but unlike deductive arguments they only provide reasons why the claim can be *presumed* to be true.

---

[1] In this paper we use only preferred extensions and so do not define grounded and stable extensions.

Such schemes follow a general pattern by which the arguments are presented as general inference rules whereby given a set of premises, a conclusion can be drawn. As noted above, however, the conclusions justified by an argumentation scheme are only presumptive, and open to question and defeat. In particular, an argument based on a particular scheme is subject to a set of critical questions characteristic of that scheme. The schemes allow arguments to be presented within a particular context but take into account that the conclusions drawn may be altered in the light of further considerations raised by the critical questions, such as new evidence or exceptional circumstances. We next illustrate the notion of argumentation schemes with an example: 'Argument from Expert Opinion'.

2.3 Argument from Expert Opinion

Several versions of this argumentation scheme have been presented. We use a recent version given in [11], which is stated as follows:

Major premise: Source E is an expert in the subject domain S containing proposition A.
Minor premise: E asserts that proposition A in domain S is true.
Conclusion: A may plausibly be taken as true.

The scheme has associated with it the following six critical questions:

CQ1: How credible is E as an expert source?
CQ2: Is E an expert in the field that A is in?
CQ3: Does E's testimony imply A?
CQ4: Is E reliable?
CQ5: Is A consistent with the testimony of other experts?
CQ6: Is A supported by evidence?

If any of these critical questions are posed, in order for the presumptive conclusion to stand, its proponent must respond satisfactorily to that critical question. What counts as a satisfactory response, however, depends upon the specific role that the critical question plays. Following [11], we recognise the following three categories of critical questions:

– Those used to question whether a *premise* of a scheme holds (e.g. CQ2 and CQ3)
– Those used to recognise *exceptions* to the use of the scheme (e.g. CQ4 and CQ5)
– Those used to question the *assumptions* used in the scheme (e.g. CQ1 and CQ6)

These categories differ: the last two contend that the presumptive conclusion does not in fact hold, whereas the first denies that the argument can be proposed at all (since its premises are false), as argued in [17]. Moreover, for assumptions the *burden of proof* is on the proponent, whereas for exceptions the burden of proof is on the opponent.

3   Abstract Argumentation Scheme Frameworks

In this section we will articulate argumentation schemes in terms of argumentation frameworks. The idea is that, as the discussion above suggested, we should not see an argumentation scheme as an atomic whole, but rather as a *process* of argumentation. We therefore need to identify the elements of an argumentation scheme and the relations between them.

We can define a general argumentation scheme as:

GAS: A proposal that a set of premises provide a reason for a conclusion: *prop = conc*, *because premises*

A set of assumptions: $assump_1 ... assump_n$
A set of exceptions: $except_1 ... except_n$
A conclusion: *conc*

For simplicity we will, without loss of generality, discuss a scheme, GAS1, with a proposal, *prop1* (i.e. *conc1*, because *prem1*), one assumption, *assump1*, one exception, *except1*, and a conclusion *conc1*.

The proponent will put forward the argumentation scheme as *prop1*. In doing so the proponent asserts that *prem1* is true, and that *conc1* should be believed on the basis of *prem1*. These claims are distinct, and although elided in normal presentation, should be considered as two claims rather than one. Since *prem1* provides a reason for *conc1* only if the assumption is satisfied and the exception does not apply, putting the scheme forward implicitly commits the proponent to *assump1* and implicitly denies *except1*. Thus the proponent of the argumentation scheme can be seen as making four claims: *prem1* being true provides a reason for *conc1*; *conc1* should be believed; *assump1* is true; and, *except1* is not true. These claims are related in the following ways.

If either *assump1* is false or *except1* is true, although *prem1* does provide a reason to believe *conc1*, *conc1* should not be accepted. Thus both *not assump1* and *except1* attack *conc1*. By asserting that *prem1* is a reason for *conc1*, however, the proponent has attacked both *not assump1* and *except1*. This enables us to see the argumentation scheme GAS1 as an AF, GASAF1, with arguments *prop1, not assump1, except1, conc1* and attacks {(*prop1, not assump1*), (*prop1, except1*), (*except1,conc1*), (*not assump1,conc1*)}. A graphical representation of this AF is shown in Figure 1.

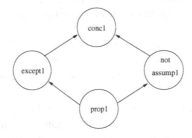

Fig. 1. AF after proponent's arguments are put forward

Considered as a standard AF in which attacks are always successful, the acceptable arguments, representing the preferred extension for the framework, are {*prop1, conc1*}, and {*not assump1, except1*} are defeated, which corresponds to the claims attributed to the proponent above.

Now consider the role of the opponent. The opponent may simply accept GAS1. If the opponent is sincere, he will do so if he believes *prem1* and *assump1* and does not believe *except1*. The opponent, however, has the right to critique the argument. Suppose opponent does not accept *prem1*. An argument for *not prem1* constitutes an attack on the proposal, *prop1*, since *prop1* holds only if *prem1* is true. If, however, the opponent does not accept the assumption, this is an assertion of *not assump1*, which does not affect the proposal, but rather denies that the conclusion should be believed on the basis of the proposal. Similarly if the opponent believes that the exception holds, this requires the assertion of *except1*, which also denies that the conclusion should be believed, rather than that the original proposal is flawed.

Nothing special is required to handle a denial of *prem1*: this is an argument attacking *prop1* which must be defeated in the usual way to reinstate *prop1*. For the moment, therefore, we will take *prop1* as unattacked, since this debate is external to GAS1. We do, however, already have arguments for *not assump1* and *except1* in GASAF1, but these are attacked by *prop1*. Here, therefore, we must be able to distinguish between attack and defeat. To reflect the opponent's *burden of production*, which reflects that the presumptive conclusion holds until challenged, we allow the opponent to mark *not assump1* and *except1* as *produced*, and say that, *within an argumentation scheme*, an attack fails to defeat an argument marked as produced, unless the attacking argument is also produced. Figure 2 shows the AF for this scenario.

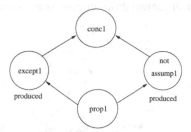

Fig. 2. AF updated with CQs produced by opponent

Suppose the opponent marks *not assump1* as produced in GASAF1. Now *not assump1* will not be defeated by *prop1*, and so will defeat *conc1*. In order to reinstate *conc1*, proponent must therefore provide an argument external to the scheme to defeat *not assump1* (i.e. justify the assumption). Suppose, however, that the opponent marks *except1* as produced in GASAF1. This will mean that *except1* is no longer defeated by *prop1*, and so will defeat *conc1*. But in this case, since here the burden of proof is on opponent, proponent should not be required to show that *except1* is false in order to reinstate *conc1*. While a sincere proponent will simply accept *except1* when produced if he has reason to believe it, proponent has the right to challenge the opponent to provide an argument for *except1*. This can be reflected by introducing another element, which we will call *challenge*, which is initially implicit and so unmarked, but which the proponent can choose to mark as *produced* if he does not believe that the exception holds. Once produced, this will defeat the exception, and so reinstate the conclusion. Figure 3 shows this scenario and the effect of the *challenge* argument on the status of *except1*.

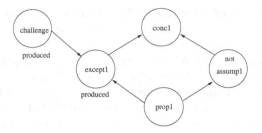

Fig. 3. AF showing *produced* exception and challenge

Now, to discharge the burden of proof, the opponent must reinstate the exception with an argument for the exception, such that this new argument attacks the challenge.

One further addition is also required. If the proposal is defeated, the conclusion must also be defeated. But it is possible that all the assumptions, and none of the exceptions are satisfied, even though the premise is defeated. A person may be credible, unbiased and so forth, but we cannot use argument from expert opinion to justify a belief in an opinion if the proposal is defeated. We therefore need an additional argument, attacked by the proposal and attacking the conclusion, representing that the proposal is unacceptable. This need not be produced: it is always defeated if the proposal is acceptable. Let us term this additional argument *unacceptable*.

This enables us to see an argumentation scheme as an AF with a particular structure, illustrated in Figure 4.

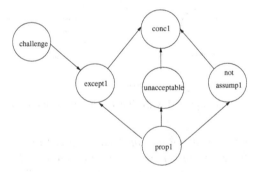

Fig. 4. AF with *unacceptable* argument

We can now define this structure as follows:

Definition 2. *Arguments in an Argumentation Scheme Framework*: An argumentation scheme framework, ASF, is a tuple <*prop, Assumps, Excepts, Challenges, unaccept, conc*> where *prop, unaccept* and *conc* are arguments, and *Assumps, Excepts* and *Challenges* are sets of arguments. All of *Assumps, Excepts* and *Challenges* contain zero or more arguments, and for every *except$_i$* ∈ *Excepts* there is a corresponding *challenge$_i$* ∈ *Challenges*. Let AS = {*prop*} ∪ *Assumps* ∪ *Excepts* ∪ *Challenges* ∪ {*unaccept*} ∪ {*conc*}.

Definition 3. *Attacks in an Argumentation Scheme Framework*:

    i) for all $assump_i \in Assumps$, attacks($prop, assump_i$)

    ii) for all $except_i \in Excepts$, attacks($prop, except_i$)

    iii) for all $assump_i \in Assumps$, attacks($assump_i, conc$)

    iv) for all $except_i \in Excepts$, attacks($except_i, conc$)

    v) for all $except_i \in Excepts$, $challenge_i \in Challenges$, attacks($challenge_i, except_i$)

    vi) attacks($prop, unaccept$)

    vii) attacks($unaccept, conc$).

Let ASatts be the set of all attacks in the ASF.

Definition 4. *Defeat in an Argumentation Scheme Framework*: Let *Produced* be the subset of AS such that an argument $ap \in Produced$ if and only if ap is marked as *produced*. Now for any $arg1, arg2 \in AS$, attacks($arg1, arg2$) succeeds if and only if attacks($arg1, arg2$) $\in$ ASatts and $arg1 \in Produced$, or it is not the case that $arg2 \in Produced$. Let ASdefs be the subset of ASatts such that $asdef \in$ ASdefs if and only if $asdef \in$ ASatts and $asdef$ succeeds.

Definition 4 uses the dialogical status of arguments in AS to determine whether the attacks in ASatts are successful. Note that success is determined entirely by the dialogical status of the arguments concerned together with the burdens of production and proof imposed by the argumentation scheme. This is therefore entirely objective, and there is no need to consider different audiences in determining whether an attack succeeds. Once, however, we have made use of the dialogical status in this way, we can abstract away from it to return to an abstract argumentation framework, but with only the successful attacks included. An argumentation scheme framework can thus be abstracted to an abstract argumentation framework, in the sense of Definition 1, with X = AS and A = ASdef. We can then determine the acceptability of arguments in this framework using any of the standard semantics applied to Dung's AFs. Note also that a deductive argument can be viewed as a degenerate argumentation scheme in which all of *Assumps*, *Excepts* and *Challenges* are empty.

4   Linking Abstract Argumentation Scheme Frameworks

Thus far we have only considered a single argumentation scheme in isolation. We now need to embed the scheme within a larger framework. First note that only conclusions can be used to attack arguments external to the scheme: if it is not established that the conclusion should be believed, it cannot attack any other argument. All of assumptions, proposals and challenges, however, can be attacked from outside the scheme: assumptions by an argument that the assumption holds, proposals by an argument that the premise does not hold, and challenges by an argument that the corresponding exception does hold.

    Suppose that we regard all arguments under consideration as representing arguments using some argumentation scheme. We will now have a set of instantiated argumentation schemes linked by the attack relation, with the conclusion of one argumentation scheme attacking the proposals, assumptions, challenges and conclusions of other argumentation schemes. We use the example of the scheme for argument from expert opinion to examine how instantiations of argumentation schemes and their associated critical questions form arguments that can be represented and evaluated in terms of the definitions just introduced.

The starting point in constructing the framework is an instantiation of an argumentation scheme. Figure 5 shows the complete ASF for an argument i and relevant fragments of ASFs for arguments j, k and m. Instantiating the scheme for argument i, an argument from Expert Opinion, creates an ASF with the following nodes: a *prop* node representing the proposal of the instantiated scheme, labelled in the graph with 'prop_i'; a *conc* node representing the conclusion of the scheme, labelled with 'conc_i'; an *assump* node representing a critical question of type assumption, labelled with 'CQ1'; and, an *except* node representing a critical question of type exception, labelled with CQ5, along with the *challenge* node on CQ5, labelled with the 'challenge_i'[2]. In accordance with our definitions, either of CQ1 and CQ5 may be answered by the conclusion of another ASF, which we indicate with the two nodes that attack the *assump* and *challenge* nodes in the framework. Furthermore, the conclusion of one ASF may attack the conclusion of another ASF, as shown in the graph by conc_i attacking conc_m. Finally, we also need to include a node in the graph for the claim that i is *unacceptable*, which is attacked by the proposal and which itself attacks the conclusion, as described previously.

Figure 5 shows a framework that includes two of the critical questions associated with the scheme from expert opinion: the remaining assumption and exception can be included similarly, whereas CQ2 and CQ3 require further ASFs with conclusions which attack the proposal.

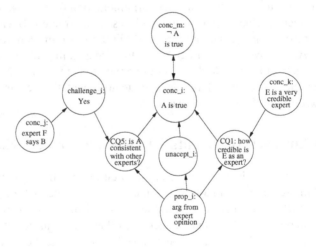

Fig. 5. AF for Argument from Expert Opinion scheme

From Figure 5, we can see that CQ1 is an argument of type *Assumps*, questioning the assumption that E is credible as an expert. As shown in the framework, the way to respond to this question is to instantiate another argumentation scheme (argument

---

[2] Due to space restrictions we include only one of the assumptions and exceptions and omit attacks on the premise.

k) to provide an argument confirming that E is indeed credible as an expert, perhaps by listing E's qualifications (as shown by the node conc_k attacking CQ1). CQ5, on the other hand, is of type *Excepts* and the response to this objection must be handled differently. The proponent of the initial argument may simply respond with a *challenge* to force the opponent to respond with some evidence to show that there are experts who disagree, best done by producing an argument from expert opinion based on a second expert.

5 Example

In this section we will make our discussion more concrete by presenting an example. The heart of the example will be two instantiations of Argument from Expert Opinion, the scheme discussed previously. We will, however, also make use of a number of other argumentation schemes for which we provide no description. We trust that the nature of these schemes will be clear from the context and the example. Identifying and classifying argumentation schemes, giving them a precise characterisation in terms of their premises and critical questions, is an area of active current investigation (e.g. [15]).

To begin the example, suppose Wilma and Bert are having breakfast and a discussion as to whether they should eat more grapefruit begins. Wilma is a grapefruit advocate, whereas Bert favours other kinds of fruit. Wilma presents her case using an argument from Expert Opinion:

W1: Jane is an expert on nutrition and she says that grapefruits are healthy.

Bert can now respond to this, and uses the critical questions characteristic of the scheme to choose his response. Suppose that he is willing to accept that Jane is indeed a credible expert on nutrition: this means that he will not question Wilma's assumption. But he recalls Jane saying something different, and so attacks the basic premise with an Argument from Testimony.

B1: Jane actually said "grapes are a healthy fruit", so she did not say that grapefruits are healthy.

Wilma needs to defeat this if she is to maintain W1, so she produces Jane's book *Eating for Life*, turns to page 69 and produces an Argument from Citation:

W2: Jane wrote *Eating for Life*, and on page 69 it says "Of all fruits, grapefruits are the most healthy." So Jane says grapefruits are healthy.

Confronted with this incontrovertible evidence, Bert must find another way to avoid grapefruit. Reviewing the critical questions, he realises he can use a dissenting expert.

B2: But other experts say different things.

Wilma is unaware of any dissenting experts and since Bert is using an exception, she can demand that he substantiate B2.

W3: Which other experts?

Bert has one, and so puts forward his own Argument from Expert Opinion:

B3: Carol is an expert on nutrition and she says that grapefruits are not good for you.

Wilma is aware that Carol is also an expert on nutrition, and now recalls that she did indeed say that. However, she can use W1 as an exception to B3. The situation is now as shown in Figure 6. Only W1 and B3 are shown in full: B1 and W2 are represented as single nodes, while B2 and W3 are included as parts of W1.

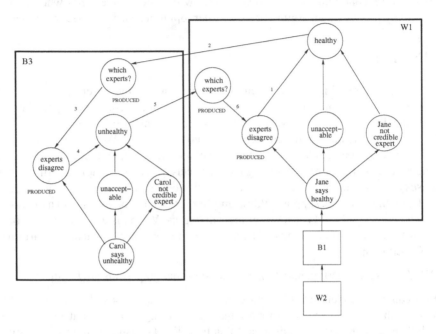

Fig. 6. Figure 6: Arguments in the Bert and Wilma debate. The cycle of six arguments, where the attacks are numbered in the graph, gives rise to two preferred extensions for the framework. This conflict is resolved for a given audience by preferring one or other of the experts . If Jane is preferred attack 5 fails and is removed: if Carol is preferred attack 2 is removed.

At this point there is something of an impasse: the six cycle in the argumentation framework means that it has two preferred extensions[3], one containing the conclusion that grapefruits are healthy, and one that they are not. How do we resolve this dilemma? It is widely recognised that the acceptability of arguments does not only depend on their intrinsic merits, but on the audience to which they are addressed [12]. In approaches such as [7] properties are ascribed to arguments and different audiences are represented by different rankings of these properties. Arguments can now be compared with their attackers on the basis of these properties, and the ranking of the relevant audience used to determine which of these attacks will succeed. Unsuccessful attacks can then be removed to provide a Dung-style argumentation framework appropriate to that audience. The question now arises as to where these properties come from. Our answer is that they come from the argumentation scheme used to generate the argument. For example, the values ascribed to arguments by [3] can be derived from their argumentation scheme for

---

[3] Only argumentation frameworks with even length cycles give rise to multiple preferred extensions [7].

practical reasoning, and in our case the Argument from Expert Opinion confers the degree of authority associated with the particular expert. This provides another important reason for paying attention to argumentation schemes. In many applications we need to supply a principled way of choosing between mutually attacking arguments.

Returning to Bert and Wilma, the successful argument in our debate will depend on whether Jane or Carol is considered the better expert. At this point, if Bert and Wilma agree on which of Jane or Carol is to be relied on, the debate can conclude. They may, however, disagree, in which case neither will be persuaded: Wilma has failed to convince Bert, but equally Bert has not forced her to change her views. They may therefore agree to disagree. Suppose, however, Bert does some searching on the internet, and finds a paper reporting a study which shows that grapefruits have some bad effects. He can return to the fray the following morning and attack the conclusion of W1 directly with an Argument from Scientific Study:

B4: A study published in *Diet Today* showed that people who eat grapefruits are significantly more likely to suffer from indigestion. This shows that grapefruits are not healthy.

Wilma may accept this, or may continue the debate using the critical questions associated with Argument from Scientific Study: for example, that *Diet Today* is not a refereed journal. We leave the example here.

The example shows: how various critical questions can be used to drive the discussion, informing the moves of the participants; the role of producing arguments in determining the dynamic status of the arguments in the course of the debate; and how argumentation schemes can be used to confer properties on the arguments and so resolve conflicts in terms of the subjective audience to which the argument is presented.

6 Discussion

So far we have shown how the presumptive style of argument supported by argumentation schemes can be integrated with abstract argumentation frameworks. The various claims made explicitly and implicitly when advancing a scheme, and the possible means of questioning those claims, all appear as nodes in the framework, related in a particular way in order to produce a structured framework characteristic of the scheme. Evaluation of these arguments can then be effected by filtering the attacks to remove those that are unsuccessful by reason of their dialogical status, and then, having used this necessary contextual information, considering them as standard, purely abstract, frameworks.

A key motivation is the use of argumentation in dialogue: the schemes enable us to identify how an opponent can respond to an argument made using a particular instantiated scheme. Consideration of the example in Figure 6 gives further pointers to how further assistance can be provided for conducting dialogues. When various critical questions are produced they need to be met by arguments to justify the assumptions or substantiate the exceptions. But because the critical questions address particular issues, the arguments answering them must also be of particular types. In our example, the conclusion of B3, which establishes the exception by showing that there are dissenting experts, is itself the conclusion of an argument using the Expert Opinion scheme. Similarly there will typically be prescribed ways of establishing that someone is indeed an expert in order to justify the assumption that the expert is credible, perhaps by pointing to qualifications or a position held in a reputable university. Thus we can expect the arguments related to a given argumentation scheme to be themselves instantiations of a limited number of particular argumentation schemes characteristic of the concern they address. In this way, a given argumentation scheme can be seen as embedded in a conversation composed from a range of argumentation schemes in a well defined manner. This has analogies with general agent communication in which it is necessary to see individual speech acts as embedded in a conversation class to provide the context necessary to their interpretation and to guide the appropriate responses as the dialogue develops [4] [5].

Of course, modelling argumentation schemes and identifying characteristic responses to their critical questions does require significant analysis of particular domains in which debates take place. What the approach in this paper provides is a means of harnessing this analysis in a way which can be made compatible with abstract argumentation frameworks for evaluation. The domain analysis guides the dialogue and provides contextual input to the evaluation. The legal domain, in particular, often states quite explicitly that certain assumptions and exceptions can only be established in specified ways. An example analysis concerning reasoning from a precedent is given in [19]. Here the critical questions of the main scheme are responded to by further specific argumentation schemes, thus allowing the whole debate to be represented as a conversation made up of a cascade of particular argumentation schemes.

As well as providing contextual information which can be used to remove unsuccessful attacks and so provide a means of abstracting these features away to reach a standard AF, argumentation schemes can provide properties which can be used to resolve conflicts between arguments that depend on the subjective interests and opinions of the audience to which they are presented. Our example in section 5 illustrated this by requiring the audience to express a preference between competing experts. Once the preference has been expressed, the unsuccessful attack can be removed, again allowing abstraction to an abstract framework appropriate to that audience.

Note that the same technique is used both to handle burden of production and audience preferences. Argumentation schemes provide properties for their constituent arguments (objective dialogical status for critical questions and subjective audience-related preferences for conflicts) and these properties are used to identify attacks that are contextually unsuccessful. Once these properties have played their part in providing the contextual and audience-related elements of evaluation, they can be discarded and the resulting AF evaluated to provide the logic of the debate.

To conclude: in this paper we have shown how argumentation schemes can guide the identification of moves to challenge claims and make appropriate responses to particular challenges. Argumentation schemes also supply contextual elements of evaluation of the arguments in a way which allows abstraction to a standard Dung-style argumentation framework. This abstract framework allows arguments to be evaluated with respect to their logical relations. In this way, we provide a means of combining the distinctive advantages of both approaches. The proposal we have presented is intended to lay the foundations for a dialogical account of argumentation whereby dialogue interactions are guided by the particular moves, corresponding to the elements of the argumentation scheme, that can be put forward and subsequently evaluated through the use of argumentation frameworks. Articulating the machinery for this dialogical setting will be the focus of the next step with this work.

Acknowledgments

We are grateful to the anonymous reviewers for their helpful comments and suggestions.

References

1. Amgoud, L., Cayrol, C.: On the acceptability of arguments in preference-based argumentation. In: Cooper, G.F., Moral, S. (eds.) Proccedings of the Fourteenth Conference on Uncertainty in Artificial Intelligence (UAI 1998), pp. 1–7. Morgan Kaufmann, San Francisco (1998)
2. Atkinson, K., Bench-Capon, T.J.M.: Action-based alternating transition systems for arguments about action. In: Proceedings of the Twenty Second Conference on Artificial Intelligence (AAAI 2007), pp. 24–29. AAAI Press, Menlo Park (2007)
3. Atkinson, K., Bench-Capon, T.J.M., McBurney, P.: Computational representation of practical argument. Synthese 152(2), 157–206 (2006)
4. Barbuceanu, M.: Coordinating agents by role based social constraints and conversation plans. In: Proceedings of the Fourteenth National Conference on Artificial Intelligence (AAAI 1997), pp. 16–21. AAAI Press, Menlo Park (1997)
5. Barbuceanu, M., Fox, M.S.: COOL: A language for describing coordination in multi agent systems. In: Proceedings of the First International Conference on Multiagent Systems, pp. 17–24 (1995)
6. Baroni, P., Giacomin, M.: A systematic classification of argumentation frameworks where semantics agree. In: Besnard, B., Doutre, S., Hunter, A. (eds.) Computational Models of Argument, Proceedings of COMMA 2008. Frontiers in Artificial Intelligence and Applications, vol. 172, pp. 37–48. IOS Press, Amsterdam (2008)
7. Bench-Capon, T.J.M.: Persuasion in practical argument using value based argumentation frameworks. Journal of Logic and Computation 13(3), 429–448 (2003)
8. Dimopoulos, Y., Nebel, B., Toni, F.: On the computational complexity of assumption-based argumentation for default reasoning. Artificial Intelligence 141(1–2), 57–78 (2002)
9. Dung, P.M.: On the acceptability of arguments and its fundamental role in nonmonotonic reasoning, logic programming and n-person games. Artificial Intelligence 77, 321–357 (1995)
10. Dunne, P.E.: Computational properties of argument systems satisfying graph-theoretic constraints. Artificial Intelligence 171(10–15), 701–729 (2007)
11. Gordon, T., Prakken, H., Walton, D.: The Carneades model of argument and burden of proof. Artificial Intelligence 171(10-15), 875–896 (2007)
12. Perelman, C., Olbrechts-Tyteca, L.: The New Rhetoric: A Treatise on Argumentation. University of Notre Dame Press, Notre Dame (1969)
13. Prakken, H.: Coherence and flexibility in dialogue games for argumentation. Journal of Logic and Computation 15, 1009–1040 (2005)
14. Prakken, H., Sartor, G.: Presumptions and burdens of proof. In: Legal Knowledge and Information Systems. JURIX 2006: The Nineteenth Annual Conference, pp. 21–30. IOS Press, Amsterdam (2006)
15. Rahwan, I.: Mass argumentation and the semantic web. Journal of Web Semantics 6(1), 29–37 (2008)

16. Reed, C.A., Rowe, G.W.A.: Araucaria: Software for argument analysis, diagramming and representation. International Journal of AI Tools 14(3–4), 961–980 (2004)
17. Verheij, B.: Dialectical argumentation with argumentation schemes. Artificial Intelligence and Law 11(2–3), 167–195 (2003)
18. Walton, D.N.: Argumentation Schemes for Presumptive Reasoning. LEA, Mahwah, NJ, USA (1996)
19. Wyner, A.Z., Bench-Capon, T.J.M.: Argument schemes for legal case-based reasoning. In: The Twentieth Annual Conference on Legal Knowledge and Information Systems. JURIX 2007, pp. 139–149. IOS Press, Amsterdam (2007)

Nested Precedence Networks with Alternatives:
Recognition, Tractability, and Models

Roman Barták[1] and Ondřej Čepek[1,2]

[1] Charles University in Prague, Faculty of Mathematics and Physics
Malostranské nám. 2/25, 118 00 Praha 1, Czech Republic
`roman.bartak@mff.cuni.cz, ondrej.cepek@mff.cuni.cz`
[2] Institute of Finance and Administration
Estonská 500, 101 00 Praha 10, Czech Republic
`cepek@mail.vsfs.cz`

Abstract. Integrated modeling of temporal and logical constraints is important for solving real-life planning and scheduling problems. Logical constrains extend the temporal formalism by reasoning about alternative activities in plans/schedules. Temporal Networks with Alternatives (TNA) were proposed to model alternative and parallel processes, however the problem of deciding which activities can be consistently included in such networks is NP-complete. Therefore a tractable subclass of Temporal Networks with Alternatives was proposed. This paper shows formal properties of these networks where precedence constraints are assumed. Namely, an algorithm that effectively recognizes whether a given network belongs to the proposed sub-class is studied and the proof of tractability is given by proposing a constraint model where global consistency is achieved via arc consistency.

Keywords: temporal networks, alternatives, constraint models, complexity.

1 Introduction

Temporal networks focus on modeling temporal relations in planning and scheduling applications. Nodes of the network describe activities (or more specifically, important time points such as start times and end times of activities) while arcs are annotated by temporal relations between the activities such as precedence relations. Traditional approaches assume that all nodes are present in the network, though the position of nodes in time may be influenced by other than temporal constraints, for example, by resource constraints. Conditional Temporal Planning (CTP) [10] introduced an option to decide which node will be present in the solution depending on a certain external condition. Hence CTP can model conditional plans where the nodes actually present in the solution are selected based on external forces. Recently, logical dependencies between the nodes were added to temporal networks to capture relations between existences of nodes. For example, the logical dependency A $\Rightarrow$ B says that if node A is present in the solution then node B must be present as well. The possibility to select nodes according to logical, temporal, and resource constraints was introduced to

D. Dochev, M. Pistore, and P. Traverso (Eds.): AIMSA 2008, LNAI 5253, pp. 235–246, 2008.
© Springer-Verlag Berlin Heidelberg 2008

manufacturing scheduling by ILOG in their MaScLib [9]. The same idea was independently formalized in Extended Resource Constrained Project Scheduling Problem (RCPSP) [8]. In the common model each node has a Boolean validity variable indicating whether the node is selected to be in the solution and these Boolean variables participate in binary logical constraints. The validity variable originates from works by Beck and Fox [4] dealing with alternative activities. *Temporal Networks with Alternatives* (TNA) [1] introduced a similar type of implicit alternatives via so called parallel and alternative branching. Basically, TNA is a directed acyclic graph with nodes corresponding to activities and arcs corresponding to temporal relations, where the input and output branchings in nodes (fan-in and fan-out sub-graphs) are annotated either as parallel or alternative branching. If only precedence relations instead of temporal ones are used, then we are talking about a *parallel/alternative graph* (*P/A graph* in short). The input parallel branching for node x means that all direct predecessors of x (branching nodes) must be processed, while the input alternative branching means that exactly one direct predecessor of x must be processed to allow processing of x (and similarly for the output branching). These branching relations describe how the processes are being joined and split (Figure 1). Formally, the branching relations implicitly define special logical constraints between the validity variables. It is possible to show that alternative branching with two or more branching nodes cannot be decomposed into binary logical constraints used in MaScLib and Extended RCPSP [3].

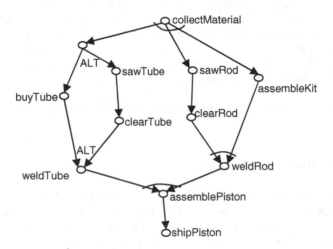

Fig. 1. A real-life manufacturing process with alternatives

The P/A graph assignment problem [1] is defined as a question whether it is possible to select a subset of nodes satisfying the above branching constraints. In practice, it corresponds to selecting a subset of activities describing a single process among alternative processes. This problem is NP-complete if some node is pre-selected [1]. Beck and Fox [4] informally proposed legal networks where logical reasoning should be easy. Nested TNAs [2] generalize the very same idea using the notion of TNA. They are motivated by real-life manufacturing processes built by decomposition of meta-processes into operations (the network in Figure 1 is nested). Hence it is not surprising that the structure of Nested TNA is identical to Temporal Planning Networks (TPN) proposed in [7] and also used in TAEMS formalism [6]. TPNs and TAEMS were proposed ad-hoc by giving a language for specifying nested networks and only the nested networks and no attention was paid to theoretical complexity of node selection. Nested TNAs are based on a more general formalism of TNA; examples in [1] show TNAs which cannot be expressed as TPNs.

This paper extends the work [2] (where Nested TNAs were first introduced) by giving formal proofs of soundness and complexity of the algorithm recognizing Nested TNAs and by proving the tractability of P/A graph assignment problem using a constraint model where global consistency is achieved via arc consistency.

2 Nested P/A Graphs

P/A graphs were proposed in [1] to formally describe processes with parallel and alternative branches. Let us first recapitulate the formal definitions of a P/A graph and a P/A graph assignment problem. Let G be a directed acyclic graph. A sub-graph of G is called a *fan-out sub-graph* if it consists of nodes $x, y_1, ..., y_k$ (for some k) such that each (x, y_i), $1 \leq i \leq k$, is an arc in G. If $y_1, ..., y_k$ are all and the only successors of x in G (there is no z such that (x, z) is an arc in G and $\forall i = 1, ..., k: z \neq y_i$) then we call the fan-out sub-graph complete. Similarly, a sub-graph of G is called a *fan-in sub-graph* if it consists of nodes $x, y_1, ..., y_k$ (for some k) such that each (y_i, x), $1 \leq i \leq k$, is an arc in G. A complete fan-in sub-graph is defined similarly as above. In both cases x is called a *principal node* and all $y_1, ..., y_k$ are called *branching nodes*.

Definition 1. A directed acyclic graph G together with its pairwise edge-disjoint decomposition into complete fan-out and fan-in sub-graphs, where each sub-graph in the decomposition is marked either as a *parallel* sub-graph or an *alternative* sub-graph, is called a *P/A graph*.

In this paper, we focus mainly on handling special logical relations imposed by the fan-in and fan-out sub-graphs – we call them *branching constraints*. In particular, we are interested in finding whether it is possible to select a subset of nodes in such a way that they form a feasible graph according to the branching constraints. Formally, the selection of nodes can be described by an *assignment* of 0/1 values to nodes of a given P/A graph, where value 1 means that the node is selected and value 0 means that the node is not selected. The assignment is called *feasible* if

- In every parallel sub-graph all nodes are assigned the same value (both the principal node and all branching nodes are either all 0 or all 1),
- In every alternative sub-graph either all nodes (both the principal node and all branching nodes) are 0 or the principal node and exactly one branching node are 1 while all other branching nodes are 0.

Notice that the feasible assignment naturally describes one of the alternative processes in the P/A graph. For example, *weldRod* is present if and only if both *clearRod* and *assembleKit* are present (Figure 1). Similarly, *weldTube* is present if exactly one of nodes *buyTube* or *clearTube* is present (but not both).

It can be easily noticed that given an arbitrary P/A graph the assignment of value 0 to all nodes is always feasible. On the other hand, if some of the nodes are required to take value 1, then the existence of a feasible assignment is by no means obvious. Let us now formulate this decision problem formally.

Definition 2. Given a P/A graph G and a subset of nodes in G which are assigned to 1, *P/A graph assignment problem* is "Is there a feasible assignment of 0/1 values to all nodes of G which extends the prescribed partial assignment?"

Intuition motivated by real-life examples says that it should not be complicated to select the nodes to form a valid process according to the branching constraints described above. The following proposition from [1] says the opposite.

Proposition 1. The P/A graph assignment problem is NP-complete.

In the rest of the paper, we will propose a restricted form of the P/A graph, a so called nested P/A graph that can cover many real-life problems while keeping the P/A graph assignment problem tractable.

When we analyzed how the P/A graphs modeling real-life processes look, we noticed several typical features. First, the process has usually one start point and one end point. Second, the graph is built by decomposing meta-processes into more specific processes until non-decomposable processes (operations) are obtained. There are basically three types of decomposition. The meta-process can split into two or more processes that run in a sequence, that is, after one process is finished, the subsequent process can start (serial decomposition). The meta-process can split into two or more sub-processes that run in parallel, that is, all sub-processes start at the same time and the meta-process is finished when all sub-processes are finished (parallel decomposition). Finally, the meta-process may consist of several alternative sub-processes, that is, exactly one of these sub-processes is selected to do the job of the meta-process (alternative decomposition). The last two decompositions have the same topology of the network (Figure 2), they only differ in the meaning of the branches. Since we focus on modeling instances of processes with particular operations that will be allocated to time, we do not assume loops (used to model abstract processes).

Fig. 2. Possible decompositions of the process

Based on above observations we propose a recursive definition of a nested graph.

Definition 3. A directed graph $G = (\{s,e\}, \{(s,e)\})$ is a *(base) nested graph*. Let $G = (V, E)$ be a graph, $(x,y) \in E$ be its arc, and $z_1,\ldots, z_k$ ($k > 0$) be nodes such that no z_i is in V. If G is a nested graph (and $I = \{1,\ldots,k\}$) then graph $G' = (V \cup \{z_i \mid i \in I\}, E \cup \{(x,z_i), (z_i,y) \mid i \in I\} - \{(x,y)\})$ is also a *nested graph*.

According to Definition 3, any nested graph can be obtained from the base graph with a single arc by repeated substitution of any arc (x,y) by a special sub-graph with k

nodes (see Figure 3). Notice that a single decomposition rule covers both the serial process decomposition ($k = 1$) and the parallel/alternative process decomposition ($k > 1$). Though this definition is constructive rather than declarative, it is practically very useful. Namely, interactive process editors can be based on this definition so the users can construct only valid nested graphs by decomposing the base nested graph.

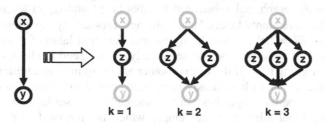

k = 1   k = 2   k = 3

Fig. 3. Arc decompositions in nested graphs

The directed nested graph defines topology of the nested P/A graph but we also need to annotate all fan-in and fan-out sub-graphs as either alternative or parallel sub-graphs. The idea is to annotate each node by input and output label which defines the type of branching. Recall that a fan-out sub-graph with principal node x and branching nodes z_i is a sub-graph consisting of nodes x, $z_1,..., z_k$ (for some k) such that each (x, z_i), $1 \leq i \leq k$, is an arc in G. Fan-in sub-graph is defined similarly.

Definition 4. *Labeled nested graph* is a nested graph where each node has (possibly empty) input and output labels defined in the following way. Nodes s and e in the base nested graph and nodes z_i introduced during decomposition have empty initial labels. Let k be the parameter of decomposition when decomposing arc (x,y). If $k > 1$ then the output label of x and the input label of y are unified and set either to PAR or to ALT (if one of the labels is non-empty then this label is used for both nodes).

Figure 4 shows how the labeled nested graph is constructed for the example from Figure 1. Notice how the labels are introduced (a semicircle for PAR label and A for ALT label) or unified in case that one of the labels already exists (see the third step). When a label is introduced for a node, it never changes in the generation process.

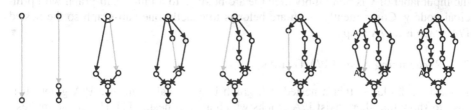

Fig. 4. Building a labelled nested graph

If an arc (x, y) is being decomposed into a sub-graph with k new nodes where $k > 1$, then we require that the output label of x is unified with the input label of y. This can be done only if either both labels are identical or at least one of the labels is empty. The following lemma shows that the second case always holds.

Lemma 1. For any arc (x, y) in the labeled nested graph, either the output label of x or the input label of y is empty.

Proof. The base nested graph contains a single arc (s, e) and labels for s and e are empty so the arc (the graph) satisfies the lemma. Assume now that graph G satisfies the lemma and we decompose some arc (x, y). During the decomposition, arc (x, y) is removed from the graph and substituted by arcs (x, z_i) and (z_i, y) for new nodes z_i, $1 \le i \le k$, which have empty labels. Hence, the new arcs satisfy the lemma. According to Definition 4 if $k > 1$ the output label of x and the input label of y are set (both or just one of them, if the other one was set already) so we need to check the other arcs going from x or going to y. If there was another arc (x, b) in G in addition to removed (x, y) then some arc (x, c) has already been decomposed to obtain two or more arcs going from x. Hence the output label of x has already been set in G and according to assumption the input label of b was empty which is preserved in the new graph. Symmetrically, if there was additional arc (b, y) in G then the output label of b is empty. So, all arcs in graph G that remain in the new graph still satisfy the lemma.   ∎

Now, we can formally introduce a nested P/A graph.

Definition 5. A *nested P/A graph* is obtained from a labeled nested graph by removing the labels and defining fan-in and fan-out sub-graphs in the following way. If the input label of node x is non-empty then all arcs (y, x) form a fan-in sub-graph which is parallel for label PAR or alternative for label ALT. Similarly, nodes with a non-empty output label define fan-out sub-graphs. Each arc (x, y) such that both output label of x and input label of y are empty forms a parallel fan-in sub-graph.

Note, that requesting a single arc to form a parallel fan-in sub-graph is a bit artificial. We use this requirement to formally ensure that each arc is a part of some sub-graph.

Proposition 2. A nested P/A graph is a P/A graph.

Proof. A nested P/A graph is a directed acyclic graph because the base nested graph is acyclic and the decomposition rule does not add a cycle. From Lemma 1, for each arc (x, y) either the output label of x or the input label of y is empty. If both labels are empty then the arc forms a separate fan-in sub-graph. If the output label of x is non-empty then the arc belongs to a fan-out sub-graph with principal node x. Similarly, if the input label of y is non-empty then the arc belongs to a fan-in sub-graph with principal node y. Consequently, each arc belongs to exactly one sub-graph so the nested P/A graph is a P/A graph.   ∎

2.1 Recognizing Nested P/A Graphs

Proposition 2 claims that a nested P/A graph is a special form of a P/A graph. It is easy to show that there exist P/A graphs which are not nested [1]. Hence, an interesting question is "Can we efficiently recognize whether a given P/A graph is nested?" In this section we will present a polynomial algorithm that can recognize nested P/A graphs by reconstructing how they are built.

First, notice that in a nested P/A graph there are no two different fan-in (fan-out) sub-graphs sharing the same principal node (Definition 5). In other words, either all arcs going to (from) a given node x belong to a single fan-in (fan-out) sub-graph with the principal node x or there is no fan-in (fan-out) sub-graph with that principal node. This feature is easy to detect so in the rest of the paper, we assume that each node participates as a principal node in at most one fan-in and at most one fan-out sub-graph. This is reflected in the following representation of P/A graphs (Figure 5). The P/A graph is represented as a set of nodes where each node x is annotated by sets of predecessors $pred(x)$ and successors $succ(x)$ in the graph and by labels $inLab(x)$ and $outLab(x)$. $inLab(x)$ = PAR if x is a principal node in a fan-in parallel sub-graph, $inLab(x)$ = ALT if x is a principal node in a fan-in alternative sub-graph. If x is not a principal node in any fan-in sub-graph then $inLab(x)$ is empty. A similar definition is done for $outLab(x)$ with relation to fan-out sub-graphs. Notice the similarity of labels to labeled nested graphs (Definition 4). The reader should realize that any nested P/A graph can be represented this way: all fan-in and fan-out sub-graphs correspond to non-empty labels and for any arc (x, y) either the label $outLab(x)$ or $inLab(y)$ is empty.

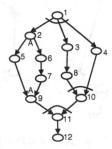

node	pred	succ	inLab	outLab
1		2,3,4	-	PAR
2	1	5,6	-	ALT
3	1	8	-	-
4	1	10	-	-
5	2	9	-	-
6	2	7	-	-
7	6	9	PAR	-
8	3	10	PAR	-
9	5,7	11	ALT	-
10	4,8	11	PAR	-
11	9,10	12	PAR	-
12	11		PAR	-

Fig. 5. Representation of a (nested) P/A Graph

The following algorithm *DetectNested* recognizes labeled nested graphs by reconstructing how they are built. Figure 6 illustrates the recognition process. Notice that some nodes in Q are not eligible for contraction because the condition in line 4 fails (e.g. node 10 in step 3, since 1 has two successors and 11 has two predecessors).

```
algorithm DetectNested(input: graph G, output: {success, failure})
1.   select all nodes x in G such that |pred(x)| = |succ(x)| = 1
2.   sort the selected nodes lexicographically according to index
         (pred(x), succ(x)) to form a queue Q
3.   while non-empty Q do
4.       select and delete a sub-sequence L of size k in Q such that
             all nodes in L have an identical index ({x}, {y}) and
             either |succ(x)| = k or |pred(y)| = k
5.       if no such L exists then stop with failure
6.       if k > 1 & outLab(x) ≠ inLab(y)  then stop with failure
7.       remove nodes z∈L from the graph
8.       remove nodes x, y from Q (if they are there)
9.       add arc (x,y) to the graph (an update succ(x) and pred(y))
10.      if |pred(x)| = |succ(x)| = 1 then insert x to Q
11.      if |pred(y)| = |succ(y)| = 1 then insert y to Q
12.  end while
13.  if the graph consists of two nodes connected by an arc then
14.      stop with success
15.  else stop with failure
```

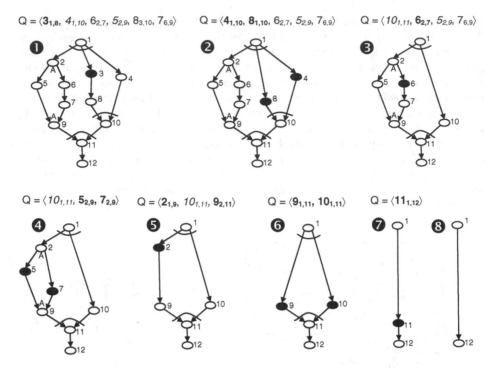

Fig. 6. Detecting nested P/A graphs by sub-graph contraction (contracted nodes are in black)

Proposition 3. Algorithm *DetectNested* always terminates and it stops with success if and only if the input P/A graph is nested.

Proof. Each line of the algorithm terminates. The body of the while loop either terminates with a failure or at least one node is removed from the graph. Because the queue Q consists of nodes that are part of the current graph, it must become empty sometime so the while loop terminates and hence the whole algorithm terminates.

We will show that the algorithm recognizes labeled nested graphs by induction on the number of decomposition steps necessary to generate a graph. The base nested graph is trivially recognized in line 13. Assume now that the algorithm can recognize all nested graphs built using *m* steps. We shall show that:

(i)   If *DetectNested* fails to find a set L of nodes to be contracted then the input graph is not a labeled nested graph, and

(ii) If *DetectNested* finds a set L of nodes and contracts them and the input graph is a labeled nested graph build using (*m*+1) steps, then the resulting graph is a labeled nested graph which can be built using *m* steps.

It is easy to see that these two claims are sufficient for the proof of the equivalence part of the proposition.

To prove (i) it is enough to realize that in any labeled nested graph constructed in accordance with Definition 3, the nodes added in the last decomposition step always fulfill the requirements on the set L in *DetectNested*. Thus if *DetectNested* fails to find a suitable set L then the input graph is not a labeled nested graph.

The proof of (ii) is more difficult because of the fact that there may be many suitable sets L = { z_1, ..., z_k } which *DetectNested* may find and contract. We have to show that any such choice produces a graph, which is labeled nested and can be built using m steps. Let us consider two cases:

a) $k > 1$. In this case a parallel or alternative sub-graph with nodes x, y, z_1, ..., z_k and arcs (x, z_i), (z_i, y) is contracted into arc (x, y). Notice, that (using the assumption that the input graph is nested) this sub-graph must be a result of an arc decomposition of (x, y) during the recursive construction and moreover no arc inside this sub-graph is further decomposed. Therefore the graph which is obtained from the input graph by contraction of L is a labeled nested graph obtainable in m steps. The sequence of decomposition steps is the same as for the input graph except that the decomposition of (x, y) is skipped.

b) $k = 1$. In this case a chain of length ≥ 2 is shortened (by one vertex and one arc) by the contraction of L. In this case there is no guarantee that the contraction can be matched to a decomposition step which built the input graph (see example below).

However, the chain of length l can be produced from a single arc by $(l-1)$ decompositions and can be contracted back into a single arc by $(l-1)$ contractions in *DetectNested* (all with $|L| = 1$). Thus, similarly as in case a) the graph which is obtained from the input graph by contraction of L is a labeled nested graph obtainable in m steps. The sequence of decomposition steps is the same as for the input graph except that the sub-sequence (not necessarily a sub-interval) of $(l-1)$ decompositions which built the chain is replaced by $(l-2)$ decompositions which build the shorter chain.   ∎

Proposition 4. The worst-case time complexity of algorithm *DetectNested* is $O(n^2)$, where n is a number of nodes in the graph.

Proof. The initial selection of nodes for the queue can be done in time $O(n)$. Time $O(n.\log n)$ is necessary to sort the queue. The sub-list for contraction can be selected in time $O(n)$ and insertion of nodes into the list can be done in $O(n)$. All other operations can be implemented in constant time. The while-loop is repeated at most n times because each time at least one node is removed from the graph. Together, the while loop takes time $O(n^2)$ so the whole algorithm takes time $O(n^2)$.   ∎

2.2 Tractability and Constraint Models

The main motivation for introducing nested P/A graphs was to make the P/A graph assignment problem tractable for this special subclass of graphs. Recall that the assignment problem consists of deciding whether it is possible to complete a partial assignment of validity variables for nodes to obtain a complete feasible assignment.

We can reformulate the P/A graph assignment problem as a constraint satisfaction problem in the following way. Each node x is represented using a Boolean validity variable v_x, that is a variable with domain $\{0,1\}$. If the arc between nodes x and y is a part of some parallel sub-graph then we define the following constraint:

$$v_x = v_y.$$

If x is a principal node and $y_1,\ldots, y_k$ for some k are all branching nodes in some alternative sub-graph then the logical relation defining the alternative branching can be described using the following arithmetic constraint:

$$v_x = \Sigma_{j=1,\ldots,k}\, v_{yj}.$$

Notice that if $k = 1$ then the constraints for parallel and alternative branching are identical (hence, it is not necessary to distinguish between them). Notice also that the arithmetic constraint for alternative branching together with the use of $\{0,1\}$ domains defines exactly the logical relation between the nodes – v_x is assigned to 1 if and only if exactly one of v_{yj} is assigned to 1. Using the arithmetic constraint simplifies a lot the formal model of the logical relation. Notice, that the task whether a completion of the partial assignment of validity variables satisfying all constraints exists is clearly equivalent to the assignment problem for the original P/A graph. Hence, if some local (polynomial) consistency such as arc consistency implies global consistency for the constraint model then the P/A graph assignment problem is trivially tractable. Recall that global consistency means that any value in the domain of a variable is part of some solution so if no domain is empty then a feasible assignment exists and can be found in polynomial time using a backtrack-free search. Unfortunately, arc consistency does not guarantee global consistency of the above-described *basic constraint model*. Assume a simple graph with two alternative branchings that are modeled using constraints $v_a = v_b + v_c$ and $v_d = v_b + v_c$. If we set v_a to 1 then arc consistency is not able to deduce that v_d must also be 1 and hence the problem is not globally consistent. Nevertheless, if we use an equivalent model $v_a = v_b + v_c$ and $v_d = v_a$ then arc consistency implies global consistency. Based on this idea we will now propose an equivalent constraint model of the nested P/A graph where arc consistency implies global consistency.

The *nested constraint model* for a nested P/A graph is designed incrementally along the process of building the nested graph. The base nested graph is modeled by a single constraint $v_s = v_e$. Assume that arc (x, y) is decomposed into a sub-graph (Figure 3) with new nodes $z_1, \ldots, z_k$ and new arcs $(x, z_i), (z_i, y)$. If the decomposition is parallel or $k = 1$ then we add constraints $y = z_i$ into the constraint model. Recall that this decomposition is allowed only if (x, y) was already a part of a serial or a parallel decomposition (or the initial arc - see Definition 4), which implies by induction on the number of decomposition steps that there is already a constraint $x = y$ in the model. Hence, the constraints $x = y$, $y = z_i$ are equivalent to the constraints of the basic model $x = z_i$, $y = z_i$. Assume now that the decomposition is alternative. It means that the arc (x, y) was a part of a serial or of an alternative decomposition (or the initial arc). In the second case, without loss of generality let us assume that x was the principal node of the alternative branching and $z_j{}'$, $j = 1,\ldots, m$ $(m \geq 0)$ are the remaining branching nodes (if any) in addition to y. Now, we add a constraint $y = \Sigma_{i=1,\ldots,k}\, z_i$. The model before decomposition implies (by induction on the number of decomposition steps) the constraint $x = y + \Sigma_{j=1,\dots,m} \, z_j'$ (this constraint is either explicitly in the model or there is a set of constraints equivalent to this constraint). The added and the implied constraints are equivalent to constraints $x = \Sigma_{i=1,\dots,k} \, z_i + \Sigma_{j=1,\dots,m} \, z_j'$, $y = \Sigma_{i=1,\dots,k} \, z_i$ which are used in the basic model to describe such a branching. Altogether, the modified constraint model of the nested P/A graph is equivalent to the original model (in terms of having the same set of feasible assignments). Note that the algorithm *DetectNested* reconstructs the decomposition process so we can define the nested constraint model for any given nested P/A graph. Figure 7 illustrates the process of building the nested constraint model and compares both nested and basic models.

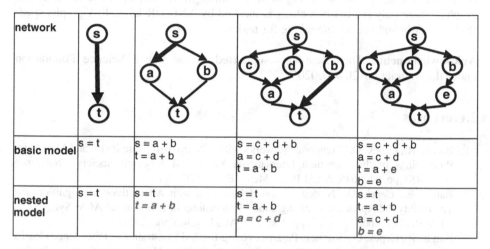

network				
basic model	s = t	s = a + b t = a + b	s = c + d + b a = c + d t = a + b	s = c + d + b a = c + d t = a + e b = e
nested model	s = t	s = t t = a + b	s = t t = a + b a = c + d	s = t t = a + b a = c + d b = e

Fig. 7. Branching constraints as arithmetic formulas over 0-1 variables (all fan-in/fan-out subgraphs are assumed to be alternative)

Proposition 5. The assignment problem for a nested P/A graph is tractable (can be solved in a polynomial time).

Proof. We shall show that the modified constraint model is Berge acyclic for which it is known that arc consistency implies global consistency [5]. The constraint model for the base nested graph consists of a single constraint so it is Berge acyclic. Any constraint added to the model after arc decomposition contains exactly one variable of the former model and new variable(s). Hence, it cannot introduce a Berge cycle. ∎

3 Conclusions

The paper studies a recursive definition of temporal networks with alternative processes, so called nested temporal networks with alternatives [2], motivated by a structure of typical manufacturing processes. Though this structure is identical to TPN [7] and TAEMS framework [6], our approach to modeling branching relations resembles more the work [4]. Surprisingly nobody so far paid attention to logical reasoning on these networks which is the main focus of this paper. We proposed an algorithm that can recognize nested TNAs, we proved its soundness, and showed its complexity. We also showed that the assignment problem for Nested TNAs is tractable by using a Berge acyclic constraint model of the problem. This constraint model applied to nested subparts of a TNA can even improve practical efficiency when solving the P/A graph assignment problem for general TNAs as showed in [3]. To solve real-life problems, the proposed model can be combined with other constraints such as the constraints modeling temporal and resource restrictions as demonstrated in [4,8,9]. Note finally, that despite a visual similarity to AND/OR graphs the concept of alternative branching differs from OR branching (the alternative branching allows two types of feasible assignment: either all nodes in the branching are invalid or the principal node is valid and exactly one branching node is valid). Moreover, opposite to AND/OR graphs, where only fan-out branching is marked by AND/OR label, P/A graphs specify both fan-in and fan-out branching for nodes.

Acknowledgments. The research is supported by the Czech Science Foundation under the contract no. 201/07/0205.

References

1. Barták, R., Čepek, O.: Temporal Networks with Alternatives: Complexity and Model. In: Proceedings of the Twentieth International Florida AI Research Society Conference (FLAIRS), pp. 641–646. AAAI Press, Menlo Park (2007)
2. Barták, R., Čepek, O.: Nested Temporal Networks with Alternatives: Recognition and Tractability. In: Applied Computing 2008 - Proceedings of 23rd Annual ACM Symposium on Applied Computing, vol. 1, pp. 156–157. ACM, New York (2008)
3. Barták, R., Čepek, O., Surynek, P.: Discovering Implied Constraints in Precedence Graphs with Alternatives. Annals of Operations Research. Springer, Heidelberg (to appear, 2008)
4. Beck, J.C., Fox, M.S.: Constraint-directed techniques for scheduling alternative activities. Artificial Intelligence 121, 211–250 (2000)
5. Beeri, C., Fagin, R., Maier, D., Yannakakis, M.: On the desirability of acyclic database schemes. Journal of the ACM 30, 479–513 (1983)
6. Horling, B., Leader, V., Vincent, R., Wagner, T., Raja, A., Zhang, S., Decker, K., Harvey, A.: The Taems White Paper, University of Massachusetts (1999), http://mas.cs.umass.edu/research/taems/white/taemswhite.pdf
7. Kim, P., Williams, B., Abramson, M.: Executing Reactive, Model-based Programs through Graph-based Temporal Planning. In: Proceedings of International Joint Conference on Artificial Intelligence (IJCAI), pp. 487–493 (2001)
8. Kuster, J., Jannach, D., Friedrich, G.: Handling Alternative Activities in Resource-Constrained Project Scheduling Problems. In: Proceedings of Twentieth International Joint Conference on Artificial Intelligence (IJCAI 2007), pp. 1960–1965 (2007)
9. Nuijten, W., Bousonville, T., Focacci, F., Godard, D., Le Pape, C.: MaScLib: Problem description and test bed design (2003), http://www2.ilog.com/masclib
10. Tsamardinos, I., Vidal, T., Pollack, M.E.: CTP: A New Constraint-Based Formalism for Conditional Temporal Planning. Constraints 8(4), 365–388 (2003)

Incorporating Learning in Grid-Based Randomized SAT Solving

Antti E. J. Hyvärinen, Tommi Junttila, and Ilkka Niemelä

Helsinki University of Technology TKK
Department of Information and Computer Science
{Antti.Hyvarinen,Tommi.Junttila,Ilkka.Niemela}@tkk.fi

Abstract. Computational Grids provide a widely distributed computing environment suitable for randomized SAT solving. This paper develops techniques for incorporating learning, known to yield significant speed-ups in the sequential case, in such a distributed framework. The approach exploits existing state-of-the-art clause learning SAT solvers by embedding them with virtually no modifications. We show that for many industrial SAT instances, the expected run time can be decreased by carefully combining the learned clauses from the distributed solvers. We compare different parallel learning strategies by using a representative set of benchmarks, and exploit the results to devise an algorithm for learning-enhanced randomized SAT solving in Grid environments. Finally, we experiment with an implementation of the algorithm in a production level Grid and solve several problems which were not solved in the SAT 2007 solver competition.

1 Introduction

In this paper we consider solving hard propositional satisfiability (SAT) problems in a grid-like, widely distributed computing environment. One realistic example of this kind of an environment is NorduGrid (http://www.nordugrid.org/) that we use in some of the experiments of this paper. Compared to, for example, a cluster of locally connected computing elements (CEs), such a widely distributed, multi-party owned environment may pose several restrictions. First, communication with the computing elements (submitting jobs and retrieving their results) can take a significant amount of time due to the widely distributed nature of such an environment. In addition, job management (for example, finding available elements by communicating with the front-end machines) causes further delays when submitting jobs. Second, communication between the CEs is not necessarily allowed at all due to security reasons; typically, all the traffic to the CEs passes through a front-end machine and one can only submit jobs, query their status, and retrieve results. Third, in order to ensure fairness between multiple users, CEs can either impose strict resource limits for jobs (for example, the maximum running time is set to four hours) or prefer jobs with predefined resource limits in a way that makes running of jobs requiring unlimited resources very slow. Fourth, computing elements (and thus jobs) are more likely to crash because they are administrated by different parties; e.g. maintenance breaks of CEs are scheduled independently, and CEs can disappear from the environment if their owner decides to prioritize local use at some time.

D. Dochev, M. Pistore, and P. Traverso (Eds.): AIMSA 2008, LNAI 5253, pp. 247–261, 2008.

In this paper we propose an approach called *Clause Learning Simple Distributed SAT* (CL-SDSAT) for solving *hard* SAT problems in such a widely distributed environment. The approach is based on running state-of-the-art, clause learning randomized SAT-solvers in a parallel environment until one of them solves the problem. In order to solve hard problems in the presence of resource limits imposed on jobs, the approach exploits the work done in *unsuccessful* jobs (i.e. those that exceeded the resource limits without finding a solution) by transferring some of the clauses learned by the solver back to the master process. When new jobs are submitted later, some of the learned clauses are passed to the jobs to constrain the search of the solver. This approach enables *parallel learning* techniques where, on one hand, learned clauses from multiple independent unsuccessful jobs are combined and, on the other hand, the clauses learned from such combinations are *cumulated*. The proposed approach can tolerate all the above mentioned restrictions related to the distributed environment as (i) a reasonable amount of data is transferred back to the master process only at the end of the execution of a job, (ii) the jobs do not communicate with each other, (iii) each job has predefined resource limits for the time and memory it is allowed to consume, and (iv) CE failures do not affect the correctness or completeness of the approach. In addition, only very modest modifications are required in a SAT solver in order to use it in the approach; thus it should be relatively easy for the approach to exploit the future improvements in clause learning SAT solvers.

Related Work. A major approach to distributed SAT solving is based on techniques relying on tight inter-process communication (for example, [1,2,3,4,5,6]). However, this approach is not directly applicable in grid environments we consider, where inter-process communication is often restricted and expensive. Methods for distributed SAT solving not based on inter-process communication have also been proposed [7,8,9]. The CL-SDSAT method developed in this work is similar to [7] as it does not involve search space partitioning like in [8,9] but it extends [7] by incorporating distributed clause learning. When the search space is partitioned, the methods are usually based on *Guiding Paths* [1,2,3]. Examples where clause learning is incorporated to the Guiding Path technique include [4,5] where the communication is performed between threads, [6] which is based on an MPI-implementation, and [10,11] where communication is performed in a more grid-like environment. The CL-SDSAT approach presented in this paper has the advantage of being able to use any clause learning SAT solver with no major changes. This is different from most approaches based on Guiding Paths and enables CL-SDSAT to exploit directly any future advances in SAT solver technology. Similar advantages are obtainable with approaches such as [8]. Distributed learning strategies are studied in [5,12,6]. In [5], the learning is based solely on the length of the clauses, whereas [6] considers several different techniques relating the learned clauses to the Guiding Paths of each solver. The learned clause distribution approaches of [6,5] exchange the learned clauses between the active jobs via a global master store dynamically, requiring modifications to SAT solvers and frequent communication. The distributed learning strategy in CL-SDSAT, on the other hand, cumulates and filters learned clauses over time. It only distributes and collects the clauses when jobs start and terminate due to resource limit, respectively, requiring much less communication and allowing jobs to have predefined resources limits.

The rest of the paper is structured as follows: Section 2 discusses the concepts used later in the presentation. In Sect. 3, basic ideas are introduced for the CL-SDSAT-framework. The issues related to the design of parallel learning are addressed in Sect. 4, and the developed ideas are implemented and evaluated in a production-level Grid environment in Sect. 5. Finally, the conclusions are presented in Sect. 6.

2 Preliminaries: Clause Learning, Randomization, and SDSAT

Most modern complete SAT-solvers, such as ZChaff [13] and MiniSAT [14] to name just two, are based on the Davis-Putnam-Logemann-Loveland (DPLL) depth-first search algorithm [15,16]. Modern implementations of the algorithm work in two alternating phases: an efficiently implementable *unit propagation*, and a heuristic selection of *decision variables* corresponding to branching in a depth-first search. To boost the search, *conflict driven clause learning* search space pruning techniques [17,18] are usually incorporated: whenever the solver reaches a conflict during the search, it analyzes the conflict and learns a new clause C that is a logical consequence of the original SAT instance $\mathcal{F}$. The solver then conjuncts the learned clause C with $\mathcal{F}$, guaranteeing that the search will not enter a similar conflict again. As C is a logical consequence of $\mathcal{F}$, a truth assignment for $\mathcal{F}$ satisfies $\mathcal{F}$ if and only if it satisfies $\mathcal{F} \wedge C$. Learned clauses usually decrease the number of decisions corresponding to the branches of the search. However, since a new clause is learned at each conflict, adding all of them into the instance $\mathcal{F}$ permanently would quickly exhaust the available memory and also slow down the unit propagation search space pruning routine forming the inner loop of the DPLL-algorithm. To avoid the exhaustion, the solvers periodically forget some of the learned clauses.

In addition to clause learning, most modern SAT solvers also apply search *restarts* and some form of *randomization* to avoid getting stuck at hard subproblems [19]. For instance, MiniSAT version 1.14 restarts the search periodically (the learned clauses are not discarded at restarts, though) and makes two percent of its branching decisions (pseudo)randomly. Despite restarts and randomness, the run times of a SAT solver can vary significantly on a single instance. As an example, observe the hundred samples based approximations of the cumulative run time distribution of the instances given in Figs. 2, 3 and 4 (the "base" plots, $q(t)$ is the probability that the instance is solved within t seconds): depending on the seed given to the pseudo-random number generator of MiniSAT 1.14, the run time varies from less than a hundred seconds to several thousands of seconds for some instances. This non-constant run time phenomenon can be exploited in a parallel environment. If we can run N randomized SAT solvers in parallel, the cumulative run time distribution $q(t)$ of an instance is improved to $q_N(t) = 1 - (1 - q(t))^N$: as an example, if $q(1000) = 0.5$ meaning that half of the runs end within 1000 seconds, then $q_8(1000) > 0.99$ and, thus, one obtains the solution almost certainly within 1000 seconds if eight parallel computing elements are available. For a more detailed analysis of running randomized SAT solvers in a parallel, distributed environment involving communication and other delays, see e.g. [7]. Although this simple strategy of running randomized SAT solvers in parallel, which we call the *Simple Distributed SAT solving* (SDSAT) approach, can reduce the expected time to solve an instance, it cannot reduce it below the minimum running time (i.e. the smallest t for which $q(t) > 0$). This is a serious drawback when solving *hard SAT problems* in a grid-like environment where the computing elements usually impose an upper limit for the computing time available for a single job. For example, if CEs only allow four hours of CPU time for each job, the basic SDSAT approach simply *cannot solve any problem with a longer minimum running time*. Notice that the "straightforward" approach of storing the memory image of a solver execution just before the time limit is reached and then continuing the execution in a new job in another CE is not a viable solution due to the amount of data that should be transferred between the jobs.

3   Clause-Learning Simple Distributed SAT Solving

The basic idea of the proposed *Clause-Learning Simple Distributed SAT* (CL-SDSAT) approach is relatively straightforward. A *master process* submits *jobs* consisting of a randomized SAT solver S and the SAT instance $\mathcal{F}$ to be solved into a grid-like distributed environment DE, which consists of computing elements (CEs) performing computations. If a job solves the problem (i.e. the satisfiability of $\mathcal{F}$ is decided) within the resource limits, the whole algorithm terminates with the solution. If the solution is not found in the job, some of the clauses the solver has learned during its search are transferred back to the master process. The master process maintains a database of such clauses, and whenever a new job is submitted, a subset of the clauses in the current database is conjuncted with $\mathcal{F}$ in the submitted SAT instance. The purpose of this is to ensure that the work done in unsuccessful jobs is not entirely wasted but can be used to prune the search in the subsequent jobs. Another advantage is that a form of parallel, history-dependent learning is enabled as clauses from several independent runs are collected together and cumulated over time. In the following we explain the proposed approach in more detail; the next sections then study different strategies for selecting the subset of learned clauses passed to the jobs and provide some experimental results.

The framework for the approach is presented in pseudo-code in Fig. 1. The learned clauses are collected to a database of clauses, denoted by *ClauseDB*, which is initially empty. The database is allowed to vary in size, but has an imposed maximum size, *MaxDBSize*. From this database, a subset of size at most *SubmSZ* is provided to each solver instance S together with the original SAT instance $\mathcal{F}$.

The main loop of the framework consists of two concurrent tasks: submitting new jobs to idle CEs in the distributed environment and receiving the results of the finished jobs.

The submission of new jobs is described on lines 2–4. If there are idle CEs in the environment (line 2), then a job $\langle S, \mathcal{F} \cup \text{Choose}(ClauseDB, SubmSZ) \rangle$ is submitted to it (line 4).[1] The function Choose selects a subset of the clauses in the current database *ClauseDB* so that the size of the subset is at most *SubmSZ*.[2] The size of the subset is restricted for two reasons: transferring data in a widely distributed environment takes non-negligible time and, as mentioned in Sect. 2, having an excessive amount of learned

---

[1] As usual, we use the notation $\mathcal{F} \cup \mathcal{C}$ to denote the conjunction $\mathcal{F} \wedge \bigwedge_{C_i \in \mathcal{C}} C_i$.

[2] By the size of a set of clauses we mean here and in the following the sum of the number of the literals in the clauses in the set.

Input: $\mathcal{F}$, a SAT instance; $\mathcal{S}$, a randomized SAT solver
   let *ClauseDB* = $\emptyset$
   let *MaxDBSize* = M
   let *SubmSZ* = N
1 **while** (True):
2   **if** there are idle CEs in DE:
3   update *SubmSZ*
4   submit the job $\langle \mathcal{S}, \mathcal{F} \cup \text{Choose}(ClauseDB, SubmSZ) \rangle$ to an idle CE
5   **if** $\langle result, \mathcal{C} \rangle$ is received from DE:
6   **if** result is in $\{\text{SAT}, \text{UNSAT}\}$:
7   **return** result
8   **else**
9   update *MaxDBSize*
10   **let** *ClauseDB* = Merge(*ClauseDB*, $\mathcal{C}$, *MaxDBSize*)

Fig. 1. A general framework for Grid-based randomized learning SAT solving

clauses can slow down the inner loop of the SAT solver $\mathcal{S}$. For the sake of completeness of the approach, the size limit *SubmSZ* may have to be increased during the search (line 3); this issue is discussed later at the end of this section.

The results received from the DE are handled on lines 5–10 with two cases.

- If the result is either SAT or UNSAT, the algorithm terminates with that result (line 7). The correctness of the result in this case, that is, the soundness of the framework, follows directly from the properties of learned clauses: a SAT instance $\mathcal{F} \cup$ Choose(*ClauseDB*, *SubmSZ*) submitted to a CE is satisfiable if and only if the original instance $\mathcal{F}$ is satisfiable because all the clauses in Choose(*ClauseDB*, *SubmSZ*) are learned clauses and, thus, logical consequences of $\mathcal{F}$.
- If the problem was not solved in the job because the resource limits were exceeded, the clause database *ClauseDB* is updated with the set $\mathcal{C}$ of learned clauses returned from the job (line 10). The function Merge takes the old clause database *ClauseDB*, the new learned clauses $\mathcal{C}$, and the maximum size *MaxDBSize* of the database, and returns a new database $ClauseDB' \subseteq ClauseDB \cup \mathcal{C}$ of size at most *MaxDBSize*. For the sake of completeness (discussed below), it may become necessary to increase the maximum size of the clause database during the search (line 9).

When instantiating the framework into a concrete implementation, of special interest are the heuristics used in the two operators Choose and Merge. We will study some natural heuristics for these operators in the next section.

Completeness. If the resource limits for the jobs as well as the size limits for the clause database (*MaxDBSize*) and submitted learned clause sets (*SubmSZ*) are fixed, the framework is not complete (that is, there are SAT instances for which the framework does not terminate). One way to ensure completeness under fixed resource limits for jobs is to increase the parameters *MaxDBSize* and *SubmSZ* periodically during the search until a solution is found. Naturally, the Choose and Merge operators must use this increased space by returning clause sets of analogously increasing size. Observe the similarity to clause learning SAT solvers: they also usually increase the limit for the number of stored learned clauses gradually during the search.

4   Parallel Learning Strategies

The CL-SDSAT algorithm presented in Sect. 3 provides a framework for parallel learning which builds on five elements: (i) the size of the clause database, *MaxDBSize*, (ii) the number of learned clauses submitted in each job, *SubmSZ*, (iii) the operator Merge for updating the clause database, (iv) the operator Choose used for selecting the learned clauses for a job, and (v) the selection of learned clauses returned by an unsuccessful job.

A key issue is the case (iv), that is, the design of the operator Choose. The heuristic devised for the operator Choose determines to a large degree the choices in other elements (i), (iii), and (v). Four potential heuristics can be identified for Choose:

- Choose$_{freq}$ prefers the most common learned clauses. Such clauses are intuitively good since they are encountered in many jobs.
- Choose$_{len}$ prefers short learned clauses. Short clauses are potentially effective in pruning the search space.
- Choose$_{123}$ returns only unary, binary, and ternary clauses. Such clauses are even more effective in pruning the search space but might be rare in practice in some cases.
- Choose$_{rand}$ returns a set of clauses which are randomly picked from the set of learned clauses so that each learned clause is returned with equal probability.

This section studies empirically the parallel learning strategies and the CL-SDSAT algorithm in Fig. 1 by (i) first studying the run time of a SAT instance $\mathcal{F}$ before and after including the learned clauses (line 4 of the algorithm), and (ii) then the cumulative effect of such inclusions after several rounds of the **while** loop.

All experiments in this section are performed using a representative set of benchmarks from the SAT 2007 Solver Competition, formed in three steps as follows. First, each of the instances in the industrial and crafted categories of the competition were attempted once with standard MiniSat v1.14 SAT solver, using a time limit of 8000 seconds, on a heterogeneous set of modern CPUs. Second, of these problems, the ones which were solved in more than 2000 seconds (but less than 8000 seconds) were collected for the third round. From the resulting set, a representative subset consisting of nine instances was formed so that there is at least one satisfiable and unsatisfiable instance from both the industrial and the crafted categories. The final benchmark set is presented in column Name of Table 1.

Heuristics for the Operator Choose. The comparison of the heuristics is performed in an idealized, zero-delay environment by running for each benchmark instance eight independent jobs which are all unsuccessful (do not return SAT or UNSAT) as for each job the run time limit is set to be 25% of the observed minimum run time (among hundred randomized runs) of the original instance. For these runs, MiniSat v1.14 was modified so that it accepts as input a seed for its internal search randomization procedure and a time limit, and upon reaching the time limit it outputs all learned clauses it is holding at that time. Based on these eight sets of learned clauses, say, $\mathcal{C}_1, \ldots, \mathcal{C}_8$, a *derived instance* is constructed by employing the respective Choose heuristic to select a subset of the learned clauses. The derived instance corresponds to the instance $\mathcal{F} \cup$ Choose(*ClauseDB, SubmSZ*) of line 4 of the algorithm in Fig. 1, with *SubmSZ* set to 100,000 literals and $ClauseDB = C_1 \cup \cdots \cup C_8$. It should be noted that the size limitation of $ClauseDB$ is ignored by effectively setting $MaxDBSize$ to infinity in these experiments.

Table 1 gives an overview of the results by comparing the expected run times of the derived instances for several heuristics and the instance without additional learned clauses (in column Base). The expected run times are computed from fifty sample runs (hundred sample runs in column Base), all ran until a solution was found in a homogeneous computing environment consisting of Intel Xeon 5130 CPUs with 2GHz clock speed and 16 gigabytes of memory. The solver used in the experiments was the MiniSAT v1.14 SAT solver modified so that the solver accepts a seed for its internal search randomization procedure as input.

Table 1. Expected run times for a selection of benchmarks from SAT 2007 competition

Name	Base	Choose$_{len}$	Choose$_{freq}$	Choose$_{123}$	Choose$_{rand}$
999999000001nc.shuffled- -as.sat05-446	2065	1436	1462	1431	1357
AProVE07-16	1563	1152	1133	1152	1647
clqcolor-10-07-09.- shuffled-as.sat05-1258	1900	1465	1447	1458	2016
cube-11-h14-sat	4832	5441	5245	5449	5164
dated-10-13-s	2278	1705	2069	1706	3283
mizh-md5-48-5	1659	1017	1436	1012	1565
mod2-rand3bip-sat-250-3- .shuffled-as.sat05-2220	1180	1205	1202	1202	1406
mod2-rand3bip-sat-280-1- .sat05-2263.reshuffled-07	2382	5231	5213	5209	5545
vmpc_28.shuffled-as.- sat05-1957	623	574	489	579	675
	18482	19226	19696	19198	22658

The results suggest that while the two length-based heuristics perform very similarly in this benchmark set, the frequency-based heuristic differs from these two. None of the three heuristics is consistently better than the other two. The total expected solving time, reported on the last row for each of the heuristics, confirms the similarity of the heuristics. Perhaps surprisingly, the results in column Base outperform others when only the sums are compared. It should be noted, however, that the effect is almost solely due to the instance mod2-rand3bip-sat-280-1.sat05- -2263.reshuffled-07. Similarity of the results can be explained by the fact that short learned clauses are also frequent and usually of length one, two or three. The results also show that there are instances for which the addition of short learned clauses increases the expected solving time. For some problems, namely the two mod2-instances, the phenomenon can be explained by the structure of the instances: they are specially crafted to be difficult for the type of solvers that are being studied here [20]. Finally, the random heuristic is usually inferior to the other heuristics. This confirms that the performance of the CL-SDSAT algorithm can be enhanced by using some specific heuristics in the operator Choose.

The run times of certain instances are examined in detail by computing the distributions $q(t)$, giving the probability that the problem is solved in less than t seconds (or decisions). The effect of the parallel learning is further studied for these instances by using an additional heuristic, Single, which selects the shortest learned clauses from only a single job. The results are given in Figs. 2 to 5, where $Choose_{rand}$, Single, $Choose_{freq}$, $Choose_{123}$, and $Choose_{len}$ are denoted by random, single, frequency, 123, and length, respectively.

In Fig. 2, the instance benefits from the learned clauses independent of the heuristic, as can also be seen from the results in Table 1. Only the heuristic Single is not useful in decreasing the run time, indicating that the parallelism can be exploited. Figure 3 illustrates a run time distribution of an instance with a different type of a distribution, where the minimum run time is two orders of magnitude smaller than the maximum run time, as opposed to the less dramatic ten-fold difference between the minimum and maximum in Fig. 2. On this instance the effects of the heuristics are similar.

Table 1 suggests further that for some instances, the introduction of new learned clauses increases the expected run time. The phenomenon is further illustrated in Figs. 4 and 5. The run times, shown in Fig. 4, increase in general. However, Fig. 5 shows that the corresponding number of decisions decreases as new clauses are included. The added clauses decrease the number of decisions the solver has to make but in some cases they increase the overhead related to solving the problem, thus increasing the expected solving time.

As stated above, we conclude that $Choose_{freq}$, $Choose_{len}$ and $Choose_{123}$ usually result in very similar run times. We will concentrate in our further analysis on the heuristic based on $Choose_{len}$ for two reasons. Firstly, the length-based heuristics are more efficiently implementable in smaller space, and secondly, $Choose_{len}$ is guaranteed to result in learned clauses and, thus, progress in the CL-SDSAT algorithm also in cases where $Choose_{123}$ performs poorly when few unary, binary or ternary clauses are discovered during the search.

Cumulative Effect of Learned Clauses. The results in Table 1 and Fig. 4 clearly show that some instances do not benefit from learned clauses obtained in the early stages of the parallel learning. In the light of this result, it would be possible that some instances are not solvable using the CL-SDSAT algorithm. In this section we study this question on instances described in Table 1 and show that ultimately the introduction of learned clauses helps in reducing the run time of these instances. We restrict our study to instances where the minimum observed run time is large. This restriction follows from the observation that if the minimum run time is small (which is the case, for example, for the two mod2-instances) the algorithm will solve the problem quickly without the learned clauses, and the question of how the learned clauses affect the run time of the instance is irrelevant as the problem is solved even before any learned clauses can be included to a job in the CL-SDSAT algorithm.

The effect of running the CL-SDSAT-algorithm for several steps is illustrated in Figs. 6, 7, and 8. The experiment is performed using the $Choose_{len}$-heuristic for both Choose and Merge to filter learned clauses, a distributed computing environment

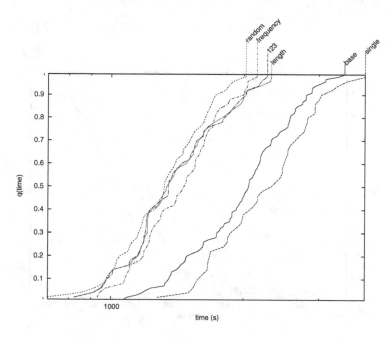

Fig. 2. Run time distributions for the derived instances of `999999000001nc.shuffled-` `-as.sat05-446` corresponding to several heuristics

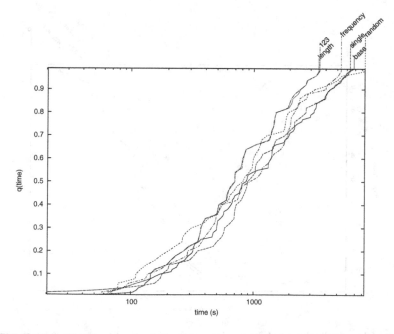

Fig. 3. Run time distributions for the derived instance of `mizh-md5-48-5` corresponding to several heuristics

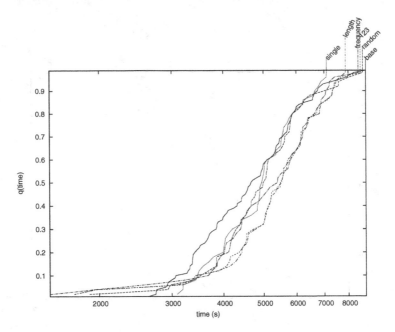

Fig. 4. Run time distributions for the derived instances of `cube-11-h14-sat` corresponding to several heuristics

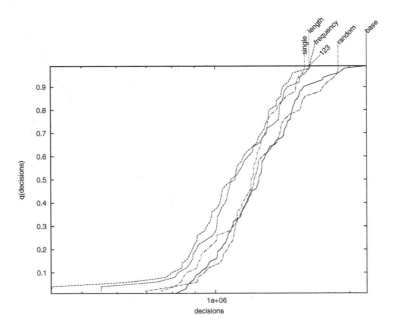

Fig. 5. Decision distribution for the derived instances of `cube-11-h14-sat` corresponding to several heuristics

consisting of 32 CEs, and for each job a run time limit of 25% of the minimum observed run time of the original instance. In all jobs the parameters *MaxDBSize* and *SubmSZ* are set to 100,000 literals. From the resulting jobs twelve are selected evenly from the range between the initial job and the final successful job. The jobs are numbered in order they are created on line 4 of the CL-SDSAT-algorithm, and the numbers are shown in the figures which present the run time distributions of the corresponding submitted SAT instances (based on fifty sample runs). The base distribution for the instance without any learned clauses is included for reference. The results in Figs. 6, 7, and 8 show that the expected run time of an instance gradually decreases when learned clauses are cumulated and added to the instance using the given heuristic for filtering the learned clauses. Notice that this happens also for the instance `cube-11-h14-sat` (Fig. 6) for which combining learned clauses from eight independent jobs only increases the expected run time (recall Fig. 4).

To sum up, the controlled experiments in this section support the conclusion that the parallel learning strategy in the CL-SDSAT-algorithm decreases the expected run time of an instance as learned clauses are iteratively cumulated and filtered when the heuristic is chosen suitably.

5   Grid Implementation

The ideas developed in this work have been implemented in a prototype of the proposed CL-SDSAT framework. The prototype uses NorduGrid, a production level Grid, as the distributed environment, and MiniSAT version 1.14 (with modifiable pseudorandom number generator seed) as the randomized SAT solver. The job management in the Grid is handled by GridJM [21], and each job has a time limit of one hour and a memory limit of one gigabyte. The implementation uses the $Choose_{len}$-heuristic preferring the shortest clauses for parallel learning; this design choice is based on the results in Sect. 4 and the fact that length-based heuristics are more efficiently implementable than frequency-based in this framework. In the prototype, the requirements needed for guaranteeing completeness are ignored by simply using a fixed clause database size of 100,000 literals. Similarly, unsuccessful jobs do not return all their learned clauses but only the shortest ones that together have at most 100,000 literals.

For the benchmark problems we selected a set of hard SAT instances for which there was little or no a priori information about the run time distribution. Such problems are available from the SAT 2007 solver competition (`http://www.satcompetition.org/`), where some of the instances were not solved by any of the competing solvers within the time bounds (10,000 seconds for the industrial and 5,000 seconds for the crafted category). Table 2 presents the results of running the CL-SDSAT prototype on a subset of these unsolved problems as well as on some other problems which were not solved by MiniSAT in the competition. Each instance was run for at most thirty hours allowing the use of at least eight and at most 64 CEs simultaneously. Column MiniSAT also reports the run times of the sequential MiniSAT v1.14 with no time limit but the memory usage restricted to two gigabytes. The MiniSAT v1.14 runs were performed using dual CPU dual-core compute nodes (Intel Xeon 5130 2GHz) with at most four processes running simultaneously in each node. It should be noted that the exact run times

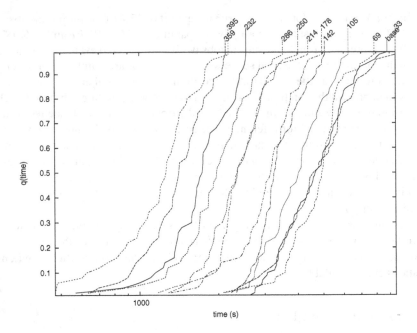

Fig. 6. Run time distributions corresponding to several jobs constructed by CL-SDSAT while solving `cube-11-h14-sat` with Choose$_{len}$

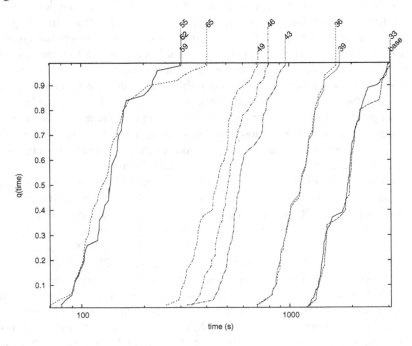

Fig. 7. Run time distributions corresponding to several jobs constructed by CL-SDSAT while solving `clqcolor-10-07-09.shuffled-as.sat05-1258`

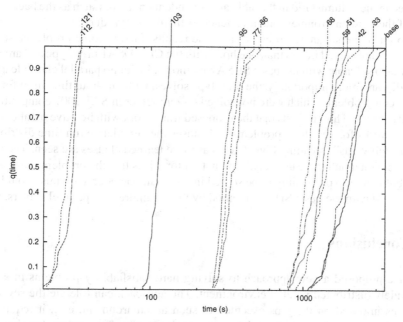

Fig. 8. Run time distributions corresponding to several jobs constructed by CL-SDSAT while solving 999999000001nc.shuffled-as.sat05-446

Table 2. Wall clock times for some difficult instances from SAT 2007 competition solved in Grid and with standard MiniSAT 1.14. Memory outs are denoted by '*', time outs by '—'

Not solved by MiniSAT in SAT 2007			
Name	Type	Grid (s)	MiniSAT (s)
ezfact64_5.sat05-452.reshuffled-07	SAT	24956	65739
vmpc_33	SAT	1072	184928
safe-50-h50-sat	SAT	25551	*
connm-ue-csp-sat-n800-d-0.02-s1542454144-.sat05-533.reshuffled-07	SAT	8269	119724
Not solved by any solver in SAT 2007			
Name	Type	Grid (s)	MiniSAT (s)
AProVE07-01	UNSAT	14567	39627
AProVE07-25	UNSAT	89206	306634
QG7a-gensys-ukn002.sat05--3842.reshuffled-07	UNSAT	40239	127801
vmpc_34	SAT	38753	90827
safe-50-h49-unsat	—		*
partial-10-13-s.cnf	SAT	7144	*
sortnet-8-ipc5-h19-sat	SAT	95497	*
dated-10-17-u	UNSAT	—	105821
eq.atree.braun.12.unsat	UNSAT	13298	59229

reported in the column Grid in the table are dependent on factors such as the background load of the Grid environment and are, hence, difficult to reproduce.

Two phenomena can be observed from the results. Firstly, some problems, such as vmpc_33, are solved in less than one hour with the CL-SDSAT prototype and are, thus, clearly also solvable with the basic SDSAT method [7] with no parallel clause learning. Secondly, and more importantly, the prototype solves, with one hour time limit for each job, several problems which were not solved by *any* solver in SAT 2007 competition in 10,000 seconds. This suggests that the proposed framework with iterative parallel learning also works for very hard problems and causes the cumulative run time distribution to "shift leftwards" (recalling Figs. 6 to 8) as more learned clauses are seen. The other, in our opinion much more unlikely, explanation for this is that the problems have a very small but non-zero probability to be solved in less than one hour and, thus, would have been solved with the basic SDSAT method by using hundreds of parallel solvers.

6   Conclusions

We have proposed a new approach to solving hard satisfiability problems in a grid-like widely distributed parallel environment. The approach can tolerate the severe restrictions imposed on the jobs executed in such an environment, e.g., it requires no inter-node communication and is inherently fault-tolerant. The approach is based on combining (i) a natural method for solving SAT in parallel by independent randomized SAT solvers, and (ii) the powerful conflict driven clause learning technique employed in many modern, sequential, DPLL-style SAT-solvers. This combination results in a novel parallel and cumulative clause learning approach. We have experimentally (i) compared different heuristics for selecting the learned clauses that are dynamically stored during the process, and (ii) demonstrated that the process can gradually make the problem easier to solve. Preliminary experimental results carried out in a production level Grid indicate that the approach can indeed solve very hard SAT problems, including several that were not solved in the SAT 2007 competition by any solver. This suggests that the developed algorithm is also useful in practical environments.

Acknowledgments. The authors wish to thank the anonymous reviewers for their valuable comments. The financial support of the Academy of Finland (projects 122399 and 112016), Helsinki Graduate School in Computer Science and Engineering, Jenny and Antti Wihuri Foundation, and Emil Aaltosen Säätiö is gratefully acknowledged.

References

1. Boehm, M., Speckenmeyer, E.: A fast parallel SAT-solver: Efficient workload balancing. Annals of Mathematics and Artificial Intelligence 17(4-3), 381–400 (1996)
2. Zhang, H., Bonacina, M., Hsiang, J.: PSATO: A distributed propositional prover and its application to quasigroup problems. Journal of Symbolic Computation 21(4), 543–560 (1996)
3. Jurkowiak, B., Li, C., Utard, G.: A parallelization scheme based on work stealing for a class of SAT solvers. Journal of Automated Reasoning 34(1), 73–101 (2005)

4. Feldman, Y., Dershowitz, N., Hanna, Z.: Parallel multithreaded satisfiability solver: Design and implementation. Electronic Notes in Theoretical Computer Science 128(3), 75–90 (2005)
5. Sinz, C., Blochinger, W., Küchlin, W.: PaSAT — Parallel SAT-checking with lemma exchange: Implementation and applications. In: SAT 2001. Electronic Notes in Discrete Mathematics, vol. 9, pp. 12–13. Elsevier, Amsterdam (2001)
6. Schubert, T., Lewis, M., Becker, B.: PaMira — a parallel SAT solver with knowledge sharing. In: MVT 2005, pp. 29–36. IEEE Computer Society Press, Los Alamitos (2005)
7. Hyvärinen, A.E.J., Junttila, T., Niemelä, I.: Strategies for solving SAT in grids by randomized search. In: 9th International Conference on Artificial Intelligence and Symbolic Computation (AISC 2008) (accepted for publication, 2008)
8. Hyvärinen, A.E.J., Junttila, T., Niemelä, I.: A distribution method for solving SAT in grids. In: Biere, A., Gomes, C.P. (eds.) SAT 2006. LNCS, vol. 4121, pp. 430–435. Springer, Heidelberg (2006)
9. Forman, S., Segre, A.: NAGSAT: A randomized, complete, parallel solver for 3-SAT. In: SAT (2002), http://gauss.ececs.uc.edu/Conferences/SAT2002/sat2002list.html
10. Blochinger, W., Westje, W., Küchlin, W., Wedeniwski, S.: ZetaSAT – Boolean satisfiability solving on desktop grids. In: CCGrid 2005, pp. 1079–1086. IEEE, Los Alamitos (2005)
11. Chrabakh, W., Wolski, R.: GridSAT: A chaff-based distributed SAT solver for the grid. In: SC 2003. IEEE, Los Alamitos (2003)
12. Inoue, K., Soh, T., Ueda, S., Sasaura, Y., Banbara, M., Tamura, N.: A competitive and cooperative approach to propositional satisfiability. Discrete Applied Mathematics 154(16), 2291–2306 (2006)
13. Moskewicz, M.W., Madigan, C.F., Zhao, Y., Zhang, L., Malik, S.: Chaff: Engineering an efficient SAT solver. In: DAC 2001, pp. 530–535. ACM Press, New York (2001)
14. Eén, N., Sörensson, N.: An extensible SAT-solver. In: Giunchiglia, E., Tacchella, A. (eds.) SAT 2003. LNCS, vol. 2919, pp. 502–518. Springer, Heidelberg (2004)
15. Davis, M., Putnam, H.: A computing procedure for quantification theory. Journal of the ACM 7(3), 201–215 (1960)
16. Davis, M., Logemann, G., Loveland, D.: A machine program for theorem proving. Communications of the ACM 5(7), 394–397 (1962)
17. Marques-Silva, J.P., Sakallah, K.A.: GRASP: A search algorithm for propositional satisfiability. IEEE Transactions on Computers 48(5), 506–521 (1999)
18. Zhang, L., Madigan, C.F., Moskewicz, M.W., Malik, S.: Efficient conflict driven learning in boolean satisfiability solver. In: ICCAD 2001, pp. 279–285. ACM Press, New York (2001)
19. Gomes, C.P., Selman, B., Crato, N., Kautz, H.A.: Heavy-tailed phenomena in satisfiability and constraint satisfaction problems. Journal of Automated Reasoning 24(1/2), 67–100 (2000)
20. Haanpää, H., Järvisalo, M., Kaski, P., Niemelä, I.: Hard satisfiable clause sets for benchmarking equivalence reasoning techniques. Journal on Satisfiability, Boolean Modeling and Computation 2, 27–64 (2006)
21. Hyvärinen, A.E.J.: GridJM. A Computer Program, http://www.tcs.hut.fi/~aehyvari/gridjm/

Solving the Course Timetabling Problem with a Hybrid Heuristic Algorithm[*]

Zhipeng Lü[1,2] and Jin-Kao Hao[1]

[1] LERIA, Université d'Angers, 2 Boulevard Lavoisier, 49045 Angers Cedex 01, France
[2] School of Computer Science and Technology, Huazhong University of Science and Technology, Wuhan, 430074, China
zhipeng.lui@gmai.com, hao@info.univ-angers.fr

Abstract. The problem of curriculum-based course timetabling is studied in this work. In addition to formally defining the problem, we present a hybrid solution algorithm (Adaptive Tabu Search–ATS), which is aimed at minimizing violations of soft constraints. Within ATS, a new neighborhood and a mechanism for dynamically integrating Tabu Search with perturbation (from Iterated Local Search) are proposed to ensure a continuous tradeoff between intensification and diversification. The performance of the proposed hybrid heuristic algorithm is assessed on two sets of 11 public instances from the literature. Computational results show that it significantly improves the previous best known results on two problem formulations.

Keywords: Timetabling, hybrid heuristic, tabu search, iterated local search, constraint solving.

1 Introduction

In recent decades, timetabling has become an area of increasing interest in the community of both research and practice [11]. In essence, timetabling consists of assigning a number of events, each with a number of features, to a limited number of timeslots and rooms subject to certain (hard and soft) constraints. In this paper, we consider one of the problems in the category of educational timetabling–the so-called curriculum-based course timetabling (CCT), the formulation of which was recently proposed as the third track of the Second International Timetabling Competition (ITC–2007)[1]. This competition aims to close the gap between research and practice within the area of educational timetabling.

The general course timetabling problem is known to be difficult and has been proved to be NP-hard [4]. In this context, exact solutions would be only possible for problems of limited sizes. Instead, algorithms based on metaheuristics have shown to be a highly effective approach to this kind of problems (see e.g. [1,3,15]).

---

[*] This algorithm is ranked the second place for the track 3 of the Second International Timetabling Competition (ITC–2007).
[1] http://www.cs.qub.ac.uk/itc2007/

D. Dochev, M. Pistore, and P. Traverso (Eds.): AIMSA 2008, LNAI 5253, pp. 262–273, 2008.

Interested readers are referred to [9] for a comprehensive survey of the automated approaches for university course timetabling presented in recent years.

In this paper, we present a hybrid Adaptive Tabu Search (ATS) algorithm for the CCT problem. We introduce a combined use of two neighborhoods including an original and very powerful double Kempe chain neighborhood. Moreover, we devise a mechanism for dynamically combining TS with a perturbation operator in order to adaptively escape from local optima and to automatically make more intensive search when promising regions of the search space are visited. Consequently, it ensures a continuous tradeoff between intensification and diversification for the search process. The performance of the ATS algorithm is assessed with the two sets of 11 instances from the literature, showing very competitive results.

The rest of this paper is organized as follows. Section 2 describes the first mathematical formulation of the CCT problem. Section 3 presents the hybrid heuristic–Adaptive Tabu Search algorithm. In Section 4 computational experiments are carried out. Conclusions are drawn in the last section.

2   Curriculum-Based Course Timetabling

2.1   Problem Description

The CCT problem of the ITC–2007 consists of scheduling all lectures of a set of courses into a weekly timetable, where each lecture of a course must be assigned a period and a room in accordance with a given set of constraints. In this problem, all hard constraints must be strictly satisfied and the number of soft constraint violations should be minimized. A feasible timetable is one in which all lectures have been scheduled at a period and a room, so that the hard constraints are satisfied. The four hard constraints H1~H4 and four soft constraints S1~S4 are defined as follows:

- **H1. Lectures.** All lectures of a course must be scheduled to a distinct period and a room.
- **H2. Room Occupancy.** Any two lectures cannot be assigned in the same period and the same room.
- **H3. Conflicts.** Lectures of courses in the same curriculum or taught by the same teacher cannot be scheduled in the same period, i.e., any period cannot have an overlapping of students or teachers.
- **H4. Availability.** If the teacher of a course is not available at a given period, then no lectures of the course can be assigned to that period.

- **S1. Room Capacity.** For each lecture, the number of students attending the course should not be greater than the capacity of the room hosting the lecture.
- **S2. Room Stability.** All lectures of a course should be scheduled at the same room. If this is impossible, the number of occupied rooms should be as few as possible.

- **S3. Minimum Working Days.** The lectures of a course should be spread into the given minimum number of days.
- **S4. Curriculum Compactness.** For a given curriculum a violation is counted if there is one lecture not adjacent to any other lecture belonging to the same curriculum within the same day, which means the agenda of students should be as compact as possible.

We present below a first mathematical formulation of the problem which is missing in the literature.

2.2 Problem Formulation

The CCT problem consists of a set of n courses $C = \{c_1, c_2, \ldots, c_n\}$ to be scheduled in a set of p periods $T = \{t_1, t_2, \ldots, t_p\}$ and a set of m rooms $R = \{r_1, r_2, \ldots, r_m\}$. Each course c_i is composed of l_i lectures (of one timeslot) to be scheduled. A period is a pair composed of a day and a timeslot and p periods are distributed in d week days and h daily timeslots, i.e., $p = d \times h$. In addition, there are a set of s curricula $CR = \{cr_1, cr_2, \ldots, cr_s\}$ where each curriculum cr_k is a group of courses that share common students.

For the solution representation, we chose a direct solution representation to make things as simple as possible. A candidate solution consists of $p \times m$ matrix X where $x_{i,j}$ corresponds to the course label assigned at period t_i and room r_j. If there is no course assigned at period t_i and room r_j, then $x_{i,j} = -1$. With this representation we assure that there will be no more than one course assigned to a room in any period, meaning that the second hard constraint H2 is always satisfied. For courses, rooms, curricula and solution representation X, a number of notations and definitions are presented in table 1.

Table 1. Table of symbols and variables

Symbol	Description
std_i	the number of students attending course c_i
l_i	the number of lectures of course c_i
tc_i	the label of the teacher instructing course c_i
mwd_i	the number of minimum working days of course c_i
cap_j	the capacity of room r_j
cr_k	the kth curriculum including a set of courses $\{c_{k1}, \ldots, c_{kv}\}$
$una_{i,j}$	whether course c_i is unavailable at period t_j. $una_{i,j} = 1$ if it is unavailable, $una_{i,j} = 0$ otherwise
$x_{i,j}$	the label of the course assigned at period t_i and room r_j
$nr_i(X)$	the number of rooms occupied by course c_i for a candidate solution X
$nd_i(X)$	the number of working days that course c_i takes place at for a candidate solution X
$cra_{k,i}(X)$	$cra_{k,i}(X) = 1$ if one lecture of any course in curriculum cr_k is scheduled at period t_i, $cra_{k,i}(X) = 0$ otherwise.

Given these notations, we redescribe the CCT problem in a formal way for a candidate solution X. The four hard constraints and four soft constraints are:

- **H1. Lectures.** $\forall c_k \in C$,

$$\sum_{i,j} sl_k(x_{i,j}) = l_k$$

$$sl_k(x_{i,j}) = \begin{cases} 1, \text{ if } x_{i,j} = c_k; \\ 0, \text{ otherwise.} \end{cases}$$

- **H2. Room Occupancy.** this hard constraint is automatically satisfied in our solution representation.
- **H3. Conflicts.** $\forall x_{i,j}, x_{i,k} \in X, x_{i,j} = c_u, x_{i,k} = c_v,$

$$(\forall cr_q, c_u \notin cr_q \lor c_v \notin cr_q) \land (tc_u \neq tc_v)$$

- **H4. Availability.** $\forall x_{i,j} \in X, x_{i,j} = c_k,$

$$una_{k,i} = 0$$

For the four soft constraints, their penalty costs are represented as:

- **S1. Room Capacity.** $\forall x_{i,j} \in X, x_{i,j} = c_k,$

$$f_{rc}(x_{i,j}) = \begin{cases} \alpha_1 \cdot (std_k - cap_j), \text{ if } std_k > cap_j; \\ 0, \hspace{3cm} \text{otherwise.} \end{cases}$$

- **S2. Room Stability.** $\forall c_i \in C,$

$$f_{rs}(c_i) = \alpha_2 \cdot (nr_i(X) - 1)$$

- **S3. Minimum Working Days.** $\forall c_i \in C,$

$$f_{md}(c_i) = \begin{cases} \alpha_3 \cdot (mwd_i - nd_i(X)), \text{ if } nd_i(X) < mwd_i; \\ 0, \hspace{3.5cm} \text{otherwise.} \end{cases}$$

- **S4. Curriculum Compactness.** $\forall x_{i,j} \in X, x_{i,j} = c_k,$

$$f_{cc}(x_{i,j}) = \alpha_4 \cdot \sum_{cr_q \in CR} c\_cr_{k,q} \cdot iso_{q,i}(X)$$

where
$$c\_cr_{k,q} = \begin{cases} 1, \text{ if } c_k \in cr_q; \\ 0, \text{ otherwise.} \end{cases}$$

$$iso_{q,i}(X) = \begin{cases} 1, \text{ if } (i\%h = 1 \lor cra_{q,i-1}(X) = 0) \land (i\%h = 0 \lor cra_{q,i+1}(X) = 0); \\ 0, \text{ otherwise.} \end{cases}$$

% is the modulo operator. One observes that in S4 the calculation is only limited within the same day. $iso_{q,i}(X) = 1$ means that there is no any course in the curriculum cr_q scheduled adjacent (before or after) to the timeslot $i\%h$ in the $[i/h]$th day (h denotes the total number of timeslots per day). More specifically, curriculum cr_q does not appear before (after) period t_i means that t_i is the first (last) timeslot of a working day or cr_q does not appear at t_{i-1} (t_{i+1}).

$\alpha_1, \alpha_2, \alpha_3$ and α_4 are the unit penalty point for each of the soft constraints. Notice that $\alpha_1 \sim \alpha_4$ are fixed in the problem formulation and should not be confused with the penalty parameters used by some solution procedures. In the literature, there are two versions of the CCT—old version and ITC–2007 competition version. We call them formulation **I** and **II** respectively. The old version ignores the second soft constraint S2 and fixes the values of $\alpha_1 \sim \alpha_4$ as:

$$\alpha_1 = 1, \alpha_2 = 0, \alpha_3 = 5, \alpha_4 = 1$$

The competition version used in ITC–2007 sets $\alpha_1 \sim \alpha_4$ as:

$$\alpha_1 = 1, \alpha_2 = 1, \alpha_3 = 5, \alpha_4 = 2$$

With the above formulation, we can then calculate the total soft penalty cost for a given candidate feasible solution X according to the cost function f defined in Eq. (1). The goal is then to find a feasible solution X that minimizes the following function.

$$f(X) = \sum_{x_{i,j} \in X} f_{rc}(x_{i,j}) + \sum_{c_i \in C} f_{rs}(c_i) + \sum_{c_i \in C} f_{md}(c_i) + \sum_{x_{i,j} \in X} f_{cc}(x_{i,j}) \qquad (1)$$

3   Hybrid Heuristic Algorithm for CCT

The basic idea of our hybrid heuristic algorithm is to combine the advantageous features of Tabu Search (TS) [8] and Iterated Local Search (ILS) [10]. Similar to the idea of [13], we devise in this work an Adaptive TS algorithm whose components and mechanisms are described in the following subsections.

TS is based on the belief that intelligent searching should be systematically based on adaptive memory and learning. TS can be used with both long and short computing budgets. In general, long computing budgets would lead to better results. However, if the total computation time is limited, it would be preferred to combine short TS runs with some robust diversification operators.

Interestingly, ILS provides such diversification mechanisms to guide the search to escape from a local optimum and move towards new regions in the solution space. When the best known solution cannot be improved any more using the TS algorithm, we employ a penalty-guided perturbation operator to destruct the obtained local optimum.

Note that starting from an empty timetable, we generate first an initial feasible solution by means of a graph coloring greedy heuristic. Because of the page limit, the details of this greedy heuristic are omitted here. We simply mention that for all the tested instances, this greedy heuristic can easily obtain feasible solutions. Once a feasible timetable that satisfies all the hard constraints is reached, our ATS algorithm is used to minimize the soft constraint cost function (Eq. (1)) without breaking hard constraints any more. Therefore, the search space of our ATS algorithm is limited to the feasible timetables.

One interesting issue concerns the influence of the initial solution on the final solution reached by ATS. Experimentations (not reported here) show that ATS is not sensitive to the quality of the initial solution.

3.1   Neighborhood Structure

In a neighborhood search procedure, applying a move mv to a candidate solution X leads to a new solution denoted by $X \oplus mv$. Let $M(X)$ be the set of all possible moves which can be applied to X and does not create any infeasibility, then the neighborhood of X is defined by: $N(X) = \{X \oplus mv | mv \in M(X)\}$. For the CCT problem, we use two distinct *moves* denoted by *SimpleSwap* and *KempeSwap*. Respectively, two neighborhoods denoted by N_1 and N_2 are defined as follows.

Neighborhood N_1: A *SimpleSwap* move consists in exchanging the hosting periods and rooms assigned to two lectures of different courses. Applying the *SimpleSwap* move to two different courses $x_{i,j}$ and $x_{i',j'}$ for the solution X consists in assigning the value of $x_{i,j}$ to $x_{i',j'}$ and inversely the value of $x_{i',j'}$ to $x_{i,j}$. Note that moving one lecture of a course to a free position is a special case of the *SimpleSwap* move where one of the swapping lectures is null and it is also included in our neighborhood N_1.

Neighborhood N_2: Our *KempeSwap* move is defined by interchanging two Kempe chains. If we focus only on courses and conflicts, each problem instance can be seen as a graph G where nodes are courses and edges connect courses with students or teacher in common. In a feasible timetable, a Kempe chain is the set of nodes that form a connected component in the subgraph of G induced by the nodes that belong to two periods. A *KempeSwap* produces a new feasible assignment by swapping the period labels assigned to the courses belonging to two specified Kempe chains. Once courses have been scheduled to periods, the room assignment can be done by solving a bipartite matching problem [15,16]. In this paper, we implement an exact algorithm–the augmenting paths algorithm introduced in [15,16].

More formally, let K_1 and K_2 be two Kempe chains in the subgraph with respect to two periods t_i and t_j, a *KempeSwap* produces an assignment by replacing t_i with $(t_i \setminus (K_1 \cup K_2)) \cup (t_j \cap (K_1 \cup K_2))$ and t_j with $(t_j \setminus (K_1 \cup K_2)) \cup (t_i \cap (K_1 \cup K_2))$. For instance, figure 1 depicts a subgraph deduced by two periods t_i and t_j and there are four Kempe chains: $K_a = \{c_1, c_2, c_7, c_8, c_{10}\}$, $K_b = \{c_3, c_6, c_9\}$, $K_c = \{c_4, c_{11}, c_{12}\}$ and $K_d = \{c_5\}$. Then, if we swap two Kempe chains K_b and K_c, a *KempeSwap* produces an assignment by moving $\{c_3, c_4, c_6\}$ to t_j and $\{c_9, c_{11}, c_{12}\}$ to t_i.

It is noteworthy to notice that our double Kempe chains interchange can be considered as a *generalization* of the single Kempe chain interchange known in the literature. In the previous definition of single Kempe chain neighborhood, each move concerns only one connected component, i.e., one of the two Kempe chains in our definition is empty [3,2,12]. Formally, it means replacing t_i with $(t_i \setminus K) \cup (t_j \cap K)$ and t_j with $(t_j \setminus K) \cup (t_i \cap K)$ where K is the non-empty Kempe chain [3,2,5,12]. Consequently, the single Kempe chain interchange is a special case of our *KempeSwap* move and it is included in our neighborhood N_2. Although not mentioned in this paper, a detailed analysis and comparison of those and other neighborhoods is conducted, showing the efficiency of the newly proposed double Kempe chains neighborhood.

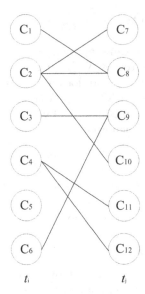

Fig. 1. Kempe chain illustrations

3.2   TS Using a Combined Neighborhood

The basic search engine of our ATS algorithm is of course based on TS. The TS procedure exploits the two neighborhoods N_1 and N_2 in a token-ring way [7]. More precisely, we start the TS procedure with one neighborhood. When the search ends with its best local optimum, we restart TS from this local optimum, but with the other neighborhood. This process is repeated until no improvement is possible and we say a TS phase is achieved. In our case, the TS procedure begins from the basic neighborhood N_1 and then neighborhood N_2: $N_1{\rightarrow}N_2{\rightarrow}N_1{\rightarrow}N_2....$

Within TS, a *tabu list* is introduced to forbid the previously visited moves. At each iteration, a best non-tabu move mv is applied to the current solution X even if $X' = X \oplus mv$ does not improve the solution quality. In our TS, when moving one lecture from one position (period-room pair) to another (N_1), or from one period to another (N_2), this lecture cannot be moved back to the original position (N_1) or period (N_2) for the next tt iterations (tt is called tabu tenure). Tabu tenure tt of a lecture x is tuned adaptively according to the current solution quality f and the frequency of the move $freq(x)$, i.e.

$$tt(x) = f + \varphi \cdot freq(x)$$

where φ is a parameter that lies in $[0, 1]$. The aspiration criterion accepts a tabu move if it improves the current best result or the set of non-tabu moves is empty in the current neighborhood. The TS procedure based on each neighborhood stops when the best solution cannot be improved within a given number of steps (denoted by θ) and we call this number the *depth of TS*. The basic TS procedure is described in algorithm 1:

Algorithm 1. Tabu Search procedure: TS(X_0, θ)

1: //X_0 is the feasible initial solution
2: //θ is *the depth of TS*
3: **repeat**
4:     $X^* = TS_{N_1}(X_0)$ based on N_1 with *depth of TS* θ
5:     $X^{*'} = TS_{N_2}(X^*)$ based on N_2 with *depth of TS* $\theta/3$
6:     $X_0 = X^{*'}$
7: **until** (no improvement is reached)

Because of the high computational effort in neighborhood evaluation of N_2, TS uses a much smaller depth (empirically fixed at $\theta/3$) (line 5) when N_2 is used. Note that the token-ring search based on TS stops when no improvement is possible. At this point, a TS phase is finished.

3.3 Perturbation

When a Tabu Search phase terminates, we employ a perturbation operator to destruct the reached local optimum in order to restart a new TS phase from this perturbed solution. Our perturbation operator consists of randomly selecting a given number of *SimpleSwap* or *KempeSwap* moves, where at least one of the moved lectures belongs to the first k highly-penalized ones. Specifically, when the current TS phase terminates, all the lectures are ranked in a non-increasing order according to their soft costs involved. Then, a certain number of lectures are selected from the first k-ranked (highly-penalized) ones. Notice that constraining the choice to highly-penalized lectures is essential because it is these lectures that contribute strongly to constraint violations (and the cost function).

Obviously, the *perturbation strength* (denoted by γ) is one of the most important ingredients of ILS and determines the quality gap between the two solutions before and after perturbation. In our case, γ is adaptively adjusted and takes values in an interval $[\gamma_{min}, \gamma_{max}]$ (set experimentally $\gamma_{min} = 4, \gamma_{max} = 15$). For acceptance criterion in the perturbation process, we use a strong exploitation technique, i.e., only better solutions are accepted.

3.4 Combination of TS with Perturbation

The *depth of TS* θ and the *perturbation strength* γ are two essential parameters which control the behavior of the ATS algorithm. On the one hand, a greater θ value ensures a more intensive search. On the other hand, greater γ corresponds to more possibility of escaping from the current local minimum. In order to get a continuous tradeoff between intensification and diversification, we devise a mechanism to dynamically and adaptively adjust these two important parameters according to the historical search records. Note that in this paper the initial values of these two and other parameters are empirically set and they are all instance-independent. It is possible that better solutions would be found by using a set of instance-dependent parameters. However, our aim is to design a robust solver which is able to solve efficiently a large panel of instances.

At the beginning of the search, we take a basic TS where the *depth of TS* θ is a small positive number, say $\theta = \theta_0$ ($\theta_0 = 10$). When TS cannot improve its best solution, perturbation is applied to this best solution with a weak strength ($\gamma = \gamma_{min}$). When the search progresses, we record the number of TS phase iterations (denoted by ξ) for which no improved solution has been found. The *depth of TS* θ and the *perturbation strength* γ are dynamically adjusted as follows: When the local minimum obtained by TS is promising, i.e., when it is close to the current best solution ($f \leq f_{best} + 2$), the *depth of TS* is gradually increased to ensure a more and more intensive search until no improvement is possible, i.e., $\theta = (1 + \eta)\theta$ at each iteration ($\eta = 0.6$). Similarly, *perturbation strength* is gradually increased so as to diversify more strongly the search if the number of non-improving TS phase iterations increases.

In this paper, we use the timeout condition required by the ITC–2007 competition rules (see next Section). Finally, our hybrid ATS algorithm is described in algorithm 2.

Algorithm 2. Adaptive Tabu Search scheme

1: X_0 is a feasible solution, X^* is the best solution found so far
2: set $\xi = 0$, $\theta = \theta_0$, $\gamma = \gamma_{min}$
3: apply TS to X_0 with *depth of TS* θ: $X^* = TS(X_0, \theta)$
4: **repeat**
5:     perturb X^* with perturbation strength γ, get X'
6:     apply TS to X' with *depth of TS* θ, get $X^{*'}$
7:     **if** the local minimum solution $X^{*'}$ is promising, i.e., $f(X^{*'}) \leq f(X^*) + 2$ **then**
8:         **repeat**
9:             call the TS procedure with a gradually increased θ: $\theta = (1 + \eta)\theta$
10:        **until** no better solution is obtained
11:     **end if**
12:     **if** better solution $X^{*'}$ has been found, i.e., $f(X^{*'}) < f(X^*)$ **then**
13:        accept $X^{*'}$ as the current best solution: $X^* = X^{*'}$
14:        reset to the basic TS ($\theta = \theta_0$) with weak perturbation ($\gamma = \gamma_{min}$)
15:     **else**
16:        reset to the basic TS: $\theta = \theta_0$
17:        $\xi = \xi + 1$
18:        update the perturbation strength: $\gamma = min\{\gamma_{min} + \lambda \cdot \xi, \gamma_{max}\}$
19:     **end if**
20: **until** (timeout condition is met)

4　Experiments and Comparisons

4.1　Problem Instances and Experimental Protocol

To evaluate the efficiency of our proposed ATS algorithm, we carry out experiments on two different data sets. The first set (4 instances named test1~test4) was previously used in the literature for the old version of the CCT problem [7].

The second set (7 instances named comp01~comp07) is from the Second International Timetabling Competition mentioned in the introduction. All these 11 instances can be downloaded from http://tabu.diegm.uniud.it/ctt/index.php.

Our algorithm is programmed in C and compiled using Dev C++ on a PC running Windows XP with 3.44GHz CPU and 2G RAM. To obtain our computational results, our ATS algorithm is run 100 times on each instance with different random seeds. The stop condition is just the timeout required by the ITC–2007 competition rules. On our PC, this corresponds to 390 seconds.

4.2   Comparative Results and Discussion

Table 2 shows the results of the ATS algorithm on the 11 instances of the two data sets for both formulations (see Section 2.2) as well as the previous best known results available in the literature [7,6,14]. For the ATS algorithm, we indicate the following information: the best score f_{min}, the average score f_{ave} and the standard deviation σ over 100 independent runs. For the reference algorithms in [7,6,14], only the best results f_{min} are available. It should be mentioned that the stop conditions of the three reference algorithms are also the timeout condition of the ITC–2007.

The algorithms in [7,6] are developed by the ITC–2007 organizers while that in [14] is the winner solver of the ITC–2007 competition. The algorithms of [7,6] employ a dynamic Tabu Search which allows unfeasible assignments during the problem solving and the neighborhood is the simple one that moves one lecture to a different period and/or a different room. One observes that this neighborhood is just the subset of our basic neighborhood N_1. The algorithm in [14] is composed of a constructive phase, a Hill Climbing algorithm and the Great Deluge technique and uses six specialized moves (neighborhoods).

From table 2, one observes that our ATS algorithm dominates the algorithms of [7,6] for 8 out of 11 instances (in bold). For the 3 remaining cases whose optimum is known, the optimum is reached by the ATS algorithm within several seconds. Moreover, the standard deviations of ATS for all the tested instances are small, showing its robustness.

If we compare ATS and the winner algorithm of the ITC–2007 in [14] on the 7 competition instances (results not available in [14] for the 11 instances of Formulation **I** and the first set of 4 instances of Formulation **II**), one observes that the results of both algorithms are quite comparable. For these 7 instances, ATS reaches better (respectively worse) results than the algorithm in [14] for 2 (respectively 3) instances, with equaling results for the 2 remaining instances. Notice that ATS is ranked the second place for the track 3 of ITC–2007 [2].

Let us mention that we also studied the behavior of the proposed ATS algorithm concerning the influence of the penalty-guided perturbation mechanism (Section 3.3) and the different adaptive mechanisms (Section 3.4). Moreover, a detailed study was conducted to analyze several neighborhoods (including those used in this paper) and the different ways of combining them. This analysis showed that the newly proposed double Kempe chains neighborhood N_4 and its

---

[2] This result is available at http://www.cs.qub.ac.uk/itc2007/winner/finalorder.htm

Table 2. Computational results and comparison on the 11 instances

Instance	Formulation I				Formulation II				
	ATS Heuristic			Best in [7, 6]	ATS Heuristic			Best in [6]	Best in [14]
	f_{min}	f_{ave}	σ	f_{min}	f_{min}	f_{ave}	σ	f_{min}	f_{min}
test1	**212**	212	0	213	**224**	229.5	1.8	234	—
test2	**8**	8	0	8	**16**	17.1	1.0	17	—
test3	**35**	35.3	0.3	36	**73**	82.9	4.1	86	—
test4	**28**	32.8	2.1	43	**76**	89.4	5.8	132	—
comp01	**4**	4	0	4	**5**	5	0	5	5
comp02	**22**	24.5	2.3	35	**34**	60.6	7.5	75	43
comp03	**41**	44.8	3.5	52	**70**	86.6	6.3	93	72
comp04	**19**	21.8	2.8	21	38	47.9	4.0	45	**35**
comp05	**224**	229.4	5.1	**244**	**298**	328.5	11.7	326	**298**
comp06	**25**	27.6	3.1	27	47	69.9	7.4	62	**41**
comp07	**4**	6.3	2.4	13	19	28.2	5.6	38	**14**

token-ring combination with the simple neighborhood N_1 contribute greatly to the efficiency of the ATS algorithm.

5   Conclusions

We have provided a mathematical formulation of the curriculum-based course timetabling problem and presented a highly effective hybrid Adaptive Tabu Search algorithm for solving this difficult problem. The effectiveness of the ATS algorithm comes from a number of original features. First, we have introduced the double Kempe chain neighborhood structure for the CCT problem. Second, for our TS procedure, we have devised a combined exploitation strategy of the Kempe chain neighborhood and the basic swap neighborhood. Third, we have proposed a mechanism for adaptively combining TS and perturbation. The computational results on 11 instances on two formulations show that our hybrid ATS algorithm dominates the reference algorithms in [7,6] and competes very well with the winner solver of the ITC–2007 in [14]. Let us comment that most of the ingredients proposed in this paper remain general and would be directly applicable or adapted to other combinatorial problems.

Acknowledgment

This work was partially supported by *"Angers Loire Métropole"* and the Region of *"Pays de la Loire"* within the MILES Project (2006-2009). The authors would like to thank the referees for their helpful comments and questions.

References

1. Burke, E., McCollum, B., Meisels, A., Petrovic, S., Qu, R.: A graph-based hyper heuristic for timetabling problems. European Journal of Operational Research 176, 177–192 (2007)
2. Casey, S., Thompson, J.: Grasping the examination scheduling problem. In: Burke, E.K., De Causmaecker, P. (eds.) PATAT 2002. LNCS, vol. 2740, pp. 232–246. Springer, Heidelberg (2003)

3. Chiarandini, M., Birattari, M., Socha, K., Rossi-Doria, O.: An effective hybrid algorithm for university course timetabling. Journal of Scheduling 9, 403–432 (2006)
4. Cooper, T.B., Kingston, J.H.: The complexity of timetable construction problems. In: Burke, E.K., Ross, P. (eds.) PATAT 1995. LNCS, vol. 1153, pp. 283–295. Springer, Heidelberg (1996)
5. Côté, P., Wong, T., Sabourin, R.: Application of a hybrid multi-objective evolutionary algorithm to the uncapacitated exam proximity problem. In: Burke, E.K., Trick, M.A. (eds.) PATAT 2004. LNCS, vol. 3616, pp. 151–168. Springer, Heidelberg (2005)
6. De Cesco, F., Di Gaspero, L., Schaerf, A.: Benchmarking Curriculum-Based Course Timetabling: Formulations, Data Formats, Instances, Validation, and Results, University of Udine (2008), http://tabu.diegm.uniud.it/ctt/DDS2008.pdf
7. Di Gaspero, L., Schaerf, A.: Multi-neighbourhood local search with application to course timetabling. In: Burke, E.K., De Causmaecker, P. (eds.) PATAT 2002. LNCS, vol. 2740, pp. 263–278. Springer, Heidelberg (2003)
8. Glover, F., Laguna, M.: Tabu Search. Kluwer Academic Publishers, Dordrecht (1997)
9. Lewis, R.: A survey of metaheuristic-based techniques for university timetabling problems. OR Spectrum 30(1), 167–190 (2008)
10. Lourenço, H., Martin, O., Stützle, T.: Iterated local search. Handbook of Metaheuristics. Springer, Berlin (2003)
11. McCollum, B.: A perspective on bridging the gap between theory and practice in university timetabling. In: Burke, E.K., Rudová, H. (eds.) PATAT 2007. LNCS, vol. 3867, pp. 3–23. Springer, Heidelberg (2007)
12. Merlot, L.T.G., Boland, N., Hughes, B.D., et al.: A hybrid algorithm for the examination timetabling problem. In: Burke, E.K., De Causmaecker, P. (eds.) PATAT 2002. LNCS, vol. 2740, pp. 207–231. Springer, Heidelberg (2003)
13. Misevicius, A., Lenkevicius, A., Rubliauskas, D.: Iterated tabu search: an improvement to standard tabu search. Information Technology and Control 35(3), 187–197 (2006)
14. Müller, T.: ITC2007: Solver Description, Technical Report, Purdue University (2008), http://www.unitime.org/papers/itc2007.pdf
15. Rossi-Doria, O., Paechter, B., Blum, C., Socha, K., Samples, M.: A local search for the timetabling problem. In: Proceedings of the 4th PATAT, pp. 124–127 (2002)
16. Sedgewick, R.: Algorithms, 2nd edn. Addison-Wesley, Reading (1988)

Heuristic Hill-Climbing as a Markov Process

Carlos Linares López

Planning and Learning Group, Computer Science Department
Universidad Carlos III de Madrid, Avda. de la Universidad, 30
28911 Leganés - Madrid, Spain
carlos.linares@uc3m.es

Abstract. The purpose of this paper is twofold: on one hand, modelling the hill-climbing heuristic search algorithm as a stochastic process serves for deriving interesting properties about its expected performance; on the other hand, the probability that a hill-climbing search algorithm ever fails when approaching the target node (i.e., it does not find a descendant with a heuristic value strictly lower than the current one) can be considered as a pesimistic measure of the accuracy of the heuristic function guiding it. Thus, in this work, it is suggested to model heuristic hill-climbing search algorithms with Markov chains in order to fulfill these goals. Empirical results obtained in various sizes of the (n, m)-Puzzle domain prove that this model leads to very accurate predictions.

1 Introduction

Any heuristic single-agent search algorithm is dramatically affected by the accuracy of the heuristic function guiding it. As a matter of fact, perfect heuristic functions $h(\cdot)$ (i.e., $h(\cdot) = h^*(\cdot)$ where $h^*(\cdot)$ denotes the optimal cost) make A^*-like search algorithms to expand only nodes that lie on the optimal path [1]. However, perfect heuristic functions are very difficult to obtain in practice, and most of them introduce an error in the estimation that make search algorithms to expand an exponentially increasing number of nodes with the depth of the search tree. In short, it is highly desirable to know in advance whether a heuristic function commits errors or not in a given domain and in case they do, whether these errors might become a concern.

Of all the heuristic search algorithms described in the specialized bibliography, heuristic hill-climbing is the usual choice in very complex systems like *automated planning* [2]. For example, the Heuristic Search Planner, HSP [3], employs a hill-climbing search algorithm guided by a heuristic function which estimates the distance between a state and a set of facts. The same authors conjectured in a different work [4] that heuristic hill-climbing is likely to be the right choice in this type of systems. Likewise, the Fast-Forward Planner, FF [5] uses a specialized version of the heuristic hill-climbing search algorithm known as *enforced* hill-climbing which is known to outperform other heuristic and non-heuristic planners. The same version of the hill-climbing search algorithm was later on adopted by Refanidis and Vlahavas in the forward state of the GRT

D. Dochev, M. Pistore, and P. Traverso (Eds.): AIMSA 2008, LNAI 5253, pp. 274–284, 2008.

heuristic planner [6]. However, though these systems (and others) do all employ quite similar search algorithms they are known to behave differently in practice. The main reason is that they do use different heuristic functions. Therefore, foretelling the performance of the hill-climbing approach when using a well-defined heuristic function is a major issue in the design of planning systems and other intelligent complex systems.

In this paper it is suggested to model the behaviour of the heuristic hill-climbing search algorithm with Markov chains with the aim of: first, determining the accuracy of any heuristic function and last, but not least, to characterize the expected performance of the heuristic hill-climbing search algorithm when it is guided by a heuristic function.

This paper is arranged as follows: next section briefly summarizes the behaviour of the heuristic hill-climbing search algorithm and discusses its suitability to Markov chains, introduced in the section after. Next, Markov chains are widely used to refer to the main concepts underlying heuristic search: accuracy and expected solution length. Finally, this paper ends with some conclusions and future work guidelines.

2 Heuristic Hill-Climbing

Table 1 shows the pseudocode of the heuristic hill-climbing search algorithm[7] where $SCS(n)$ is the set comprising all descendants of node n, and $k(n, n_k)$ is the cost of the edge from node n to its descendant n_k. The heuristic hill-climbing search algorithm strives to reach the goal state t from any node s by selecting any of the most promising descendants n_i of the current node n, i.e., it commits to any of the descendants with the lowest heuristic estimation to the target node, t, underestimating all the other successors. Thus, the hill-climbing search algorithm evolves by selecting locally the next node to expand in spite of the nodes previously expanded (in other words, it does not use any lists as best-first does) so that it is known as a *"memory-less"* search algorithm in the specialized bibliography.

Table 1. Pseudocode of the Heuristic Hill-Climbing search algorithm

```
HILL-CLIMBING (n:NODE): COST
    if (n=t) return 0
    compute n_k = argmin{h(n_i)}, ∀n_i ∈ SCS(n)
    return k(n,n_k)+HILL-CLIMBING (n_k)
```

This actually means that the heuristic hill-climbing approach exhibits the *Markovian property* which states that the conditional probability distribution function of future states depends only upon the current state. According to this observation, the stochastic theory of Markov processes [8] deserves some attention when studying this type of algorithms.

Before moving further, it shall be noted that the analysis is restricted to domains which meet the following requirements:

– The cost of all arcs, $k(n, n_i)$, with n_i being a successor of n, is always equal to 1.
– The only allowed changes of the heuristic function, $h(\cdot)$, from node n to any of its descendants, n_i, are $\{-1, +1\}$ when estimating its distance to the same node m: $|h(n, m) - h(n_i, m)| = 1$.

Note that the second assumption amounts to assuming that $h(\cdot)$ is *consistent* when the first requirement is met. Also, *consistency* leads to *admisibility* of the heuristic funcion or, in other words, that $h(\cdot)$ will never overestimate the cost to reach the target node [1].

3   Markov Chains

Let $X(t)$ be a family of random variables which denote the current state of a hill-climbing search algorithm at time t. Since the evolution of the search algorithm can be actually sampled at equally sized intervals, $i = 1, 2, \ldots$, $X(t)$ is said to be a *discrete-time parameter* stochastic process. As mentioned before, since the hill-climbing search algorithm exhibits the Markovian property, the state $X(t_i)$ is said to contain all the relevant information necessary to decide what state to visit next. Moreover, these transitions do not depend upon the current time t_i and the resulting process is said to be *homogeneous* and it is denoted just as X. When the state space of a Markov process is discrete, it is referred as a *Markov chain*.

According to [8], let X_i denote the successive observations of a homogeneous discrete-time Markov chain at time steps $i, i + 1$, etc. The Markovian property can now be written down more specifically as:

$$P(X_{n+1} = x_{n+1}|X_0 = x_0, X_1 = x_1, \ldots, X_n = x_n) = P(X_{n+1} = x_{n+1}|X_n = x_n)$$

where $P(X_{n+1} = x_{n+1}|X_n = x_n)$ are known as the *single-step transition probabilities*, i.e., the conditional probability of making a transition from markvovian state x_n to state x_{n+1}. These probabilities will also be denoted as:

$$p_{ij} = P(X_{n+1} = j|X_n = i)$$

The matrix P, formed by placing p_{ij} in row i and column j is called the *transition probability matrix* and it is said to be stochastic since:

1. $p_{ij} \geq 0$ for all i and j
2. $\sum_j p_{ij} = 1$ for all i
3. No column contains only zeroes

3.1   Markovian States

Obviously, it would make no sense to consider states x_i traversed by a Markov chain X (also called *markovian states*) to be the same than the nodes expanded by a hill-climbing search algorithm, n_j. Instead, some sort of reductionism (from nodes to Markov states) shall be practiced in order to allow a realistic statistical analysis.

Clearly, the more specific the description of markovian states, the more number of markovian states would be required to represent the whole state space to be traversed by the hill-climbing search algorithm —leading to a problem of efficiency. Conversely, the more general the description of markovian states, the less truthful the statistical results, hence, leading to a problem of accuracy. Ideally, two nodes n_i and n_j shall belong to the same markov state x_k if they both have the same chances to evolve towards the goal state. In this work, a representation of markov states, called the τ-representation is introduced with the hope of being a good trade-off between efficiency and accuracy.

A good contribution to the statistical characterization of nodes in a state space was provided by Korf et al. [9]. The *equilibrium distribution* they introduced served for predicting very accurately the number of generated nodes by a heuristic single-agent search algorithm, and it resulted from splitting nodes having the same heuristic estimation to the target node, $h(n, t)$, according to their branching factor, b.

However, taking into account only the heuristic estimation, h, and branching factor, b, lacks of some important information for guessing where to step next. In other words, it is not expected that nodes having the same values of h and b have also the same probability to get to the goal state in the same number of steps. As a matter of fact, the probability to find a path from node n to the target node t shall be somewhat related to the number of alternatives in the surroundings of node n: the more descendants with the least heuristic value, the more likely to get to the target node from the current state. According to this principle, it is defined the *class* c of a node n, $c(n)$, as the number of descendants of node n with minimal heuristic value, $h(n_i)$, where $h(n_i)$ is less than $h(n)$. Clearly, $c \leq b$ always holds.

Obviously, the hill-climbing search algorithm will deter from finding a solution in a number of steps equal to the heuristic value of a node n, $h(n, t)$, if its class equals 0. Thus, nodes belonging to the null class (this is, with $c(\cdot) = 0$) do represent errors for the heuristic function that shall have ranked them with a higher score, since its real cost shall never be less than:

$$\min_{n_i \in \text{SCS}(n)} \{k(n, n_i) + h(n_i, t)\}$$

provided that $h(\cdot)$ is admissible, where $\text{SCS}(n)$ stands for the successors of node n and $k(n, n_i)$ is the cost of the operator from n to n_i. Therefore, if no descendant have a heuristic estimation lower than the heuristic value of its parent, it has to be necessarily concluded that the parent is wrongly classified by the heuristic function. In conclusion, since $h(\cdot)$ is assumed to be admissible (according to section 2), the heuristic value of a node n, $h(n)$, can be recognized as being wrong everytime the hill-climbing search algorithm reaches the goal in a number of steps exceeding the heuristic value, $h(n)$, and this happens everytime a node with $c = 0$ is found along the path.

From a general point of view there is no reason to expect that nodes with the same heuristic estimation h and branching factor b belong to the same class c with the same likelihood. Henceforth, it is suggested to classify nodes within the state

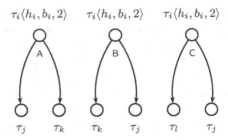

Fig. 1. Example of the computation of the single-step transition probabilities p_{ij}

space to be traversed by the hill-climbing search algorithm according to these parameters resulting in the so-called τ representation and denoted as $\tau\langle h, b, c\rangle$.

3.2   Sampling the State Space

Given a mapping function $\Theta : n_i \mapsto \tau_i$ which transforms any state n_i in the state space $\mathcal{E}$ into its τ-representation $\tau_i\langle h(n_i), b(n_i), c(n_i)\rangle$ in the Markov state space[1], where $h(n_i)$, $b(n_i)$ and $c(n_i)$ are the heuristic value, branching factor and class of node $n_i \in \mathcal{E}$, it is feasible to sample the probability p_{ij} of stepping from markovian state τ_i to τ_j by examining the descendants of node n_i and reasoning locally as follows: for all descendants, n_j, of node n_i, whose heuristic value $h(n_j) < h(n_i)$ equals the minimum of the heuristic value of all descendants, the likelihood it will be selected to be expanded next is $1/c(n_i)$, all its other siblings getting a likelihood equal to 0. However, if the minimum heuristic estimation of all the descendants is strictly greater than $h(n_i)$, the likelihood of any descendant been expanded equals[2] $1/b$. Let q_{ij} denote this local probability, thus:

$$p_{ij} = \frac{Q_{ij}}{Q_i}, \quad Q_i = \sum_j Q_{ij}$$

and Q_{ij} is equal to the summation of all the individual q_{ij} computed for every sampled transition from state i to state j.

For the sake of clarity, consider a node A with a τ-representation $\tau_i\langle h_i, b_i, c_i = 2\rangle$ which is observed to have exactly $c_i = 2$ descendants with a heuristic value equal to $(h_i - 1)$, all the other $(b_i - c_i)$ descendants having heuristic values larger than $(h_i - 1)$. Assume that the eligible children of A to be expanded by the heuristic hill-climbing search algorithm have τ-representations $\tau_j\langle h_i - 1, b_j, c_j\rangle$ and $\tau_k\langle h_i - 1, b_k, c_k\rangle$. Say other node B with the same τ-representation than node A (i.e., belonging to the same markov state than node A) is observed to choose equally likely between two descendants with the same τ-representations than the

---

[1] In the forthcoming discussions, 'state' is generally referred to markov states unless otherwise explicitly stated.

[2] In fact, this probability is $1/m$ where m is the number of nodes with minimal heuristic estimation. However, according to the assumptions in section 2, all descendants shall be scored in this case equally, resulting in $1/b$.

descendants of node A. Finally, a third node C with the same τ-representation than nodes A and B is sampled and transitions are observed to either one descendant belonging to the markov state represented by $\tau_j \langle h_i - 1, b_j, c_j \rangle$ or a new one termed $\tau_l \langle h_i - 1, b_l, c_l \rangle$. Figure 1 shows graphically this example.

In this case, $Q_{ij} = \frac{1}{2} + \frac{1}{2} + \frac{1}{2} = \frac{3}{2}$; $Q_{ik} = \frac{1}{2} + \frac{1}{2} = 1$ and $Q_{il} = \frac{1}{2}$. Thus, $Q_i = Q_{ij} + Q_{ik} + Q_{il} = 3$ leading to the following single-step transition probabilities:

$$p_{ij} = \frac{Q_{ij}}{Q_i} = \frac{\frac{3}{2}}{3} = \frac{1}{2}$$

$$p_{ik} = \frac{Q_{ik}}{Q_i} = \frac{1}{3} = \frac{1}{3}$$

$$p_{il} = \frac{Q_{il}}{Q_i} = \frac{\frac{1}{2}}{3} = \frac{1}{6}$$

In any case, after sampling a large number of nodes n_i and filling in matrix P computing p_{ij} according to the previous expression, it might happen that the resulting matrix P is not stochastic since for some rows its summation equals 0 or, in other words, no transition from it has ever been observed. Thus, after sampling the state space, matrix P has to be refined in order to ensure it is a properly defined stochastic matrix.

In case $Q_k = 0$ (i.e., no transition has ever been observed from state k) all transitions from any state i to state k, $p_{ik} > 0$, shall be deleted from the matrix P and also the state k. This is, row and column k shall be deleted from matrix P. Considering there are up to n transitions from state i, this process actually involves distributing the probability p_{ik} among all the $(n-1)$ remaining probabilities $p_{ij} \neq 0, j \neq k$.

One way consists of equally distributing p_{ik} among all the legal transitions from i so that $p_{ij}, j \neq k$, becomes $p_{ij} + \delta$ where $\delta = p_{ik}/(n-1)$. However, it shall lead to better results to weigh this over-assignment according to each likelihood $p_{ij}, j \neq k$, making p_{ij} to become $p_{ij}(1 + \delta')$ such that:

$$\sum_{j=1}^{n-1} (p_{ij} + p_{ij}\delta') = 1$$

so that:

$$\sum_{j=1}^{n-1} p_{ij}\delta' = 1 - \sum_{j=1}^{n-1} p_{ij} = p_{ik}$$

leading to:

$$\delta' = \frac{p_{ik}}{\sum\limits_{j=1}^{n-1} p_{ij}}$$

since $\sum\limits_{j=1}^{n-1} p_{ij} = (1 - p_{ik})$ the following value results:

$$\delta' = \frac{p_{ik}}{1 - p_{ik}}$$

In the preceding example, assume that state l has never been observed to transit to any other state, so that row and column l of matrix P shall be deleted. In this case, $\delta' = \frac{p_{il}}{1-p_{il}} = \frac{1}{5}$ so that p_{ij} and p_{ik} are updated to $\frac{1}{2}(1 + \frac{1}{5}) = \frac{3}{5}$ and $\frac{1}{3}(1 + \frac{1}{5}) = \frac{2}{5}$, respectively, so that $p_{ij} + p_{ik} = \frac{3}{5} + \frac{2}{5} = 1$, as stated above.

Finally, after re-computing the probabilities p_{ij} for all states i that step to any state k from which no transition has ever been sampled (this is, $Q_k = 0$), row and column k can be safely deleted from the matrix P. This process shall be run iteratively until no ilegal transitions are left leaving P as a properly defined stochastic matrix.

4   Heuristic Hill-Climbing as a Markov Process

Nonetheless, the preceding procedure served only for deriving the so-called *single-step transition matrix* P. Fortunately, the so-called Chapman-Kolmogorov equations [8] generalize the single-step transition matrix P to a λ-step transition probability matrix $P^{(\lambda)}$ whose element $p_{ij}^{(\lambda)} = P(X_{m+\lambda} = j | X_m = i)$. For a homogeneous discrete-time Markov chain, the following results from the Markov property:

$$p_{ij}^{(\lambda)} = \sum_{\forall k} p_{ik}^{(l)} p_{kj}^{(\lambda-l)}, \text{ for } 0 < l < \lambda$$

and, in general terms, the Chapman-Kolmogorov equations are written as $P^{(\lambda)} = PP^{(\lambda-1)} = P^{(\lambda-1)}P$.

If 0 denotes a Markovian state whose heuristic estimation to the target node is 0 (i.e., it is a *goal state*), then $p_{i0}^{(\lambda)}$ is the probability that a goal state can be reached from state i in λ steps. More specifically, column 0 of $P^{(\lambda)}$ contains the probabilities of finding a goal state from any Markovian state in exactly λ transitions.

Hence, the key observation here results from rephrasing the notion of error discussed at the end of section 3.1 and it is that for all states (rows) whose heuristic value equals $h(\cdot) = x$, column 0 of the stochastic matrix $P^{(x)}$ contains the probability that heuristic hill-climbing will find a path from node n to the goal state in exactly x steps or, similarly, *the probability that $h(\cdot)$ provided perfectly informed values all along the path to the target node.*

Obviously, if $p_{x0}^{(x)} = 1$ for all states (rows) with $h(\cdot) = x$ then $h(\cdot)$ can be tagged as a perfect heuristic function. Instead, if $0 \leq p_{x0}^{(x)} < 1$, there is a positive likelihood $(1 - p_{x0}^{(x)})$ of being confused in the path to the goal state t, i.e., of $h(\cdot)$ to provide misleading estimations along the path to the goal state t. In brief, $p_{x0}^{(x)}$ can be actually considered as a *pesimistic* measure of the accuracy of the heuristic function $h(\cdot)$.

In most cases, however, deriving accurate figures about the accuracy of the heuristic hill-climbing search algorithm might not be enough, and one might still wonder whether the error margins are too large or not. The answer relies on the analysis of the mean number of steps necessary to reach the target node t from every state and more precisely, from every heuristic value, e.g., being x the current heuristic estimation what is the expected number of steps to reach t?

In order to answer this question, one can compute the *mean first passage times*:

$$M_{ij} = \sum_{\lambda=1}^{\infty} \lambda f_{ij}^{(\lambda)}$$

where $f_{ij}^{(\lambda)}$ is the probability that, starting from state i, the first passage to state j occurs in exactly λ steps. Obviously, $f_{ij}^{(1)} = p_{ij}$, and for arbitrary values of λ, this probability can be computed as

$$f_{ij}^{(\lambda)} = p_{ij}^{(\lambda)} - \sum_{l=1}^{\lambda-1} f_{ij}^{(l)} p_{jj}^{(\lambda-l)}$$

that can be intuitively explained as the probability of arriving at state j in λ steps considering even cycles minus the probability for these cycles to happen.

Thus, M_{i0} is the mean number of steps to reach the goal state from state i or, to put it this way, the mean solution length when applying the heuristic hill-climbing search algorithm from nodes with heuristic value equal to i.

5   Results

The experiments have been conducted on various sizes of the sliding tile puzzle: 2×4, 3×3, 2×5 and 4×4 using the Manhattan distance. The first three domains were selected because they are small enough as for being fully enumerable so that a high accuracy is expected there if the model proposed herein is as good as expected. The last domain has been chosen to test the accuracy when sampling large state spaces.

For testing the accuracy of the predicted performance of the heuristic hill-climbing search algorithm, the expected number of times that hill-climbing would succeed in finding the goal state from every node n, in a number of steps equal to its heuristic value, was sampled as:

$$\bar{p}(n) = \frac{1}{c(n)} \sum_{i=1}^{c(n)} \bar{p}_i(n_i) \quad , n_i \in \text{SCS}(n) \tag{1}$$

This probability, $\bar{p}(n)$, is computed for every sampled node just running the heuristic hill-climbing search algorithm on each descendant that decrements the heuristic distance of its immediate ancestor, and returning 1 every time the target node is found or 0 if a node with class 0 (or, equivalently, an error of the heuristic function) has been found. The backed-up values are joined as shown in equation (1). Clearly, $\bar{p}(n)$ stands for the probability that the heuristic hill-climbing search algorithm will find the goal state from node n, in a number of steps equal to its heuristic value.

Table 2. Results in different sliding tile puzzles

h	2×4 (20160 nodes)		3×3 (181440 nodes)		2×5 (1814400 nodes)		4×4 (10^{13} nodes)	
	Obs. ($\bar{p}_x$)	Pred. ($\hat{p}_x$)	Obs. ($\bar{p}_x$)	Pred. ($\hat{p}_x$)	Obs. ($\bar{p}_x$)	Pred. ($\hat{p}_x$)	Obs. ($\bar{p}_x$)	Pred. ($\hat{p}_x$)
0	1	1	1	1	1	1	1	1
1	1	1	1	1	1	1	1	1
2	1	1	1	1	1	1	1	1
3	0.75	0.75	0.8	0.8	0.75	0.75	0.714286	0.714286
4	0.142857	0.142857	0.13913	0.13913	0.097345	0.097345	0.062015	0.062015
5	0.081761	0.077268	0.073170	0.068879	0.065934	0.062834	0.052	0.050238
6	0.044025	0.038978	0.041726	0.036011	0.036710	0.031671	0.026440	0.025341
7	0.028130	0.026218	0.030864	0.025866	0.021269	0.021315	0.016626	0.016534
8	0.016414	0.013524	0.015230	0.011262	0.008424	0.009189	0.006601	0.006764
9	0.010649	0.008110	0.010228	0.007179	0.005080	0.005606	0.004788	0.004858
10	0.007405	0.004935	0.006235	0.003555	0.002983	0.002978	0.004843	0.004806
11	0.005022	0.002939	0.004706	0.002504	0.001833	0.001826		
12	0.003513	0.001775	0.002991	0.001266	0.001044	0.001026		
13	0.002069	0.001016	0.002023	0.000873	0.000630	0.000624		
14	0.001298	0.000600	0.001432	0.000470	0.000381	0.000371		
15	0.000712	0.000325	0.001081	0.000319	0.000217	0.000223		
16	0.000254	0.000178	0.000912	0.000178	0.000123	0.000134		
17	9.6×10^{-5}	9.8×10^{-5}	0.000765	0.000119	6.5×10^{-5}	8.0×10^{-5}		
18	3.1×10^{-5}	5.0×10^{-5}	0.000821	7.0×10^{-5}	3.2×10^{-5}	4.8×10^{-5}		
19	3.2×10^{-5}	2.8×10^{-5}	0.000790	4.4×10^{-5}	1.5×10^{-5}	2.8×10^{-5}		
20	0	1.5×10^{-5}	0.001358	2.9×10^{-5}	7.0×10^{-6}	1.6×10^{-5}		
21	0	1.1×10^{-5}	0.001213	1.6×10^{-5}	3.7×10^{-6}	1.0×10^{-5}		
22			0.001469	1.5×10^{-5}	2.0×10^{-6}	5.8×10^{-6}		
23					1.2×10^{-6}	3.3×10^{-6}		
24					5.3×10^{-7}	1.9×10^{-6}		
25					2.9×10^{-7}	1.1×10^{-6}		
26					1.8×10^{-7}	6.3×10^{-7}		
27					2.1×10^{-8}	3.8×10^{-7}		
28					1.8×10^{-8}	2.2×10^{-7}		
29					0	1.4×10^{-7}		
30					0	8.6×10^{-8}		
31					0	6.2×10^{-8}		

So far, the empirical value for the performance of the heuristic hill-climbing search algorithm is finally computed for every heuristic distance x as:

$$\bar{p}_x = \frac{1}{N_x} \sum_{\forall n, h(n)=x} \bar{p}(n)$$

where N_x is the number of nodes with $h(n) = x$. This value, $\bar{p}_x$, simply binds all nodes with the same heuristic value and is used for reporting purposes. Thus, it stands for the *observed* probability that the heuristic hill-climbing search algorithm will find the goal state from an arbitrary node with $h(n) = x$ in exactly x steps.

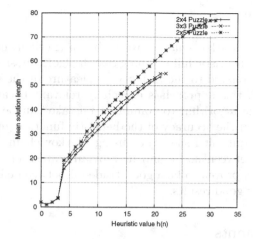

Fig. 2. Mean solution length in the 2×4, 3×3 and 2×5 puzzles

On the other hand, the *theoretical* value predicted by the Markov chain results from the theorem of total probability:

$$\hat{p}_x = \sum_{\forall i, h(i)=x} p_{i0}^{(x)} \frac{Q_i}{\sum_{\forall i} Q_i}$$

where i are the markov states whose heuristic distance to the target node equals x. Obviously, $\bar{p}_0 = \hat{p}_0 = 1$, this is, the target node, when being inmediately available has a probability equal to 1 to be chosen.

The results of comparing $\hat{p}_x$ vs. $\bar{p}_x$ are shown in tabular form since they result indistinguishable in a figure. Table 2 shows the probabilities either observed (sampled) or computed (predicted) by the Markov chain with up to 6 decimals, for the 2×4, 3×3 and 2×5 puzzles for all plausible heuristic values —note that every puzzle have different maximum heuristic values. It also shows the differences between the predicted and observed values for the 4×4 puzzle up to depth 10. As it can be seen, the precision of the mathematical model proposed herein is very accurate. In the 2×4, 3×3 and 2×5 puzzles, the whole state space was enumerated and all the transitions among their respective markovian states were computed. In the case of the 4×4 puzzle, all states with heuristic value up to 10 were considered.

Finally, the second class of experiments that were ran are relative to the mean solution length, as computed by the Markov chain, of the heuristic hill-climbing search algorithm using the Manhattan distance in the 2×4, 3×3 and 2×5 puzzles. Although these results are very hard (or even impossible) to contrast exactly with empirical data[3], the results obtained are very sound. It shall be noted that the mean solution length grows dramatically for $h(n) = 4$, exactly the heuristic value where the error, as shown in table 2, grows steeply.

---

[3] The reason is that this implies enumerating all paths from every node n in the state space $\mathcal{E}$ to the goal state, t and taking the mean length over all paths.

6   Conclusions

Intuitively speaking, Markov chains have to be an excellent mean for studying the behaviour of the heuristic hill-climbing search algorithm since the markovian property is directly related to the memory-less feature of the search algorithm. On the other hand, this approach is especially relevant because the probability for this heuristic algorithm to commit a mistake can be understood as a pesimistic measure of the accuracy of the heuristic function guiding it.

One of the most difficult issues when using Markov chains is to find a good mapping function between nodes and markovian states. The τ representation shown here has been proven to be a good trade-off between efficiency and accuracy leading to very good results.

Acknowledgements

Thanks to the reviewers who improved significantly the paper with their comments —especially to the first reviewer who made an outstanding review both in terms of quantity and quality.

This work has been partially supported by the Spanish MEC project TIN2005-08945-C06-05 and UC3M-CAM project CCG06-UC3M/TIC-0831.

References

1. Pearl, J.: Heuristics. Addison-Wesley, Reading (1984)
2. Ghallab, M., Nau, D., Traverso, P.: Automated Planning: Theory and Practice. Morgan Kaufmann, San Francisco (2004)
3. Bonet, B., Geffner, H.: Planning as heuristic search. Artificial Intelligence 129(1–2), 5–33 (2001)
4. Bonet, B., Geffner, H.: Planning as Heuristic Search: New Results. In: Biundo, S., Fox, M. (eds.) ECP 1999. LNCS, vol. 1809, pp. 359–371. Springer, Heidelberg (2000)
5. Hoffmann, J., Nebel, B.: The FF planning system: Fast plan generation through heuristic search. Journal of Artificial Intelligence Research 14, 253–302 (2001)
6. Refanidis, I., Vlahavas, I.: The grt planning system: Backward heuristic construction in forward state-space planning. Journal of the Artificial Intelligence Research 15, 115–161 (2001)
7. Korf, R.E.: Search. In: Encyclopedia of Artificial Intelligence, vol. 2, pp. 994–998 (1987)
8. Stewart, W.J.: Introduction to the Numerical Solution of Markov Chains. Princeton University Press, Princeton (1994)
9. Korf, R.E., Reid, M., Edelkamp, S.: Time complexity of iterative-deepening-A*. Artificial Intelligence 129(1–2), 199–218 (2001)

DynABT: Dynamic Asynchronous Backtracking for Dynamic DisCSPs

Bayo Omomowo, Inés Arana, and Hatem Ahriz

School of Computing,
The Robert Gordon University, Aberdeen
{bo,ia,ha}@comp.rgu.ac.uk

Abstract. Constraint Satisfaction has been widely used to model static combinatorial problems. However, many AI problems are dynamic and take place in a distributed environment, i.e. the problems are distributed over a number of agents and change over time. Dynamic Distributed Constraint Satisfaction Problems (DDisCSP) [1] are an emerging field for the resolution of problems that are dynamic and distributed in nature. In this paper, we propose DynABT, a new Asynchronous algorithm for DDisCSPs which combines solution and reasoning reuse i.e. it handles problem changes by modifying the existing solution while re-using knowledge gained from solving the original (unchanged) problem. The benefits obtained from this approach are two-fold: (i) new solutions are obtained at a lesser cost and; (ii) resulting solutions are stable i.e. close to previous solutions. DynABT has been empirically evaluated on problems of varying difficulty and several degrees of changes has been found to be competitive for the problem classes tested.

Keywords: constraint satisfaction, distributed AI, dynamic problems.

1 Introduction

A Constraint Satisfaction Problem (CSP) can be defined as a triple $Z = (X, D, C)$ containing a set of variables $X = \{x_1....x_n\}$, for each variable x_i, a finite set $D_i \in D$ of possible values (its domain), and a set of constraints C restricting the values that the variables can take simultaneously. A solution to a CSP is an assignment to all the variables such that all the constraints are satisfied.

Dynamic Constraint Satisfaction problems (DCSPs) were introduced in [2] to handle problems that change over time. Loosely defined, a DCSP is a sequence of CSPs, where each one differs from the previous one due to a change in the problem definition. These changes could be due to addition/deletion of variables, values or constraints.Since all these changes can be represented as a series of constraint modifications [3], in the remainder of this paper we will only consider constraint addition and retraction. Several algorithms have been proposed for solving DCSPs e.g Dynamic Backtracking for Dynamic Constraint Satisfaction Problems [4] and Solution Reuse in Dynamic Constraint Satisfaction Problems [5].

A Distributed Constraint Satisfaction Problem (DisCSP) is a CSP in which variables, domains and constraints are distributed among autonomous agents [6]. Formally, a DisCSP can be described as a four tuple Z = (X, D, C, A) where

D. Dochev, M. Pistore, and P. Traverso (Eds.): AIMSA 2008, LNAI 5253, pp. 285–296, 2008.
© Springer-Verlag Berlin Heidelberg 2008

- X, D and C remain as described in CSPs and
- A is a set of agents with the mapping assigning variables to agents

Agents are only aware of their local constraints and the inter-agent constraints they are involved in and do not have a global view of the problem due to privacy, security issues and communication costs [7]. Solving a DisCSP consist of finding an assignment of values to variables by the collective and coordinated action of these autonomous agents which communicate through message passing. A solution to a DisCSP is a compound assignment of values to all variables such that all constraints are satisfied.

Various algorithms have been proposed for solving DisCSPs e.g Asynchronous Backtracking algorithm (ABT) [8], Asynchronous weak-Commitment search Algorithm (AWCS) [9] and Distributed Breakout algorithm (DBA) [10]. In DisCSPs, the following assumptions are usually made: (i) There is one variable per agent (ii) Agents are aware of their neighbours and constraints they share with them, (iii) Message delays are finite though random and messages arrive in the order they are sent between two related agents [8] and we shall also be making these assumptions in this paper.

Many hard practical problems can be seen as DisCSPs. Most DisCSP approaches however assume that problems are static. This has a limitation for dynamic problems that evolve over time e.g timetabling shifts in a large hospital where availability of staff changes over time. In order to handle this type of problems, traditional DisCSP algorithms naively solve from scratch every time the problem changes which may be very expensive or inadequate, i.e. there may be a requirement for the solution to the new (changed) problem to remain as close as possible to the original solution.

Distributed and Dynamic Constraint Satisfaction Problems (DDisCSPs) can be described as a five tuple (X,D,C,A,δ) where

- X, D, C and A remain as described in DisCSPs and
- δ is the change function which introduces changes at different time intervals

This definition is different from that of DisCSPs only in the introduction of the change function δ, which is a representation of changes in the problem over time [1]. DDisCSPs can be used to model problems which are distributed in nature and change over time.

Problem changes which have been widely modelled as a series of constraint additions and removals can be episodic (where changes occur after each problem has been solved) or occur while a problem is being solved. In this paper, we shall assume that changes shall be episodic.

Amongst the DDisCSP algorithms is the Dynamic Distributed Breakout Algorithm (DynDBA) [1] which is the dynamic version of DBA - a distributed local search algorithm inspired by the breakout algorithm of [11]. In DBA, agents assign values to their variables and communicate these values to neighbouring agents by means of messages. Messages passed between agents are in the form of *OK* and *Improve* messages. When agents discover inconsistencies they compute the best possible improvement to their violations and exchange it with neighbouring agents. Only the agent with the best possible improvement among neighbours is allowed to implement it. When an inconsistent state cannot be improved, i.e. a quasi local minimum is reached, the weights on violated constraints are increased [10], thus prioritising the satisfaction of these constraints.

In DynDBA, agents solve problems just like in the DBA algorithm but have the ability to react to changes continuously in each cycle with the aid of *pending lists* for holding new neighbours and messages.

In this paper we introduce our Dynamic Asynchronous Backtracking Algorithm (DynABT) which is based on the Asynchronous Backtracking Algorithm (ABT) [6] to handle DDisCSPs.

The remainder of this paper is structured as follows: section 2 describes ABT; next, section 3 introduces DynABT; this algorithm is evaluated in section 4 and; finally conclusions are presented in section 5.

2   Asynchronous Backtracking Algorithm (ABT)

Asynchronous Backtracking (ABT) is an asynchronous algorithm for DisCSPs in which agents act autonomously based on their view of the problem. ABT places a static ordering amongst agents and each agent maintains a list of higher priority agents and their values in a data structure known as the *agentview*. Constraints are directed between two agents: the *value-sending* agent (usually higher priority agent) and the *constraint-evaluating* agent (lower priority agent). The value-sending agents make their assignments and send them to their lower priority (constraint-evaluating) neighbours who try to make consistent value assignments. If a constraint-evaluating agent is unable to make a consistent assignment, it initiates backtracking by sending a *nogood* message to a higher priority agent, thus indicating that it should change its current value assignment. Agents keep a *nogood list* of backtrack messages and use this to guide the search. A solution is found if there is quiescence in the network while unsolvability is determined when an empty nogood is discovered. The correctness and completeness of ABT has been proven in [8].

ABT sends a lot of obsolete messages and uses a lot of space for storing nogoods. Therefore, various improvements to ABT have been proposed [12,13,14,15] which either reduce the number of obsolete messages or the space required for storing nogoods. In addition there is a version of ABT which uses just one nogood per domain value [15] which is of interest to us. This version uses the nogood recording scheme of Dynamic Backtracking [16] when recording and resolving nogoods but maintains the static agent ordering of ABT. Thus, a nogood for an agent x_k with value a is represented in the form $x_i = b \cap x_j = c \Rightarrow x_k \neq a$, where x_i and x_j are neighbouring agents with values b and c. In the remainder of this paper, we will use ABT to refer to the version which keeps just one nogood per eliminated value.

3   DynABT

DynABT is an asynchronous, systematic algorithm for dynamic DisCSPs. Based on ABT, it repairs the existing solution when the problem changes. DynABT combines solution reuse, reasoning reuse and justifications where a justification for the removal of a value states the actual constraint causing the removal in the explanation set recorded for the removed value.

Like in ABT, DynABT agents maintain a list of higher priority agents and their values in their *agentview* and a list of neighbouring agents' values inconsistent with their *agentview* in the nogood store. Higher priority agents send their value assignments to lower priority agents in the form of *info messages*. When an *info message* is received, the agent updates its *agentview* and checks for consistency. When its value is inconsistent, the agent composes a nogood but, unlike ABT nogoods, these are coupled with a set of justifications (actual constraints causing the violations). A nogood in DynABT is now of the form $x_i = b \cap x_j = c\{C_1, ..C_n\} \Rightarrow x_k \neq a$, where x_k currently has value a. Thus, the justification included in the nogoods acts as a pointer to which nogoods should become obsolete when constraints are retracted. We shall call the ABT with this new form of nogood recording ABT^+.

In DynABT (see Algorithms 1 to 5), each agent initialises its variables, starts the search and solves the problem like in ABT. However agents monitor the system to see if there are any changes and if so, react appropriately. Problem changes are handled in a two phase manner namely the *Propagation phase* (see Algorithm 2) and the *Solving phase* (ABT^+). In the propagation phase, agents are informed of constraint addition/retraction and they promptly react to the situation by updating their *constraint lists*, *neighbour lists*, *agentview* and *nogoods* where necessary. After all changes have been propagated, the new problem is at a consistent starting point, the *canProceed* flag is set to true and the agents can move on to the *Solving phase* and solve the new problem in a way similar to the ABT algorithm.

Three new message types (*addConstraint*, *removeConstraint* and *adjustNogood*) are used in order to handle agent behaviour during the propagation phase. When an agent receives an *addConstraint* message, the agent updates its *constraint* and *neighbour* lists where necessary (see Algorithm 3). When a *removeConstraint* message is received the agent modifies its *neighbour list* by excluding neighbours that only share the excluded constraint from its neighbour list and removing them from its agentview. The constraint is then removed and the nogood store is updated by removing nogoods whose justification contains the retracted constraint (see algorithm 4).

When a constraint is removed, an *adjustNogood* message is broadcasted to agents that are not directly involved in this constraint. The agents receiving this message update their nogoods store by removing the nogoods containing the retracted constraint as part of its justification and returning the values to their domains (see Algorithm 5). This step ensures that values that have been invalidated by retracted constraints are returned and made available since the source of inconsistency is no longer present in the network. Performing these processes during the propagation stage ensures that the new problem starts at a consistent point before the search begins.

3.1   Sample Execution

Figure 1a represents a DisCSP involving four agents (a, b, c, d) each with its own variable and domain values enclosed in brackets and having 3 *Not Equal* constraints ($C_1, C2, C3$) between them. Let us assume that the initial DisCSP was solved with the solution ($a = 1, b = 0, c = 0, d = 0$) and the following nogoods were generated:

Algorithm 1. DynABT

$changes \leftarrow 0$; $changeBox \leftarrow empty$; $canProceed \leftarrow true$
ABT^{+} (ABT with nogoods containing justifications)
repeat
   $changes \leftarrow monitorChanges$
   if (changes) **then**
     $canProceed \leftarrow false$
     PropagateChange(changeBox)
     current value $\leftarrow$ value from the last solution
     $ABT^{+}()$
   end if
until termination condition met

Algorithm 2. PropagateChanges

PropagateChange(changebox)
while $changeBox \neq empty \cap canProceed \leftarrow false$ **do**
   $con \leftarrow getChange$; $changeBox \leftarrow changeBox - con$
   Switch (con.msgType)
   con.removeConstraint : removeConstraint(con);
   con.addConstraint : includeConstraint(con);
   con.adjustNogood : incoherentConstraint(con);
end while

Algorithm 3. IncludeConstraint

IncludeConstraint(con)
$newCons \leftarrow$ con.getConstraint()
add new neighbours in newCons to neighbour list
$constraintList \leftarrow constraintList \cup newCons$

Algorithm 4. ExcludeConstraint

ExcludeConstraint(con)
incoherentConstraint(con)
$constraint \leftarrow$ con.getConstraint()
Remove unique neighbours in constraint from neighbour list
Delete unique neighbours from agentView
Remove constraint from constraintlist

Algorithm 5. AdjustNogoods

IncoherentConstraint(con)
$constraint \leftarrow$ con.getConstraint()
for each nogood in nogoodstore **do**
   if $contains(nogood, constraint)$ **then**
     return eliminated value in nogood to domain
     remove nogood from nogoodStore
   end if
end for

- Agent a : $(() \{C_1\} \Rightarrow a \neq 0)$
- Agent c : $((a = 1) \{C_2\} \Rightarrow c \neq 1)$
- Agent d : $((a = 1) \{C_3\} \Rightarrow d \neq 1)$

In Figure 1b, we assume that the solved problem has now changed and the constraint between a and d (C_3) has been retracted and a new constraint between c and d (C_4)has been added. At this stage, DynABT goes into the *propageChanges* mode in which agents c and d are informed of a new constraint between them and also agent a and d are made aware of the loss of the constraint between them. In addition to this set of messages, agents b and c are also sent adjustNogood messages, informing them of the loss of constraint C_3 and the need for them to adjust their nogoods if it is part of their justification sets. When these messages have been fully propagated (agent d will adjust its nogood and regain the value 0 back in its domain), the nogood store of the agents will now be in the form below:

- Agent a : $(() \{C_1\} \Rightarrow a \neq 0)$
- Agent c : $((a = 1) \{C_2\} \Rightarrow c \neq 1)$

The agents can now switch back to the solving mode because the problem is at a consistent starting point and the algorithm can now begin solving again. A new solution to the problem will be $(a = 1, b = 0, c = 0, d = 1)$ with d having to change its value to 1 in order for the new problem to be consistent.

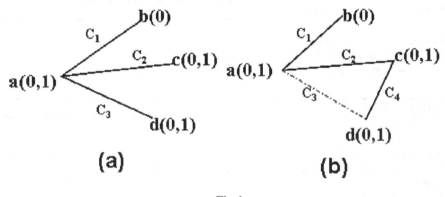

(a)   (b)

Fig. 1.

In our implementation, we have used a system agent for detecting quiescence just as been done in [15], in addition to this, we have also used it to communicate changes in the problem to the agents and also set the *canProceed* flag of agents to true when it determines that all propagation has been done. Completion of the propagation stage is determined in the following way: every time an agent receives any of the three messages (*addConstraint*, *removeConstraint* and *adjustNogood*) and performs the appropriate computation, the agent sends a dummy message back to the system agent indicating that it has received and treated a propagation message. The system agent can determine the total number of such messages to receive when all agents have received messages and acted on them in the *propagateChanges* and can therefore set the *canProceed* flag of all agents to true. This total number of messages can be calculated in the following way: Let x represents the number of constraints of a certain arity r added to the new problem and let N be the total number of agents in the network and y be the total number of constraint removed from the problem. The total messages to receive can be computed as $tot = (\sum(x_i * r_i)) + N * y$. In our implementation, we have reported these messages as part of the cost incured by DynABT.

3.2 Theoretical Properties

DynABT is sound, since whenever a solution is claimed, there is quiescence in the network. If there is quiescence in the network, it means that all agents have satisfied their constraints. If not all constraints have been satisfied, then there will be at least an agent unsatisfied with its current state and at least one violated constraint in the network. In this case, the agent involved would have sent at least a message to the culprit agent closest to it. This message is not obsolete and the culprit agent involved on receiving the message, will act on it and send out messages thus breaking our quiescence claim. It therefore follows that whenever there is quiescence in the network, agents are satisfied with their current state and whatever solution inferred is sound.

In the DynABT, agents update their nogood list when they receive *info messages* and evaluate constraints, during domain wipe out and also when changes are introduced. Nogoods are always generated in two ways: (1) when a constraint is violated because of an *info message* (this knowledge is explicitly enclosed in the constraint) and (2) When a domain wipe-out occurs and all nogoods are resolved into one. In essence, all the nogoods that can be generated are logical extension of the constraint network, therefore the empty nogood cannot be inferred if the network is satisfiable.

Also because every nogood discovered by an agent will always involve higher priority agents, which are eventually linked to the agent through the *addLink Message*, it follows that agents will not keep obsolete nogoods forever, since they will be informed of value changes by higher priority agents and thus update their nogood store, ensuring that the algorithm will terminate. We now need to show that when changes occur, these properties are still preserved.

When constraints are added to the problem, previous nogoods invalidating domain values remain consistent and since the nogood stores remain unchanged during constraint addition, these nogoods are preserved. Therefore when constraints are added to the problem, the soundness property of the algorithm is preserved.

When a constraint is retracted, nogoods are updated to exclude the retracted constraint and the associated values are returned to the agent's domain. If these values are still useless, this inconsistency will be rediscovered during search since they will violate constraints with some other agents and, therefore, solutions are not missed.

The *adjustNogoods* method ensures that all agents (whether participating in a retracted constraint or not) update their nogoods store and all nogoods containing retracted constraints as part of their justification are removed and the associated values returned to their domain.

Because retraction triggers the updating of the nogood store in a cautious manner in which nogoods are quickly forgotten but can be rediscovered if necessary during search, DynABT is complete and does terminate.

4 Experimental Evaluation

In order to evaluate DynABT, ABT, DynABT and DynDBA have been implemented in a simulated environment. The implementations of DynABT and ABT use the Max Degree heuristic.

Two sets of experiments were conducted using both randomly generated problems and graph colouring problems: (i) Comparing DynABT with ABT; (ii) Comparing DynABT and DynDBA. In all our comparisons with DynDBA, we have modified the DynDBA algorithm to make it react to changes episodically and also improved it by increasing the weight of a newly added constraint within a neighbourhood to the maximum constraint weight within that neighbourhood. This encourages DynDBA to satisfy the newly added constraints quicker.

Table 1. DynABT (Alg1) vs ABT (Alg2) on Random Problems, Density 0.2

t	Avg Messages		Avg CCC		Avg Stability		Median Msgs		Median CCC		Median Stability	
	Alg1	Alg2	Alg1	Alg2	Alg1	Alg2	Alg1	Alg2	Alg1	Alg2	Alg1	Alg2
Constraint changes 2(%)												
0.1	152	106	40	34	0.09	0.99	151	104	38	32	0	0
0.2	152	133	43	58	0.27	3.59	151	131	42	54	0	3
0.3	161	221	53	163	0.96	7.53	156	198	49	117	0	6
0.4	283	1262	140	1019	2.41	10.99	189	650	61	437	0	10
0.5	45868	83071	25129	45882	5.96	11.90	8566	63325	3998	35946	1	7
0.6	6879	27778	2442	10069	-	-	1194	23200	360	8439	-	-
0.7	1482	12204	373	3413	-	-	65	10134	14	2916	-	-
0.8	495	5301	103	1193	-	-	65	4769	13	992	-	-
0.9	92	1964	15	413	-	-	65	1776	12	386	-	-
Constraint Changes 6(%)												
0.1	280	106	44	33	0.32	2.35	279	105	42	31	0	2
0.2	281	132	53	59	0.76	6.17	279	130	52	56	1	5
0.3	293	208	80	153	1.65	11.33	285	192	65	109	1	11
0.4	547	1084	270	946	5.28	15.69	327	732	81	453	2	16
0.5	83746	86068	44929	47981	12.39	16.58	56532	60330	30847	35965	14	19
0.6	17797	27951	6075	10168	-	-	12986	24057	4462	8552	-	-
0.7	3952	12487	971	3619	-	-	1771	10337	356	3179	-	-
0.8	1320	5052	251	1116	-	-	193	4389	34	997	-	-
0.9	351	1944	58	427	-	-	193	1570	33	375	-	-
Constraint Changes 32(%)												
0.1	986	105	77	34	1.5	5.73	984	104	74	33	1	6
0.2	991	134	120	62	3.92	12.95	988	131	119	57	4	13
0.3	1023	201	198	119	7.15	18.72	1005	193	170	105	7	19
0.4	1670	987	718	1020	13.87	21.89	1214	644	331	403	14	22
0.5	138979	120871	84532	80589	23.13	24.03	91770	87106	58606	55678	24	25
0.6	39080	38525	15397	17323	-	-	31594	32527	11341	14132	-	-
0.7	10788	13216	2903	4286	-	-	8312	10413	2124	3598	-	-
0.8	3733	5255	741	1329	-	-	3298	4508	582	1103	-	-
0.9	1929	1781	329	406	-	-	1436	1479	207	372	-	-

In all our experiments, we have introduced a rate of change δ as a percentage of the total constraints/edges in the problem ($\delta \in \{2, 6, 32\}$). These changes [1] were made to

---

[1] All constraints/edges have equal probability of being selected for retraction.

Table 2. DynABT (Alg1) vs ABT (Alg2) on Graph Colouring Problems

deg	Avg Messages		Avg CCC		Avg Stability		Median Msgs		Median CCC		Median Stability	
	Alg1	Alg2	Alg1	Alg2	Alg1	Alg2	Alg1	Alg2	Alg1	Alg2	Alg1	Alg2
Constraint changes 2(%)												
4.1	**1107**	1556	**141**	691	**8.31**	26.41	**937**	1358	**72**	564	**3**	25
4.2	**3908**	4987	**668**	1544	**20.37**	33.5	**1133**	3737	**87**	1310	**4**	39
4.3	**6598**	9089	**1422**	2411	**20.67**	31.13	**1278**	5199	**91**	1777	**5**	34
4.4	**15173**	22450	**2682**	5246	**36.29**	52.42	**7921**	18607	**742**	4187	**54**	62
4.5	**73051**	84269	**14949**	17969	**16.33**	39.94	**1377**	25921	**95**	6199	**2**	44
4.6	177267	**170370**	39241	**35248**	-	-	205565	**188612**	42964	**35633**	-	-
4.7	**96149**	129664	**20166**	26006	-	-	**108513**	123541	**20957**	24526	-	-
4.8	**112436**	129634	**24159**	26271	-	-	**123294**	125345	**24648**	24999	-	-
4.9	**85917**	99803	**17027**	19782	-	-	98236	**92975**	**18319**	19395	-	-
Constraint Changes 6(%)												
4.1	2234	**1657**	**225**	648	**17.1**	36.58	1953	**1452**	**100**	398	**14**	37
4.2	6779	**6775**	**1230**	1919	**31.07**	46.3	**3685**	4380	**381**	1343	**36**	51
4.3	**9601**	10861	**1868**	2706	**33.07**	46.53	**5753**	7574	**947**	1841	**39**	48
4.4	**26788**	29292	**4915**	6708	**52.37**	61.86	**16683**	20158	**3205**	4904	**61**	66
4.5	**75872**	77264	**14510**	16812	**34.54**	49.08	**9213**	22545	**1433**	5540	**36**	58
4.6	159183	**139501**	32148	**28569**	-	-	107662	**94313**	20425	**18939**	-	-
4.7	171735	**163474**	34317	**32337**	-	-	152500	**150572**	30855	**30847**	-	-
4.8	163775	**140212**	32315	**28694**	-	-	154104	**127484**	28749	**25688**	-	-
4.9	131197	**117767**	23826	**22452**	-	-	118803	**106355**	21776	**20325**	-	-
Constraint Changes 32(%)												
4.1	12408	**6993**	**1305**	2050	**42.18**	56.23	8295	**2689**	**339**	1007	**43**	57
4.2	19706	**13531**	**2826**	3613	**50.90**	60.03	12011	**5855**	**895**	2175	**52**	61
4.3	53886	**45458**	**9984**	10446	**51.11**	60.43	15515	**10114**	**1890**	3139	**52**	62
4.4	77395	**68769**	**14230**	16216	**57.16**	65.76	23579	**20745**	**3707**	4843	**59**	67
4.5	183139	**158873**	35975	**34616**	**55.48**	62.03	**53710**	66718	**11238**	15459	**56**	62
4.6	267015	**215258**	51554	**44995**	-	-	139890	**109348**	25002	**23862**	-	-
4.7	345887	**302694**	62451	**61463**	-	-	272262	**211478**	51923	**43559**	-	-
4.8	375866	**344063**	71460	**68691**	-	-	215529	**177088**	40890	**35357**	-	-
4.9	285956	**236785**	49152	**45935**	-	-	242464	**193879**	40676	**37038**	-	-

be uniform between restriction and retraction. For example, if 4 changes are introduced, 2 are constraint additions and 2 are constraint retractions, thus ensuring that the overall constraint density remains unchanged.

In our experiments with randomly generated problems, we used with parameters (n, d, p_1, p_2) where n = number of variables = 30, d = domain size = 10, p_1 = density = 0.2, p_2 = tightness with values 0.1 - 0.9 step of 0.1. The range of tightness 0.1 - 0.4 contains solvable problems, 0.5 contains a mixture of both solvable and unsolvable (52% - 48%) and tightness 0.6 - 0.9 problems are unsolvable. For the unsolvable region, stability cannot be measured, as there is no solution to the problem. Each problem was solved and the solution obtained was kept for future reuse. Constraint changes were introduced and the new problem was solved. In all, 100 trials were made for each tightness value and a total of 1800 problems (900 original problems + 900 changed problems) were solved for each rate of change.

For our evaluation with graph colouring problem, we generated graph colouring problems with nodes = 100, d = 3 and degree k (4.1 - 4.9 step 0.1). These problems ranges from solvable through phase transition to unsolvable problems. In all, 100 trials were made per degree and a total of 1800 problems (900 original problems + 900 changed problems) were solved for each rate of change.

We measured the number of messages sent, Concurrent Constraint Checks (CCC) as defined[2] in [17] and the solution stability. For solution stability, we measure the total

---

[2] Cost of transmitting a message is zero in our implementation.

Table 3. DynABT (Alg1) vs DynDBA(Alg3) on Random Problems, Density 0.2

t	Avg Msgs		Avg CCC		Avg Stability		Median Msgs		Median CCC		Median Stability	
	Alg1	Alg3	Alg1	Alg3	Alg1	Alg3	Alg1	Alg3	Alg1	Alg3	Alg1	Alg3
Constraint changes 2(%)												
0.1	**152**	403	**40**	121	**0.09**	0.1	151	401	**38**	120	0	0
0.2	**152**	402	**43**	125	0.27	**0.21**	151	402	42	120	0	0
0.3	**161**	485	**53**	151	0.96	**0.45**	156	404	49	120	0	0
0.4	**283**	2291	**140**	735	2.41	**1.99**	189	408	61	130	0	0
0.5	**17941**	42675	**10442**	12515	5.96	**1.79**	225	408	72	120	1	**0**
Constraint Changes 6(%)												
0.1	**280**	415	**44**	121	0.32	**0.28**	279	410	**42**	120	0	0
0.2	**281**	430	**53**	131	0.76	**0.63**	279	411	52	120	1	1
0.3	**293**	582	**80**	187	1.65	**0.97**	285	416	65	125	1	1
0.4	**547**	4497	**270**	1511	5.28	**4.31**	327	520	81	180	2	2
0.5	**51970**	209324	**28851**	64411	12.39	**3.7**	**15807**	20235	8783	**6380**	14	2
Constraint Changes 32(%)												
0.1	986	**523**	**77**	139	1.5	**1.25**	984	**457**	**74**	120	1	1
0.2	991	**673**	**120**	193	3.92	**2.51**	988	**476**	**119**	150	4	2
0.3	**1023**	1387	**198**	426	7.15	**4.32**	1005	**811**	**170**	270	7	4
0.4	**1670**	8097	**718**	2691	13.87	**10.34**	**1214**	4624	**331**	1540	14	10
0.5	**122384**	608790	**75763**	193657	23.13	**18.33**	**63667**	586656	**38303**	172440	24	**15**

distance between successive solutions when both exists (the number of variables which get different values in both solutions). All the results reported are the mean and median of the observed parameters and we have only presented results of observed parameters when resolving. We also measured CPU time (not reported here) and it correlated to the trends observed with messages and concurrent constraint checks.

4.1 Comparison with ABT

For random problems, results obtained in table 1 show a reduction in the cost incured when a new problem is solved using previous solution, i.e. DynABT significantly outperforms ABT on small and intermediate changes while on large problems, ABT peforms better than DynABT: this is due to the fact that the new problem is substantially different from the previous one because of the quantity of changes involved and also because DynABT incurs more cost as changes increase during the propagation phase.

For Graph Colouring problems, the results obtained in table 2 are mixed between DynABT and ABT. With small and intermediate changes DynABT performs better than ABT on messages and CCC in the solvable region betwee 4.1 - 4.4 and the Phase transition region of 4.5, while in the unsolvable region from 4.6 - 4.9, ABT performs better. This behaviour is due to the fact that more cost is incured during the propagation stage of DynABT, when agents are modifying their nogoods before the new search starts. With large changes, ABT performs better than DynABT. However, with both problems, DynABT outperforms ABT on solution stability for all degrees of changes, which suggests that reusing solution, improves stability.

4.2 Comparison with DynDBA

In order to compare DynABT with DynDBA the latter algorithm was allowed a cut-off of at least 50% more cycles than DynABT when solving a problem because DynDBA is a two-phased algorithm (it takes an agent two cycles to make a value change compared to DynABT in which values can be changed in one cycle).

For our comparison with DynDBA, we have only presented results for solvable problems for both algorithms because DynDBA being an incomplete algorithm, cannot determine a problem is unsovable. For our Comparison with DynDBA on random problems, results from table 3 shows that DynABT outperforms DynDBA in terms of messages sent and concurrent constraint checks.

DynDBA outperforms DynABT in terms of solution stability. Our take on this is the fact that DynDBA with its min-conflict heuristic helps the algorithm find a stable solution. However it was shown that both algorithms depreciate in terms of solution stability as changes increase. This could be due to the fact that as more changes are introduced, the difference between the inital problem and the new problem is more pronounced, therefore new solutions are needed.

Table 4 presents results of our experiment on Graph Colouring problems. DynABT also outperforms DynDBA on messages and CCCs while DynDBA performs better on solution stability.

Table 4. DynABT (Alg1) vs DynDBA (Alg3) on Graph Colouring Problems

t	Avg Msgs		Avg CCC		Avg Stability		Median Msgs		Median CCC		Median Stability	
	Alg1	Alg3	Alg1	Alg3	Alg1	Alg3	Alg1	Alg3	Alg1	Alg3	Alg1	Alg3
Constraint changes 2(%)												
4.1	1107	42481	141	1553	8.31	14.82	937	3650	72	120	3	3
4.2	3908	116632	668	3989	20.37	14.41	1133	1449	87	60	4	2
4.3	6598	171520	1422	5923	20.67	16.17	1278	2439	91	90	5	3
4.4	15201	208067	2697	6967	36.29293	21.82	7907	10592	717	360	54	5
4.5	14217	449156	2742	14685	16.32558	10.03	1355	4414	81	135	2	2
Constraint Changes 6(%)												
4.1	2234	70844	225	2539	17.1	24.43	1953	30074	100	1080	14	17
4.2	6779	213290	1230	7511	31.07	31.37	3685	75800	381	2700	36	26
4.3	9497	305758	1856	10498	33.07	33.21	5640	109373	914	3810	39	30
4.4	23708	462501	4305	15322	52.37	35.71	15843	235621	3114	8025	61	35
4.5	19548	528244	3661	17237	34.54	24.46	7334	429983	983	14340	36	19
Constraint Changes 32(%)												
4.1	9617	196977	804	6761	42.18	46.9	8255	98926	331	3552	43	50
4.2	15210	434704	1725	15188	50.9	48.96	11885	234240	850	8145	52	51
4.3	22483	445826	3677	14872	51.11	54.05	14419	273725	1601	9540	52	54
4.4	33362	581442	5372	18767	57.16	54.81	21107	474626	3121	14634	59	56
4.5	54785	1090806	10535	35821	55.48	56.43	30829	980126	6647	32670	56	58

5   Summary and Conclusions

We have presented DynABT, an asynchronous, systematic search algorithm for DDisCSPs. An empirical comparison between DynABT and ABT on dynamic random problems shows a significant reduction in computational effort and a substantial gain in solution stability. Comparison with ABT on dynamic graph colouring problems however produces mixed results with DynABT outperforming ABT on messages and concurrent constraint checks on problems with small changes. With Intermediate changes, ABT performs better than DynABT in the unsolvable region while DynABT outperforms ABT in terms of stability for all categories of problem changes. Experimental results also show that DynABT requires less messages and constraint checks than DynDBA. However, the latter produces more stable solutions.

We have also shown that both DynABT and DynDBA cope well with problems where the rate of change is small but, as the number of changes increases, performance decreases. This is unsurprising since, with a high rate of change, the new problem is substantially different from the previous one. Future work will investigate other ways of improving the performance of DynABT in terms of solution stability.

References

1. Mailler, R.: Comparing two approaches to dynamic, distributed constraint satisfaction. In: AAMAS 2005: Proceedings of the fourth international joint conference on Autonomous agents and multiagent systems, pp. 1049–1056. ACM Press, New York (2005)
2. Dechter, A., Dechter, R.: Belief maintenance in dynamic constraint networks. In: Seventh National Conference on Artificial Intelligence (AAAI), pp. 37–42. St. Paul, MN (1988)
3. Verfaillie, G., Jussien, N.: Constraint solving in uncertain and dynamic environments: A survey. Constraints 10(3), 253–281 (2005)
4. Verfaillie, G., Schiex, T.: Dynamic backtracking for dynamic constraint satisfaction problems. In: Proceedings of the ECAI 1994 Workshop on Constraint Satisfaction Issues Raised by Practical Applications, Amsterdam, The Netherlands, pp. 1–8 (1994)
5. Verfaillie, G., Schiex, T.: Solution reuse in dynamic constraint satisfaction problems. In: National Conference on Artificial Intelligence, pp. 307–312 (1994)
6. Yokoo, M.: Distributed constraint satisfaction: foundations of cooperation in multi-agent systems. Springer, London (2001)
7. Meisels, A.: Distributed constraints satisfaction algorithms, performance, communication. In: Tenth International Conference on Principles and Practice of Constraint Programming, Canada, pp. 161–166 (2004)
8. Yokoo, M., Durfee, E.H., Ishida, T., Kuwabara, K.: The distributed constraint satisfaction for formalizing distributed problem solving. In: Proc. of the 12th. DCS, pp. 614–621 (1992)
9. Yokoo, M.: Asynchronous weak-commitment search for solving distributed constraint satisfaction problems. In: Montanari, U., Rossi, F. (eds.) CP 1995. LNCS, vol. 976, pp. 88–102. Springer, Heidelberg (1995)
10. Yokoo, M., Hirayama, K.: Distributed breakout algorithm for solving distributed constraint satisfaction problems. In: Second International Conference on Multiagent Systems, pp. 401–408 (1996)
11. Morris, P.: The breakout method for escaping from local minima. In: AAAI, pp. 40–45 (1993)
12. Zivan, R., Meisels, A.: Dynamic ordering for asynchronous backtracking on discsps. Constraints 11(2-3), 179–197 (2006)
13. Silaghi, M.C., Sam-Haroud, D.B.: Generalized dynamic ordering for asynchronous backtracking on discsps. In: Second Asia-Pacific Conference on Intelligent Agent Technology (IAT), Maebashi, Japan (2001)
14. Silaghi, M.C., Sam-Haroud, D., Hybridizing, B.: abt and awc into a polynomial space, complete protocol with reordering. TR4 EPFL-TR-01/36, Swiss Federal Institute of Technology, Lussane, Switzerland (May 2001)
15. Bessière, C., Maestre, A., Brito, I., Meseguer, P.: Asynchronous backtracking without adding links: a new member in the abt family. Artif. Intell. 161(1-2), 7–24 (2005)
16. Ginsberg, M.L.: Dynamic backtracking. Journal of Artificial Intelligence Research 1, 25–46 (1993)
17. Meisels, A., Kaplansky, E., Razgon, I., Zivan, R.: Comparing performance of distributed constraints processing algorithms. In: Proceedings of the AAMAS 2002 Workshop on Distributed Constraint Reasoning, Bologna, July 2002, pp. 86–93 (2002)

Centralized Indirect Control of an Anaerobic Digestion Bioprocess Using Recurrent Neural Identifier

Ieroham S. Baruch[1], Rosalba Galvan-Guerra[1], and Boyka Nenkova[2]

[1] CINVESTAV-IPN, Department of Automatic Control,
Ave. IPN No 2508, A.P. 14-470, Mexico D.F., C.P. 07360, Mexico
{baruch,rgalvan}@ctrl.cinvestav.mx
[2] Institute of Information Technologies, BAS, 1113 Sofia, Bulgaria

Abstract. The paper proposed to use a Recurrent Neural Network Model (RNNM) and a dynamic Backpropagation learning for centralized identification of an anaerobic digestion bioprocess, carried out in a fixed bed and a recirculation tank of a wastewater treatment system. The anaerobic digestion bioprocess represented a distributed parameter system, described by partial differential equations. The analytical model is simplified to a lumped ordinary system using the orthogonal collocation method, applied in three collocation points, generating data for the neural identification. The obtained neural state and parameter estimations are used to design an indirect sliding mode control of the plant. The graphical simulation results of the digestion wastewater treatment indirect control exhibited a good convergence and precise reference tracking.

Keywords: Recurrent neural network model, backpropagation learning, distributed parameter system, systems identification, sliding mode control, anaerobic digestion bioprocess, wastewater treatment bioprocess.

1 Introduction

In the last two decades the number of Neural Network (NN) applications for identification and control of biotechnological plants, [1], has been incremented. Among several possible network architectures the most widely used are the Feedforward NN (FFNN) and the Recurrent NN (RNN), [1]-[2]. The main NN property to approximate complex non-linear relationships without prior knowledge of the model structure made them very attractive alternative to the classical modeling and control techniques. This property has been proved for both types of NNs by the universal approximation theorem, [2]. The preference given to NN identification with respect to the classical methods of process identification is clearly demonstrated by the "bias-variance dilemma" [2]. The FFNN and the RNN have been applied for Distributed Parameter Systems (DPS) identification and control too. In [3], a RNN is used for system identification and process prediction of a DPS dynamics - an adsorption column for wastewater treatment of water contaminated with toxic chemicals. In [4] and [5], a spectral-approximation, based on intelligent modeling approach, is proposed for

D. Dochev, M. Pistore, and P. Traverso (Eds.): AIMSA 2008, LNAI 5253, pp. 297–310, 2008.
© Springer-Verlag Berlin Heidelberg 2008

the distributed thermal processing of the snap curing oven DPS that is used in semi-conductor packaging industry. After finding a proper approximation of the complex boundary conditions of the system, the spectral methods can be applied to time–space separation and model reduction, and NNs are used for state estimation, and system identification. Then a neural observer has been designed to estimate the states of the Ordinary Differential Equations (ODE) model from measurements taken at specified locations in the field. In [6], it is presented a new methodology for the identification of DPS, based on NN architectures, motivated by standard numerical discretization techniques used for the solution of Partial Differential Equations (PDE). In [7], an attempt is made to use the philosophy of the NN adaptive-critic design to the optimal control of distributed parameter systems. In [8], the concept of proper orthogonal decomposition is used for the model reduction of DPS to form a reduced order lumped parameter problem. The optimal control problem is then solved in the time domain, in a state feedback sense, following the philosophy of adaptive critic NNs. The control solution is then mapped back to the spatial domain using the same basis functions. In [9], measurement data of an industrial process are generated by solving the PDE numerically using the finite differences method. Both centralized and decentralized NN models are introduced and constructed based on this data. The models are implemented on FFNN using Backpropagation (BP). Unfortunately, all these works suffered of the same inconvenience that the FFNNs used are multilayer and of higher dimension, having great complexity, which made difficult their application. Furthermore, in all papers the problem of identification and control of DPS is stated in particular and not in a universal way.

In [10]-[13], Baruch et al. defined a new canonical Recurrent Trainable NN (RTNN) architecture, possessing simple hybrid two layer structure with minimum number of learning parameters, and a dynamic BP learning algorithm, applied for centralized systems identification and control of multi-input multi-output square nonlinear plants. In the present paper this RTNN model will be used for identification, state and parameter estimation of an anaerobic digestion wastewater treatment distributed parameter plant, [14]. The analytical DPS model, generating simulation data for RTNN learning is derived by PDE/ODE and it is simplified using the orthogonal collocation technique in three collocation points of the bioreactor fixed bed, [14]. So, the distributed parameter plant could be viewed as a plant with excessive multi-point measurements. Motivated for this higher order ODE nonlinear rectangular model we decided to approximate it in a universal way, using the RTNN by simply augmenting the number of plant outputs, taking into account the process variables measured in different points of the bioreactor. The issued by the RTNN state and parameter information is used to design a linear equivalent sliding mode control law. The means squared solution for the rectangular local plant dynamics compensation caused an offset of reference tracking which is compensated introducing a learnable threshold in the control law. The paper is organized as follows. Section 4 is dedicated to the bioprocess plant description and control problem formulation. Section 2 contained a short description of the RTNN topology and the algorithm of weight learning. Section 3 described the identification and control methodology used. It is devoted to the design of the sliding mode control law, based on the issued by RTNN state and parameter information, considering both the square and the rectangular case. Section 5 contained the simulation results and Section 6 represented the concluding remarks.

2 Simplified Analytical Model of the Anaerobic Digestion System

The anaerobic digestion system, formed by a fixed bed bioreactor and a recirculation tank, is depicted on Fig. 1. The figure contained a fixed bed bioreactor, a recirculation tank and a pump to mix and move the activated slurry back to the bioreactor (the pump dynamics is omitted in the process model). The physical meaning of all variables and constants is given in Table 1. The given values of all constants are taken from [14]. For practical purpose, the full PDE anaerobic digestion process model, [14], could be reduced to an ODE system using an early lumping technique and the orthogonal collocation method, [15], in three points (0.25H, 0.5 H, 0.75H), obtaining the following system of nonlinear ordinary differential equations:

$$\frac{dX_{1,i}}{dt}=\left(\mu_{1,i}-\varepsilon D\right)X_{1,i}\ , \quad \frac{dX_{2,i}}{dt}=\left(\mu_{2,i}-\varepsilon D\right)X_{2,i}\ , \tag{1}$$

$$\frac{dS_{1,i}}{dx}=\frac{E_z}{H^2}\sum_{j=1}^{N+2}B_{i,j}S_{1,j}-D\sum_{j=1}^{N+2}A_{i,j}S_{1,j}-k_1\mu_{1,i}X_{1,i}\ , \tag{2}$$

$$\frac{dS_{1T}}{dt}=\frac{Q_T}{V_T}\left(S_1\left(1,t\right)-S_{1T}\right), \quad \frac{dS_{2T}}{dt}=\frac{Q_T}{V_T}\left(S_2\left(1,t\right)-S_{2T}\right). \tag{3}$$

$$\frac{dS_{2,i}}{dx}=\frac{E_z}{H^2}\sum_{j=1}^{N+2}B_{i,j}S_{1,j}-D\sum_{j=1}^{N+2}A_{i,j}S_{2,j}+k_2\mu_{1,i}X_{2,i}-k_3\mu_{2,i}X_{2,i}\ , \tag{4}$$

$$\frac{dS_{1T}}{dt}=\frac{Q_T}{V_T}\left(S_{1,N+2}-S_{1T}\right), \quad \frac{dS_{2T}}{dt}=\frac{Q_T}{V_T}\left(S_{2,N+2}-S_{2T}\right), \tag{5}$$

$$S_{k,1}=\frac{1}{R+1}S_{k,in}\left(t\right)+\frac{R}{R+1}S_{kT}\ , \quad S_{k,N+2}=\frac{K_1}{R+1}S_{k,in}\left(t\right)$$
$$+\frac{K_1 R}{R+1}S_{kT}+\sum_{i=2}^{N+1}K_i S_{k,i} \tag{6}$$

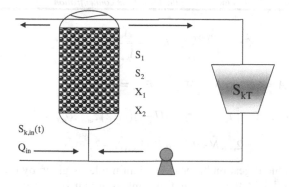

Fig. 1. Block-diagram of an anaerobic digestion bioreactor

Table 1. Summary of the variables in the plant model

Variable	Units	Name	Value
z	$z \in [0,1]$	Space variable	
t	D	Time variable	
E_z	m^2/d	Axial dispersion coefficient	1
D	$1/d$	Dilution rate	0.55
H	m	Fixed bed length	3.5
X_1	g/L	Concentration of acidogenic bacteria	
X_2	g/L	Concentration of methano-genic bacteria	
S_1	g/L	Chemical Oxygen Demand	
S_2	$mmol/L$	Volatile Fatty Acids	
ε		Bacteria fraction in the liquid phase	0.5
k_1	g/g	Yield coefficients	42.14
k_2	$mmol/g$	Yield coefficients	250
k_3	$mmol/g$	Yield coefficients	134
μ_1	$1/d$	Acidogenesis growth rate	
μ_2	$1/d$	Methanogenesis growth rate	
μ_{1max}	$1/d$	Maximum acidogenesis growth rate	1.2
μ_{2s}	$1/d$	Maximum methanogenesis growth rate	0.74
K_{1s}	g/g	Kinetic parameter	50.5
K_{2s}	$mmol/g$	Kinetic parameter	16.6
K_{12}	$mmol/g$	Kinetic parameter	256
Q_T	m^3/d	Recycle flow rate	0.24
V_T	m^3	Volume of the recirculation tank	0.2
S_{1T}	g/L	Concentration of Chemical Oxygen Demand in the recirculation tank	
S_{2T}	$mmol/L$	Concentration of Volatile Fatty Acids in the recirculation tank	
Q_{in}	m^3/d	Inlet flow rate	0.31
V_B	m^3	Volume of the fixed bed	1
V_{eff}	m^3	Effective volume tank	0.95
$S_{1,in}$	g/l	Inlet substrate concentration	
$S_{2,in}$	$mmol/L$	Inlet substrate concentration	

$$K_1 = -\frac{A_{N+2,1}}{A_{N+2,N+2}}, \quad K_i = -\frac{A_{N+2,i}}{A_{N+2,N+2}}, \tag{7}$$

$$A = \Lambda \phi^{-1}, \quad \Lambda = \left[\varpi_{m,l} \right], \quad \varpi_{m,l} = (l-1) z_m^{l-2}, \tag{8}$$

$$B = \Gamma \phi^{-1}, \quad \Gamma = \left[\tau_{m,l} \right], \quad \tau_{m,l} = (l-1)(l-2) z_m^{l-3}, \phi_{m,l} = z_m^{l-1} \tag{9}$$

$$i = 2,...,N+2, \quad m,l = 1,...,N+2. \tag{10}$$

The reduced anaerobic digestion bioprocess plant model is given by the equations (1)-(10), where the model of the recirculation tank is given by the equations (5). This ODE model described the plant in a universal way as a general plant model with excessive multi-point measurements where the dimension of the plant output is higher than the dimension of the plant input (representing a rectangular system). The complete simplified dynamical plant model could be used as a target model generating input/output process data for centralized recurrent neural identification and indirect adaptive control system design.

3 Description of the RTNN Topology and Learning

The block-diagram of the RTNN topology is depicted in Fig. 2. The RTNN topology could be described analytically by the following vector-matrix equations:

$$X(k+1) = AX(k) + BU(k); A = \text{block-diag } (Ai), |Ai| < 1; \tag{11}$$

$$B = [B_1 ; B_0]; U^T = [U_1 ; U_2]; \tag{12}$$

$$Z_1(k) = G[X(k)]; Z^T(k) = [Z_1(k) ; Z_2(k)]; \tag{13}$$

$$V(k) = CZ(k); C = [C_1 ; C_0]; Y(k) = F[V(k)]; \tag{14}$$

where X, Y, U are state, augmented output and input vectors with dimensions n, (l+1), (m+1), respectively; Z_1 and U_1 are the (nx1) output and (mx1) input of the hidden layer; the constant scalar threshold entries are $Z_2 = -1$, $U_2 = -1$, respectively; V is a (lx1) pre-synaptic activity of the output layer; A is a (nxn) block-diagonal weight matrix; B and C are [nx(m+1)] and [lx(n+1)]- augmented weight matrices; B_0 and C_0 are (nx1) and (lx1) threshold weights of the hidden and output layers; F[.], G[.] are vector-valued tanh(.)-activation functions with corresponding dimensions. Based on the RTNN topology and using the diagrammatic method rules of [16], we could build an adjoint error predicting RTNN model, depicted in Fig. 3. The dynamic BP algorithm of RTNN learning could be derived using the error predictions and applying the simple delta rule, [2], which yields:

$$W(k+1) = W(k) + \eta \Delta W(k) + \alpha \Delta W_{ij}(k-1); \tag{15}$$

$$E(k) = T(k) - Y(k); E_1(k) = F'[Y(k)] E(k); F'[Y(k)] = [1 - Y^2(k)]; \tag{16}$$

$$\Delta C(k) = E_1(k) Z^T(k); \tag{17}$$

$$E_3(k) = G'[Z(k)] E_2(k); E_2(k) = C^T(k) E_1(k); G'[Z(k)] = [1 - Z^2(k)]; \tag{18}$$

$$\Delta B(k) = E_3(k) U^T(k); \tag{19}$$

$$\Delta A(k) = E_3(k) X^T(k); \tag{20}$$

$$\text{Vec}(\Delta A(k)) = E_3(k) \circ X(k); \tag{21}$$

where T is the (lx1) plant output vector, considered as a RNN reference; F'[.], G'[.] are the derivatives of the tanh(.) activation functions; W is a general weight, denoting each weight matrix (C, A, B) in the RTNN model, to be updated; ΔW (ΔC, ΔA, ΔB), is the weight correction of W; η, α are learning rate parameters; ΔC is a weight correction of the learned matrix C; ΔA is an weight correction of the learned matrix A; ΔB is an weight correction of the learned matrix B; the diagonal of the matrix A is denoted by Vec(.) and equation (21) represented its learning as an element-by-element product; E, E_1, E_2, E_3, are error vectors with appropriate dimensions, predicted by the adjoint RTNN model, given on Fig.3. The stability of the RTNN model is assured by the tanh (.) activation function bounds (-1, 1) and by the local stability weight bound condition, given by the right equation of (11).

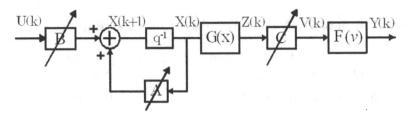

Fig. 2. Block-diagram of the RTNN model

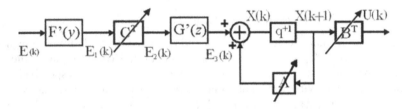

Fig. 3. Block-diagram of the adjoint RTNN model

4 Indirect Adaptive Control Using a RTNN Identifier

The indirect adaptive control using the RTNN as plant identifier has been described in [11], [12], [13] and applied for aerobic fermentation bioprocess plant control. Here the same control approach will be represented as a Sliding Mode Control (SMC) and applied for wastewater anaerobic digestion bioprocess, modeled by a simplified rectangular ODE system. The block diagram of this control is shown on Fig. 4. The stable nonlinear plant is identified by a RTNN identifier with topology, given by equations (11)-(14) and learned by the stable BP-learning algorithm, given by equations (15)-(21), where the identification error tends to zero. Let us first to admit that l= m (square system). The identified local linear plant model described in both state-space and input-output form is given by the equations:

$$X (k+1) = A X (k) + B U(k) \tag{22}$$

$$Y (k) = C X(k) \tag{23}$$

$$Y(k+1) = C X(k+1) = C [AX(k) + BU(k)] \tag{24}$$

In [17], the sliding surface is defined with respect to the state variables and the SMC objective is to move the states from an arbitrary state-space position to the sliding surface in finite time. In [18], the sliding surface is also defined with respect to the states but the states of the SISO system are obtained from the plant outputs by differentiation.

In [19], the sliding surface definition and the control objectives are the same. The equivalent control systems design is done with respect to the plant output, but the reachability of the stable output control depended on the plant structure. Here, the sliding surface is defined directly with respect to the plant outputs which facilitated the equivalent SMC systems design and decoupled the closed-loop system. Let us define the following sliding surface as an output tracking error function:

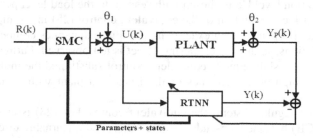

Fig. 4. Block diagram of the closed-loop system using RTNN identifier and a SMC

$$S(k+1)=E(k+1)+\sum_{i=1}^{p} \gamma_i\, E(k-i+1); \ |\gamma_i| < 1 \tag{25}$$

where S (.) is the Sliding Surface Error Function (SSEF) defined with respect to the plant output; E(.) is the systems output tracking error; γ_i are parameters of the desired stable SSEF; p is the order of the SSEF. The tracking error for k, k+1 is defined as:

$$E(k) = R(k) - Y(k); E(k+1) = R(k+1) - Y(k+1) \tag{26}$$

Here the l-dimensional vectors R (k), Y (k) are the system reference and the output of the local linear plant model. The objective of the sliding mode control systems design is to find a control action which maintained the system error on the sliding surface, assuring that the output tracking error reached zero in p steps, where p < n. So, the control objective is fulfilled if the SSEF is S (k+1) = 0. From (24)-(26), we obtained:

$$R(k+1) - CAX(k) - CBU(k) + \sum_{i=1}^{p} \gamma_i\, E(k-i+1) = 0 \tag{27}$$

As the local approximation plant model (22), (23), is controllable, observable and stable, [10], the matrix A is diagonal, and l = m, then the matrix product (CB), representing the plant model static gain, is nonsingular, and the plant states X (k) are smooth non-increasing functions. Now, from (27) it is easy to obtain the equivalent control capable to lead the system to the sliding surface which yields:

$$U_{eq}(k) = (CB)^{-1} [- CAX(k) + R(k+1) + \sum_{i=1}^{p} \gamma_i\, E(k-i+1)] \tag{28}$$

Following [17], the SMC avoiding chattering is taken using a saturation function instead of sign one. So the SMC took the form:

$$U^*(k) = \begin{cases} U_{eq}(k), & \text{if } \|U_{eq}(k)\| < Uo \\ -Uo\, U_{eq}(k)/\|U_{eq}(k)\|, & \text{if } \|U_{eq}(k)\| \geq Uo \end{cases} \tag{29}$$

Here the saturation level Uo is chosen with respect to the load level perturbation. It is easy to see that the substitution of the equivalent control (28) in the input-output linear plant model (24) showed an exact complete plant dynamics compensation which avoided oscillations, so that the chattering effect is not observed. Furthermore, the designed plant output sliding mode equivalent control substituted the multi-input multi-output coupled high order dynamics of the linearized plant with desired decoupled low order one.

If $l > m$ (rectangular system) the equivalent control law (28) is changed. The inverse of the (CB) becomed a pseudo-inverse $(CB)^+$ and a learnable threshold Of with dimension (mx1) is added to the right hand side of (28), so to obtain:

$$U_{eq}(k) = (CB)^+ [- CAX(k) + R(k+1) + \sum_{i=1}^{p} \gamma_i E(k-i+1)] + Of \tag{30}$$

The threshold Of is trained using the input control error, applying the BP algorithm, given by (15). The input control error is obtained from the output control error (26) backpropagating it through the RTNN adjoint model (see Fig.3). An approximate way to obtain the input error from the output error is to multiply it by the same (CB)-pseudo-inverse. So, the threshold correction is obtained as:

$$\Delta Of\,(k) = (CB)^+ E(k); \quad (CB)^+ = [(CB)^T (CB)]^{-1} (CB)^T \tag{31}$$

5 Simulation Results of Sliding Mode Control

The plant contained 14 output process variables and two input variables, representing itself a rectangular distributed parameter system. The output variables are as follows: 4 variables for each collocation point of the fixed bed (z=0.25H, z=0.5H, z=0.75H, H= 3.5 m- the fixed bed length), which are X_1 (acidogenic bacteria), X_2 (methanogenic bacteria), S_1 (chemical oxygen demand), S_2 (volatile fatty acids) and two variables for the recirculation tank, which are S_{1T} (chemical oxygen demand), S_{2T} (volatile fatty acids). The input variables are: S_{1in} (substrate concentration of the acidogenic bacteria), S_{2in} (substrate concentration of the methanogenic bacteria). The topology of the RTNN identifier is (2-number of inputs, 16-number of states, 14-number of outputs). The number of states is chosen as a sum of inputs and outputs, the activation functions are tanh (.) for both layers, and the learning rate parameters used are α=0.001, η=0.1. The graphical simulation results of SMC are obtained on-line in real-time during 100 days with a step of 0.1 day (To = 0.1 sec., N_t = 1000 iterations). The Figs. 5-9 compared the respective plant outputs with the reference signals during N_t = 1000 iterations of RTNN learning and sliding mode bioprocess control.

Fig. 5 b, c, d showed the variable X_1 (dotted line) and the respective reference (continuous line) in three measurement points of the fixed bed bioreactor (z=0.25H, z=0.5H, z=0.75H, H= 3.5 m). The Fig. 5 a show a 3-D view which gave a spatiotemporal representation of the output variable X_1 for the bioreactor fixed bed length z and the time. The interpolation of the variable X_1 between the space measurement points (z=0.25H, z=0.5H, z=0.75H, H= 3.5 m) is linear. The graphical results of RTNN learning and SMC, given on Fig. 6 a, b, c, d, Fig. 7 a, b, c, d, Fig. 9 a, b, c, d showed similar results for the variables of the fixed bed X_2, S_1, S_2. The graphical results of RTNN learning and SMC, given on Fig. 8 a, b showed temporal results for the variables of the recirculation tank S_{1T}, S_{2T}. The 3-D views showed a decrease of the variables with the increase of the fixed-bed length z. The temporal graphics of RTNN learning and SM control showed some bigger discrepancies between the plant variables and the respective references at the beginning of the learning which decreased faster on time. In order to see the accuracy of the system identification and SM control, the values of the Mean Squared Error (MSE%) of all plant output variables for the fixed bed and the recirculation tank at the end of the 1000 iterations of learning are shown on Table 2. The value of the MSE% obtained for the worse case (the variable S_1 for z=0.25H) is $3.2284 \cdot 10^{-8}$. For sake of comparison similar results of MSE% for proportional linear optimal control of the same bioprocess variables are given on Table 3. The value of the MSE% obtained for the worse case (the variable S_1 for z=0.25H) is $1.1978 \cdot 10^{-5}$.

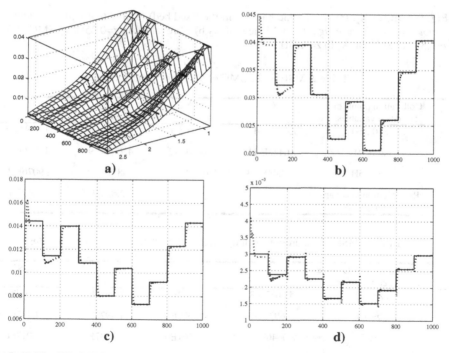

Fig. 5. SMC of X_1 (acidogenic bacteria in the fixed bed) (*dotted line -plant output, continuous line- reference signal*); a) 3-D view of X_1; b) SMC of X_1 in z=0.25H; c) SMC of X_1 in z=0.5H; d) SMC of X_1 in z=0.75H

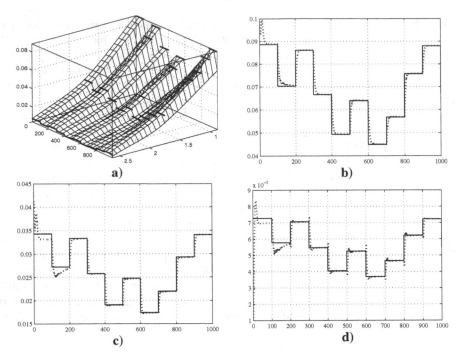

a) b) c) d)

Fig. 6. SMC of X_2, (methanogenic bacteria in the fixed bed) (*dotted line – plant output, continuous line –reference signal*); a) 3d view of X_2 b) SMC of X_2 in z=0.25H; c) SMC of X_2 in z=0.5H, d) SMC of X_2 in z=0.75H

Table 2. MSE of the SMC of all output plant variables

Collocation points of the fixed bed variables	X1	X2	S1	S2
z=0.25H	1.4173E-10	1.1231E-9	3.2284E-8	1.4994E-8
z=0.5H	1.6442E-11	8.0341E-12	4.7359E-9	1.8789E-9
z=0.75H	1.9244E-13	3.1672E-13	4.3831E-11	2.9076E-12
Recirculation tank variables			2.6501E-9	2.4932E-9

Table 3. MSE of the proportional optimal control of all output plant variables

Collocation points of the fixed bed variables	X1	X2	S1	S2
z=0.25	5.3057E-8	1.7632E-7	1.1978E-5	2.1078E-5
z=0.5	6.6925E-9	4.2626E-8	1.4922E-6	4.4276E-6
z=0.75	3.0440E-10	2.0501E-9	6.8737E-8	2.0178E-7
Recirculation tank variables			2.7323E-7	6.0146E-7

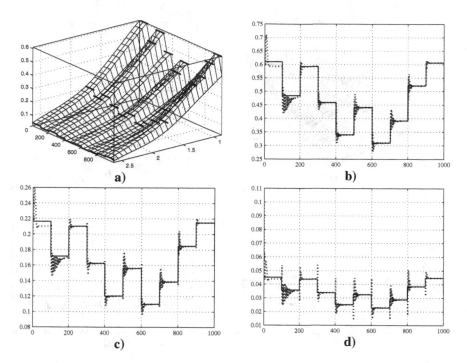

Fig. 7. SMC of S_1 (chemical oxygen demand in the fixed bed) (*dotted line-plant output, continuous line –reference signal*); a) 3d view of S_1; b) SMC of S_1 in z=0.25H; c) SMC of S_1 in z=0.5H; d) SMC of S_1 in z=0.75H

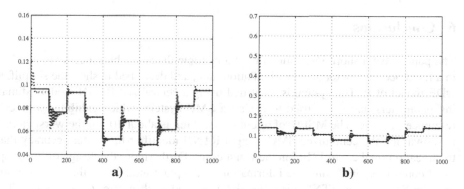

Fig. 8. SMC of the recirculation tank variables; a) SMC of S_{1T} (chemical oxygen demand in the recirculation tank) (*dotted line –plant output, continuous line –reference signal*); b) SMC of S_{2T} (volatile fatty acids in the recirculation tank) (*dotted line- plant output, continuous line – reference signal*)

The MSE% results of bioprocess identification and control showed a very good RTNN convergence and precise reference tracking with slight priority of the indirect adaptive control over the linear optimal control due to the adaptation ability of the first one.

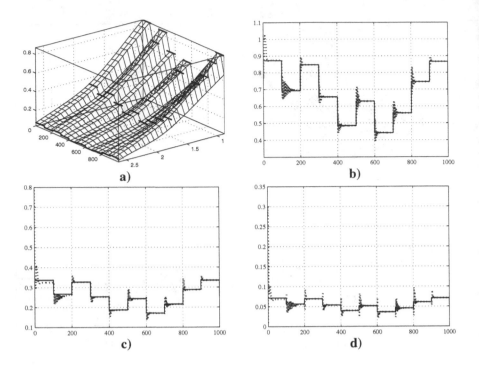

Fig. 9. SMC of S_2 (volatile fatty acids in the fixed bed) (*dotted line- plant output, continuous line –reference signal*); a) 3d view of S_2 ; b) SMC of S_2 in z=0.25H; c) SMC of S_2 in z=0.5H; d) SMC of S_2 in z=0.75H

6 Conclusions

The paper gave a short description of the anaerobic digestion bioprocess taking place in a fixed bed bioreactor and a recirculation tank, and described in short the simplified ODE model of it. The system is identified using a centralized RTNN model estimating parameters and states used to design a SMC system, which exhibited a precise reference tracking. The MSE% of reference tracking is about 10^{-8} in the worst case. It is shown that the centralized SMC using a RTNN identifier has better accuracy than the linear optimal control (10^{-5} in the worst case). Furthermore the design of linear optimal control required complete information of the plant model plus the cost function, and the SMC using RTNN identifier did not need such a-priory information. The practical implementation of the indirect adaptive control system required the measurement of all participating process variables.

Acknowledgments. The Ph.D. student Rosalba Galvan-Guerra is thankful to CONACYT for the scholarship received during her studies at the Department of Automatic Control, CINVESTAV-IPN, Mexico.

References

1. Boskovic, J.D., Narendra, K.S.: Comparison of Linear, Nonlinear, and Neural Network-Based Adaptive Controllers for a Class of Fed-Batch Fermentation Processes. Automatica 31(6), 817–840 (1995)
2. Haykin, S.: Neural Networks, a Comprehensive Foundation. Second Edition. Section 2.13, 84–89; Section 4.13, pp. 208–213. Prentice-Hall, Upper Saddle River (1999)
3. Bulsari, A., Palosaari, S.: Application of Neural Networks for System Identification of an Adsorption Column. Neural Computing and Applications 1, 160–165 (1993)
4. Deng, H., Li, H.X.: Hybrid Intelligence Based Modeling for Nonlinear Distributed Parameter Process with Applications to the Curing Process. IEEE Trans. on Systems, Man and Cybernetics 4, 3506–3511 (2003)
5. Deng, H., Li, H.X.: Spectral-Approximation-Based Intelligent Modeling for Distributed Thermal Processes. IEEE Transactions on Control Systems Technology 13, 686–700 (2005)
6. Gonzalez-Garcia, R., Rico-Martinez, R., Kevrekidis, I.: Identification of Distributed Parameter Systems: A Neural Net Based Approach. Computers and Chemical Engineering 22 (4-supl. 1), 965–968 (1998)
7. Padhi, R., Balakrishnan, S., Randolph, T.: Adaptive Critic based Optimal Neuro Control Synthesis for Distributed Parameter Systems. Automatica 37, 1223–1234 (2001)
8. Padhi, R., Balakrishnan, S.: Proper Orthogonal Decomposition Based Optimal Neurocontrol Synthesis of a Chemical Reactor Using Approximate Dynamic Programming. Neural Networks 16, 719–728 (2003)
9. Pietil, S., Koivo, H.N.: Centralized and Decentralized Neural Network Models for Distributed Parameter Systems. In: Proc. of the Symposium on Control, Optimization and Supervision, CESA 1996. IMACS Multi-conference on Computational Engineering in Systems Applications, pp. 1043–1048. IMACS Press, Lille (1996)
10. Baruch, I.S., Flores, J.M., Nava, F., Ramirez, I.R., Nenkova, B.: An Advanced Neural Network Topology and Learning, Applied for Identification and Control of a D.C. Motor. In: Proc. 1st Int. IEEE Symp. Intelligent Systems, Varna, Bulgaria, vol. 1, pp. 289–295 (2002)
11. Baruch, I.S., Georgieva, P., Barrera-Cortes, J., Feyo de Azevedo, S.: Adaptive Recurrent Neural Network Control of Biological Wastewater Treatment. International Journal of Intelligent Systems 20(2), 173–194 (2005)
12. Baruch, I.S., Hernandez, L.A., Barrera-Cortes, J.: Recurrent Neural Identification and Sliding Mode Adaptive Control of an Aerobic Fermentation Plant. Cientifica, ESIME-IPN 11(2), 55–62 (2007)
13. Baruch, I.S., Hernandez, L.A., Mariaca-Gaspar, C.R., Nenkova, B.: An Adaptive Sliding Mode Control with I-term Using Recurrent Neural Identifier. Cybernetics and Information Technologies 7(1), 21–32 (2007)
14. Aguilar-Garnica, E., Alcaraz-Gonzalez, V., Gonzalez-Alvarez, V.: Interval Observer Design for an Anaerobic Digestion Process Described by a Distributed Parameter Model. In: Proc. of the Second International Meeting on Environmental Biotechnology and Engineering (2IMEBE), Mexico City, Mexico, paper 117, 26-29 Sept. 2006, pp. 1–16 (2006)
15. Bialecki, B., Fairweather, G.: Orthogonal Spline Collocation Methods for Partial Differential Equations. Journal of Computational and Applied Mathematics 128, 55–82 (2001)

16. Wan, E., Beaufays, F.: Diagrammatic Method for Deriving and Relating Temporal Neural Networks Algorithms. Neural Computations 8, 182–201 (1996)
17. Young, K.D., Utkin, V.I., Ozguner, U.: A Control Engineer's Guide to Sliding Mode Control. IEEE Trans. on Control Systems Technology 7(3), 328–342 (1999)
18. Levent, A.: Higher Order Sliding Modes, Differentiation and Output Feedback Control. International Journal of Control 76(9/10), 924–941 (2003)
19. Eduards, C., Spurgeon, S.K., Hebden, R.G.: On the Design of Sliding Mode Output Feedback Controllers. International Journal of Control 76(9/10), 893–905 (2003)

Logical Analysis of Mappings between Medical Classification Systems

Elena Cardillo[1], Claudio Eccher[1], Luciano Serafini[2], and Andrei Tamilin[2]

[1] e-Health Applied Research Unit, FBK-IRST
[2] Data & Knowledge Management Research Unit, FBK-IRST
Via Sommarive 18, 38050 Povo di Trento, Italy
{cardillo,cleccher,serafini,tamilin}@fbk.eu

Abstract. Medical classification systems provide an essential instrument for unambiguously labeling clinical concepts in processes and services in healthcare and for improving the accessibility and elaboration of the medical content in clinical information systems. Over the last two decades the standardization efforts have established a number of classification systems as well as conversion mappings between them. Although these mappings represent the agreement reached between human specialists who devised them, there is no explicit formal reference establishing the precise meaning of the mappings. In this work we close this semantic gap by applying the results that have been recently reached in the area of AI and the Semantic Web on the formalization and analysis of mappings between heterogeneous conceptualizations. Practically, we focus on two classification systems which have received great widespread and preference within the European Union, namely ICPC-2 (*International Classification of Primary Care*) and ICD-10 (*International Classification of Diseases*). The particular contributions of this work are: the logical encoding in OWL of ICPC-2 and ICD-10 classifications; the formalization of the existing ICPC-ICD conversion mappings in terms of OWL axioms and further verification of its coherence using the logical reasoning; and finally, the outline of the other semantic techniques for automated analysis of implications of future mapping changes between ICPC and ICD classifications.

Keywords: Medical Informatics, Medical Classification Systems, Mapping, Reasoning, ICPC-ICD.

1 Motivation and Approach

Coding and classification systems have a long history in medicine, tracing their origin to the early attempts to classify the causes of death in the eighteenth century. These systems, in fact, allow to systematically classify the medical concepts by assigning them labels that can be used for many purposes: unambiguous communication of diseases and their causes, statistical analysis for epidemiological studies, etc. In the last few decades these terminological systems have started to attract wide-spread attention and resources, mainly due to the ever growing

D. Dochev, M. Pistore, and P. Traverso (Eds.): AIMSA 2008, LNAI 5253, pp. 311–321, 2008.

need to manage, analyze and exchange clinical data stored in clinical information systems, no longer just for epidemiological purposes, but to provide effective tool for managing the entire care process in and between healthcare organizations. There exist many coding systems, each developed for a particular medical domain and with a particular purpose. In this context, there also raised the necessity to established unambiguous mappings between different coding systems to guarantee their interoperability.

The mappings between different medical systems are defined on clinical bases; typically, however, there is no clear formal reference establishing the precise meaning of mappings between the coding systems. In addition, the interpretation of all mappings can not be done coherently — different groups of mappings can hold different implicit semantics. As such, formal analysis of mappings and their clear logical encoding is an important prerequisite for the correct integration of different systems. Clear logical model of mappings can advance the whole mapping assessment process by allowing the formal evaluation of the quality of mappings, verification of coherence of the newly added mappings or modifications done to the mappings, debugging the causes of contradictions, and others.

In this paper we focus on two classification systems which have received great widespread and preference within the European Union, namely ICPC-2 and ICD-10. The two systems address fundamental parts of healthcare process: ICPC is used by general practitioners for encoding symptoms and diagnosis, while ICD is used mainly by clinicians in hospitals for encoding diseases. A technical conversion between ICPC and ICD-10 has been made in [15,12], allowing primary care physicians to implement ICD-10 as a reference nomenclature within the classification structure of ICPC and leading to a substantial increase of the diagnostic potential of ICPC. In this paper we perform the logical analysis of the existing technical conversion between the two systems, which further allows to design the formal mapping model between them.

The concrete contributions of this work are: (1) the formal encoding of ICPC and ICD systems using OWL language; (2) formalization, in terms of OWL axioms, of ICPC-ICD mapping and practical validation of its coherence and soundness using Pellet OWL Reasoner[1]; and finally (3) discussion of applicability and usefulness of other recently developed semantic techniques to the task of formal assessment of ICPC-ICD mapping.

2   Background

2.1   ICPC-2 and ICD-10 Classification Systems

In this section we briefly recall the main characteristics of the two classification systems ICP-2 and ICD-10.

The International Classification of Diseases. ICD-10 is the tenth revision of the International Classification of Diseases published by the World Health

---
[1] http://pellet.owldl.org

Organization (WHO)[2]. The goal of the system is to allow the systematic collection and statistical analysis of morbidity and mortality data from different countries around the world. Due to its importance, ICD acts as a *defacto* reference point for many healthcare terminologies.

The ICD-10 is a multiple-axis classification system structured into 21 chapters. At its core, the basic ICD is a single list of rubrics, labeled by a three alphanumeric character code, organized by category, from A00 to Z99 (excluding U codes which are reserved for research, and for the provisional assignment of new diseases of uncertain aetiology). The first character of the ICD code is a letter associated with a particular chapter. The 3-character code allows to divide each chapter into homogeneous blocks reflecting different axes of classification. Each ICD category can be further divided into up to 10 subcategories, if more details are required, using a fourth numeric character after a decimal point. This is used, for example, to classify histological varieties of neoplasms. A few ICD chapters adopt five or more characters to allow further sub-classification along different axes. For example, in Chapter II – Neoplasms (letters C and D) the first axis is the behavior of the neoplasm, and the next is its site. A neoplasm of pancreas, belonging to the class *Malignant neoplasms of digestive organs* (code range C15-C26), has code C25.

The International Classification of Primary Care. In primary care, many symptoms and non-disease conditions are difficult to code in ICD, which has a disease-based structure. For this reason, the International Classification of Primary Care (ICPC) was created to allow to codify many of vague and ill-defined conditions for which patients contact their general practitioner. ICPC was published in 1987 by WONCA[3], the world organization of general practitioners, to allow the classification of three important elements of the healthcare encounter: reasons for encounter (RFE), diagnoses or problems, and process of care. In problem-oriented medical records, these elements allow to classify an episode from the beginning with an RFE to its conclusion with a more defined problem. The current revision is ICPC-2, issued in 1998; an electronic version of ICPC-2, referred to as ICPC-2-E, was released in 2000 [12], and updated in 2005 to ICPC-2-R.

ICPC has a biaxial structure with 17 chapters, identified by a single alpha code, divided into seven components, identified by a range of two-digit numeric codes that are not always uniform across chapters. The structure of ICPC is reported in Table 1.

Differently from ICD-10, which includes separate chapters for different diseases, in ICPC problems are distributed among chapters, depending on the body system to which they belong. Component 7 (*Diagnoses, disease*) is further organized in five subgroups, which are not numerically uniform across chapters: infectious diseases, neoplasms, injuries, congenital anomalies, and other diseases. For example, the rubric corresponding to malignant neoplasms of pancreas (code

---

[2] http://www.who.int/classifications/icd
[3] http://www.globalfamilydoctor.com/wicc/icpcstory.html

Table 1. The structure of ICPC

Components	Chapters																
	A	B	D	F	H	K	L	N	P	R	S	T	U	W	X	Y	Z
1. Symptoms																	
2. Diagnostic, screening, prevention																	
3. Treatment, procedures, medication																	
4. Test results																	
5. Administrative																	
6. Other																	
7. Diagnosis, disease																	

A. General
B. Blood, blood forming
D. Digestive
F. Eye
H. Ear
K. Circulatory

L. Musculoskeletal
N. Neurological
P. Psychological
R. Respiratory
S. Skin
T. Metabolic, endocrine, nutrition

U. Urinary
W. Pregnancy, family planning
X. Female genital
Y. Male genital
Z. Social

D76) belongs to component 7 (*Diagnoses, disease*), subgroup *neoplasms*, and chapter D (*Digestive*).

2.2 ICPC-ICD Relationship

A technical conversion between ICPC and ICD-10 has been made, allowing primary care physicians to implement ICD-10 as a reference nomenclature within the classification structure of ICPC and leading to a substantial increase of the diagnostic potential of ICPC [15]. This conversion was revised with the release of ICPC-2 and its electronic edition ICPC-2-E [12]. Because ICD-10 and ICPC were designed for different purposes, in many cases the conversion could not be done on a one-to-one basis, considering also that the diagnostic ICD-10 classes at the three digit level are far more specific than any primary care classification. In making the mapping, three situations arose.

1. A set of three-digit ICD-10 rubrics were compatible on a one-to-one basis with a three-digit rubric in the first or seventh component of ICPC.
2. A set of ICD-10 three-digit rubrics had to be broken open into four-digit-rubrics for at least one compatible conversion to one or more ICPC rubrics.
3. To allow the compatible conversion to the remaining rubrics of ICPC, the rest of ICD-10 rubrics, either on the three- or four-digit level, had to be grouped into a combination of classes.

In Tables 2, 3 and 4 we show an example of the three situation and the corresponding mappings. In general, the mapping result is that one ICPC rubric may be mapped to n ICD-10 (three or four-digit) rubrics and one ICD-10 (three or four-digit) rubric may be mapped to m ICPC rubrics.

3 OWL Encoding of Medical Classifications

To support further formal analysis of ICPC-ICD mappings, we encode ICPC-2 and ICD-10 classifications using recently developed Web Ontology Language

(OWL) [1]. This additionally allows to re-use a range of techniques and implemented tools for reasoning with OWL.

In conversion of classifications to OWL we have to preserve two important properties of classifications: the disjointness of nodes and the closed world assumption. The class disjointness reflects the principle that any given individual should be classified to a single classification code. To guarantee this, we explicitly define OWL sibling classes to be disjoint. The closed world assumption reflects the exhaustiveness of classifications, which is guaranteed by introducing a residual class "Other" at every stage of subdivision. This later property is modeled in OWL through a closure definition of any subdivision class as to be equivalent to a disjunction of all its child classes including "Other". In fact, ICPC-2 contains many rag-bag rubrics, which are further mapped to multiple ICD-10 rubrics, e.g., Y99 (Genital disease male other) is mapped to a good 31 ICD rubrics.

Encoding of ICD-10. Starting from the XML format of ICD-10, we automatically created an OWL class for each chapter, a class for each range of subchapters and a class for each three or four-digit ICD-10 rubric. We defined only the subsumption relations, in such a way that each range is subclass of the corresponding chapter, the three digit code class is subclass of the corresponding range class, and each four-digit code class is subclass of the corresponding three-digit code class. For example, we created a class for Chapter II (Neoplasms) with a subclass for the range C15-C26 (Malignant neoplasms of digestive organs), which has a subclass for code C25 (Malignant neoplasm of pancreas), which, in its turn, has a subclass for code C25.0 (Head of pancreas). Name of the classes corresponds to the rubric (or rubric range) code. For text strings associated to the rubric (e.g., description) we created datatype properties.

Encoding of ICPC-2. In encoding ICPC-2 we reproduced in OWL its biaxial structure. To this end, we created a class for each component and a class for each chapter. At the highest level we created two disjoint sibling classes; one superclass *Chapter* and the other superclass *Component*. In particular, the *Component* includes three subclasses: *Symptom and Complaint*; *Procedure*; and *Diagnosis and Disease*. The first class comprehends all the classes which represent the symptoms in ICPC-2, from code 00 to 29, prefixed by the code of the chapter which the symptom refers to (from A to Z). The *Procedure* class is composed by five subclasses (*Administrative Procedure, Diagnostic screening and prevention, Test Result, Treatment and Medication,* and *Other Reason for Encounter*). The ICPC-2 procedures (from 30 to 69) are the same for all chapters;

Table 2. One to one conversions between ICPC and ICD-10

ICPC	ICD-10
A71 Measles	B05 Measles
D76 Malignant neoplasm pancreas	C25 Malignant neoplasm of pancreas
S81 Angiomatous birthmark, portwine stain	D18 Haemangioma and lymphangioma, any site

Table 3. Breaking open of ICD-10 rubrics and conversion to one ICPC rubric

ICPC	ICD-10
Y14 Family planning male, other	Z30.0 General counselling and advice on contraception
	Z30.8 Other contraceptive management
	Z30.9 Contraceptive management, unspecified
W14 Contraception female, other	Z30.0 General counselling and advice on contraception
	Z30.8 Other contraceptive management
	Z30.9 Contraceptive management, unspecified
X10 Postponement of menstruation	Z30.9 Contraceptive management, unspecified

Table 4. Conversion of a group of ICD-10 classes to one ICPC rubric

ICD-10	ICPC
K80 Cholelithiasis	D98 Cholecystitis/cholelithiasis
K81 Cholecystitis	
K82 Other diseases of gallbladder	
K83 Other diseases of biliary tract	
K87.0 Disorders of gallbladder and biliary tract in diseases classified elsewhere	

consequently their codes don't bear any information on chapters. The third component *Diagnosis and Disease* is divided into five subclasses according to the type of disease: *Infections, Injuries, Neoplasms, Congenital Anomalies*, and *Other Diseases*. These subclasses include all the ICPC-2 rubrics from 70 to 99 prefixed by the chapter, as for *Symptom and Complaint*. The addition of disjointness statements between siblings was a straightforward task except for several particular situations. In the chapter *Skin* some diagnoses coincide with symptoms, reflecting the fact that some diseases of skin are immediately evident. This is the case of the symptom S12 (Insect bite/sting) or the symptom S09 (Infected finger/toe), which are also considered an injury disease and an infection disease, respectively. Consequently, the disjointness of all siblings in OWL leads to an inconsistency. To solve this problem, we defined the disjointness by means of the followings axiom between these two components:

$$SymptomAndComplaint \sqcap \neg(\exists isSymptomOfKind.S\_Skin)$$

and the corresponding axiom for the *Symptom and Complaint* class. To connect components and chapters, we created axioms at the level of components. For instance, to say that the class A08 (Swelling) is a symptom of kind General (chapter A), we added to the class A08 the existential and closure restrictions:

$$\exists isSymptomOfKind.A\_General \sqcap \forall isSymptomOfKind.A\_General$$

ICPC provides a field "Consider", which relates a rubric to a (set of) rubric(s) that physician has to consider when classifying a clinical element. Consequently, to each class having this relation, we added as many restrictions as the related classes. For instance, to the class A71 (Measles), which considers S07, A76, A03, we added the following restriction:

$$\exists Consider.S07 \sqcap \exists Consider.A76 \sqcap \exists Consider.A03$$

This means that a diagnosis of measles involves the symptoms Rash Generalized (S07) and Fever (A03), and the diagnosis Viral Exantem, Other (A76).

4 OWL Encoding and Analysis of Mappings

Having constructed the OWL formalization of the ICPC-2 and ICD-10 classifications, we can now pass to the question of formal encoding of mappings between them. A number of approaches has been reported in the literature for representing mappings between heterogeneous representations [6]. In this work, we pursue the simplest approach, which is of expressing mappings as OWL axioms.

Standardly, given two heterogeneous representations, a mapping can be viewed as a triple $\langle e, e', r \rangle$, where e, e' are the entities (e.g., formula, terms, classes, etc.) belonging to the different representations, and r is the relation asserted by the mapping. Due to the idea of encoding ICPC-ICD mappings as OWL axioms, the entities in the mapping correspond to ICPC-2 and ICD-10 classes and expressions, while the relation r is given a set-theoretic meaning by using subsumption and equivalence.

Since historically the goal of establishing ICPC-ICD mapping was in allowing primary care physician to implement ICD-10 as a reference nomenclature within the classification structure of ICPC-2 we decided to start formalization having exactly this vision in the mind. Taking the ICPC-ICD mapping, as a first trial we performed the following encoding: for each ICPC-2 class we selected all ICD-10 classes to which it can be translated; further we declared the mapping axiom by stating the OWL equivalence between the ICPC-2 class and the disjunction of the ICD-10 classes. For example, for ICPC-2 class X70 (Syphilis Female) related to ICD-10 classes A50 (Congenital syphilis), A51 (Early syphilis), A52 (Late syphilis), A53 (Other and unspecified syphilis), and A75 (Nonvenereal syphilis) we constructed the following axiom:

$$X70 \equiv A50 \sqcup A51 \sqcup A52 \sqcup A53 \sqcup A75$$

In total 686 such mapping axioms have been constructed. At the next step, we integrated the mapping axioms and OWL formalizations of ICPC-2 and ICD-10 together into a combined knowledge base and analyzed it. The idea behind the analysis consists in loading it to the OWL reasoner and further see possible logical shortcomings. Practically, we used the state of the art Semantic Web reasoner Pellet, which has been asked to detect whether there are any unsatisfiable classes coming out of the proposed mapping encoding. As a result, 423 classes in the integrated representation has been derived as unsatisfiable. This later meant that the selected vision was not appropriate for correct representation of ICPC-ICD mappings. At the next step, by applying the recently proposed in [9], and implemented in Pellet, OWL debugging technique, we have diagnosed the unsatisfiable classes. During the diagnosis, for each of the unsatisfiable classes Pellet has been asked to detect sets of conflicting axioms causing together the unsatisfiability. Each such set of axioms has been further analyzed in order to understand (1) what kinds of interactions between axioms standing for the mappings and axioms of ICPC and ICD are problematic, and (2) how the mapping axioms can be repaired to eliminate the unsatisfiability.

In overall, three groups of problematic interactions between mappings and axioms of medical classifications have been identified. In these groups the unsatisfiability has been caused by (1) declared disjointness of siblings in the classifications, (2) considered terms, and (3) "Other and Not Specified" terms. For illustration, let us have a look at some typical examples:

- ICPC-2 classes Y14 and W14 in Table 3 map both to the same ICD-10 classes. The reasoner here found that disjoint Y14 and W14 can't be mapped to the same terms in ICD-10. While in the reality this is perfectly legal because ICD-10 does not distinguish between different aspects of contraception distinguished by general practitioners in ICPC-2.
- "Consider" restriction forms an additional channel for propagating already derived unsatisfiability. For example, ICPC-2 classes N06 (Sensation disturbance other) and S01 (Pain/tenderness of skin) map both to the ICD-10 class R20.8 (Other and unspecified disturbances of skin sensation). Since S01 is found to be unsatisfiable the "consider" additionally renders the connected S70 (Herpes zoster) as unsatisfiable.
- ICPC-2 classes Y06 (Prostate symptom/complaint) and Y99 (Genital disease male other) map both to two ICD-10 classes: N42.8 (Other specified disorders of prostate) and N42.9 (Disorder of prostate, unspecified). This is a m:n map, in which the unsatisfiability is due to the disjointness of the symptom Y06 and the rag-bag disease Y99.

To repair the above problems, a number of relaxations for encoding the problematic mappings have been done. First of all, we saw that many ICPC-2 classes can not be mapped on the 1:n basis via equivalence relation. As such, the problematic cases have been converted to m:n versions by taking the union of the corresponding ICPC-2 classes and stating weaker subsumption relation to mapping; additionally the other way round subsumption axioms from ICD-10 to ICPC-2 classes have been introduced. For example, given the mappings in Table 3, the following axioms have been constructed:

$$Y14 \sqcup W14 \sqsubseteq Z30.0 \sqcup Z30.8 \sqcup Z30.9$$
$$X10 \sqsubseteq Z30.9$$
$$Z30.9 \sqsubseteq Y14 \sqcup W14 \sqcup X10$$
$$Z30.0 \sqcup Z30.8 \sqsubseteq Y14 \sqcup W14$$

After performed relaxations, the reasoner did not fire any unsatisfiability any more. As such, we constructed the sound version of ICPC-ICD mappings.

5 Related Work

The problem of establishing and formal analysis of mappings between heterogeneous representations is hard and challenging task addressed in many areas of information technologies. The situation in the healthcare domain is not exceptional in this sense. Besides the investigation of ICPC-2 and ICD-10 classification systems done in this work, there are a number of approaches focused on the analysis of other medical systems. Although many of the techniques and formalizations for these systems have been criticized as violating core ontological modeling principles (see Ceuster and Smith [2]), they have been found as practically useful in different application scenarios.

On the one side, a large amount of the proposed algorithms and heuristics for mapping discovery between medical classifications are based on the extensive use of the Metathesaurus UMLS (Unified Medical Language System) as a Knowledge Resource for semantic mapping of concepts belonging to different classifications and terminologies. The notable works in this direction are the one proposed by Fung and Bodenreider [7]; Elkin and Brown [5] on mapping between SNOMED RT (Systematized Nomenclature of Medicine-Reference Terminology) and ICD9-CM (International Classification of Diseases version 9 - Clinical Modifications); Wang et al. [14] on mapping ICPC2-Plus to SNOMED CT, and others.

From the other side, a number of formally grounded mapping approaches have been proposed. For example, Dolin et al. [4] applied the description logics for a semantically-based mapping, evaluating a LALR (lexically assign, logically refine) strategy for merging overlapping healthcare terminologies. De Keizer and Abu-Hanna [3] provided a representation formalism based on Entity Relationship Diagrams (ERD) and First Order Logic (FOL), as part of a framework for representing the structure of 5 terminological systems: ICD, SNOMED, NHS clinical terms, UMLS, and GALEN. Lomax and McCray [11] described their experiences with mapping the Gene Ontology (GO) to UMLS, while Lee and Geller [10] performed a Semantic Enrichment for medical ontologies using the two level structure of UMLS and WordNet SUMO.

To conclude, we mention Heja et al. [8] that, to our knowledge, is the only one attempt of a formal representation of the first two chapters of ICD-10 coding system, based on GALEN CMR.

6 Concluding Remarks

In this paper we have argued on the benefits of logic-based formalization and analysis of mappings between medical classification systems. We showed that by application and adoption of recent advances developed in the Semantic Web one can fruitfully reuse available logical methods and system implementations for the analysis of mappings between medical classification systems.

Practically, we focused on the task of devising the explicit formal representation of the mappings between ICPC-2 and ICD-10 medical classification systems, which have a particular importance in healthcare processes in countries of the European Union. As a formal language for expressing mappings we used the semantic web standard - Web Ontology Language (OWL). Starting from the existing conversion tables between ICPC-2 and ICD-10 systems proposed in [12] our task was to find the logically coherent encoding of the conversions in terms of OWL axioms. To assess the logical coherence and debug the reasons of logical incoherence the open-source OWL Reasoner Pellet has been applied. As a result of this work, we have constructed and validated OWL representations of ICPC-2 and ICD-10 systems and mappings between them.[4]

Recently WHO and Wonca have decided to issue the updated versions of ICD and ICPC classification systems, namely ICD-11 and ICPC-3. Consequently, the question of introducing changes into mappings between them will arise. The experience with mapping current ICPC-2 and ICD-10 systems have demonstrated that the process of establishing mappings and their further correctness analysis is a question of years involving multiple human expert efforts. We think that by pushing the use of the Semantic Web developments the efforts for upgrading the mappings between ICPC-3 and ICD-11 can be noticeably reduced. As we demonstrated in this work, the OWL encoding of mappings allows to advantageously reuse existing reasoning tools for checking mappings correctness, as well as for debugging and fixing errors in mappings utilizing techniques proposed in [9,13].

References

1. Bechhofer, S., van Harmelen, F., Hendler, J., Horrocks, I., McGuinness, D.L., Patel-Schneider, P.F., Andrea Stein, L.: OWL Web Ontology Language Reference. W3C Recommendation (February 2004), http://www.w3.org/TR/owl-ref
2. Ceusters, W., Smith, B., Flanagan, J.: Ontology and Mapping Terminology: Why Description Logics Are Not Enough. In: Proceedings of TEPR 2003 (2003)
3. de Keizer, N.F., Abu-Hanna, A.: Understanding Terminological System II: Experience with Conceptual and Formal Representation of Structure. Methods Inf. Med. 39(1), 22–29 (2000)
4. Dolin, R.H., Huff, S.M., Rocha, R.A., Spackman, K.A., Campbell, K.E.: Evaluation of a lexically assign, logically refinestrategy for semi-automated integration of overlapping terminologies. J. Am. Med. Inform. Assoc. 5(2), 203–213 (1998)
5. Elkin, P.L., Brown, S.H.: Automated enhancement of decription logic-defined terminologies to facilitate mapping to ICD9-CM. Journal of Biomedical Informatics 35, 5–6 (2002)
6. Euzenat, J., Scharffe, F., Serafini, L.: Specification of the delivery alignment format. Deliverable 2.2.6, KnowledgeWeb (2006)
7. Fung, K.W., Bodenreider, O.: Utilizing the UMLS for Semantic Mapping between Terminologies. In: Proceedings of AMIA Annual Symposium 2005, pp. 266–270 (2005)

---

[4] http://ehealth.fbk.eu/resources/medical-terminologies.html

8. Heja, G., Surján, G., Lukácsy, G., Pallinger, P., Gergely, M.: GALEN based formal representation of ICD10. International Journal of Medical Informatics 76(2-3), 118–123 (2007)
9. Kalyanpur, A., Parsia, B., Sirin, E., Hendler, J.: Debugging unsatisfiable classes in owl ontologies. Journal of Web Semantics 3(4), 268–293 (2005)
10. Lee, Y., Geller, J.: Semantic enrichment for medical ontologies. Journal of Biomedical Informatics 39, 206–226 (2006)
11. Lomax, J., McCray, A.T.: Mapping the Gene Ontology into the Unified Medical Language System. Comparative and functional genomics 5(4), 354–361 (2004)
12. Okkes, I.M., Jamoullea, M., Lamberts, H., Bentzen, N.: ICPC-2-E: the electronic version of ICPC-2. Differences from the printed version and the consequences. Family Practice 17, 101–107 (2000)
13. Schlobach, S., Cornet, R.: Non-standard reasoning services for the debugging of description logic terminologies. In: Proceedings of the 18th International Joint Conference on Artificial Intelligence, Acapulco, Mexico, pp. 355–362 (2003)
14. Wang, Y., Patrick, J., Miller, G., O'Halloran, J.: Linguistic Mapping of Terminologies to SNOMED CT. In: Proceedings of SMCS 2006 (2006)
15. Wood, M., Lamberts, H., Meijer, J.S., Hofmans-Okkes, I.M.: The Conversion Between ICPC and ICD-10. Requirements for a Family of Classification Systems in the Next Decade. Family Practice 9, 340–348 (1992)

Modeling Reuse on Case-Based Reasoning with Application to Breast Cancer Diagnosis

Carles Pous[1], Pablo Gay[1], Albert Pla[1], Joan Brunet[2], Judit Sanz[3,4],
Teresa Ramon y Cajal[3], and Beatriz López[1]

[1] Universitat de Girona, Campus Montilivi, edifice P4, 17071 Girona
carles.pous@udg.edu,beatriz.lopez@udg.edu
http://eXiT.udg.edu/
[2] Institut Català d'Oncologia, Girona
Jbrunet@iconcologia.net
http://www.ico.net/
[3] Hospital de la Santa Creu i Sant Pau
juditsanzbuxo@gmail.com,tramon@santpau.cat
http://www.santpau.es
[4] Hospital Universitari Sant Joan de Reus

Abstract. In the recent years, there has been an increasing interest on the use of case-based reasoning (CBR) in Medicine. CBR is characterized by four phases: retrieve, reuse, revise and retain. The first and last phases have received a lot of attention by the researchers, while the reuse phase is still in its infancy. The reuse phase involves a multi-facet problem which includes dealing with the closeness to the decision threshold used to determine similar cases, among other issues. In this paper, we propose a new reuse method whose decision variable is based on the similarity ratio. We have applied the method and tested in a breast cancer diagnosis database.

1 Introduction

Case-Based Reasoning (CBR) is experiencing a rapid grow and development in Medicine [3], mainly due to their relationship with evidence-based practice: case-based reasoning focuses on the reuse of previous experiences or cases to solve new problems [4,10]. A case-based system accompanies a conclusion of a problem with the cases found similar in a case base, thus presenting compelling justification for the decision.

Case-based reasoning consists of four phases, that are repeated for each new situation [1]. The first step is to seek for past situation or situations similar to the new one (*Retrieval*). In this stage it is necessary to define a metric function and decide how many cases to retrieve. *Reuse* is the second phase and it consists in using the extracted cases to propose a possible solution. Once the solution is proposed it has to be revised (*Revise*) by a human expert or automatically. After the revision process, it has to be decided if it is useful to keep the new situation in the case base in order to help on the future diagnosis of new situation. This last stage is known as *Retain*. Case-based reasoning has evolved in the last years, and there are currently a big number of available techniques for each phase of the system.

D. Dochev, M. Pistore, and P. Traverso (Eds.): AIMSA 2008, LNAI 5253, pp. 322–332, 2008.

Nevertheless, most of the advances have been achieved at the retrieval and retain phase [14,18,6,15]. In the reuse phase, advances have been obtained depending on the system purpose: diagnosis, classification, tutoring and planning (such as therapy support). Regarding diagnosis and classification, most of the systems rely on adaptation methods that consist on copying the solution of the most similar case or a combination of them (the class of the majority of them, for example). More elaborated methods have been defined for therapy, guidelines and protocols (planning tasks). For example, in [12] a reuse phase for breast cancer treatment protocol decision is based on reformulating the problem in subparts (paths) and then elaborate an adaption for each part. In [9] a particular adaptation method is proposed for deciding the reuse of an antibiotic treatment, that depends on the features used in the retrieval phase. Adaptation methods for planning are influenced by the work done in other application domains, as software design [8]. However, most of the methods are problem specific [16,17]. For example, in [11] the method proposed is based on a library of feature adaptation plans.

According to [12], adaptation in Medicine is complex, needs to deal with the lack of relevant information about a patient, the applicability and consequences of the decision, the closeness to the decision thresholds and the necessity to consider patients according to different viewpoints. In fact, [17] points out about the difficulty on giving autonomy to this CBR step in Medicine. So, the authors propose a user-interactive approach for this CBR stage. In this sense, we believe that all of the approaches found in the literature are complementary and necessary: the use of generic approaches (as the majority rule), specific knowledge (as a library of feature adaptation plans), and user interaction.

In this paper, we introduce a new a simple generically applicable approach to perform case adaptation. Our research concerns reuse for classification, since our goal is to diagnose whether a patient has cancer or not. Our method, takes into account the closeness of the retrieved cases, refining the approach provided by the majority rule. The simplicity of our approach means that the development cost required to apply it is low. We have experimentally tested our method with two cancer data bases, and we have got uncourageous results.

The paper is structured as follows. First, a brief description of an elemental CBR system is given. Section 3 describes the main contribution of the paper, the proposed reuse methodology. Afterwards, the description of the experiments done are reported, following with the results achieved. Finally, we end with the conclusions and future work.

2   Case-Based Reasoning Approach

We are faced with the problem of building a CBR system to support breast cancer diagnosis. As a first approach, we have developed a system similar to the one described in [5],which includes the basic functions. So we have the complete CBR cycle covered with simple functions that enable the study of the reuse stage performance.

Concerning to the retrieve stage, we have used the distance based on the Mathematical distance for numeric attributes and the Hamming distance for categorical ones. Regarding missing values, statisticians have identified three situations [20]. Missing completely at random (MCAR) is the case when the probability of missing a value is the same for all variables. On the other hand, not missing at random (NMAR) values occurs when the probability of missing a value is also dependent on the value of the another missing variable. At last, the missing values classified as missing at random (MCR) happens when the probability of missing a value is only dependent on other variables. We do not distinguish between MCAR,NMAR or MCR values. So, when the attribute of the test case and the memory case are both missed, the distance is considered 0. Second, when the value of the attribute either the test case or the memory case are missed, the distance is also taken as 0. Finally, when both are missed, the corresponding distance function is applied (Mathematical or Hamming).

Thus, the *local distance* between two cases concerning attribute a with a value of x_a and y_a for each case is given by the following equation:

$$d(x_a, y_a) = \begin{cases} 0 & if\ either\ x\ or\ y\ are\ unknown \\ Hamming(x_a, y_a) & if\ x_a = y_a\ and\ a\ categorical \\ |x_a - y_a| & if\ x_a = y_a\ and\ a\ numerical \end{cases} \tag{1}$$

where

$$Hamming(x_a, y_a) = \begin{cases} 1 & if\ x_a \neq y_a \\ 0 & if\ x_a = y_a \end{cases} \tag{2}$$

The *global distance* is the average of local distances obtained for all the attributes. Finally the similarity between a test case T and a memory case C, $sim(C, T)$, is computed as the inverse of the global distance as follows:

$$sim(C, T) = 1 - \frac{\sum_{i=1}^{n} d(x_i, y_i)}{n} \tag{3}$$

where n is the amount of attributes in the application.

Note, then, that we are using a very simple data retrieval phase. In this paper we do not focus on optimize the number of features to use, or on investigating different similitude functions and attributes weighting. We are focussing on the reuse methods.

3   Reuse Method

The retrieval phase returns a set of k similar cases, $C_1, \ldots, C_k$ to the current test case T. Often they are ranked according to the similarity degree (i.e. $sim(C_1, T) > sim(C_2, T) > \ldots > sim(C_k, T)$). Then, in the reuse phase the solution to the problem posed by T should be computed. Particularly, when dealing with a classification problem, the class corresponding to T should be determined.

In the particular case of breast cancer diagnosis, two classes conforms the solution space of a problem: cancer (+) and no cancer (-). According to the *majority rule*, the new problem is classified in the same class than the majority of the retrieved cases. However, in domains like the one we are dealing with, it is the case that in most situations the amount of positive and negative cases retrieved are the same. So, the majority rule is not a valid classification criteria.

An alternative approach is the one proposed by Bilska-Wolak and Floyd who define a decision variable, DV_{bw}, as the ratio of "+" cases in $C_1, \ldots, C_k$ to all similar cases [5]. This definition represents an intuitive approach for describing the likelihood of cancer in a case. Thus,

$$DV_{bw} = \frac{\text{Number of + cases in } \{C_1, \ldots, C_k\}}{k} \qquad (4)$$

Then, the decision variable is compared against a given threshold τ. Whenever the current value of the decision variable equals or goes beyond this threshold, the test case T is classified as positive or not.

This approach presents several disadvantages regarding the closeness degree of the similar cases to the test case, as shown in the following example. Let us suppose that we have four similar cases, two positives and two negatives: $C_1^+, C_2^+, C_3^-, C_4^-$. In this case, the decision variable according to [5] is $DV_{bw} = \frac{2}{4} = 0.4$. If τ is set to 0.5, the test case T will have the "+" class assigned. This decision is independent of the similarity degree of the cases. For example, in Table 1 several possibilities are given. In the first one, when all similarities are the same, the results cannot be improved. In the second situation, it is clear that positive cases are closer to the test case that the negative ones, and the decision variable should catch that. Conversely, in the third analyzed scenario, negative cases are more likely. Finally, in the last situation, as in the first scenario, there is no room for precision on the decision variable.

Table 1. Different situations with the same DV_{bw} value

$sim(C_1^+,T)$	$sim(C_2^+,T)$	$sim(C_3^-,T)$	$sim(C_4^-,T)$	DV_{bw}	T class ($\tau = 0.5$)
0.5	0.5	0.5	0.5	0.5	+
0.8	0.8	0.5	0.5	0.5	+
0.4	0.4	0.5	0.5	0.5	+
0.4	0.5	0.4	0.5	0.5	+

To take into account the closeness degree of similarity between the test and a memory case in the decision variable we propose an alternative definition as the ratio of similarities of + cases to the addition of the similarities of all similar cases. Formally,

$$DV_{pous} = \frac{\sum_{C_i^+} sim(C_i^+, T)}{\sum_{i=1}^{k} Sim(C_i, T)} \qquad (5)$$

where C_i^+ are the cases in $C_1, \ldots, C_k$ which belong to the + class (suffering cancer).

As it is possible to observe in Table 2, our definition of the decision variable captures the closeness of the similarity to the test case involved in positive and negative cases. So, with the same threshold $\tau = 0.4$ the results are slightly different than the ones obtained with the DV_{bw} (see Table 1).

Table 2. Different situations with the same DV_{pous} value

$sim(C_1^+, T)$	$sim(C_2^+, T)$	$sim(C_3^-, T)$	$sim(C_4^-, T)$	DV_{pous}	T class ($\tau = 0.5$)
0.5	0.5	0.5	0.5	0.5	+
0.8	0.8	0.5	0.5	**0.6**	+
0.4	0.4	0.5	0.5	**0.4**	-
0.4	0.5	0.4	0.5	0.5	+

4   Experimental Set-Up

We have implemented our simple CBR system with the reuse method in Java. In order to test our methodology, we have used two data sets: The Breast Cancer Wisconsin (Diagnostic) Data Set [19] and our own cancer data set (HSCSP). The former is composed by 699 instances, 100 attributes each (integer values). Features are computed from a digitized image of a fine needle aspirate of a breast mass. They describe characteristics of the cell nuclei present in the image. There are some missing values.

The other data set used was provided by the Hospital de la Santa Creu i Sant Pau (HSCSP) from Barcelona. It consists of 871 cases, with 628 corresponding to healthy people and 243 to women with breast cancer. There are 1199 attributes for each case. They correspond to people habits (smoker or not, diet style, sport habits,...), desease characteristics (type of tumor,...), and gynaecological history among others. Since there are redundant, wrong and useless information a preprocess was carried out. Also, the preprocess was used to obtain data corresponding to independent individuals, since there are patients in the database that are relatives. After this operations the final database was constituted of 612 independent cases, with 373 patients and 239 healthy people. From 1199 attributes, 680 are considered useless, redundant or too much incomplete. About the rest of attributes, 192 are discrete, 279 numeric and 34 text (such as postal address, etc.). We have used discrete and numeric attributes for our experimentation (471). Another preprocessing step has also been applied to normalize numerical values.

As our method depends on two parameters, the number of cases retrieved (k) and the threshold used to classify the case according to the corresponding decision variable (τ), we have defined several experiments to be carried out varying them. Particularly we have varied each parameter as follows:

- k: from 1 to 10, step 1
- τ: from 0.1 to 1.0, step 0.1.

A cross-validation methodology has been followed. Figure 1 illustrates the process when using a stratified cross validation methodology with a fixed length of

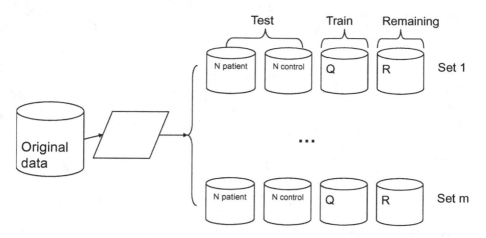

Fig. 1. Data sets generated for stratified cross validation on HSCSP data

test files (N cases per file) and a fixed number of data sets (m). So, from each original data set, up to 10 different training and testing sets have been generated. So, a run have been computed with each data set and each pair of parameter values.

5   Results

We have analyzed the four different possible outcomes: true positives (TP), true negatives (TN), false positive (FP) and false negative (FN). True positive and negative are the correct classifications. A false positive occurs when the outcome is incorrectly predicted as cancer (+); conversely, a false negative occurs when the outcome incorrectly predicts a no-cancer value (-). Next, the true positive rate (tp), and the false positive rate (fp) have been computed as follows:

$$tp = \frac{TP}{TP+FN} \tag{6}$$

$$fp = \frac{FP}{FP+TN} \tag{7}$$

The true and false positive rates have been used to visualize the results in ROC graphs. These graphs have been used in cost-sensitive learning because of the ease with which class skew and error cost information can be applied to them to yield cost-sensitive decisions [7]. In the x-axis, the fp values are plotted, while in the y-axis, the tp ones. Any point in the two-dimensional graph represents an experiment result. The ideal situation would be a system with tp=1 and fp=0, that is a point located just at the right upper part of the graphic.

In order to average the results of the cross validation process, we have followed the threshold averaging methodology explained in [7]. Thus, since the Pous method have two parameters, first we have averaged the results obtained for each k value given a threshold τ. Next, we have averaged the results of 10

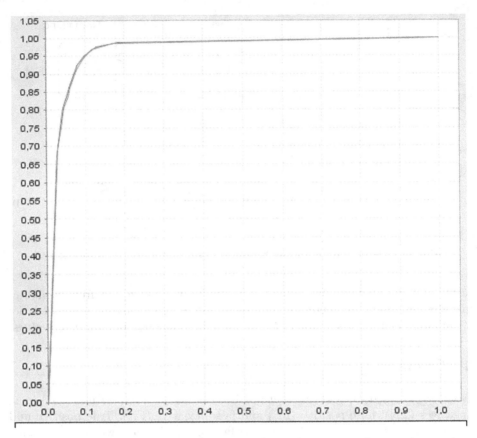

Fig. 2. Results on the Winsconsin data set. fp on the x-axis and tp on the y-axis.

runs for every possible τ. So, any point at the ROC graph corresponds to the results of the method according to the a given τ value.

The results on the Wisconsin data set are shown in figure 2. As it is possible to observe, the tp rate is above the 95% in almost all the ocassions. Regarding the HSCSP data set, results are shown in figure 3. With this data set, we have obtained a lower tp rate than with the Wisconsin data set. However, the complexity of the data set is different (10 attributes versus 471). As a first approach to the case-based breast cancer diagnosis, we believe that these results are good enough: With a $\tau = 0.7$, we got a tp (sensitivity) of 70 % while maintaining the fp (1-specificity) rate to 20 %. Of course, there is room for improvement with the incorporation in a future work of more appropriate retrieval, attribute selection and other methods.

6   Related Work

This section describes the main papers related to the adaption stage of a CBR cycle, when used in several medicine domains. Our starting point has been the

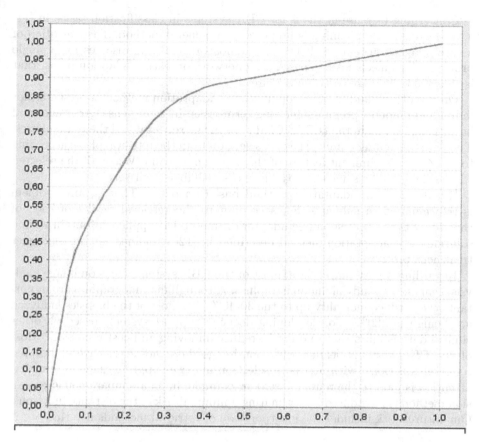

Fig. 3. Results on the HSCSP data set. fp on the x-axis and tp on the y-axis.

work of Bilska-Wolak and Floyd on breast cancer [5]. The authors propose the probabilistic measure presented in equation 4. As a previous step, they use a similarity threshold SIM, empirically estimated, in order to decide if a case is similar or not to the current case. The estimation is based on the goodness of the distance function used in the retrieval phase according to an exhaustive analysis of all the features. In their case, this approach is feasible, since they are dealing with a reasonable number of variables (10). However, in other scenarios, as it is our, this exhaustive approach is unfeasible. Regarding the reuse phase, the differences between our method is expressed in equations 4 and 5.

Our methodology is tightly related to weighted knn-methods. Such methods follow a voting scheme to decide the labeling of a test case, as the majority rule, but using a weight function for every vote [13]. When taking the weight function as the similarity one, we have the same results. Note, however, that we are not just assigning the class with the high aggregated weight, but computing the ratio of the weights of cancer that matched the test case to the total number of k matched cases.

Concerning to reuse methods for CBR systems in classification tasks, in [11] a knowledge-based planning is proposed as a general method. The execution of a case adaptation plan leads to the diagnosis of multiple diseases for a single problem. This mechanism is based on a network of feature adaption plans, that is, specific knowledge costly to acquire.

On the other hand, in [2] a sophisticated adaptation method is presented for individual diabetes prognosis for long-term risks. In some sense, Armengol and her colleagues mix in the DIRAS CBR system the retrieval and the reuse phases. They are continuing refining the case selection from the retrieval phase according to the most discriminant feature of the problem at hand. When all the selected cases belong to the same class, the new case is labeled with it.

[12] develops an adaptation method based on paths. That is, any problem is reformulated in subparts. So, according to the retrieval results of each of the subparts,the corresponding adaptation method is applied to obtain a final solution. The adaptation method described in [12] is defined for breast cancer treatment protocols.

Regarding breast cancer data, most of the CBR systems focus on image analysis, that is, the study of mammographies. Although mammographies are important (they reduce mortality up to the 30-40% [5]), most of the biopsies analyzed are benign (75-80%). So, developing decision support systems based on other clinical data is important in order to reduce annoying and costly biopsies. Most of the CBR systems dealing with non-image data, however, have elaborated retrieval functions, with attribute selection and weighted mechanism, and few attention is paid to the reuse phase. For example, in [6] a nonparametric regression method is introduced for estimating a minimal risk of the distance functions. Our future work includes the incorporation of retrieval methods like this.

7   Conclusions and Discussion

There is an increasing interest on the use of CBR techniques for medical diagnosis. However, most of the advances concentrate on the retrieval phase, in which similar cases are selected from memory. Regarding the reuse phase, researchers have focalized on planning tasks, as therapy or protocols, while for diagnosis the majority rule seems to be the most simple and general method utilized.

In this paper, we have presented a new general method for reuse, that is also simple to compute. It takes into account the similarity degree of the selected cases when determining the class of the new case. The method has been tested in two cancer databases: Wisconsin (from the UCI repository) and HSCSP (our own), and the results obtained are encouraging.

As a future work, we wish to complement our general reuse method, with other knowledge specific methods and user interaction abilities, according to the requirements of the medical applications. From the complementarity of all of these techniques should arise a reuse stage for CBR useful for physicians. We are also planning to integrate more accurate retrieval functions, as well as other attribute selection and weighing methods in our initial CBR prototype.

Acknowledgments. This research project has been partially funded by the Spanish MEC project DPI-2005-08922-CO2-02, Girona Biomedical Research Institute (IdiBGi) project GRCT41 and DURSI AGAUR SGR 00296 (AEDS).

References

1. Agnar, A., Plaza, E.: Case-based reasoning: Foundational issues, methodological variations, and system approaches. AI Communications 7(1), 39–59 (1994)
2. Armengol, E., Palaudáries, A., Plaza, E.: Gindividual prognosis of diabetes long-term risks: a cbr approach. Methods on Information in Medicine 40(1), 46–51 (2001)
3. Bichindaritz, I., Marling, C.: Case-based reasoning in the health sciences:what's next? Artificial Intelligence in Medicine (36), 127–135 (2006)
4. Bichindaritz, I., Montani, S., Portinale, L.: Special issue on case-based reasoning in the health sciences. Appl. Intelligence (28), 207–209 (2008)
5. Bilska-Wolak, A.O., Floyd Jr., C.E.: Development and evaluation of a case-based reasoning classifier for prediction of breast biopsy outcome with bi-radstm lexicon. Medical Physics 29(9), 2090–2100 (2002)
6. Dippon, J., Fritz, P., Kohler, M.: A statistical approach to case based reasoning, with application to breast cancer data. Computational Statistics and Data Analysis 40, 579–602 (2002)
7. Fawcett, T.: Roc graphs: Notes and practical considerations for data mining researchers. Technical Report HPL-2003-4, HP Labs (2003)
8. Gomes, P., Pereira, F.C., Carreiro, P., Paiva, P., Seco, N., Ferreira, J.L., Bento, C.: Case-based adaptation for uml diagram reuse. In: Negoita, M.G., Howlett, R.J., Jain, L.C. (eds.) KES 2004. LNCS (LNAI), vol. 3215, pp. 678–686. Springer, Heidelberg (2004)
9. Heindl, B., Schmidt, R., Schmid, R.G., Haller, M., Pfaller, P., Gierl, L., Pollwein, B.: A case-based consiliarius for therapy recommendation (icons): computer-based advice for calculated antibiotic therapy in intensive care medicine. Computer Methods and Programs in Biomedicine 52(2), 117–127 (1997)
10. Holt, A., Bichindaritz, I., Schmidt, R., Perner, P.: Medical applications in case-basd reasoning. The Knowledge Engineering Review 20(3), 289–292 (2005)
11. Hsu, C.-C., Ho, C.-S.: A new hybrid case-based architecture for medical diagnosis. Information Sciences 166, 231–247 (2004)
12. Lieber, J., d' Aquin, M., Badra, F., Napoli, A.: Modeling adaptation of breast cancer treatment decision protocols in the KASIMIR project. Applied Artificial Intelligence (28), 261–274 (2008)
13. Macleod, J.E.S., Luk, A., Titterington, D.M.: A re-examination of the distance-weighted k-nearest neighbor classification rule. IEEE Transactions on Systems, Man and Cybernetics 17(4), 689–696 (1987)
14. Sollenborn, M., Nilsson, M.: Advancements and trends in medical case-based reasoning: An overview of systems and system development. In: American Association for Artificial Intelligence, Proceedings of the 17th International FlAIRS Conference, Special Track on Case-Based Reasoning, pp. 178–183 (2004)
15. Perner, P.: Concepts for novelty detection and handling based on a case-based reasoning process scheme. In: Perner, P. (ed.) ICDM 2007. LNCS (LNAI), vol. 4597, pp. 21–33. Springer, Heidelberg (2007)

16. Salem, A.-B.M.: Case-based reasoning technology for medical diagnosis. In: Proc. World Academy of Science, Engineering and Technology, vol. 25, pp. 9–13 (2007)
17. Montani, S., Bellazzi, R., Portinale, L., Gierl, L., Schmidt, R.: Casebased reasoning for medical knowledge-based systems. International Journal of Medical Informatics 64, 355–367 (2001)
18. Shie, J.-D., Chen, S.-M.: Feature subset selection based on fuzzy entropy measures for handling classification problems. Applied Artificial Intelligence (28), 69–82 (2008)
19. Wolberg, W., Street, N., Mangasariam, O.: UCI machine learning repository. breast cancer wisconsin (diagnostic) data set (2007)
20. Zhang, S., Qin, Z., Ling, C.X., Sheng, S.: "missing is useful": Missing values in cost-sensitive decision trees. IEEE Transactions on Knowledge and Data Engineering, 17(12), 1689–1693 (2005)

Toward Expert Knowledge Representation for Automatic Breast Cancer Detection

Marina Velikova[1], Maurice Samulski[1], Nico Karssemeijer[1], and Peter Lucas[2]

[1] Department of Radiology, Radboud University Nijmegen Medical Centre
Nijmegen, The Netherlands
{m.velikova,m.samulski,n.karssemeijer}@rad.umcn.nl
[2] Institute for Computing and Information Sciences, Radboud University Nijmegen
Nijmegen, The Netherlands
peterl@cs.ru.nl

Abstract. In reading mammograms, radiologists judge for the presence of a lesion by comparing at least two breast projections (views) as a lesion is to be observed in both of them. Most computer-aided detection (CAD) systems, on the other hand, treat single views independently and thus they fail to account for the interaction between the breast views. Following the radiologist's practice, in this paper, we develop a Bayesian network framework for automatic multi-view mammographic analysis based on causal independence models and the regions detected as suspicious by a single-view CAD system. We have implemented two versions of the framework based on different definitions of multi-view correspondences. The proposed approach is evaluated and compared against the single-view CAD system in an experimental study with real-life data. The results show that using expert knowledge helps to increase the cancer detection rate at a patient level.

Keywords: Bayesian network, causal independence model, mammography, multi-view breast cancer detection.

1 Introduction

The goal of breast cancer screening programs is to detect cancers at an early (pre-clinical) stage, by using periodic mammographic examinations in asymptomatic women. At a screening mammographic exam each breast is usually scanned in two views–mediolateral oblique (MLO), taken under 45° angle, and craniocaudal (CC), taken top-down (Figure 1). In evaluating cases, radiologists judge for the presence of a lesion by comparing both views as a lesion tends to be observed in different breast projections.

Most computer-aided detection (CAD) systems, on the other hand, treat each view independently. Hence, the interaction between the breast views is ignored and thus the breast cancer detection can be obscured. This limits the usability and the trust in the performance of such systems.

To tackle this problem, in this paper, we exploit multi-view dependencies by developing a Bayesian network framework to improve the breast cancer detection

D. Dochev, M. Pistore, and P. Traverso (Eds.): AIMSA 2008, LNAI 5253, pp. 333–344, 2008.

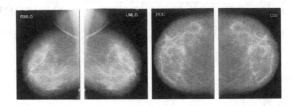

Fig. 1. MLO and CC projections of a left and right breast

rate at a patient level. The main idea is to use the context information about each breast, represented as the regions detected by a single-view CAD system in MLO and CC, to obtain a single likelihood measure for a patient being cancerous.

We first introduce the main domain terminology. By *lesion* we refer to a physical cancerous object detected in a patient. We call a contoured area on a mammogram a *region*. A region detected by a CAD system is described by a number of continuous (real-valued) *features* (e.g., size, location, contrast). By *link* we denote matching (established correspondence) between two regions in MLO and CC views. The term *case* refers to a patient who has undergone a mammographic exam.

The remainder of the paper is organized as follows. In the next section we review earlier works in automatic multi-view breast cancer detection. In Section 3 we introduce basic definitions related to Bayesian networks and causal independence models. Then, we describe a Bayesian network framework for multi-view breast cancer detection on mammograms. We implement two versions of the framework following different definitions of multi-view correspondences. The proposed approach is evaluated and compared against a single-view CAD system in a real case with breast cancer screening data presented in Section 4. Section 5 concludes the paper and discusses further directions for research.

2   Previous Research

There has been previous research on the automatic multi-view breast cancer detection. In [1], Good et al. have proposed a probabilistic method for true matching of lesions detected in both views, based on Bayesian network and multi-view features. The results from experiments demonstrate that their method can significantly distinguish between true and false positive links of regions. Van Engeland et al. have developed another linking method in [2] based on Linear Discriminant Analysis (LDA) classifier and a set of view-link features to compute a correspondence score for every possible region combination. The proposed approach demonstrates an ability to discriminate between true and false links. In [3], Van Engeland and Karssemeijer have extended this matching approach by building a cascaded multiple-classifier system for reclassifying the initially detected region based on the linked candidate region in the other view. Experiments show that the lesion based detection performance of the two-view detection system is significantly better than that of the single-view detection method. Paquerault et al.

have also considered established correspondence between suspected regions in both views to improve lesion detection based on LDA ([4]). By combining the resulting correspondence score with its one-view detection score the lesion detection improves and the number of false positives reduces. Only in this study, however, the authors report improvement in the case-based performance based on multi-view information. Therefore more research is required to build CAD systems that discriminate well between normal and cancerous cases–the ultimate goal of breast cancer screening programs.

In our earlier work ([5]) we have already discussed certain aspects on modelling multi-view information in the domain of breast cancer. We suggested a possible knowledge representation scheme using probabilistic first-order languages where the information for one view allows us to update our belief for the information on the other view. The proposed scheme, however, has not been supported with experimental results. Therefore, in this paper, we extend our ideas of probabilistic multi-view modelling of breast cancer based on a Bayesian network framework, as presented in the next section. Furthermore, we provide empirical evidence using real-life data that the developed framework has a potential to support screening radiologists in their practice.

3   Bayesian Multi-view Detection

3.1   Basic Definitions

Consider a finite set $\mathbf{U}$ of random variables, where each variable U in $\mathbf{U}$ takes on values from a finite domain $dom(U)$. Let P be a joint probability distribution of $\mathbf{U}$ and let X, Y, Z be subsets of $\mathbf{U}$. We say that X and Y are conditionally independent given Z, denoted by $X \perp\!\!\!\perp_P Y \mid Z$, if for all $x \in dom(X)$, $y \in dom(Y)$, $z \in dom(Z)$, the following holds:

$$P(x \mid y, z) = P(x \mid z), \text{ whenever } P(y, z) > 0.$$

In short, we have $P(X \mid Y, Z) = P(X \mid Z)$.

A Bayesian network is defined as a pair $\mathrm{BN} = (G, P)$ where G is an acyclic directed graph $G = (\mathbf{V}, \mathbf{E})$ with a set of nodes $\mathbf{V}$ corresponding to the random variables in $\mathbf{U}$ and a set of edges (arcs) $\mathbf{E} \subseteq (\mathbf{V} \times \mathbf{V})$ corresponding to direct causal relationships between the variables. We say that G is an $I\text{–}map$ of P if any independence represented in G, denoted by $A \perp\!\!\!\perp_G B \mid C$ with $A, B, C \subseteq \mathbf{V}$ mutually disjoint sets of nodes, is satisfied by P, i.e.,

$$A \perp\!\!\!\perp_G B \mid C \quad \Longrightarrow \quad X_A \perp\!\!\!\perp_P X_B \mid X_C,$$

where A, B and C are sets of nodes of G and X_A, X_B and X_C are the corresponding sets of random variables. A Bayesian network BN allows a compact representation of independence information about the joint probability distribution P by specifying a *conditional probability table* (CPT) for each random variable. This table describes the conditional distribution of the node given each possible combination of parents' values. The joint probability can be computed by simply multiplying the CPTs. For a more recent detailed description of Bayesian networks, the reader is referred to [6].

3.2 Causal Independence Models

One way to specify interactions among statistical variables in a compact fashion is offered by the notion of *causal independence* ([7]). Causal independence arises in cases where multiple causes (parent nodes) lead to a common effect (child node). Here we present the formal definition of the notion of causal independence as given in [8].

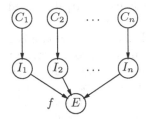

Fig. 2. Causal-independence model

The general structure of a causal-independence model is shown in Figure 2; it expresses the idea that causes $C_1, \ldots, C_n$ influence a given common effect E through intermediate variables $I_1, \ldots, I_n$. We denote the assignment of a value to a variable by lower-case letter, e.g., i_k stands for $I_k = \top$ (true) and $\bar{i}_k$ otherwise. The *interaction function* f represents in which way the intermediate effects I_k, and indirectly also the causes C_k, interact. This function f is defined in such way that when a relationship between the I_k's and $E = \top$ is satisfied, then it holds that $f(I_1, \ldots, I_n) = e$; otherwise, it holds that $f(I_1, \ldots, I_n) = \bar{e}$. It is assumed that if $f(I_1, \ldots, I_n) = e$ then $P(e \mid I_1, \ldots, I_n) = 1$; otherwise, if $f(I_1, \ldots, I_n) = \bar{e}$, then $P(e \mid I_1, \ldots, I_n) = 0$. Note that the effect variable E is conditionally independent of $C_1, \ldots, C_n$ given the intermediate variables $I_1, \ldots, I_n$ and that each variable I_k is only dependent on its associated variable C_k. Using information from the topology of the network, the notion of causal independence can be formalised for the occurrence of effect E, i.e. $E = \top$, in terms of probability theory as follows:

$$P(e \mid C_1, \ldots, C_n) = \sum_{f(I_1, \ldots, I_n) = e} \prod_{k=1}^{n} P(I_k \mid C_k)$$

Finally, it is assumed that $P(i_k \mid \bar{c}_k) = 0$ (absent causes do not contribute to the effect); otherwise, $P(I_k \mid C_k) > 0$.

An important subclass of causal-independence models is obtained if the deterministic function f is defined in terms of separate binary functions g_k; it is then called a *decomposable* causal-independence model [7]. Usually, all functions $g_k(I_k, I_{k+1})$ are identical for each k. Typical examples of decomposable causal-independence models are the noisy-OR [9] models, where the function g represents a logical OR. The formulation of this function states that the presence of any of the causes C_k with absolute certainty will cause the effect e $(E = \top)$. Causal independence model with the logical OR is used in the theoretical model presented next.

3.3   Model Description

An example of a multi-view breast cancer detection scheme is depicted in Figure 3. We have a left breast projected in MLO and CC view. A lesion (represented by the circles in both projections) is present in the breast and thus, the whole breast is cancerous. In both views an automatic single-view system detects potential cancerous regions described by a number of extracted real-valued features. In the figure regions A_1 and B_1 are correct detections of the cancer, i.e., these are true positive (TP) regions whereas A_2 and B_2 are false positive (FP) regions. Since we deal with projections of the same breast there are certain dependencies between the detected regions. To account for them we introduce *links* (L_{ij}) between the regions in both views, A_i and B_j as shown in Figure 3.

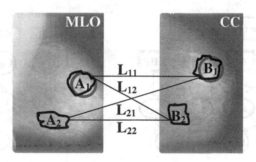

Fig. 3. Multi-view mammographic analysis with automatically detected regions

Next to every link we assign a binary class with values of "false" and "true". Given the definition of a lesion, it is logical to define the class link $L_{ij} = \ell_{ij}$ as follows:

$$\ell_{ij} = \begin{cases} \text{true} & \text{if } A_i \textbf{ AND } B_j \text{ are TP,} \\ \text{false} & \text{otherwise.} \end{cases} \tag{1}$$

In other words, the cancer must be detected by at least one region in each view in order to define the link between these regions as true. In the example depicted in Figure 3 the only "true" link corresponding to the lesion is L_{11}. However, in practice there are cases where the cancer may not be observed in one of the two views due to obscurity of the lesion, the very small lesion size or undetected region by the single-view CAD system. Therefore in this work we adopt a second definition of a class link where we refer to a link as "true" if at least one of the detected regions is TP, otherwise it is "false". More formally we have:

$$\ell_{ij} = \begin{cases} \text{true} & \text{if } A_i \textbf{ OR } B_j \text{ are TP,} \\ \text{false} & \text{otherwise.} \end{cases} \tag{2}$$

This definition allows us to maintain information about the tumor character-istics of the breast even if there is no observed cancerous region in one of the views. Given this definition, in our example we obtain 3 "true" links: L_{11}, L_{12}, and L_{21}, and 1 "false" link: L_{22}. The advantage of the first definition is that it distinguishes between true links related to lesions and those containing FP re-gions. However the second definition better represents the reality and thus may facilitate the modeling process. In our experimental study we compare models based on data using both definitions.

A region, view, breast and case has also a binary class with values of "normal" and "cancerous", determined by a radiologist or pathology.

In any case, multiple views corresponding to the same canceorus area contain correlated characteristics whereas normal views tend to be less correlated. To account for view interaction, we propose a two-step Bayesian network framework where all the region links from corresponding views are considered simultane-ously to compute a single measure for the breast (case) being cancerous. Figure 4 represents the framework.

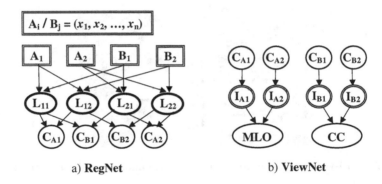

a) **RegNet**   b) **ViewNet**

Fig. 4. Bayesian network framework for representing the dependencies between mul-tiple views. Every variable (region) A_i and B_j is described by a real-valued feature vector $(x_1, x_2, \ldots, x_n)$.

In the first step we build a Bayesian network to estimate the probabilities that a detected region is cancerous given its links to the regions in the other view. A straightforward way to model a link L_{ij} is to use the corresponding regions A_i and B_j as causes for the link class, i.e., $A_i \longrightarrow L_{ij} \longleftarrow B_j$. Since the link variable is discrete and each region variable is represented by a vector of real-valued features x_n extracted from the CAD system, we apply logistic regression to compute $P(L_{ij} = \ell_{ij}|A_i, B_j)$:

$$P(L_{ij} = \ell_{ij}|A_i, B_j) = \frac{\exp\left(\beta_0^{\ell_{ij}} + \beta_1^{\ell_{ij}} x_1 + \cdots + \beta_{2n}^{\ell_{ij}} x_{2n}\right)}{1 + \exp\left(\beta_0^{\ell_{ij}} + \beta_1^{\ell_{ij}} x_1 + \cdots + \beta_{2n}^{\ell_{ij}} x_{2n}\right)}$$

where β's are the model parameters to be estimated.

Since one region participates in more than one link it is natural and easy to construct a causal model where all the links with their respective causes (regions) are considered together (see the first top layer in the network in Figure 4a).

Next we compute the probabilities $P(C_{Ai} = 1|L_{ij} = \ell_{ij})$ and $P(C_{Bj} = 1|L_{ij} = \ell_{ij})$ where C_{Ai} and C_{Bj} are the classes of regions A_i and B_j, respectively. Given the definitions in (1) and (2), we can easily model the relations between links and regions through a causal independence model using the logical OR. This network is referred to as RegNet.

At the second step of our Bayesian network framework we simply combine the computed region probabilities from RegNet by building a causal independence model with the logical OR to obtain the probability of the respective view being cancerous. We call this network ViewNet.

Finally, we combine the view probabilities obtained from ViewNet into a single probabilistic measure for the breast as whole. This is done by considering the class of the breast and applying a logistic regression model with the view probabilities as input variables. Since we have two definitions for the class links, given in (1) and (2), our whole multi-view detection scheme thus described corresponds to two models: MV-TPTP and MV-FPFP.

4 Experiments with Breast Cancer Data

4.1 Data Description

The data set we use contain 1063 cases (2126 breasts) from which 383 are cancerous. All cases contained both MLO and CC views. All cancerous breasts had one visible lesion in at least one view, which was verified by pathology reports to be cancerous. Lesion contours were marked by an experienced screening radiologist.

As input for our multi-view detection scheme we use the regions detected by a single-view CAD system ([2]). For every region, based on extracted real-valued features and supervised learning with a neural network (NN), the CAD system computes a likelihood for cancer that is converted into *normality score* (NormSc): the average number of normal regions in an image (mammogram) with the same or higher cancer likelihood. For each image we took the first 5 regions with the lowest NormSc. In total there were 10478 MLO regions and 10343 CC regions. Every region is described by 11 real-valued features–NN's output from the single-view CAD and lesion characteristics such as star-like shape, size, contrast, linear texture and location. For each breast, every region from MLO was linked with every region from CC, leading to 51088 links in total. Finally we add the binary link classes following the definitions in (1) and (2). Out of all links in the data 583 links correspond to TP-TP links (as cancer can be hit by more than one region on a mammogram). Thus based on our first link class definition only around 1% of all links are true. Based on the second definition, the number of true links is considerably larger: 2698, accounting also for 39 cancerous breasts with a detected TP region only in one view.

4.2   Training and Evaluation

Both `RegNet` and `ViewNet` networks have been trained and tested using the Bayesian Network Toolbox in Matlab ([10]). The learning is based on the EM algorithm, used to approximate a probability function with incomplete data (in our models the logical OR-nodes are not observed) ([11]). Each network performance is evaluated using two-fold stratified cross validation: the dataset is split into two subsets with approximately equal number of observations and proportion of cancerous breasts. Each half is used as a training set and a test set.

Next, as we discussed in the description of our model, we apply logistic regression to compute the probability for a breast being cancerous. The input vector V contains view information represented by the probabilities for MLO and CC obtained from `ViewNet` and the minimum NormScs for each view, which are also indicators for view suspiciousness. The output is the breast probability $P(Br|V)$, which is estimated using two-fold cross validation: the breast data is split in two halves and each half is used for training and testing. This splitting is repeated 10 times. As a result for each breast we obtain 10 probabilities and for each case we get 20 breast probabilities in total. Out of these 20 probabilities we first choose the maximum probability and assign it to the respective breast. Then consider only the 10 probabilities for the other breast and take the minimum as a final probability for being cancerous.

Finally we compute the likelihood of a case (patient) being cancerous by applying two combining schemes based on the computed right and left breast probabilities, $P(RBr|V_R)$ and $P(LBr|V_L)$, given their respective view information V_R and V_L. The first most straightforward scheme (Case-MAX) is to take the maximum out of both probabilities:

$$Lik_{MAX}^{MV}(Case) = \max\big(P(RBr|V_R), P(LBr|V_L)\big).$$

Our second combining scheme (Case-$DIFF$) is based on the intuition that the absolute probability difference between the left and right breast for cancerous cases should be larger than that for normal cases and thus allowing for a better case distinction. Furthermore we use the normality score for each breast/case (NormSc$_{Br}$/NormSc$_C$), computed as the minimum out of the scores for all regions in the breast/case, as a weighting parameter for breast/case suspiciousness. Since the normality score has negative correlation with the breast/case likelihood for cancer, we reverse its values by applying a simple linear transformation (NormSc$_{Br}^R$/NormSc$_C^R$). Thus, we define the case likelihood by:

$$Lik_{DIFF}^{MV}(Case) =$$
$$\left| P(RBr|V_R) \cdot \text{NormSc}_{RBr}^R - P(LBr|V_L) \cdot \text{NormSc}_{LBr}^R \right| \cdot \text{NormSc}_C^R,$$

where $|\cdot|$ denotes an absolute value.

The performance of our multi-view models is compared with that of the single-view CAD system (`SV-CAD`). For the latter, we compute the likelihood for MLO view, CC view and breast being cancerous by taking the minimum NormSc out of all the regions in MLO view, CC view and both views, respectively. Then, the case likelihood for cancer is computed following the two combining schemes discussed above:

$$Lik^{SV}_{MAX}(Case) = \max(\text{NormSc}^R_{RBr}, \text{NormSc}^R_{LBr}),$$

$$Lik^{SV}_{DIFF}(Case) = \left|\text{NormSc}^R_{RBr} - \text{NormSc}^R_{LBr}\right| \cdot \text{NormSc}^R_C.$$

The comparison analysis between the single-view and the multi-view systems is based on the Receiver Operating Characteristic (ROC) curve and the Area Under the Curve (AUC) as a performance measure ([12]). The significance of the differences obtained in the AUC measures is tested in LABROC4 ([13]).

4.3 Results

Table 1 presents the respective AUCs (with standard deviations) from our multi-view models and SV-CAD system as well as the one-sided p-values obtained from the tests on the differences with respect to the single-view system. The results indicate that our models tend to show better distinction between normal and cancerous breasts (cases) with respect to the SV-CAD. However, the difference in the AUC measures between MV-TPTP and SV-CAD is statistically insignificant at a breast level and a case level using the Case-MAX combining scheme. On the other hand MV-FPFP outperforms significantly the single-view CAD system in detecting cancerous breasts and cases. This confirms our expectation that allowing for more true links by MV-FPFP leads to better learning.

We also observe interesting outcome with respect to the two combining schemes for computing the case likelihood for cancer. While the results for both multi-views models support the hypothesis that Case-$DIFF$ can better distinguish between normal and cancerous cases, for the single-view CAD system this is not the case. This finding indicates that using multi-view information helps to improve the individual breast classification in comparison to the independent single-view approach.

Table 1. AUC (with std. dev.) obtained from the single- and multi-view systems at a breast and a case level

Method	Breast	p-value	Case–MAX	p-value	Case–$DIFF$	p-value
SV-CAD	0.850 (0.011)	–	0.797 (0.014)	–	0.697 (0.017)	–
MV-TPTP	0.863 (0.011)	0.168	0.801 (0.014)	0.097	0.812 (0.014)	0.000
MV-FPFP	0.875 (0.010)	0.001	0.832 (0.013)	0.014	0.840 (0.013)	0.000

To get more insight into the areas of improvement we plotted ROC curves for each of the three models with the best performance; see Figure 5. It is interesting to observe that both multi-view models have the same tendency to detect more cancers at (very) low FP rates (< 0.1) in comparison with the single-view CAD

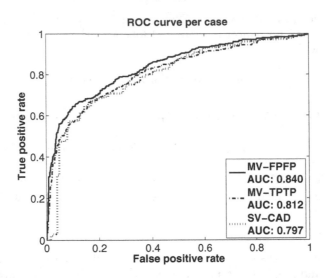

Fig. 5. ROC analysis per case

system–a result ultimately desired at the screening practice where the number of normal cases exceeds that of the cancerous ones.

Although the main goal of our multi-view systems is to distinguish between cancerous and normal cases, it is interesting to check to what extent MV-TPTP and MV-FPFP can detect the correct location of cancer on a breast. The intuition suggests that a cancerous lesion, which is related to a true link between both breast views, should have the highest probability within a case. Therefore, a straightforward performance measure is to count in how many cancerous cases the link with the maximum probability within a case corresponds to a true link. Given our link definitions in (1) and (2), we expect that MV-TPTP would find more TP-TP links with a maximum probability within the case, whereas for MV-FPFP this number would be lower since the latter does not distinguish between links related to lesions (TP-TP links) and those containing 1 FP region. Indeed the results confirm our hypothesis. For MV-TPTP in 247 cases out of all 383 (64.5%), the maximum link probability was obtained for a true (TP-TP or lesion) link. For MV-FPFP in 331 cases a link with at least one TP region obtained the maximum probability, i.e. 86.4% out of all cases. Out of these links 94 correspond to lesion links. However, in the screening practice the correct detection of the lesion in at least one of the views would suffice to refer the patient for a further examination.

5   Conclusions

In this paper we proposed a Bayesian network framework for multi-view breast cancer detection on mammograms. In comparison to a single-view CAD system we showed that the former performs better in terms of increased detection rate of breast cancer, especially for low false positive rates at a case level. This improvement can be explained by a number of factors. First, the Bayesian multi-view scheme is built upon the single-view CAD system that already achieves a relatively good cancer detection rate. However, the biggest advantage of our method is that it is largely based on domain knowledge. Following the mammography analysis, we consider multi-view dependencies between MLO and CC views to obtain a single measure for the breast and case being cancerous. This is done by: (1) defining links between the regions detected by the single-view CAD system in MLO and CC views (2) building a causal model where all detected regions with their feature vectors and the established region links are considered simultaneously, and (3) using the logical OR to compute the region and view probability for cancer. Furthermore, our multi-view scheme benefits from its Bayesian nature that allows to handle noisy or incomplete information such as the lack of detected (or not visible) lesions in one of the views. As a possible extension of the network structure, one can consider adding the so-called leak node as an intermediate variable to account for causes that are not directly observed (or measured) but influence the effect variable, e.g., the quality of the mammogram produced: high image quality increases the chance of detecting the cancer.

In our study we use two link class definitions—one following the lesion definition and the other accounting for the fact that a lesion might not be visible/detected in one of the views. The results from our experimental study showed that the second, less restrictive definition allows the model to better learn from data where information about tumor characteristics for the breast is present. An interesting direction for future research is to consider a link class definition with four values corresponding to TP-TP, TP-FP, FP-TP and FP-FP links. In this case, we expect to better capture and represent the real relationships between the detected regions.

Finally we note that the straightforward nature of the proposed methodology allows for easy extensions. For example, one can consider including correlation (view combining) features based on the output of the single-view CAD system or breast cancer characteristics obtained by a human expert such as risk factors. Thus, the developed Bayesian network framework can be also applied to any domain where the goal is computerized multi-view (object) detection.

Acknowledgments. This research has been funded by the Netherlands Organisation for Scientific Research under BRICKS/FOCUS grant number 642.066.605.

References

1. Good, W., Zheng, B., Chang, Y., Wang, X., Maitz, G., Gur, D.: Multi-image cad employing features derived from ipsilateral mammographic views. Proceedings of SPIE, Medical Imaging 3661, 474–485 (1999)
2. van Engeland, S., Timp, S., Karssemeijer, N.: Finding corresponding regions of interest in mediolateral oblique and craniocaudal mammographic views. Medical Physics 33(9), 3203–3212 (2006)

3. van Engeland, S., Karssemeijer, N.: Combining two mammographic projections in a computer aided mass detection method. Medical Physics 34(3), 898–905 (2007)
4. Paquerault, S., Petrick, N., Chan, H., Sahiner, B., Helvie, M.A.: Improvement of computerized mass detection on mammograms: Fusion of two-view information. Medical Physics 29(2), 238–247 (2002)
5. Velikova, M., de Carvalho Ferreira, N., Lucas, P.: Bayesian network decomposition for modeling breast cancer detection. In: Bellazzi, R., Abu-Hanna, A., Hunter, J. (eds.) AIME 2007. LNCS (LNAI), vol. 4594. Springer, Heidelberg (2007)
6. Jensen, F.V., Nielsen, T.D.: Bayesian Networks and Decision Graphs. Springer, Heidelberg (2007)
7. Heckerman, D., Breese, J.S.: Causal independence for probability assessment and inference using Bayesian networks. IEEE Transactions on Systems, Man and Cybernetics, Part A 26(6), 826–831 (1996)
8. Lucas, P.J.F.: Bayesian network modelling through qualitative pattern. Artificial Intelligence 163, 233–263 (2005)
9. Diez, F.: Parameter adjustment in Bayes networks: The generalized noisy or-gate. In: Proceedings of the Ninth Conference on UAI, San Francisco, CA (1993)
10. Murphy, K.: Bayesian Network Toolbox (BNT),
    http://www.cs.ubc.ca/~murphyk/Software/BNT/bnt.html
11. Dempster, A., Laird, N., Rubin, D.: Maximum likelihood from incomplete data via the EM algorithm. Journal of the Royal Stat. Soc., Series B 39(1), 1–38 (1977)
12. Hanley, J.A., McNeil, B.J.: The meaning and use of the area under a Receiver Operating Characteristic (ROC) curve. Radiology 143, 29–36 (1982)
13. Metz, C.E.: Some practical issues of experimental design and data analysis in radiological ROC studies. Investigative Radiology 24, 234–245 (1988)

Hybrid Wavelet-RBFNN Model for Monthly Anchovy Catches Forecasting

Nibaldo Rodriguez[1], Broderick Crawford[1], Carlos Castro[2,*],
and Eleuterio Yañez[1]

[1] Pontificia Universidad Católica de Valparaíso, Chile
FirstName.Name@ucv.cl
[2] Universidad Técnica Federico Santa María, Valparaíso, Chile
FirstName.Name@inf.utfsm.cl

Abstract. A hybrid method to forecast 1-month ahead monthly anchovy catches in the north area of Chile is proposed in this paper. This method combined two techniques, stationary wavelet transform (SWT) and radial basis function neural network (RBFNN). The observed monthly anchovy catches data is decomposed into two subseries using 1-level SWT and the appropriate subseries are used as inputs to the RBFNN to forecast original anchovy catches time series. The RBFNN architecture is composed of linear and nonlinear weights, which are estimates using the least square method and Levenberg-Marquardt algorithm; respectively. Wavelet-RBFNN based forecasting performance was evaluated by comparing it with classical RBFNN model. The benchmark results shown that a 99% of the explained variance was captured with a reduced parsimony and high speed convergence.

Keywords: wavelet transform, neural network, forecasting.

1 Introduction

The forecasting of monthly anchovy catches in the north area of Chile is a basic topic, because it play a central role in management of stock. In fisheries management policy the main goal is to establish the future catch per unit of effort (CPUE) values in a concrete area during a know period keeping the stock replacements. To achieve this aim lineal regression methodology has been successful in describing and forecasting the fishery dynamics of a wide variety of species [1,2]. However, this technique is inefficient for capturing both nonstationary and nonlinearities phenomena in anchovy catch forecasting time series. Recently there has been an increased interest in combining nonlinear techniques and wavelet theory to model complex relationship in nonstationary time series. Nonlinear model based on Neural networks have been used for forecasting model due to their ability to approximate a wide range of unknown nonlinear functions [3]. Gutierrez *et. al.* [4], propose a forecasting model of monthly anchovy catches

---

* The third author has been partially supported by the Chilean National Science Fund through the project FONDECYT 1070268.

D. Dochev, M. Pistore, and P. Traverso (Eds.): AIMSA 2008, LNAI 5253, pp. 345–352, 2008.
© Springer-Verlag Berlin Heidelberg 2008

based on a sigmoidal neural network, whose architecture is composed of an input layer of 6 nodes, two hidden layers having 15 nodes each layer, and a linear output layer of a single node. Some disadvantages of this architecture is its high parsimony as well as computational time cost during the estimation of linear and nonlinear weights. As shown in [4], when applying the Levenberg Marquardt (LM) algorithm [5], the forecasting model achieves a determination coefficient of 82%. A better result of the determination coefficient can be achieved if sigmoidal neural network is substituted by a reduced multivariate polynomial (MP) combined with wavelet transform based on translation-invariant wavelet transform. Coifman and Donoho [6] introduced translation-invariant wavelet denoising algorithm based on the idea of cycle spinning, which is equivalent to denoising using the discrete stationary wavelet transform (SWT) [7,8]. Beside, Coifman and Donoho showed that SWT denoising achieves better root mean squared error than traditional discrete wavelet transform denoising. Therefore, we employ the SWT for monthly anchovy catches data preprocessing.

In this paper, we propose a RBFNN combined with stationary wavelet transform for forecasting of monthly anchovy catches.The anchovy catches data considered is decomposed into two wavelet subseries by using SWT. Then, RBFNN model is constructed with appropriate wavelet subseries as inputs and original anchovy catches time series as putput. The RBFNN's architecture consists of two components [9]: a linear weights subset and a nonlinear hidden weights subset. Both components are estimated by using the separable nonlinear least squares (SNLS) minimization procedures [10].The SNLS scheme consists of two phases. In the first phase, the hidden weights are fixed and output weights are estimated with a linear least squares method. In a second phase, the output weights are fixed and the hidden weights are estimated using the LM algorithm [5]. For anchovy catches forecasting, advantages of the proposed model are reducing the parsimony, improvement of convergence speed,and increasing accuracy precision. On the other hand, Wavelet denoising algorithm employs the stationary wavelet transform with universal threshold rule.

The layout of this paper is as follows. In section 2, the forecasting scheme based on both stationary wavelet transform and RBFNN with hybrid algorithm for adjusting the linear and nonlinear weights are presented. The performance evaluation curves of the forecaster effect are discussed in Section 3. Finally, the conclusions are drawn in the last section.

2   Wavelet-Neuro Based Forecasting

This section presents the proposed forecasting model for monthly anchovy catches in north area of Chile, which is based on stationary wavelet transform and RBF neural network.

2.1   Stationary Wavelet Preprocessing

A signal can be represented at multiple resolutions by decomposing the signal on a family of wavelets and scaling functions [6,7,8]. The scaled (approximation) signals are computed by projecting the original signal on a set of orthogonal scaling functions of the form

$$\phi_{jk}(t) = \sqrt{2^{-j}}\phi(2^{-j}t - k) \tag{1}$$

or equivalently by filtering the signal using a low pass filter of length r, $h = [h_1, h_2, ..., h_r]$, derived from the scaling functions. On the other hand, the detail signals are computed by projecting the singal on a set of wavelet basis functions of the form

$$\psi_{jk}(t) = \sqrt{2^{-j}}\psi(2^{-j}t - k) \tag{2}$$

or equivalently by filtering the signal using a high pass filter of length r, $g = [g_1, g_2, ..., g_r]$, derived from the wavelet basis functions. Finally, repeating the decomposing process to any scale J, the original signal can be represented as the sum of all detail coefficients and the last approximation coefficient.

In time series analysis, discrete wavelet transform (DWT) often suffers from a lack of translation invariance. This problem can be tacked by mean of the stationary wavelet transform. The SWT is similar to the DWT in that the high-pass and low-pass filters are applied to the input signal at each level, but the output signal is never decimated. Instead, the filters are upsampled at each level.

Consider the following discrete signal $x(n)$ of length N where $N = 2^J$ for some integer J. At the first level of SWT, the input signal $x(n)$ is convolved with filter h to obtain the approximation coefficients $a(n)$ and with filter $g(n)$ to obtain the detail coefficients $d(n)$, so that

$$a(n) = \sum_k h(n-k)x(k) \tag{3a}$$

$$d(n) = \sum_k g(n-k)x(k) \tag{3b}$$

Because no subsampling is performed, $a(n)$ and $d(n)$ are of length N instead of $N/2$ as in the DWT case. The output of the SWT $s(n) = [a(n) \; d(n)]$ is then the approximation coefficients and the detail coefficients. In this paper, the $h(n)$ and $g(n)$ filters are based on Haar wavelet filter and are given as

$$h = \begin{bmatrix} \dfrac{1}{\sqrt{2}} & \dfrac{1}{\sqrt{2}} \end{bmatrix} \tag{4a}$$

$$g = \begin{bmatrix} \dfrac{-1}{\sqrt{2}} & \dfrac{1}{\sqrt{2}} \end{bmatrix} \tag{4b}$$

2.2   RBFNN Model

The forecasted signal $s(n)$ can be decomposed in a low frequency component and a high frequency component. The low frequency component is approximated using a autoregression model and the high frequency component is approximated using a RBFNN. That is,

$$\hat{y} = \sum_{j=1}^{N_h} b_j \phi_j(u_k, v_j) + \sum_{i=1}^{m} c_i u_i \tag{5}$$

where N_h is the number of hidden nodes, m is the number input nodes, u denotes the regression vector $u = (u_1, u_2, \ldots u_m)$ containing lagged m values, $w = [b_0, b_1, \ldots b_{N_h}, c_1, c_2, \ldots c_m]$ are the linear output parameters, $v = [v_1, v_2, \ldots v_{N_h}]$ are the nonlinear hidden parameters, and $\phi_j(\cdot)$ are hidden activation functions, which is derived as [9]

$$\phi_j(u_k) = \phi(\|u_k - v_j\|^2) \tag{6a}$$

$$\phi(\lambda) = (\lambda + 1)^{-1/2} \tag{6b}$$

In order to estimate the linear parameters $\{w_j\}$ and nonlinear parameters $\{v_j\}$ of the forecaster an hybrid training algorithm is proposed, which is based on least square (LS) method and Levenberg-Marquardt (LM) algorithms. The LS algorithm is used to estimate the parameters $\{w_j\}$ and the LM algorithm is used to adapts the nonlinear parameters $\{v_j\}$.

Now suppose a set of training input-output samples, denoted as $\{u_{i,k}, d_i, i = 1, \ldots, N_s, k = 1, \ldots, m\ )$. Then we can perform N_s equations of the form of (5) as follows

$$\hat{Y} = W\Phi \tag{7}$$

where the desired output d_i and input data (wavelet coefficients) u_i are obtained as

$$d_i = [x(t)] \tag{8a}$$

$$u_i = [s(t-1)s(t-2) \cdots s(t-m)] \tag{8b}$$

where $x(t)$ represent origin anchovy catches data. For any given representation of the nonlinear parameters $\{v_j\}$, the optimal values of the linear parameters $\{\hat{w}_j\}$ are obtained using the LS algorithm as follows

$$\hat{W} = \Phi^\dagger D \tag{9}$$

where $D = [d_1\ d_2\ \cdots\ d_{N_s}]$ is the desired output patter vector and $\Phi^\dagger$ is the Moore-Penrose generalized inverse [10] of the activation function output matrix Φ.

Once linear parameters $\hat{W}$ are obtained, the LM algorithm adapts the nonlinear parameters of the hidden activation functions minimizing mean square error, which is defined as

$$E(v) = \sum_{i=1}^{N_s} (d_i - y_i)^2 \tag{10a}$$

$$\hat{Y} = \hat{W}\Phi \tag{10b}$$

Finally, the LM algorithm adapts the parameter $v = [v_1 \cdots v_{Nh}]$ according to the following equations [5]

$$v = v + \Delta v \tag{11a}$$

$$\Delta v = (JJ^T + \alpha I)^{-1}J^T E \tag{11b}$$

where J represent Jacobian matrix of the error vector $e(v_i) = d_i - y_i$ evaluated in v_i, I is the identity matrix. The error vector $e(v_i)$ is the error of the RBFNN for i patter. The parameter μ is increased or decreased at each step.

3   Experiments and Results

The monthly anchovy catches data was conformed by historical data from January 1963 to December 2005, divided into two data subsets as shown in Fig.1. In the first subset, the data from January 1963 to December 1985 was chosen for the training phase (weights estimation), while the remaining was used for the validation phase.

In the training process the hidden weights v_i were initialized by a gaussian random process with a normal distribution $N(0, 1)$. The number of input and hidden nodes were set to 2 and the stopping criterion was a maximum number

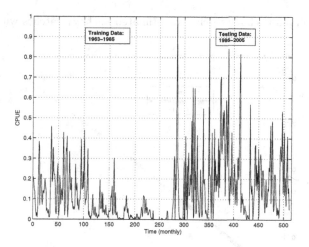

Fig. 1. Monthly anchovy catches data

Table 1. Statistic of the forecasting model with LS-LM algorithm

Model	Run	MSE	R-Squared
	1	6.74E-11	99.99
	2	9.40E-11	99.99
	3	5.01E-11	99.99
	4	7.06E-11	99.99
	5	7.00E-11	99.99
RBFNN(2,2,1)	6	4.71E-11	99.99
	7	5.42E-11	99.99
	8	6.62E-11	99.99
	9	6.80E-11	99.99
	10	7.06E-11	99.99
Std. Dev.		**1.33E-11**	**0.0**

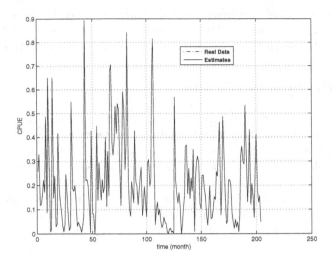

Fig. 2. Observed anchovy catches vs estimated anchovy catches with testing data

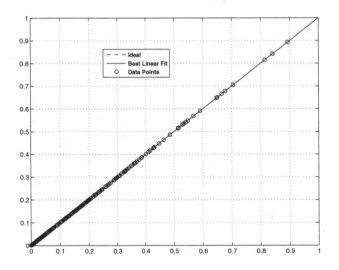

Fig. 3. Observed anchovy catches

iterations set to 2. Due to random initialization of the hidden nodes was used ten runs to calculate the standard deviation of the RBFNN. The RBFNN architecture according to mean square error (MSE) and determination coefficient R^2 are given in Table 1. From Tables 1, we conclude that RBFNN forecasting model based on hybrid algorithm finds a little MSE value using smaller number of parameters, so that seven weights.

Fig.2 describes the performance evaluation of the validation phase with testing data for the RBFNN forecasting model. This plot shows the original time series and the one predicted by proposed forecasting model. The difference is so tiny that it is impossible to tell one from another by eye inspection. That is why you probably see only the RBFNN(2,2,1) forecasting curve. From Fig.2 it can be observed that the forecasting model according to its parsimony and precision is the architecture composed by 2 input nodes $u = [a(n - 1), d(n - 1)]$, $v = 2$ hidden nodes and one output node.

Fig.3 illustrates the determination coefficient estimation of the validation phase between the observed and estimated anchovy catches data with the best forecasting model. Please note that the forecasting model shown that a 99.99% of the explained variance was captured by the RBFNN(2,2,1) model.

Also, monthly anchovy catches forecasting based on classical RBFNN with gaussian activation function has been evaluated for the purpose of comparison. The classical RBFNN architecture were CRBFNN(2,2,1), so that 1 input node, 2 hidden nodes and 1 output node. The CBFNNN training were stopped after 2 iterations using gradient descent algorithm. The RBFNN and CRBFNN are compared with respect to MSE and R^2 in Table 2. From Tables 2, can be seen that the CRBFNN achieved a poor forecasting performance.

Table 2. Statistic of each model in test phase

Model	Run	MSE	R-Squared
RBFNN(2,2,1)	1	4.71E-11	99.99
CRBFNN(2,2,1)	1	1.10E-02	66.70

4   Conclusions

In this paper, one-step-ahead forecasting of monthly anchovy catches based on stationary wavelet transform and radial basis function neural network with hybrid algorithm (LS-LM) was presented. The LS-LM based forecasting model can predict the future CPUE value based on previous values $a(t - 1), d(t - 1)$ and the results found shown that a 99.99% of the explained variance was captured with a reduced parsimony and high speed convergence.

References

1. Stergiou, K.I.: Prediction of the Mullidae fishery in the easterm Mediterranean 24 months in advance. Fish. Res. 9, 67–74 (1996)
2. Stergiou, K.I., Christou, E.D.: Modelling and forecasting annual fisheries catches: comparison of regression, univariate and multivariate time series methods. Fish. Res. 25, 105–138 (1996)
3. Hornik, K., Stinchcombe, M., White, H.: Multilayer feedforward networks are universal approximators. Neural Network 2(5), 359–366 (1989)
4. Gutierrez, J.C., Silva, C., Yañez, E., Rodriguez, N., Pulido, I.: Monthly catch forecasting of anchovy engraulis ringens in the north area of Chile: Nonlinear univariate approach. Fisheries Research 86, 188–200 (2007)
5. Hagan, M., Menhaj, M.: Training feed-forward networks with marquardt algorithm. IEEE Trans. Neural networks 5(6), 1134–1139 (1994)

6. Coifman, R.R., Donoho, D.L.: Translation-invariant denoising, Wavelets and Statistics. Springer Lecture Notes in Statistics (103), pp. 125–150 (1995)
7. Nason, G., Silverman, B.: The stationary wavelet transform and some statistical applications, Wavelets and Statistics. Springer Lecture Notes in Statistics (103), pp. 281–300 (1995)
8. Pesquet, J.-C., Krim, H., Carfantan, H.: Time-invariant orthonormal wavelet representations. IEEE Trans. on Signal Processing 44(8), 1964–1970 (1996)
9. Karayiannis, N.B.: Reformulated radial basis neural networks trained by gradient descent. IEEE Trans. Neural networks 10(3), 188–200 (1999)
10. Serre, D.: Matrices: Theory and applications. Springer, New York (2002)

Trilateration Analysis for Movement Planning in a Group of Mobile Robots

Svetoslav Genchev[1], Pencho Venkov[2], and Boris Vidolov[3]

[1] Technical University of Sofia, 8, Kliment Ohridski St.,
Sofia-1000, Bulgaria
genchevs@abv.bg
[2] Technical University of Sofia, 8, Kliment Ohridski St.,
Sofia-1000, Bulgaria
pven@tu-sofia.bg
[3] Université de Technologie de Compiègne, rue Personne de Roberval,
60200 Compiègne, France
bvidolov@hds.utc.fr

Abstract. This paper presents a real-time, low-computation trilateration model, designed for teams of cooperating mobile robots, working in unknown in-door environments without GPS signal accessibility. It allows to analytically build precision maps of the working area, giving information for the potential localization errors in each point of the space around the robots and thus simplifying movement planning and tracking. The technique relies on Time-Of-Flight measurements between several members of the team, used as static beacons, and the rest of the robots currently in motion. Simple parameter analysis is proposed for choosing the best suited robots to act as beacons for a given task. The performances of the mathematical and the statistical models have been validated numerically by computer simulations.

1 Introduction

The main assertion in mobile robotics, prior to making any decisions related to movement planning, is the good knowledge of the actual position of the robot itself. Thus the most frequent failures in mobile robot operation are due to incorrect position tracking, which could occur even in well-structured industrial environments [1], where the correct localization is crucial for the safety of the manipulated objects or working personnel. In practice, determining one's current position could be achieved by a set of measurements relative to its previous known position, or relative to other objects from the same environment, the locations of which are already known [2]. In each case, the required information is collected with appropriate interceptive or exteroceptive sensors, which firstly induce some error related to the sensor quality. Afterwards, additional error could be easily appended because of the mathematical model used to manipulate the measurements and obtain the required position estimate. Finally there is a whole universe of potential errors related to unpredictable phenomena in the working area. All these uncertainties could be greatly reduced when multiple sensors with different kind of operation are used in parallel and their multiple

D. Dochev, M. Pistore, and P. Traverso (Eds.): AIMSA 2008, LNAI 5253, pp. 353–364, 2008.
© Springer-Verlag Berlin Heidelberg 2008

outputs fused into a single one, obtaining unique and more precise data. The most common approach is to combine absolute and relative localization techniques via some accuracy based probabilistic algorithm like the Kalman filter [3], thus combining the speed and the court-term precision of the dead-reckoning methods with absolute, not time-depending information. Typical example is the combination of odometry or inertial navigation, with GPS [4][5]. However, the Global Positioning System is not appropriate for small-scale or in-door applications. In these cases, the navigation could be helped by artificial vision [19] or techniques like triangulation or trilateration, but all requiring special landmarks and therefore more or less structured and known environments. Another sensor-fusion approach, getting around this limitation, is the cooperative localization, recently emerged from the multi-robot applications. Here the sensing capabilities of a single robot are very restricted, but there are multiple robots working together in the same environment, and sharing sensing information [6][7]. A team of small robots, with limited capabilities, could perform easily tasks, like exploration and map building [8][9], otherwise very complex for a single, spatially limited robot, no matter how advanced are its sensors. The localization in the team of robots is achieved by measuring their relative position and/or orientation and communicating with each other. A lot of solutions exist in the literature, coping in different manners with the localization problem and using varied kind of sensors, but mostly designed for specific tasks [10][11][12][13]. Furthermore, most of the proposed techniques for statistical analysis are either too complex or not appropriate to facilitate movement planning. That's why our objective was firstly to build a methodology not depending on the topology of the working area, in order to be applied for almost any application, and secondly to establish faster ways for its error analysis, based rather on heuristic approaches. It uses inertial navigational system (INS) as a relative source of position information [14][24] for each robot and a trilateration procedure based on time-of-flight (TOF, TOA) measurements [15][16] between the robots in the team, in order to calculate a second position estimate for sensor fusion by Kalman filter. In this paper we will analyse our trilateration model and its application for aiding movement planning. While the triangulation uses angles to calculate the unknown coordinates, the trilateration requires distance measurements to previously placed beacons in the working area, which location is very accurate. In our case, the role of the beacons is given to some of the robots, thus creating a movable trilateration framework. Major problems would be the fluctuations in the wave speed during its travel, the multi-paths and the clock synchronization measuring the wave time of flight. Furthermore, the mobile beacon's accuracy is not as good as the accuracy of static beacons, and therefore our statistical model has been designed to deal with that feature as well. But, as shown in the experimental section, the importance of these factors diminishes when increasing the number of beacons used to perform the trilateration. Another common approach, where there is no synchronization between the wave emission and reception, is to measure the Time-Difference-of-Arrival (TDOA) [17][24] but it's highly non-linear technology which cannot be modelled with our mathematical model. In order to avoid additional errors, our non linear, multi-agent system for localisation has to be well modelled mathematically. Iterative algorithms [17][21] are often used in navigation, performing several iterations to linearly approximate the trilateration solution, starting from the last calculated point. Unfortunately, it could be time and computation costly as the number of beacons grows up. Moreover it's not certain that the system would converge towards the real solution in very noisy environments. In such cases, maximum likelihood estimators [16][22] are often applied for fitting the model into the measurements. That could be done by using their uncertainties into likelihood functions that are then optimized to retrieve the most accurate solution. But for our particular case where there are many noisy beacons, the solution of such tasks is too complex to be resolved analytically, and iterative algorithms are needed. In the other hand, in an unpredictable active environment, it's not suitable to rely on statistical knowledge for trilateration computing, because it could be totally wrong due to some external not-predicted perturbations, e.g. multi-paths. A more appropriate and faster fitting technique in this scenario would be the method of lest squares (LSM). It does not need any statistical information (in the case when no preliminary classification is performed), but if enough beacons are in use, after applying LSM the wrong sensor data for one of them would not influence the final result, making it more robust. Finally, the obtained solution that best satisfies the measurements has to be analysed statistically before fused with other position information (i.e. that provided periodically by the INS). The main task of the error analysis is to find an analytical law that best describes the output error, given knowledge of the input one. Advanced statistical analyses are found in [15][17][18][20][22].

In the next chapter a linear model for solving trilateration with large number of wrongly positioned beacons is presented. Three issues are faced – the initial system non-linearity, the beacon redundancy and the model error analysis, and more specifically the linear error transfer law, which allows us to easily describe the statistical properties of our results. The next section continues this thematic by studying alternative ways to estimate the trilateration precision a priori for each point of the space, thus giving precious information for movement planning in real-time. After that, in the last one, we have performed several experimentations by simulation in order to validate certain theoretical assertions in our work.

2 Multi-Beacon Trilateration and Uncertainty Estimation

Let's consider the case where we want to find a point R with coordinates x, y and z, given the knowledge of the distances l_i to n known positions R_i with coordinates x_i, y_i, z_i, in the same three-dimensional Cartesian coordinate system. If the radius vectors from the reference frame's origin to R and R_i are denoted respectively r and r_i, then the mathematical relation between each distance, each radius vector and the unknown coordinates may be expressed by the quadratic equation (1):

$$\left(x - x_i\right)^2 + \left(y - y_i\right)^2 + \left(z - z_i\right)^2 = l_i^2; \quad i = 1, 2, \dots, n \tag{1}$$

$$\Rightarrow xx_i + yy_i + zz_i = \frac{r^2 + r_i^2 - l_i^2}{2} \tag{2}$$

When having three beacons for resolving a two-dimensional position and four, when in tree-dimensions, the solution of the obtained set of non-linear equations (2) is trivial. Therefore let's proceed to a specific case, where there are many more possible beacons to use, with coordinates that are not precisely defined, but have some statistically-known uncertainties. By substitution and subtraction of the equations, we get rid of the square of r, thus obtaining linear relationships for x, y and z of the form:

$$x\left(x_i - x_j\right) + y\left(y_i - y_j\right) + z\left(z_i - z_j\right) = \frac{1}{2}\left(r_i^2 - l_i^2 - r_j^2 + l_j^2\right) \tag{3}$$

This kind of linearization is not precision costly, but at the expense of a part of the redundant input information instead, i.e. the subtracting equation. In our case it works very well and the obtained system of linear equations does not present any abnormalities. However, this is not a straightforward methodology, and cannot be applied successfully for any similar redundant non-linear problem, because of the potential singularities. For example, if TDOA trilateration technique is used instead of TOF, it would result in abnormal behaviour for any symmetric beacon configuration. In the general case, where all beacons have wrong information about their locations, we have to build a symmetric model that gives no priority to the precision of any of them. By choosing i and j to be consecutive indexes and using matrix formulation for (3), we obtain the following linear matrix equation:

$$\underbrace{\begin{bmatrix} x_{12} & y_{12} & z_{12} \\ x_{23} & y_{23} & z_{23} \\ \cdots & \cdots & \cdots \\ x_{n-1n} & y_{n-1n} & z_{n-1n} \\ x_{n1} & y_{n1} & z_{n1} \end{bmatrix}}_{A_{n\times3}} \begin{bmatrix} x \\ y \\ z \end{bmatrix}_{R_{3\times1}} = \frac{1}{2} \underbrace{\begin{bmatrix} r_1^2 - l_1^2 - r_2^2 + l_2^2 \\ r_2^2 - l_2^2 - r_3^2 + l_3^2 \\ \cdots \\ r_{n-1}^2 - l_{n-1}^2 - r_n^2 + l_n^2 \\ r_n^2 - l_n^2 - r_1^2 + l_1^2 \end{bmatrix}}_{B_{n\times1}} \tag{4}$$

$$A^* = \left(A^T A\right)^{-1} A^T \tag{5}$$

$$\Rightarrow R = A^* B \tag{6}$$

where A^* is the pseudo-inverse of A. Because of the redundant input information, the matrix A in (4) is not square and in order to find a solution for R we are going to use pseudo-inversion, which is equivalent to the Least Square Method (LSM). We do not agree that (5) is computation costly procedure, as implied in some references [20]. It involves two matrix multiplications and one inversion of a 3x3 matrix. There are no square roots or any other non-linear function. It should be noted also, that here the matrix A^* depends only on the beacon configuration and therefore has to be calculated only when the beacons change their positions. The final expression (6) gives us a solution for R, in function of the actual beacon's coordinates and the measured distances. But just obtaining a solution is not enough to perform data fusion with other sources of information. We need to know its statistical properties first, e.g. what is the probability distribution that best characterizes its uncertainty and what are the corresponding parameters. There are two main sources of errors. First, the beacons R_i are not exactly where they are supposed to be, and second, the range measurements l_i in any TOF trilateration system could be very wrong being based on wave traveling times measurements. Finally, a bad geometrical configuration of beacons (e.g. all of them are laying on a line or are too close to each other) could further amplify these errors. Besides the main geometrical factor, the configuration quality could be improved by using more beacons than the minimum needed to perform the trilateration (i.e. three for a two-dimensional problem and four in three dimensions). Even erroneous, the usage of more of them could diminish the final error to a desired threshold value. Also, as already mentioned, any no-predicted phenomena in the information of a single beacon would not influence the final result so much. Now, supposing that the input errors have been estimated, we have to determine how our model projects them to the output. It has been found that the distribution of the input variables is mostly transferred to the output, i.e. the character of the trilateration error is very similar to the beacon's position and range measurement errors, suggesting a relationship close to linear. For example if the input errors are supposed to be distributed normally, with zero mean, we shall expect such a distribution to the output. Considering the above premises, our analysis is based on a first-order error propagation law of the type:

$$P_{OUT} = J P_{IN} J^T = \sum_{i=1}^{n} J_i P_i J_i^T \tag{7}$$

where P_{IN} and P_{OUT} are the full input and output covariance matrices and J denotes the full Jacobian matrix, relating the output variations with the variations of each input parameter calculated for the analytic solution (6), which is assumed to be the expected value of the real solution [23]. Considering the fact, that the beacons are not correlated statistically with each other, allows us to separate J and P_{OUT} thus obtaining a sum of little, easier to calculate sub-products where P_i is the covariance matrix corresponding to a single beacon, given the assumption that the distance measurements are independent statistically and the errors in the three dimensions are possibly correlated, and J_i is the corresponding Jacobian matrix calculated at the point of interest R. After differentiation of (6) we obtain solutions for J_i, using information from the already calculated pseudo-inverse A^*:

$$dR = A^* \left(dB - dAR \right) \tag{8}$$

$$J_i = \left[\frac{\partial R}{\partial x_i} \quad \frac{\partial R}{\partial y_i} \quad \frac{\partial R}{\partial z_i} \quad \frac{\partial R}{\partial l_i} \right] = \left(C_{i-1} - C_i \right) L_i \tag{9}$$

$$L_i = \left[x - x_i \quad y - y_i \quad z - z_i \quad l_i \right]; \quad C_i = \left[a_{1i}^* \quad a_{2i}^* \quad a_{3i}^* \right]^T \tag{10}$$

where C_i is the i-th column in the matrix A^*, and L_i is a quantity containing information about both the measured distance l_i and its calculated value. Once we have a solution for R, then both sets of vectors C and L (10) are already known and it is a matter of multiplications to calculate the Jacobian matrices, thus estimating the output covariance matrix after applying (7).

However, the presented methodology has a remarkable drawback – it does not consider the precision of each beacon when computing its information. It is a fast and efficient solution when the beacons have similar uncertainties, but otherwise, i.e. there are some beacons introducing far greater errors, the system would not ignore them and they will dominate the final result, which is clearly seen from (7). A way around this problem may be the preliminary classification of the beacons for elimination of such outliners.

3 Error Analysis for Movement Planning

The information obtained from the error transfer law proposed in the previous chapter is required for sensor fusion. However, we still need simpler and faster analysis in order to make decisions related to robot guidance and trajectory building in the space easier. Moreover, except their position accuracy, we don't have any quantitative evaluation about which robots to use as beacons. Let us define an error transfer factor as a quantitative characteristic how each beacon uncertainty influences the resulting error. As such a measure, we are using the *Frobenius norm* of the associated Jacobian matrix, which leads to a square root of the sum of the squares of its elements, or a square root of the sum of the diagonal elements of the product JJ^T. Finally a substitution from (9) yields:

$$\|J_i\| = \sqrt{trace\left(J_i J_i^T\right)} = \sqrt{\sum_k \sum_l j_{i\,kl}^2} = \sqrt{2} l_i \|C_{i-1} - C_i\| \tag{11}$$

These quantities explain how much the computed solution R depends on the associated beacon's information – if one factor is bigger than another, it means that all changes in the corresponding beacon's position or distance measurements, i.e. all possible errors, would be reflected more strongly on R. Now given the fact that C, being a column of A^*, contains only information about the beacon locations, it turns out that there is a linear relationship between this factor value and the distance l_i, which means that the importance of the beacon increases as R is moving away from it. This fact allows us to isolate l_i and to define a new, purely geometrical influence factor, associated only with the importance of the beacon's location in the current configuration:

$$m_i = \frac{1}{2}\left(\frac{\|J_i\|}{l_i}\right)^2 = \frac{1}{2l_i^2}\sum_k \sum_l j_{i\,kl}^2 = \|C_{i-1} - C_i\|^2 \tag{12}$$

From now on we will refer to this quantity as beacon geometrical mass. The mass analysis allows us to estimate the importance of each beacon in the current geometrical configuration, without calculating any Jacobian matrices. The mass is a function of the actual geometrical arrangement only, and shall be constant until the beacons relocate. In case that arrangement is good (bigger covered area), the common mass of all beacons would be smaller. It would be smaller also if more beacons are used to perform the trilateration. The more important property of the mass values is that they are used to analytically find the point in the space with minimum trilateration error, i.e. the best place for a robot to be localized:

$$R_{\min} = \left(\sum_{i=1}^n m_i P_i\right)^{-1} \sum_{i=1}^n m_i P_i R_i \tag{13}$$

As error estimate we are using the trace of the output covariance matrix, also referred as *full variance* in the remaining of this presentation. It has only one minimum in (13), the knowledge of which is of practical interest, but is not enough to describe the quality of a given beacon configuration. We have found by experimentation, that the full variance is governed in the space by a very simple geometrical law:

$$\sigma^2 = \sigma_{min}^2 + \left(R - R_{min} \right)^T U \left(R - R_{min} \right) \tag{14}$$

where U is a *symmetric, positive-definite* matrix, depending on the geometrical and statistical properties of the beacons. The elements in its main diagonal (*slope* parameters) dictate how fast the error grows in the three dimensions as R is moving away from R_{min}. The remaining correlation parameters modify that behavior in different directions. If the variances and correlations are all equal, the slope parameters in the main diagonal of U are also equal. If there are no correlations, the other correlation parameters are null. The main advantage from (14) is the possibility to calculate an estimate for the localization precision in any point of the space around the robots just by computing a simple quadratic equation. It allows to determinate areas with variance below a given threshold value and to choose appropriate beacons for navigating a given trajectory. Equation (14) is somewhat similar to a Mahalanobis distance between the points R and R_{min}, with U being the inverse of a covariance matrix expressing their dispersion in all directions. Presently we don't have a real-time procedure to determine the contents of U, that's why we are deriving them indirectly from several variance values calculated in different points in the space, i.e. at least three points for the two-dimensional case, and six for the tree dimensional. For that purpose we are using the beacon positions R_i, assuming that their number satisfy this condition.

4 Simulation Results

In order to visualize better our results, without loosing objectivity, we have chosen to present two-dimensional problems here. Let's consider an area with given dimensions, containing a configuration of a randomly chosen number of beacons, with random positions and with random uncertainties governed by a normal law. The exact experimental setup is shown on Table 1:

Table 1. Experimental Setup

Area dimensions	100x100
Beacon position range	(0,100)
Position variance range (for each dimension)	(0,2]
Correlation range (between dimensions)	[-0.3,0.3]
Distance variance range	(0,2]
Position and distance error means	0

There aren't any specified metric units because it's the relative size of the quantities that matters. The setup information concerning each beacon has been randomly generated in the given ranges. In the first experiment we are analyzing the movement of a robot when following the same predefined trajectory for different number of beacons.

Let's begin with the minimum case of three available beacons, shown on Fig. 1. The chosen trajectory is visible on Fig. 1a, while Fig. 1b presents the corresponding precision map (darker zones mean lower variance levels) build with relation (14) for all points in the space enclosing the beacons. The two crossing lines mark the error minimum (13) while the doted line passing trough the minimum is a transverse section of the map shown on Fig. 1c:

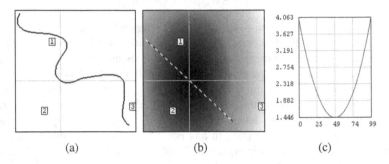

(a)   (b)   (c)

Fig. 1. a) Configuration of three beacons and trajectory. b) Corresponding precision map for navigation (darker zones mean lower variance levels). Crossed lines mark variance minimum. Doted line marks a transversal section; c) Transversal section passing through the minimum.

In the next two scenarios (Fig 2a, b), we are adding each time three more noisy beacons. The term *noisy* here means that the beacon positions and the distance measurements between the beacons and the robot are contaminated with statistically known noise. In the last fourth test (Fig 2c), we are leaving active only those beacons which are lying in a line.

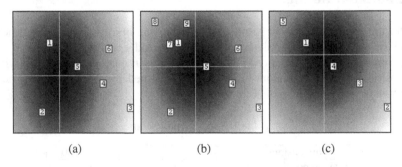

(a)   (b)   (c)

Fig. 2. a) Configuration increased by three more noisy beacons; b) Three more beacons appended; c) Configuration reduced to a line formation

The contents of the U matrix, the error minimum location R_{min}, its variance (σ^2_{min}) and the common mass, obtained for each of the four scenarios, are presented in Table 2. As we can see from the second and third configurations, the usage of more beacons, even noisy, may improve the overall quality of the configuration – the error slope, error minimum and common mass are diminishing. Still, their improper arrangement in the fourth case yields extremely high variance levels and therefore

Table 2. Error Geometrics

Config.	I	II	III	IV
Beacons	3	6	9	5 (line)
R_{min}	(37.845, 58.436)	(38.746, 53.335)	(44.555, 45.842)	(45.953, 36.177)
σ^2_{min}	1.446	1.115	0.934	320.96
U_{11}	1.747071e-3	1.528731e-3	7.069159e-4	0.601296
U_{22}	5.839934e-4	4.918705e-4	4.052344e-4	0.345402
U_{12}, U_{21}	2.839096e-5	1.612481e-5	1.824650e-5	1.887162e-2
Mass	9.236e-4	7.541e-4	5.432e-4	0.235

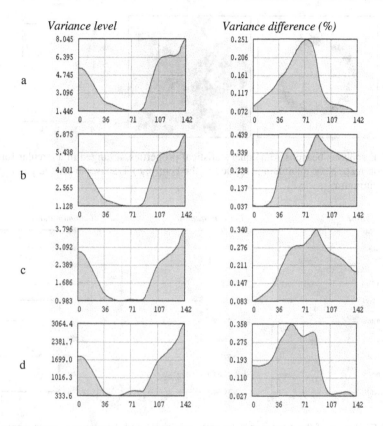

Fig. 3. Analytically estimated variance levels and variance differences between the analytic and the numeric estimators (measured in percents) for the trajectory in each configuration: a) 3 beacons; b) 6 beacons; c) 9 beacons; d) 5 beacons lying in a line

totally wrong trilateration results. This fact is reflected to the common beacon mass which is about thousand times bigger than its previous values. Now let's observe the evolution of the variance estimated with relation (14) and the difference between its value and the numerically estimated one (in percents) for each of the four configurations while following the trajectory shown on Fig. 1a. The population of errors for the numerical analysis is build by adding normally distributed, zero mean white noise (generated using the Box-Muller method) to the beacon position and distance measurements, and then calculating already known coordinates of points laying on the experimental trajectory. It has to be noted that since the numerical estimator is working with samples of considerable, but still finite number of erroneous trajectories, its output is slightly variable between experiments. However, in all of the considered cases, the observed differences between the two estimators do not exceed 0.5 percents even for the worst, fourth configuration, where the variance levels are very high (Fig. 3d). This fact affirms that the variance of our solution (6) may be very well modeled by relation (14) which is far more simple and easy to compute than (7).

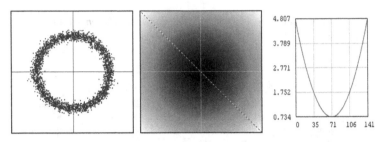

Fig. 4. a) Erroneous beacons with equal statistical properties, arranged in a circular formation; b) Precision map; error minimum is located in the centre; c) Transverse section passing trough the error minimum;

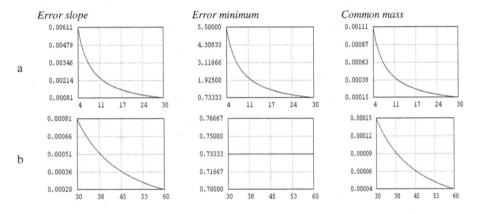

Fig. 5. a) Evolution of the error slope, the error minimum and the common mass when increasing the number of beacons from 4 to 30, with radius 30 units; b) The same quantities captured when increasing the radius from 30 to 60 units, with 30 active beacons

In the second experiment we are trying to demonstrate how the trilateration precision is increasing when using more beacons or increasing the area covered by their geometrical configuration. For that purpose, in this test we need beacons with absolutely equal properties, i.e. they will have equal uncertainties, without any correlations, and will be positioned in a circle (Fig. 4a), therefore expecting equal slope parameters (U_{11}, U_{22}), null correlation parameters (U_{12}, U_{21}), equal masses and an error minimum located in the centre (Fig. 4b). The evolutions of the error slope, error minimum and the common beacon mass, when increasing the number of beacons in the formation from 4 to 30 with circle radius of 30 units are displayed on Fig 5a, while Fig. 5b shows the same quantities for increasing the radius from 30 to 60 units with 30 beacons. In both cases we obtain a global precision improvement (low errors in bigger area), although the error minimum is decreasing only in the first case. It turns out that the mass parameter is closely related to the error slope and do not influence the value of the error minimum (but do define its position, according to relation (13) in the previous chapter).

5 Conclusion

This paper presents a fast and simple real-time algorithm for time-of-flight trilateration, designed for mobile robots navigated in-door. It is based entirely on linear methodologies, requires small computation resources which allows the implementation in low cost robot platforms. It does not require static beacons in the working area because it uses some of the robots in the team as temporal beacons. The possibility of quick precision inspection of the area around the robots may help movement planning and trajectory building. Several simulation results have been presented to validate its theoretical assertions in different beacon configurations. Future work will include constructing the entire hybrid system combined with inertial navigation for further increasing the movement precision. The combination between an inertial navigation system and the proposed trilateration technique would render the robots independent of its environment and thus very versatile.

References

1. Dhillon, B.S.: Robot Reliability and Safety. Springer, New York (1991)
2. Borenstein, J., Everett, H.R., Feng, L., Wehe, D.: Mobile Robot Positioning: Sensors and Techniques. Invited paper for the Journal of Robotic Systems, Special Issue on Mobile Robots 14(4), 231–249
3. Brown, R.G., Hwang, P.Y.C.: Introduction to Random Signals And Applied Kalman Filtering, 2nd edn. John Wiley & Sons, Chichester (1992)
4. Kim, J., Jee, G., Lee, J.: A complete GPS/INS integration technique using GPS carrier phase measurements. In: IEEE Position Location and Navigation Symposium 1998, April 20-23, 1998, pp. 526–533 (1998)
5. Ohno, K., Tsubouchi, T., Shigematsu, B., Yuta, S.: Differential GPS and odometry-based outdoor navigation of a mobile robot. Advanced Robotics 18(6), 611–635
6. Parker, L.E.: Current state of the art in distributed autonomous mobile robotics. In: Parker, L.E., Bekey, G., Barhen, J. (eds.) Distributed Autonomous Robotic Systems, vol. 4, pp. 3–12. Springer, Tokio (2000)
7. Cao, Y.U., Fukunaga, A.S., Kahng, A.B.: Cooperative Mobile Robotics: Antecedents and Directions. Autonomous Robots 4, 1–23 (1997)
8. Hayes, A.T., Martinoli, A., Goodman, R.M.: Distributed Odor Source Localisation, Special Issue on Artificial Olfaction. In: Nagle, H.T., Gardner, J.W., Persaud, K. (eds.) IEEE Sensors Journal, vol. 2 (3), pp. 260–271 (2002)
9. Rekleitis, I.M., Dudek, G., Milios, E.E.: Multi-robot collaboration for robust exploration. In: Proceedings of the 2000 IEEE International Conference on Robotics and Automation, ICRA 2000, vol. 4, pp. 3164–3169 (2000)

10. Spletzer, J., Das, A.K., Fierro, R., Taylor, C.J., Kumar, V., Ostrowski, J.P.: Cooperative Localization and Control for Multi-Robot Manipulation. In: Proceedings of the 2001 IEEE/RSJ International Conference on Intelligent Robots and Systems 2001, vol. 2, pp. 631–636 (2001)
11. Tinós, R.: Navarro-Serment and Christiaan J.J. Paredis, Fault Tolerant Localization for Teams of Distributed Robots. In: Proceedings of the 2001 IEEE/RSJ International Conference on Intelligent Robots and Systems, Maui, Hawaii, USA, October 29 - November 03 (2001)
12. Bisson, J., Michaud, F., Letourneau, D.: Relative positioning of mobile robots using ultra-sounds. In: Proceedings on 2003 IEEE/RSJ International Conference on Intelligent Robots and Systems, 2003 (IROS 2003), October 27-31, 2003, vol. 2, pp. 1783–1788 (2003)
13. Roumeliotis, S.I., Rekleitis, I.: Analysis of multirobot localization uncertainty propagation. In: Proc. IEEE Int. Workshop on Intelligent Robots and Systems, Las Vegas, NV, USA, pp. 1763–1770 (2003)
14. Oliver, J.: Woodman, An introduction to inertial navigation, Technical report UCAM-CL-TR-696 (ISSN 1476-2986), University of Cambridge, Computer Laboratory (August 2007)
15. Manolakis, D.E.: Efficient solution and performance analysis of 3-D position estimation by trilateration. IEEE Transactions on Aerospace and Electronic Systems 32(4), 1239–1248 (1996)
16. Grabowski, R., Khosla, P.: Localization techniques for a team of small robots. In: Proceedings of the 2001 IEEE/RSJ International Conference on Intelligent Robots and Systems, vol. 2, pp. 1067–1072 (2001)
17. Manolakis, D.E., Cox, M.E.: Effect in range difference position estimation due to stations position errors. IEEE Transactions on Aerospace and Electronic Systems 34(1), 329–334 (1998)
18. Thomas, F., Ros, L.: Revisiting Trilateration for Robot Localization. IEEE Transactions on Robotics 21(1), 93–101 (2005)
19. Stella, E., Lovergine, F.P., Caponetti, L., Distante, A.: Mobile robot navigation using vision and odometry. In: Proceedings of the Intelligent Vehicles 1994 Symposium, 24-26 October 1994, pp. 417–422 (1994)
20. Cheng, X., Thaeler, A., Xue, G., Chen, D.: TPS: A time-based positioning scheme for outdoor wireless sensor networks. In: INFOCOM 2004. Twenty-third Annual Joint Conference of the IEEE Computer and Communications Societies, 7-11 March 2004, vol. 4, pp. 2685–2696 (2004)
21. Foy, W.H.: Position-Location Solutions by Taylor-Series Estimation. IEEE Transactions on Aerospace and Electronic Systems 12(2), 187–194 (1976)
22. Pages-Zamora, A., Vidal, J.: Evaluation of the improvement in the position estimate accuracy of UMTS mobiles with hybrid positioning techniques. In: 55th IEEE Vehicular Technology Conference 2002, vol. 4, pp. 1631–1635 (2002)
23. Arras, K.O.: An introduction to error propagation: derivation, meaning and examples, Technical Report N° EPFL-ASL-TR-98-01 R3 of the Autonomous Systems Lab, Institute of Robotic Systems, Swiss Federal Institute of Technology Lausanne (EPFL) (September 1998)
24. Genchev, S., Venkov, P.: Hybrid System for Localizion of Robots Moving in Group. In: Proceedings of the International Conference on Automatics and Informatics 2007, VI International Symposium Robotics Systems, Sofia, Bulgaria, 3-6 October 2007, vol. 2, pp. 653–657 (2007) ISSN: 1313 – 1850

Dynamic and Neuro-Dynamic Optimization
of a Fed-Batch Fermentation Process

Tatiana Ilkova and Mitko Petrov

Centre of Biomedical Engineering - Bulgarian Academy of Sciences
Sofia 1113, 105 Acad. George Bonchev Str., Bulgaria
{Tanja,Mpetrov}@clbme.bas.bg

Abstract. A fed-batch fermentation process is examined in this paper for experimental and further dynamic optimization. The optimization of the initial process conditions is developed for to be found out the optimal initial concentrations of the basic biochemical variables – biomass, substrate and feed substrate concentration. For this aim, the method of dynamic programming is used. After that, these initial values are used for the dynamic optimization carried out by neuro-dynamic programming. The general advantage of this method is that the number of the iterations in the cost approximation part is decreased.

Keywords: Dynamic programming, Neuro-dynamic programming, Neural network.

1 Introduction

This work focuses on optimization and optimal control of a laboratory *E. coli* fed-batch fermentation process.

One approach for solving problems of optimization is Dynamic programming (DP), which is successfully applied to fermentation processes. However, the approach is largely considered impractical as due to the analytical solution of resulting dynamic program is seldom possible and the numerical solution suffers from the "curse of dimensionality" [1].

Neuro-dynamic programming (NDP) is a relatively new class of DP methods for control and further decision making under uncertainties [1]. These methods have the potential to deal with problems that for a long time were thought to be intractable due to either a large state space or the lack of an accurate model. The name NDP expresses the reliance of the methods on both DP and neural networks (NN) concepts. In this case, in the artificial intelligence community, from where the methods originated the name reinforcement learning is also used [2].

The aim of this paper is optimization of the initial conditions of the basic kinetic variables of a fed-batch fermentation and further future dynamic optimization.

2 Model of the Process

The model of the process (*E. coli* fermentation) is expressed by the following biochemical variables: biomass concentration, substrate concentration, oxygen concentration in liquid phase, and oxygen concentration in gas phase [3]:

D. Dochev, M. Pistore, and P. Traverso (Eds.): AIMSA 2008, LNAI 5253, pp. 365–369, 2008.

$$\frac{dX}{dt} = \mu(S, C_L)X - \frac{F}{V}X \tag{1}$$

$$\frac{dS}{dt} = \frac{F}{V}(S_0 - S) - K_1\eta(\mu)X \tag{2}$$

$$\frac{dC_G}{dt} = K_G a(C_L - C_G) - \frac{F}{V}C_G \tag{3}$$

$$\frac{dC_L}{dt} = K_L a(C_G - C_L) - K_2\mu(S, C_L)X - \frac{F}{V}C_L \tag{4}$$

$$\frac{dV}{dt} = F \tag{5}$$

where: $\mu(S, C_L) = \mu_{max}\dfrac{S^2}{(K_s + S^2)}\dfrac{C_L^2}{(K_c + C_L^2)}$, $\eta = \dfrac{\mu(S, C_L)}{Y} + K_m$;

X – biomass concentration, g/l; S – substrate concentration, g/l; S_0 – feed substrate concentration, g/l; C_L – oxygen concentration in liquid phase, %; C_G – oxygen concentration in gas phase, %; $K_G a$ – oxygen mass-transfer coefficient in gas phase, h^{-1}; $K_L a$ – oxygen mass-transfer coefficient in liquid phase, h^{-1}; F – feeding rate, l/h; μ – specific growth rate, h^{-1}; η – specific substrate utilization rate; h^{-1}; $g.g^{-1}$; V – bioreactor volume, l; Y, K_s, K_c, K_m, K_1 and K_2 – yield coefficients.

The statistical investigations with comparison between the experimental and the model data show high degree of adequacy of the model. This model will be used for static optimization and further dynamic optimization of the process aiming maximum of biomass concentration at the end of the bioprocess.

3   Optimization and Optimal Control of the of Process

3.1   Optimization of the Initial Conditions

The optimization problem can be formulated in the following way: to be found such initial values of the biochemical variables X(0), S(0) and S_0 that maximizing the biomass quantity at the end of the process:

$$\max_u J = \int_{t_0}^{t_f} X(t)V(t)dt \tag{6}$$

where: t_0 – initial time, t_f – final time of the fermentation.

The vector of control variables has a type: $u=u[u_1, u_2, u_3]$, where: $u_1=X(0)$ g/l, $u_2=S(0)$ g/l and $u_3=S_0$ g/l.

The intervals of variation of variables, based on a few years' experiences it was found that the best reasonable limits of changing of the initial conditions are as follows [3]: $(0.08 \leq u_1 \leq 5.00)$ g/l; $(2.6 \leq u_2 \leq 3.8)$ g/l; $(85.0 \leq u_3 \leq 135.0)$ g/l.

The choice of the ranges of the control variables was accepted depending on the bioreactor structure.

To solve this problem DP method was applied. It was established that this method has been successfully applied for finding of initial conditions for desired maximum. An algorithm is synthesized and a program is developed and the following initial conditions of the biochemical variables have been found:

$u_1=0.135$ g/l; $u_2=3.1$ g/l; $u_3=112$ g/l.

Using these initial values, the biomass quantity increases with 5.12% in comparison with the original data. Fig. 1 shows the results before and after static optimization for the biomass concentration.

3.2 Optimal Control

For determination of the optimal control problem of fed-batch fermentation processes maximizing of the optimization criterion at the end of the process max J (in this case biomass) on the used substrate S is accepted. Thus, the optimization problem is reduced to find a profile of the control variable that maximizes the criterion [3]:

$$\max_F J = \int_{t_0}^{t_f} X(t)\, V(t)\, dt \qquad (7)$$

The feeding rate F is accepted as a control variable.

The optimization problem will be solved by NDP. It aims to approximate the optimal cost-to-go function by the cost of some reasonably good suboptimal policy, called base policy. Depending on the context, the cost of the base policy may be calculated either analytically, or more commonly by simulation [1].

With this method, the model of the process (1) –(5) and the vector of the control variable are examined as developing in time processes and they are analyzed of consecutive stages. Admissible values for this bioprocess for the control variable are taken in the interval $0 \leq F \leq 1$, with discrete step $\Delta F=0.01$. In the fed-batch fermentation process the control variable (F) is limited by [3]:

$$\sum_{i=1}^{N} F_i = F \qquad (8)$$

then the total quantity feeding substrate is:

$$V_i = \sum_{i=1}^{N} F_i h_i \qquad (9)$$

where $h_i=t_i-t_{i-1}$ - is the step of discretization.

The control objective is, therefore, to drive the reactor from the low biomass steady state to the desirable high biomass yield state. It may be considered as a step change in the set point at time t=0 from the low biomass to the high biomass yield steady state.

4 NDP Algorithm

The following steps describe the general procedure of the developed method of NDP algorithm.

The following steps describe the general procedure of the developed method of NDP algorithm:

1. Starting with a given policy (some rule for choosing a decision u at each possible state i), and approximately evaluate the cost of that policy (as a function of the current state) by least-squares-fitting a scoring function $\tilde{J}^j(X)$ to the results of many simulated system trajectories using that policy [1];
2. The solution of one-stage-ahead cost plus cost-to-go problem, results in improvements of the cost values [1];
3. The resulting deviation from optimality depends on a variety of factors, principal among which is the ability of the architecture $\tilde{J}^j(X)$ to approximate accurately the cost functions of various policies;
4. Cost-to-go function is calculated using the simulation data for each state visited during the simulation, as for each closed loop simulation (simulation part).
5. A new policy is then defined by minimization of Bellman's equation, where the optimal cost is replaced by the calculated scoring function, and the process repeats. This type of algorithm typically generates a sequence of policies that eventually oscillates in a neighbourhood of an optimal policy;
6. Fit a neural network function approximator to the data to approximate cost-to-go function as a smooth function of the states;
7. The improved costs are again fitted to a neural network, as described above, to obtain subsequent iterations $\tilde{J}^1(X)$, $\tilde{J}^2(X)$, and so on ..., until convergence.

5 Results

A functional approximation relating cost function with augmented state was obtained by using neural network with eight hidden neurons, nine input and one output neurons. NN is trained with "back-propagation" method.

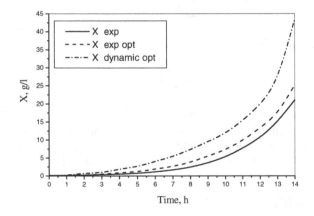

Fig. 1. Optimization results

Cost is said to be "converged" if the sum of the absolute error was less than 10% of the maximum cost. The general advantage of this method is that the number of iterations in the cost approximation part is decreased. In this case the number of iterations is diminished from 1560 to 1340 for one step. In addition, the CPU time needed for NDP is decreased in comparison with the classical DP (without the use of neural approximator). When NDP is used the optimization time is 720 seconds and when DP is used the optimization time is 6126 seconds. The results show generally an increase of 22.37 % amount of biomass production after the static and dynamic optimization. These results are shown in Fig. 1.

6 Conclusion

A static optimization of an *E. coli* fed-batch fermentation process is performed in order to be found the optimal initial values of the basic biochemical variables. After that these initial values are used for the dynamic optimization. For this aim a new method – NDP is developed and is applied for dynamic optimization of the process. With this method, the number of iterations in the cost approximation part are decreased using of the simulation part. In this way number of iterations is diminished. The results show that NDP approach is particularly simple to implement and is applicable for on-line implementation.

References

1. Bertsekas, D., Tsitsiklis, J.: Neuro-Dynamic Programming. Athena Scientific, Belmont (1996)
2. Driessens, K., Dzeroski, S.: Integrating Guidance intro Relational Reinforcement Learning. Mach. Learn. 57, 217–304 (2004)
3. Petrov, M.M., Ilkova, T.S.: Modeling and Fuzzy Optimization of Fed-Batch Fermentation Process. CABEQ 16, 173–178 (2002)

Classification Model for Estimation of Risk Factors in Cardiological Diagnostics

Plamena Andreeva

Knowledge Based Systems Department at the Institute of Control and System Research, Bulgarian Academy of Sciences, Acad. G. Bonchev Str, Bl.2, P.O.Box 79, 1113 Sofia, Bulgaria
plamena@icsr.bas.bg

Abstract. The proposed classification model for risk factor estimation makes semi-automatic data analysis based on advanced machine-learning methods. The objective is to provide intelligent computer-based support for medical diagnostics. The developed fuzzy boundary classification determines risk factors importance and adjusts the threshold. Experimental results are presented.

Keywords: classification model, fuzzy boundaries, data mining, risk estimation, diagnostics.

1 Introduction

Intelligent systems based on data mining make easier the acquisition of knowledge [3], and the diagnosis based on it. Various systems of multiple classifier ensembles have been proposed in the literature [6]. The base classifiers influence the generated decision boundaries. In this paper is consi-dered the semi-automatic data mining analysis. A fuzzy decision boundary classification model is developed for determining the risk factor importance. Our model uses fuzzy membership weights to adjust the importance of the classifier to predict the class, based on previous training error. The objective is to provide intelligent computer-based support for medical specialists.

A graded diagnosis allows the treatment of patients to be adapted to the individual needs. The principle of fuzzy control will be the key to the success of many future fuzzy medical systems [2]. Recently, a great deal of work has been done in cardiology [5]. Fuzzy diagnosis is the original application domain of fuzzy set theory in medicine, where vagueness, linguistic uncertainty, measurement imprecision, and subjectivity are present in medical diagnosis. The treatment of non-statistical uncertainty is still a challenge. One way to use fuzzy sets is to express degree of abnormality of a certain ECG. This is considered here as the basis for the conducted research.

The acute coronary syndromes comprise a variety of clinical scenarios ranging from unstable angina pectoris to non-ST-segment elevation myocar-dial infarction, leading to the direct cause of more than 4,3 million deaths in Europe each year, and over 2 million people in the EU (43% of all deaths) (www.ehnheart.org).

2 Classification with Fuzzy Boundary

The problem of medical diagnosis can be formalized as follows. Let $\mathbf{C}= \{C_1, C_2, \ldots ,C_M\}$ is a set of M diagnoses possible in the context of a certain medical problem. C

D. Dochev, M. Pistore, and P. Traverso (Eds.): AIMSA 2008, LNAI 5253, pp. 370–374, 2008.
© Springer-Verlag Berlin Heidelberg 2008

can be: a list of disorders, types of blood cells, etc. We call **C** a set of class labels. Let **x** is the description of an object (e.g. a patient) in the form of an n-dimensional real vector $\mathbf{x} = [x_1,..x_n]^T \in R^n$. The components of **x** encode the features (clinical measurements and findings; risk factors - smoking, blood pressure, etc.). So if $x \in R^n$ is a feature vector and $\{1, 2,.. c\}$ is the label set of c classes, then a classifier is every mapping $\mathbf{D}: R^n \rightarrow [0,1]^c - \{\mathbf{0}\}$, where $\mathbf{0} = [0,0,...,0]^T$ is the origin of R^C. The classifier output is a "class label", or decision to which class the particular data **x** belongs. The output of **D** is denoted as: $\mu_D(x) = [\mu_D^1(x),..., \mu_D^c(x)]^T$, where μ is the membership degree for class C_i, given **x**. The class is chosen by the maximum rule (1).

$$\mu_D(\mathbf{x}) = \max_{i=1,...c}\{\mu_D^i(\mathbf{x})\} \tag{1}$$

For every object $x \in R^n$ the classifier specifies a single class label C_i (diag-nosis). Fuzzy diagnosis is characterized by the fact that the classifier relies on fuzzy sets. The concept of *fuzzy boundary* classification adopted here is used for calculating the objective function for separating the classes.

The rationale of this model leads to a fuzzy rule-based system with a fuzzy output **D(x)**. The membership degrees determine the action of the system (alerts and alarms), and the thresholds can be tuned by the user.

2.1 Classification Model

Due to the complexity of the medical classification problems, it is impossible to build a perfect classifier, but realistic target is to build an effective decision support system. The weight for the measured risk factors is based on confusion matrix - of size c. Once the matrices have been obtained, we select the pair $\{i, j\}$ with maximum value according to the following expression:

$$\{i, j\} = \arg\max_{i \neq j}(C_{train}(i, j) + C_{train}^T(i, j) + C_{test}(i, j) + C_{test}^T(i, j)\} \tag{2}$$

for all (i,j) in $[1,..C]$, where C^T is transposed matrix. Given C classes and a coding matrix, for the ith iteration, first the pair of classes i, j (with highest error) is selected, then the weight based on (1) is updated. The weight is: $w=0,5.log(1-e)/e$ where e_i is the error of the particular class. So we focus on the difficult partitions, trying to increase the distance between the classes.

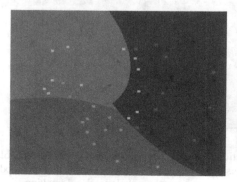

Fig. 1. The classification fuzzy boundary for two classes with tolerance ε

The curves plot on fig. 1 shows the separation boundary for the points predicted within the tolerance ε. In the experiments these curves provide more compelling presentation of the results than alternatives such as tables of MSE.

3 Related Work

The implemented Support Vector Machines algorithm [7] assign relative weights to various datasets. Kernel choice is of greatest influence on the clas-sifier being most relevant to the input distribution. In Sniderman and Furberg [8] a crucial distinction is made about the modifiable causal factors that are used to predict risk. Other than age and gender, classic modifiable causative factors for cardiovascular disease seem to affect the individual risk. For this reason we propose our classification model to assist the decision for the boun-daries for age and abnormality. Results show that the decision based on fuzzy boundary classification with fuzzy measures of similarity is superior to the ordinary schemes. The proposed new bounding technique provides transpa-rent understanding of parameter estimation as *min-max* problem. The bounding technique guaranties the convergence to a locally optimal solution [7].

4 Case Study and Experimental Results

The performance of the classification model on a heart dataset is investigated empirically. For the purposes of this study the (www.cs.waikato.ac.nz/ml/weka) package WEKA is used. The goal is to predict the presence or absence of heart disease in a predefined risk group given the results of various medical tests. The dataset data from a publicly available cardiology database (UCI Repository) contains 13 attributes and two classes: presence and absence. For each risk factor, the corresponding accuracy on the training set was recorded. Then all classifiers with accuracy of 95% or higher on the training set were tested on the 1/3 of 270 independent testing samples. The

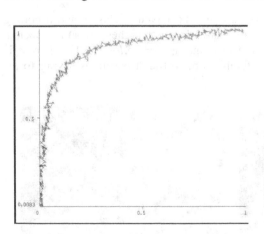
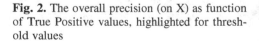
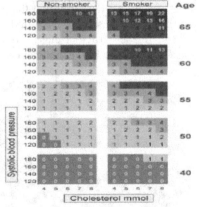

Fig. 2. The overall precision (on X) as function of True Positive values, highlighted for threshold values

Fig. 3. The risk (as numbers in percent) of cardiovascular diseases depending on blood pressure, age, cholesterol and smoking

final error estimate for the model built on the 2/3 dataset is calculated with fuzzy boundary for the classification function. This research uses a recently proposed framework [1], to analyze the relationship between structural dimensions and classifiers. After the test evaluation, training risk factors are reweighed according to the total cost assigned to each class.

The obtained averaged result for correctly classified data is 87.67%. All cases for which the classifier output exceeds the threshold are labeled as positive and the remaining objects are labeled as negative. On fig. 2 is shown the true positive rate (sensitivity) versus false positive rate (1-specificity) for the proposed classifier. The medical user is then able to decide on a compromise between sensitivity and specificity achievable simultaneously by the classifier. When adding fuzzy threshold the accuracy increases. The correct interpretation depends on the understanding of the risk factor relationship. A good visualization tool is of useful assistance for the doctor. Visual results are shown on fig. 3.

The key idea in our approach is not primarily to define more complex functions, but to deal with more complex output spaces by extracting structural information. The differences in the clinical manifestation of cardiovascular diseases and the response to therapy depend also on age and gender [4]. The classification of risk factors can help to delineate the strategies for changing the misperceptions of cardiovascular diseases. Risk factors of particularly importance in older age are modeled differently from younger age, e.g. when combined with smoking and bad cholesterol. The presence of every risk factor shifts the age boundary to a higher risk level.

5 Conclusion and Discussion

This paper considers the use of classification models for estimation of risk factor importance by applying fuzzy boundaries. We tested different machine learning models where the formed boundary is sensitive to the given preference of the user. The use of the overall classification accuracy as evaluation metric is adequate for constant class distribution among examples. The basic conclusion from our experiments is that the proposed model is superior to other model techniques, needing many parameters and leading to over-training. The used classifier with fuzzy decision boundary appears to be simple and intuitive, and do not require heavy calculations. The understanding of the differences of risk factor influences is of crucial importance for the improvement of the diagnostic management of cardiovascular diseases.

Acknowledgments. Many thanks to the referees for their valuable com-ments. This work is partially supported by NSRF *Project MI - № 1509/2005*.

References

1. Andreeva, P.: Data Modelling and Specific Rule Generation via Data Mining Techniques. Journal of E-learning and Knowledge Society (in press, 2008), http://www.je-lks.it
2. Bartlett, P., Sh-Taylor, J.: Generalization performance of support vector machines and other pattern classifiers. In: Schölkopf, B., Burges, C.J.C., Smola, A.J. (eds.) Advances in Kernel Methods, pp. 43–54. MIT Press, MA (2002)

3. Brito, P., Bertarand, P., Cucumel, G., De Carvalho, F.: Studies in classification, data analysis and knowledge organization, Selected Contributions in Data Analysis and Classification. Springer, Heidelberg (2007)
4. Bezdek, J.C., Keller, J.M., Krishnapuram, R., Pal, N.R.: Fuzzy Models and Algorithms for Pattern Recognition and Image Processing. Kluwer, Dordrecht (1999)
5. European Society of Cardiology (2008), http://www.escardio.org/knowledge/guidelines/
6. Kuncheva, L.: Combining Pattern Classifiers: Methods and Algorithms. Wiley, Chichester (2004)
7. Schölkopf, B., Smola, A.J.: Learning with Kernels: Support Vector Machines, Regularization, Optimization and Beyond. MIT Press, Cambridge (2002)
8. Hernàndez, A., Gil, D.: A Deterministic-Statistical Strategy for Adventitia Segmentation in IVUS images, CVC Tech. Rep. 89 (2005), http://cvc.uab.es/aura/
9. Gil, D., Radeva, P.: Extending anisotropic operators to recover smooth shapes. Comput. Vis. Imag. Under. 99, 110–125 (2005)

A Hybrid Approach to Distributed Constraint Satisfaction

David Lee, Inés Arana, Hatem Ahriz, and Kit-Ying Hui

School of Computing,
The Robert Gordon University,
Aberdeen, United Kingdom
{dl,ia,ha,khui}@comp.rgu.ac.uk

Abstract. We present a hybrid approach to Distributed Constraint Satisfaction which combines incomplete, fast, penalty-based local search with complete, slower systematic search. Thus, we propose the hybrid algorithm PenDHyb where the distributed local search algorithm DisPeL is run for a very small amount of time in order to learn about the difficult areas of the problem from the penalty counts imposed during its problem-solving. This knowledge is then used to guide the systematic search algorithm SynCBJ. Extensive empirical results in several problem classes indicate that PenDHyb is effective for large problems.

Keywords: Constraint Satisfaction, Distributed AI, Hybrid Systems.

1   Introduction

Constraint satisfaction is an increasingly important paradigm for the resolution of combinatorial problems. A Constraint Satisfaction Problem (CSP) [1] is a triple (X, D, C) where $X = \{x_1, ..., x_n\}$ is a set of variables, $D = \{D_1, ..., D_n\}$ is a set of domains and C is a set of constraints which restrict the values that variables can take simultaneously. A solution to a CSP is an assignment of values to variables which satisfies all constraints.

A Distributed Constraint Satisfaction Problem (DisCSP) [2], is a CSP where the problem (variables, domains and constraints) is distributed amongst a number of agents each of which has only a partial view of the problem due to privacy issues, communication costs or security concerns. Thus, in order to solve the problem, agents must communicate and cooperate whilst disclosing as little information as possible. Assumptions commonly shared by researchers in the field [2] are: (i) message delays are finite and for any pair of agents, messages are received in the order that they are sent; (ii) each agent is responsible for one variable.

Existing solution methods for DisCSPs can be classified as systematic search and local search methods. In systematic algorithms, ordered agents sequentially instantiate their variable, backtracking to the previous agent if no consistent value is found. Zivan and Meisels [3] devised SynCBJ, a distributed version of conflict-directed backjumping [4] combined with dynamic backtracking [5]. Systematic search algorithms are complete[1].

---

[1] They either find a solution to a problem or detect that the problem is unsolvable.

D. Dochev, M. Pistore, and P. Traverso (Eds.): AIMSA 2008, LNAI 5253, pp. 375–379, 2008.

Consequently, distributed local search algorithms have been devised which, for large problems, converge quicker to a solution but are generally incomplete. Distributed local search approaches iteratively improve an initial set of values until a set of values which is a solution is found. However, it may find one set of non-optimal values (local optima) always appears more promising than moving to other combinations of values (the neighbourhood) and get stuck. Therefore, local search approaches rely heavily on strategies for escaping local optima, e.g. weights on constraints [2] or penalties on values [6]. For large problems, they are faster than the systematic approaches.

There are only two distributed hybrid approaches which combine both types of search to produce hybrid algorithms which are 'fast' and complete. DisBOBT [7] uses Distributed Breakout [2] as its main problem-solver and, if Distributed Breakout fails to solve the problem, its weight information orders the agents for an SBT [3] search. LSDPOP [8] is an optimisation algorithm running the systematic algorithm DPOP [9], until the maximum inference limit is exceeded when local search guided by DPOP is run to find the best solution to the problem.

We introduce a hybrid approach, PenDHyb, combining penalty-based local search, DisPeL [6], with systematic search, SynCBJ, for distributed constraint satisfaction.

The remainder of this paper is structured as follows. Our hybrid approach, PenDHyb, is explained in Section 2; Section 3 presents empirical results on both solvable and unsolvable problems. Finally, Section 4 concludes the paper.

2   PenDHyb: Penalty-Based Distributed Hybrid Algorithm

We propose a new approach, PenDHyb, for Distributed Constraint Satisfaction which combines penalty-driven local search (DisPeL) with systematic search (SynCBJ) in order to speed-up the latter. In the former type of search, when a quasi-local optimum is reached, penalties are imposed on 'current' variable values causing constraint violations. Penalties therefore indicate values that, though looking promising, fail to lead to a solution. The higher the penalties accumulated by a value, the less desirable it becomes. The penalty and value information learnt during penalty-driven problem solving can be used to guide systematic search as follows:

– **Difficult variables.** Penalties on values are used to learn which variables are difficult to assign during problem solving. A variable which has many heavily penalised values is seen as more troublesome than a variable whose values have few or no penalties. Variables are ordered according to their degree and difficulty (penalties) and this order is used to drive the systematic search algorithm.
– **Best variable values.** The best solution found (the one with the least constraint violations) in the penalty-based algorithm, is used for value ordering in the systematic search.

DisPeL [6] is an iterative improvement algorithm where agents take turns to improve a random initialisation in a fixed order. In order to resolve deadlocks (quasi-local-optima where an agent's *view* remains unchanged for 2 iterations), DisPeL applies penalties to variable values which are used in a 2-phased strategy as follows: (i) First the values are penalised with a *temporary penalty* in order to encourage agents to assign other values and; (ii) If the temporary penalties fail to resolve a deadlock *incremental penalties* are imposed on the culprit values. In the more efficient Stoch-DisPeL [6], agents decide randomly to either impose a temporary penalty or to increase the incremental penalty. In the remainder of this paper we refer to Stoch-DisPeL as DisPeL.

SynCBJ is a synchronous systematic search algorithm where each agent keeps track of the reasons why values have been eliminated from their variable's domain. When a bactrack step is required, the agent is able to determine the variable responsible for the conflict and backjumps to the agent holding that variable. This increases the performance of the algorithm very substantially when compared to SBT.

In order to learn penalty information we modified DisPeL by adding:

- A *penalty counter* pc_i for each variable v_i. pc_i is incremented whenever a penalty is imposed on any of v_i's values. Unlike penalties on values, penalty counters are never reset and, therefore, highlight repeated penalisation of variables, i.e. *troublesome* variables.
- A *best value* bv_i store for each variable v_i which keeps the value participating in the best solution found by DisPeL so far [2].

Our hybrid algorithm, PenDHyb, combines DisPeL and SynCBJ. After agent initialisation, a standard DisPeL search runs (as described for Stoch-DisPeL in [6]) but only for a very small number of cycles e.g. 21 cycles for randomly generated problems with 40 variables. We ran a series of experiments and determined that a very small number which steadily increases as the number of variables increase was optimal. If the problem is solved, the solution is returned. Otherwise, variables are arranged, in descending order, according to their degree and penalty count before SynCBJ is run. In addition to the variable ordering information, SynCBJ makes use of value ordering information as follows: for each variable v_i, the first value to be tried is the best value bv_i found by DisPeL, i.e. the one participating in the best instantiation found.

PenDHyb is complete since either DisPeL reports a solution within the small number of cycles (typically DisPeL solves 5% of problems) or SynCBJ runs. Since SynCBJ is complete, completeness of PenDHyb is guaranteed. PenDHyb is sound since both DisPeL and SynCBJ are sound [6,3].

3   Empirical Evaluation

Our SynCBJ implementation was verified with the distributed randomly generated problems described in [3] with the results at least as good as those reported by SynCBJ's authors. We use SynCBJ with max-degree variable ordering which obtains substantially better results than lexicographic ordering.

We evaluated PenDHyb on solvable and unsolvable distributed randomly generated problems measuring the two established costs for DisCSPs: (i) the number of messages sent; (ii) the number of constraint checks performed. Although CPU time is not an established measure for DisCSPs [10], we also measured it and the results obtained were consistent with the other measures used.

---

[2] Note that determining the best solution does not incur any additional messages.

Table 1. Performance of SynCBJ and PenDHyb on randomly generated problems

N. messages	solvable problems				unsolvable problems			
n	30	40	50	60	30	40	50	60
SynCBJ	2301	22590	262178	1897645	5307	55692	557359	3069301
PenDHyb	2115	18566	100417	486301	5546	51916	451218	2686295
N. cnstr. checks	solvable problems				unsolvable problems			
n	30	40	50	60	30	40	50	60
SynCBJ	11489	119209	1344941	10421510	24790	285591	2924331	17153383
PenDHyb	12041	116573	714961	2975784	27913	284847	2424052	15077229

Table 2. SynCBJ and PenDHyb on graph colouring problems for $degree = 5$

N. messages	solvable problems				unsolvable problems			
n	125	150	175	200	125	150	175	200
SynCBJ	18781	75778	191988	722256	127054	660334	1957622	6793331
PenDHyb	18577	60005	161213	463601	113590	557434	1849564	5357801
N. cnstr. checks	solvable problems				unsolvable problems			
n	125	150	175	200	125	150	175	200
SynCBJ	46234	178942	477713	1750199	309383	1587410	4518670	15694031
PenDHyb	52534	162748	416520	463601	281142	1327274	4498886	12527968

We evaluated PenDHyb against SynCBJ on a wide variety of randomly generated problems ($n \in \{30, 40, 50, 60\}$, $d \in \{8, 9, 10, 11, 12\}$, $p1 \in \{0.1, 0.15, 0.2, 0.25, 0.3\}$ and $p2 \in \{0.1, 0.15, ..., 0.95\}$) where n is the number of variables, d is the domain size, $p1$ is the constraint density and $p2$ is the constraint tightness. We present the results at the phase transition point which represents hard problems for SynCBJ. Other tightness values showed similar performance for both algorithms. The results, shown in Table 1 for problems with ($n \in \{30, 40, 50, 60\}$, $d = 10$, $p1 = 0.15$ and $p2 = 0.6(30), 0.5(40), 0.45(50), 0.4(60)$), are median values over 100 problems. For solvable problems, PenDHyb is significantly more efficient than SynCBJ with performance difference increasing with the number of variables. For unsolvable problems SynCBJ is marginally better on problems with 30 variables but PenDHyb is substantially better on problems with 40 or more variables.

We also evaluated the performance of PenDHyb against SynCBJ on distributed graph colouring problems ($nodes \in \{125, 150, 175, 200\}$, domain size $d = 3$ and degree $k \in \{4.6, ..., 5.3\}$). These problems are of similar size to the ones used for the experiment on randomly generated problems above. Median values over 100 solvable problems and 100 unsolvable problems are shown in Table 2 for problems with a degree of 5. The results showed that PenDHyb is significantly more efficient for both solvable and unsolvable problems. This efficiency becomes more profound as the number of nodes increase and thereby mirrors the performance of PenDHyb with randomly generated problems namely that PenDHyb is substantially more efficient on medium to large-sized problems. Experiments for other degrees (not shown here) gave similar results, i.e. PenDHyb performed better, especially for graphs with a large number of nodes.

4 Conclusions

We have presented, PenDHyb, a hybrid approach to Distributed Constraint Satisfaction using penalty-based local search algorithm, DisPeL, to learn about the problem with the knowledge gained guiding a systematic search algorithm, SynCBJ.

We also evaluated other methods of exploiting the knowledge gained from running DisPeL to provide variable and value ordering for SynCBJ. We found that the best performing method was the one used in PenDHyb, where variables are sorted using max-degree and penalties and values are prioritised using sticking values.

We have shown that PenDHyb's performance is significantly better than systematic search for large, difficult problems in two problem classes, randomly-generated problems and graph colouring problems.

Our future work with PenDHyb will investigate the effectiveness of our approach on non-binary problems and the approach's applicability to coarse-grained problems, where agents are responsible for more than one variable.

References

1. Rossi, F., van Beek, P., Walsh, T.: Handbook of Constraint Programming. Elsevier, Amsterdam (2006)
2. Yokoo, M., Hirayama, K.: Algorithms for Distributed Constraint Satisfaction: A Review. Autonomous Agents and Multi-Agent Systems 3(2), 185–207 (2000)
3. Zivan, R., Meisels, A.: Synchronous vs asynchronous search on DisCSPs. In: Proceedings of the First European Workshop on Multi-Agent Systems (EUMA), Oxford (December 2003)
4. Prosser, P.: Hybrid Algorithms for the Constraint Satisfaction Problem. Computational Intelligence 9(3), 268–299 (1993)
5. Ginsberg, M.L.: Dynamic backtracking. Journal of Artificial Intelligence Research 1, 25–46 (1993)
6. Basharu, M., Arana, I., Ahriz, H.: Stoch-DisPeL: Exploiting randomisation in DisPeL. In: Proceedings of 7th International Workshop on Distributed Constraint Reasoning, Hakodate, Japan (May 2006)
7. Eisenberg, C.: Distributed Constraint Satisfaction for Coordinating and Integrating a Large-Scale Heterogeneous Enterprise. PhD thesis, Ecole Polytechnique Federale De Lausanne (2003)
8. Petcu, A., Faltings, B.: A hybrid of inference and local search for distributed combinatorial optimization. In: Proceedings of 2007 IEEE/WIC/ACM International Conference on Intelligent Agent Technology, pp. 342–348. IEEE Computer Society, Los Alamitos (2007)
9. Petcu, A., Faltings, B.: A scalable method for multiagent constraint optimization. In: Proceedings of the 19th International Joint Conference on Artificial Intelligence (IJCAI 2005), Edinburgh, Scotland (August 2005)
10. Meisels, A., Kaplansky, E., Razgon, I., Zivan, R.: Comparing performance of distributed constraints processing algorithms. In: Proceedings of the AAMAS-2002 Workshop on Distributed Constraint Reasoning, Bologna, July 2002, pp. 86–93 (2002)

Interleaved Alldifferent Constraints: CSP vs. SAT Approaches

Frédéric Lardeux[3], Eric Monfroy[1,2], and Frédéric Saubion[3]

[1] Universidad Técnica Federico Santa María, Valparaíso, Chile
[2] LINA, Université de Nantes, France
[3] LERIA, Université d'Angers, France

Abstract. In this paper, we want to handle multiple interleaved Alldiff constraints from two points of view: a uniform propagation framework with some CSP reduction rules and a SAT encoding of these rules that preserves the reduction properties of CSP.

1 Introduction

When modeling combinatorial problems, a suitable solution consists in expressing the problem as a constraint satisfaction problem (CSP), which is classically expressed by a set of decision variables whose values belong to finite integer domains. Constraints are used to model the relationships that exist between these variables. Another possibility is to encode the problem as a Boolean satisfiability problem (SAT), where a set of Boolean variables must satisfy a propositional formula.

The two paradigms share some common principles [2] and here, we focus on complete methods that aim at exploring a tree by enumerating variables and reducing the search space using propagation techniques.

On the CSP side, the identification of global constraints that arise in several real-world problems and the development of very specialized and efficient algorithms have considerably improve the resolution performances. The first example was certainly the Alldiff constraint [4] expressing that a set of n variables have all different values. Furthermore, one may notice that handling sets of distinct variables is often a more general problem and that, in some cases, such Alldiff constraints could be interleaved, leading to a high computational complexity. Many Alldiff constraints overlap in various problems such as Latin squares and Sudoku games. On the SAT side, no such high level modeling feature is offered to the user, who has to translate its problem into propositional logic. Systematic basic transformations from CSP to SAT have been proposed [3,5] to ensure some consistency properties to the Boolean encodings.

Here, we want to provide, on the CSP solving side, a uniform propagation framework to handle Alldiff constraints and, in particular, interleaved Alldiff. We also want to generalize possible encodings of Alldiff and multiple Alldiff constraints in SAT (i.e., by a set of CNF formulas).

D. Dochev, M. Pistore, and P. Traverso (Eds.): AIMSA 2008, LNAI 5253, pp. 380–384, 2008.

2   Encoding CSP vs. SAT

A CSP (X, C, D) is defined by a set of variables $X = \{x_1, \cdots, x_n\}$ taking their values in their respective domains $D = \{D_1, \cdots, D_n\}$. A constraint $c \in C$ is a relation $c \subseteq D_1 \times \cdots \times D_n$. A tuple $d \in D_1 \times \cdots \times D_n$ is a solution if and only if $\forall c \in C, d \in c$.

Usual resolution processes [1,2] consists in enumerating the possible values of a given variable in order to progressively build a variables assignment and reach a solution. Reduction techniques are added at each node to reduce the search tree (local consistency mechanisms) by removing values of variables that cannot satisfy the constraints. We recall a basic consistency notion (the seminal arc consistency is the binary subcase of this definition).

Definition 1 (Generalized Arc Consistency (GAC)). *A constraint[1] c on variables $(x_1, \cdots, x_m)$ is generalized arc-consistent iff $\forall k \in 1..m, \forall d \in D_k$, $\exists (d_1, \cdots, d_{k-1}, d_{k+1}, \cdots d_m) \in D_i \times \cdots \times D_{k-1} \times D_{k+1} \times \cdots \times D_m$, s.t. $(d_1, \cdots, d_m) \in c$.*

Domain Reduction Rules

Inspired by [1], we abstract constraint propagation as a transition process over CSPs:

$$\frac{(X, C, D)|\Sigma}{(X, C, D')|\Sigma'}$$

where $D' \subseteq D$ and Σ and Σ' are first order formulas (i.e., conditions of the application of the rules) such that $\Sigma \wedge \Sigma'$ is consistent. We canonically generalize $\subseteq$ to sets of domains as $D' \subseteq D$ iff $\forall x \in X$ $D'_x \subseteq D_x$. Given a set of variables V, we also denote D_V the union $\bigcup_{x \in V} D_x$. $\#D$ is the set cardinality.

Given a CSP (X^k, C^k, D^k), a transition can be performed to get a reduced CSP $(X^{k+1}, C^{k+1}, D^{k+1})$ if there is an instance of a rule (i.e., a renaming without variables' conflicts):

$$\frac{(X^k, C^k, D^k)|\Sigma^k}{(X^{k+1}, C^{k+1}, D^{k+1})|\Sigma^{k+1}}$$

such that $D^k \models \bigwedge_{x \in X} x \in D_x^k \wedge \Sigma^k$, and D^{k+1} is the greatest subset of D^k such that $D^{k+1} \models \bigwedge_{x \in X} x \in D_x^{k+1} \wedge \Sigma^{k+1}$.

In the conclusion of a rule (in Σ), we use the following notations: $d \notin D_x$ means that d can be removed from the domain of the variable x (without loss of solution); similarly, $d \notin D_V$ means that d can be removed from each domain variables of V; and $d_1, d_2 \notin D_x$ (resp. D_V) is a shortcut for $d_1 \notin D_x \wedge d_2 \notin D_x$ (resp. $d_1 \notin D_V \wedge d_2 \notin D_V$).

Since we only consider here rules that does not affect constraints and variables, the sets of variables will be omitted and we highlight the constraints that are required to apply the rules by restricting our notation to $< C, D >$. We will say that $< C, D >$ is GAC if C is GAC w.r.t. D. The transition relation using a

---

[1] This definition is classically extended to a set of constraints.

rule R is denoted $< C, D > \rightarrow_R < C, D' >$. $\rightarrow_R *$ denotes the reflexive transitive closure of $\rightarrow_R$. It is clear that $\rightarrow_R$ terminates due to the decreasing criterion on domains in the definition of the rules (see [1]). This notion can be obviously extended to sets of rules $\mathcal{R}$. Note also that the result of $\rightarrow_{\mathcal{R}} *$ is independent from the order of application of the rules [1]: from a practical point of view, it is thus generally faster to first sequence less complicated rules (or rules that execute faster).

An instance of the SAT problem can be defined by a pair (Ξ, ϕ) where Ξ is a set of Boolean variables $\Xi = \{\xi_1, ..., \xi_n\}$ and ϕ is a Boolean formula $\phi \colon \{0, 1\}^n \rightarrow \{0, 1\}$. The formula is said to be satisfiable if there exists an assignment $\sigma \colon \Xi \rightarrow \{0, 1\}$ satisfying ϕ and unsatisfiable otherwise. The formula ϕ is in conjunctive normal form (CNF) if it is a conjunction of clauses (a clause is a disjunction of literals and a literal is a variable or its negation).

In order to transform our CSP (X, D, C) into a SAT problem, we must define how the set Ξ is constructed from X and how ϕ is obtained. Concerning the variables, we use the direct encoding [5]: $\forall x \in X, \forall d \in D_x, \exists \xi_{x,d} \in \Xi$ ($\xi_{x,d}$ is true when x has the value d, false otherwise).

To enforce exactly one value for each variable, we use the next clauses:

$$\bigwedge_{x \in X} \bigvee_{d \in D_x} \xi_{x,d} \quad \text{and} \quad \bigwedge_{x \in X} \bigwedge_{\substack{d_1, d_2 \in D_x \\ d_1 \neq d_2}} (\neg \xi_{x,d_1} \vee \neg \xi_{x,d_2})$$

Our purpose is to define uniform transformation rules for handling multiple Alldiff constraints, which are often involved in many problems.

From the resolution point of view, complete SAT solvers are basically based on a branching rule that assign a truth value to a selected variable and unit propagation (UP) which allows to propagate unit clauses in the current formula [2]. We will say that a SAT encoding preserves a consistency iff all variables assigned to false by UP have their corresponding values eliminated by enforcing GAC.

3 Alldiff Constraints: Reduction Rules and Transformation

In the following, we classically note $Alldiff(V)$ the Alldiff constraint on a subset of variables V, which semantically corresponds to the conjunction of $n*(n-1)/2$ pairwise disequality constraints $\bigwedge_{x_i, x_j \in V, i \neq j} x_i \neq x_j$.

A Single $Alldiff$ constraint. We first reformulate a well known consistency property [4] w.r.t. the number of values remaining in the domain of the variables. This case corresponds of course to the fact that if a variable has been assigned then the corresponding value must be discarded from other domains.

$$[O1] \quad \frac{< C \wedge Alldiff(V), D > | x \in V \ \wedge \ D_x = \{d_1\}}{< C \wedge Alldiff(V), D' > | d_1 \notin D'_{V \setminus \{x\}}}$$

Property 1. If $< Alldiff(V), D > \rightarrow^*_{[O1]} < Alldiff(V), D' >$, then the corresponding conjunction $\bigwedge_{x_i, x_j \in V} x_i \neq x_j$ is GAC w.r.t. $< D' >$. Note that enforcing GAC on the disequalities with $[O1]$ reduces less the domains than enforcing GAC on the global Alldiff constraint.

This rule can be generalized when considering a subset V' of m variables with m possible values, $1 \leq m \leq (\#V - 1)$:

$$[Om] \quad \frac{< C \wedge Alldiff(V), D > |V' \subset V \ \wedge \ D_{V'} = \{d_1, \ldots, d_m\}}{< C \wedge Alldiff(V), D' > |d_1, \ldots, d_m \notin D'_{V \setminus V'}}$$

Consider $m = 2$, and that two variables of an Alldiff only have the same two possible values. Then, these two values cannot belong to the domains of the other variables.

Property 2. Given $< Alldiff(V), D > \rightarrow^{*}_{[Om]_{1 \leq m \leq (\#V-1)}} < Alldiff(V), D' >$, then $< Alldiff(V), D' >$ has the GAC property.

Now, the Alldiff constraints can be translated in SAT, by encoding $[O1]$ with a set of $\#V * (\#V - 1)$ CNF clauses:

$$[SAT - O1] \quad \bigwedge_{x \in V} \bigwedge_{y \in V \setminus \{x\}} (\neg \xi_x^d \vee (\bigvee_{f \in D_x \setminus \{d\}} \xi_x^f) \vee \neg \xi_y^d)$$

This representation preserves GAC. Indeed, if $\neg \xi_x^d$ is false (i.e. when the variable x is valued to d) and $\bigvee_{f \in D_{x_1} \setminus \{d\}} \xi_{x_1,f}$ is false (i.e., when the variable x_1 is valued to d) then $\neg \xi_{x_2,d}$ must be true to satisfy the clause (x_2 cannot be valued to d). Generalized to a subset V' of m variables $\{x_1, ..., x_m\}$ with m possible values $\{d_1, ..., d_m\}$, $1 \leq m \leq (\#V - 1)$, the corresponding $\#(V \setminus V') * m^{m+1}$ clauses are:

$$[SAT - Om]$$

$$\bigwedge_{y \in V \setminus V'} \bigwedge_{k=1}^{m} \bigwedge_{p_1=1}^{m} \cdots \bigwedge_{p_m=1}^{m} [(\bigvee_{s=1}^{m} \neg \xi_{x_s}^{d_{p_s}}) \vee$$
$$(\bigvee_{i=1}^{m} \bigvee_{f \in D_{x_i} \setminus \{d_1, ..., d_m\}} \xi_{x_i}^f) \vee \neg \xi_y^{d_k}]$$

Property 3. $\bigcup_{1 \leq m \leq \#V-1} [SAT - Om]$ preserves the GAC property.

Multiple Overlapping *Alldiff* **Constraints.** In presence of several overlapping Alldiff constraints, specific local consistency properties can be enforced according to the number of common variables, their possible values, and the number of overlaps. To simplify, we consider Alldiff constraints $Alldiff(V)$ such that $\#V = \#D_V$.

If a value appears in variables of the intersection of two Alldiff, and that it does not appear in the rest of one of the Alldiff, then it can be safely removed from the other variables' domains of the second Alldiff.

$$[OI2] \quad \frac{< C \wedge Alldiff(V_1) \ \wedge \ Alldiff(V_2), D > |d \in D_{V_1 \cap V_2} \ \wedge \ d \notin D_{V_2 \setminus V_1}}{< C \wedge Alldiff(V_1) \ \wedge \ Alldiff(V_2), D' > |d \notin D'_{V_1 \setminus V_2}}$$

$[OI2]$ is coded in SAT as $\#D_{V_1 \cap V_2} * \#(V_1 \cap V_2) * \#(V1 \setminus V_2)$ clauses:

$$[SAT - OI2] \quad \bigwedge_{d \in D_{V_1 \cap V_2}} \bigwedge_{x_1 \in V_1 \cap V_2} \bigwedge_{x_2 \in V_1 \setminus V_2} \bigvee_{x_3 \in V_2 \setminus V_1} (\neg \xi_{x_1}^d \vee \xi_{x_2}^d \vee \neg \xi_{x_3}^d)$$

[$OI2$] can be extended to [OIm] to handle m ($m \geq 2$) $Alldiff$ constraints connected by one intersection. Let denote by V the set of variables appearing in the common intersection: $V = \bigcap_{i=1}^{m} V_i$

$$[OIm] \quad \frac{< C \bigwedge_{i=1}^{m} Alldiff(V_i), D > |d \in D_V \ \wedge \ d \notin D_{V_1 \setminus V}}{< C \bigwedge_{i=1}^{m} Alldiff(V_i), D' > |d \notin \bigcup_{i=1}^{m} D'_{V_i \setminus V}}$$

Note that this rule can be implicitly applied to the different symmetrical possible orderings of the m Alldiff.

[OIm] is translated in SAT as $\#D_V * \#V * \sum_{i=2}^{m} (\#(V_i \setminus V))$ clauses:

$$[SAT - OIm] \quad \bigwedge_{d \in D_V} \bigwedge_{x_1 \in V} \bigwedge_{i=2}^{m} \bigwedge_{x_3 \in V_i \setminus V} \bigvee_{x_2 \in V_1 \setminus V} (\neg \xi_{x_1}^d \vee \xi_{x_2}^d \vee \neg \xi_{x_3}^d)$$

Property 4. Consider $m > 2$ Alldiff with a non empty intersection. Given $< C, D >\rightarrow^*_{[OIm]}< C, D' >$ and $< C, D >\rightarrow^*_{[OI2]}< C, D'' >$, then $D'' \subseteq D'$.

Other rules can be defined for several Alldiff connected by several intersections.

4 Conclusion

We have defined a set of consistency rules for general Alldiff constraints that can be easily implemented in usual constraint solvers. Recent works deal with the combination of several global constraints. Nevertheless, theses approaches require some specialized and complex algorithms for reducing the domains, while our approach allows us to simplify and unify the presentation of the propagation rules and attempts at addressing a wider range of possible combinations of Alldiff. The basic encodings of CSP into SAT have been fully studied to preserve consistency properties and induce efficient unit propagation in SAT solvers. Our transformation is based on the reduction rules and extended to multiple connected Alldiff. As some of these works we proved that it is correct w.r.t. GAC. This work provides then an uniform framework to handle interleaved Alldiff and highlights the relationship between CSP and SAT in terms of modeling and resolution when dealing with global constraints.

References

1. Apt, K.: Principles of Constraint Programming. Cambridge University Press, Cambridge (2003)
2. Bordeaux, L., Hamadi, Y., Zhang, L.: Propositional satisfiability and constraint programming: A comparative survey. ACM Comput. Surv. 38(4), 12 (2006)
3. Gent, I.: Arc consistncy in SAT. Technical Report APES-39A-2002, University of St Andrews (2002)
4. Régin, J.C.: A filtering algorithm for constraint of difference in csps. In: National Conference of Artificial Intelligence, pp. 362–367 (1994)
5. Walsh, T.: SAT v CSP. In: Dechter, R. (ed.) CP 2000. LNCS, vol. 1894, pp. 441–456. Springer, Heidelberg (2000)

Adaptation of Personality-Based Decision Making to Crowd Behavior Simulations

Fatemeh Alavizadeh, Caro Lucas, and Behzad Moshiri

Control and Intelligent Processing Center of Excellence
School of ECE, University Of Tehran, Tehran, Iran
f.alavi@ece.ut.ac.ir,
lucas@ut.ac.ir,
moshiri@ut.ac.ir

Abstract. Simulation of crowd behaviors in emergency contexts is a critical issue concerning human lives. However, lack of human behavioral data in most of current simulation tools made them unrealistic. This is partially because of instability in the theoretical issues due to complexity of human behavior, and partially because of the gap between theories and what that can be implemented in agent environments. This research is an effort to investigate a system for realistic and intelligent crowd simulations based on personality, age, and gender of individuals. The model demonstrates some emergent behaviors, such as competitive pushing, queuing, and herding.

1 Introduction

Numerous incidents are reported regarding overcrowding, stampeding, pushing, rushing, and herding during emergency situations which are consequences of human behavior. Therefore, understanding human social behavior in disaster is crucial. But, this is a difficult issue, since it often requires exposing real people to actual dangerous environments. A realistic computational tool that considers human behavior could serve as a viable alternative.

On the other hand, human behavior choice is highly dependent on individual characteristics such as personality, gender, age, and so on. Among them, decision makers' personality traits are crucial sources of difference between individuals, and accounting for personality can make human decision model more natural. In this research we will propose a multi-agent framework for simulation of human behaviors and emergent crowd behaviors during emergency evacuations based on individual differences.

Fortunately, there are some researches on geographical distribution of big five personality traits. Thus the personality distribution of a society of agents based on their region and culture is known; consequently the emergent social behavior would be predictable. Therefore, this project can effectively enhance disaster response systems; based on the distribution of personality types in a culture disaster response systems can predict social behavior and train their forces effectively. Similarly, in urban planning, buildings can be constructed with

D. Dochev, M. Pistore, and P. Traverso (Eds.): AIMSA 2008, LNAI 5253, pp. 385–389, 2008.

more effective exit plans. More generally, this framework would be a step forward to development of realistic models of human decision process.

2   Related Works

Many works are done on evacuation simulations. Very few number of these simulation tools [1] account for behaviors based on individual differences; yet, none of them accounts for human personality trait. For instance, MASSEgress [3] is a multi-agent simulation environment for egress analysis which is similar to this work in accounting for individual differences. However, MASSEgress decision making process is rule based and it is not concerned with personality. In Legion [11], occupants are intelligent individuals with social, physical, and behavioral characteristics. Simulex [4] is another example which assigns a certain set of attributes to each person, such as physical motions and gestures, gender and age. But, the program assumes the presence of a rational agent able to assess the optimal escape route [1], which is not realistic. Regularly, untrained people don't act rationally under pressure and in public buildings most people are unaware of exit paths. Although Legion and Simulex consider individual differences, they did not account for personality. DrillSim is another simulation environment [2] which has the potential for defining user profiles and decision making process based on this profile. Yet, DrillSim is a development environment for developers to write their own decision making process and evaluate it.

On the other hand, personality has been conceptualized from variety of theoretical perspectives. One of available taxonomies is the "Big Five" personality model derived from the natural-language terms people use to describe themselves. This model first mentioned in 1933; but in 1981, available personality tests were reviewed and concluded that tests with the most promise measured a subset of five factors. This event was followed by widespread acceptance of this model among personality researchers. These factors are often called *Openness, Conscientiousness, Extraversion, Agreeableness,* and *Neuroticism.* Agreeableness concerns with cooperation and social harmony. While agreeable individuals are considerate and helpful, disagreeable individuals place self-interest above getting along with others. Conscientiousness concerns the way in which people control their impulses. Highly conscientiousness people have strong control of their impulses; they persistently follow their goals by planning. Extraversion is marked by engagement with the external world. Extraverts enjoy being with people and often experience positive emotions. But introverts tend to be less dependent on the social world. Neuroticism refers to the tendency to experience negative emotions. People high in Neuroticism respond emotionally to events that would not affect most people. Openness distinguishes imaginative and creative people from conventional people. Open people are intellectually curious with unconventional and individualistic beliefs [5,6].

Different studies have confirmed the predictive value of the big five across a wide range of behaviors. For instance, in the area of job performance in [8] authors found that conscientiousness showed consistent relations with performance

criteria, and extraversion was a predictor of success in occupations involving social interaction. Although no direct research was found on the predictive value
of big five on emergent behavior in disasters, risk taking behavior has a strong
impact in action selection in such contexts. In [9] it is argued that openness and
extraversion have the most impact on risk taking.

3 Proposed Model

The proposed model intends not only to simulate human behavior at an individual level but also to capture the emergent social behaviors. At individual
level, each agent based on defined profile choose between possible behaviors including "Moving to nearest exit door", "following the crowd", and "moving to
the less crowded door". Each agent has a profile indicting his personality vector
on facets of big five, age and gender. This profile will be used in individual's
decision making process. On the other hand, for each behavior a selection likelihood vector is defined. An individual would select the behavior that has the
least weighted distance with his own personality vector. Behavior vectors and
dimension weights are defined based on [9] which discussed the predictive value
of big five model on risk taking behavior. For instance, finding your own way
compared to moving with the crowd has a significant level of risk taking which
means higher levels of openness and extraversion [9]. Yet, we should consider
that in emergency context neuroticism plays a major rule since it makes people
panic and leads to irrational behaviors. So we should consider this dimension as
well. Although finding exit path is a kind of planning that needs conscientiousness, emergency context decreases the impact of conscientiousness on decision
making, so we assigned a lower weigh to conscientiousness. Moreover, in [9] it is
argued that men have higher levels of risk taking than women, and risk taking
behavior decreases with age.

Besides defined actions, individuals decide on two other major behaviors. The
first one appears at the doors where they decide whether to push others in a
competitive manner or to queue for exit. Obviously, in highly crowded buildings,
queuing can enhance evacuation significantly. The second behavior is concerned
with helping others. Based on personality some people tend to help others with
announcing the exit directions.

4 Implementation

The simulation environment is a grid world that shows the building plan consisting of obstacles, doors, and exit signs. Population will be generated based
on mean and standard deviation of personality distribution of a culture which
are provided in [7,10]. The simulation executes in several iterations in which all
individuals decide on their next move and it will end whenever all people exit
the building. Similar to cellular automata approaches, each agent decide on his
next move by getting information from surrounding cells and neighbors. The
agent can scan neighbor cells looking for exits. If he couldn't find an exit, he will

decide whether to move with the crowd or to look for the exit signs. In first case, he will ask neighbors for their direction. Based on personality neighbors may answer or not. Then the agent will follow the direction that most of neighbors are following. If the agent decides to follow the signs, he will look for a sign in surrounding cells. If moving with crowd or following the sign was impossible, the agent would walk randomly. At the doors, people have the chance to queue or push. Since each cell has a specific capacity (4 agents) individuals would know whether their front cell is crowded or not. So they can decide whether to enter the cell, wait in their current location, or take a detour. Door's performance is related to the amount of people in surrounding cells. If the cell at a door has 4 people in it only one can exit. But if 2 people are in that cell both of them can exit. This can simulate the negative impact of pushing at the doors. Figure 1 shows one screenshot of the system.

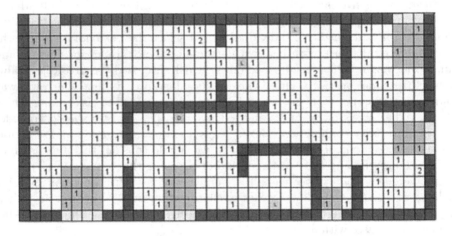

Fig. 1. The population of each cell is written in the cell. Doors are light gray cells and exit signs are shown by letters: U, D, L, R.

5   Results

Here we present results of personality traits for Japanese and Germans. Table 1 shows the average values of personality dimensions which are drown from [7]. Japanese are scored higher on neuroticism and scored lower in extraversion and openness [9]. This make us expect Japanese to have tendency of staying with the crowd, since they get nervous easily and they are not open to risk taking. The average values of 10 simulations shows that about 15 percent of people tend

Table 1. Average in each dimension on 0 to 99 scale

Nation	Extraversion	Agreeableness	Conscientiousness	Neuroticism	Openness	SD
Japan	46	42	37	57	41	8.8
Germany	50	57	46	47	47	8.99

to follow their own paths by searching for signs. Also about 50 percent tend to help others. These simulations show herding behavior which can lead to under utilization of some exits and over utilization of the others. On the other hand, the average values of 10 simulations shows that about 70 percent of Germans tend to follow their own paths by searching for signs; which was expected due to higher levels of agreeableness and extraversion and lower level of neuroticism.

6 Conclusions and Future Works

This work is an effort to develop a model for realistic simulations of humans based on individuality. A tool has been developed in which each human perceives the surroundings, decides how to behave according to its individuality and acts on the environment. The model demonstrates some emergent behaviors, such as pushing, queuing, and herding. As a part of future work, in order to simulate more complicated behaviors such as "Helping" and "Guiding" collaborative behaviors can be added to the model. In addition relations between humans are important in their decision process. For instance, people who would decide to follow the sign may change decision if their relatives are going in another way. Thus addition of relations is a good extension to current model.

References

1. Santos, G., Aguirre, B.: Critical Review of Emergency Evacuation Simulation Models. In: Workshop on Occupant Movement During Emergencies, pp. 27–52 (2005)
2. Balasubramanian, V., Massaguer, D., Mehrotra, Sh., Venkatasubramanian, N.: DrillSim: A Simulation Framework for Emergency Response Drills, ISI (2006)
3. Pan, X., Han, C.S., Law, K.H.: A Multi-agent Based Simulation Framework for the Study of Human and Social Behavior in Egress Analysis. In: International Conference on Computing in Civil Engineering, Cancun (2005)
4. Thompson, P.A., Wu, J., Marchant, E.W.: Modelling Evacuation in Multi-storey Buildings with Simulex. Fire Engineering 56, 7–11 (1996)
5. Lim, M.Y., Aylett, R., Jones, C.M.: Emergent affective and Personality Model. In: Panayiotopoulos, T., Gratch, J., Aylett, R.S., Ballin, D., Olivier, P., Rist, T. (eds.) IVA 2005. LNCS (LNAI), vol. 3661, pp. 371–380. Springer, Heidelberg (2005)
6. John, O., Srivastava, S.: The Big-Five trait taxonomy: History, measurement, and theoretical perspectives. Handbook of personality 2, 102–138 (1999)
7. Reips, S., et al.: Geographic distribution of Big Five personality traits: Patterns and profiles of human self-description across 56 nations. Journal of Cross-Cultural Psychology 38(2), 173–212 (2007)
8. Barrick, M.R., Mount, M.K.: The big five personality dimensions and job performance: A meta-analysis. Personnel Psychology 44, 1–26 (1991)
9. Maryann, G.: High Rolling Leaders: The Big Five Model of Personality and Risk Taking During War. International Studies Association, USA (2006)
10. McCrae, R.R., Terracciano, A.: Personality profiles of cultures: Aggregate personality traits. Journal of Personality and Social Psychology 89(3), 407
11. Legion International L, http://www.legion.biz/system/research.cfm

Conceptualizing Interactions with Matchmakers and Front-Agents Using Formal Verification Methods

Amelia Bădică and Costin Bădică

University of Craiova, A.I. Cuza 13, Craiova, 200585, Romania
ameliabd@yahoo.com, badica_costin@software.ucv.ro

Abstract. In this note we propose a rigorous analysis of interactions of requesters and providers with matchmakers and front-agents based on formal specification and verification using FSP process algebra and FLTL temporal logic.

1 Introduction

Connecting requesters and providers for dynamic sharing of information, resources and services on a global scale is a key function of modern information society. This problem requires the use of middle-agents – replacement of middle-men in a virtual environment ([5]). For example, middle-agents are encountered in e-commerce applications ([2,7]) as well as in general-purpose agent platforms ([9,8]).

Based on what it is initially known by the requesters, middle-agent, and providers about requester preferences and provider capabilities, authors of [5] proposed 9 types of middle-agents including *Matchmaker* and *Front-agent*. However, we observed that "notions of middle-agents, matchmakers [...] are used freely in the literature [...] without necessarily being clearly defined" (see survey from [10]). In our work we focused on improving this state-of-affair by defining formal models of middle-agents with the goal of understanding their similarities and differences. We have considered *finite state process algebra* – FSP ([11]) models of *Recommender* ([3]), of *Arbitrator* for a single-item English auction ([4]), and of *Matchmaker*, *Front-agent* and *Broker* ([1]).

In this paper we reconsider *Matchmaker* and *Front-agent* models and show how their qualitative properties can be formally defined and verified using *fluent linear temporal logic* – FLTL ([11]). Our work can be seen as an initial attempt for defining an FLTL-based framework for formal verification of systems with middle-agents.

2 Background

Matchmaker. The block diagram of a system composed of requesters, providers and a *Matchmaker* is shown in figure 1. For the FSP model please consult [1].

Provider agent registers its capability offer (action *offer*) with the *Matchmaker* and then enters a loop where it receives requests from *Requester* agents via action *receive_request* and processes and replies accordingly via action *send_reply*. Note that a *Provider* can also withdraw a registered capability offer and while its capability is not registered it always refuses to serve a request (action *refuse_request*).

D. Dochev, M. Pistore, and P. Traverso (Eds.): AIMSA 2008, LNAI 5253, pp. 390–394, 2008.

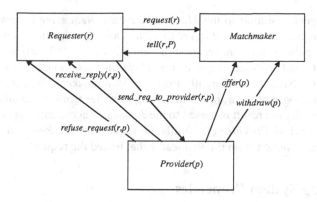

Fig. 1. System with *Matchmaker* middle-agent

Requester agent submits a request to the *Matchmaker* (action *send_request*) and then waits for a reply. The *Matchmaker* replies with a set of matching providers (action *tell* with argument *P* representing the set of matches). Next *Requester* may choose what provider $p \in P$ to contact for performing the service (action *send_request_to_provider*). Finally, *Requester* waits for a reply from the contacted provider (action *receive_reply*).

Matchmaker agent registers and deregisters *Provider* offers and answers *Requester* requests for matching offers. *Matchmaker* informs *Requester* about the available registered offers (action *tell*). Note that *Requester* has the responsibility to choose an appropriate matching offer from the available matching offers (action *send_request_to_provider*). This complicates a bit *Requester* behavior as compared with *Front-agent* case.

A critical situation may occur when a match is found but the matching *Provider* withdrawn its offer before it was actually contacted by the *Requester*. This ability of the *Provider* is modeled with action *refuse_request*. Note that this situation cannot occur with the *Front-agent* (*Front-agent* intermediates the request on behalf of the *Requester*) so we did not have to model this ability of the *Provider*.

Front-agent. The block diagram of a system composed of requesters, providers and a *Front-agent* is shown in figure 2. For the FSP model please consult [1].

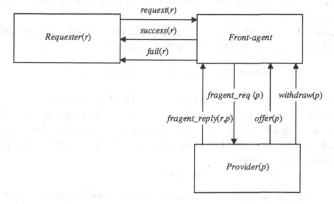

Fig. 2. System with *Front-agent* middle-agent

Provider agent is similar to the *Matchmaker* case. *Requester* is however simpler: it submits a request to the *Front-agent* (action *send_request*) and then waits for the *Front-agent* to either resolve that request (action *success*) or fail (action *fail*).

Front-agent agent processes requests from *Requester* agents and registers offers from *Provider* agents. Note that, differently from the *Matchmaker*, *Front-agent* has the responsibility to choose an appropriate matching provider from the available matching providers *P*. Finally the result is passed to the *Requester* agent using action *success*. In conclusion, the actual *Provider* that fulfils the request for the *Requester* on behalf of the *Front-agent* is hidden from the *Requester* that issued the request.

3 Modeling System Properties

Formal modeling of agent systems has the advantage that models can be systematically checked against user-defined properties. A property is defined by a statement that is true for all the possible execution paths of the system. A property is used to describe a desirable feature of the system behavior. Definition and analysis of properties of middle-agents have the advantage that they enable a formal and concise, rather than informal and speculative comparison of various types of middle-agents.

Properties are usually expressed as formulas in a temporal logic language. A property holds if the associated formula is true for all the possible executions of the system, as it is described by the model of the system. For system models captured using FSP it has been argued that a very convenient temporal logic for property expression is *fluent temporal logic* – FLTL ([11]).

In FLTL primitive properties are expressed using *fluents*. A fluent is a property whose truth is triggered by an initiating event and that holds until the signalling of a terminating event. In FSP it is natural to model initiating and terminating events by execution of specific actions. Therefore, following [11], a fluent is defined as a triple:

$$\textbf{fluent } F = \langle \{i_1, \ldots, i_m\}, \{t_1, \ldots, t_n\} \rangle \textbf{ initially } B$$

where (i) $\{i_1, \ldots, i_m\}$ and $\{t_1, \ldots, t_n\}$ are disjoint sets of *initiating events* and *terminating events* and (ii) *B* is true or false and represents the initial value of *F*. When any of the initiating actions is observed, *F* becomes true and stays true until any of the terminating actions is observed.

Every action *a* defines a singleton fluent $F(a)$ having *a* as the single initiating action and the rest of all actions as terminating actions, as follows:

$$\textbf{fluent } F(a) = \langle \{a\}, A \setminus \{a\} \rangle \textbf{ initially false}$$

A singleton fluent $F(a)$ is usually written as *a* in FLTL formulas.

FLTL formulas are built over fluent propositions (including singleton fluents) using the usual logical operators $\wedge, \vee, \rightarrow, \neg$ and temporal operators **X** (next), **U** (until), **W** (weak until), **F** (eventually) and **G** (always) (see [11]). A property *P* is specified using an FLTL formula Φ as follows:

$$\textbf{assert } P = \Phi$$

Property specification was recognized as a difficult task requiring expert knowledge in formal methods, especially in temporal logic. However, according to the rigorous analysis performed in [6], most of the specifications fall into a category of specification patterns. Therefore in our work we have initially looked into the application of property specification patterns to the verification of *Matchmaker* and *Front-agent* middle-agents. In this note we show how *response properties* that check if a given situation must always be followed by another situation, can be applied for the verification of our system.

For the *Matchmaker* middle-agent we can define a fluent that holds while the *Matchmaker* is processing a request from a given *Requester*.

fluent $MM\_PROC\_REQ(r \in \mathcal{R}) = \langle\{request(r)\}, \{tell(r, P \subseteq \mathcal{P})\}\rangle$ **initially false**

Similarly we can define a fluent that holds while the *Front-agent* is processing a request from a given *Requester*.

fluent $FA\_PROC\_REQ(r \in \mathcal{R}) = \langle\{request(r)\}, \{success(r), fail(r)\}\rangle$ **initially false**

We can also describe a fluent that holds while a *Provider* is registered with either the *Matchmaker* or the *Front-agent*.

fluent $REGISTERED(p \in \mathcal{P}) = \langle\{offer(r)\}, \{withdraw(r)\}\rangle$ **initially false**

Using the $REGISTERED(p \in \mathcal{P})$ properties we can define an indexed set of properties $MATCHES(r \in \mathcal{R})$ that are true when there is a registered *Provider* that matches the given *Requester* $r \in \mathcal{R}$.

assert $MATCHES(r \in \mathcal{R}) = \vee_{p \in M(r)} REGISTERED(p)$

Using the defined fluents and formulas we can express two characterizing response properties of a system with *Front-agent*: (i) "If a *Requester* issues a request to the *Front-agent* then the *Front-agent* with eventually reply either with success or failure" and (ii) "If a *Requester* issues a request to the *Front-agent* and there is at least one matching *Provider* registered with the *Front-agent* then the *Front-agent* with eventually reply with success". These properties can be formally defined using FLTL as follows:

assert $FA\_RESPONSE(r \in \mathcal{R}) =$
  $\mathbf{G}(request(r) \rightarrow (FA\_PROC\_REQ(r) \mathbf{U} (success(r) \vee fail(r))))$
assert $FA\_MATCHING\_RESPONSE(r \in \mathcal{R}) =$
  $\mathbf{G}((request(r) \wedge MTACHES(r)) \rightarrow (FA\_PROC\_REQ(r) \mathbf{U} success(r)))$

Different response properties can be defined for the *Matchmaker* middle-agent: "If a *Requester* issues a request to the *Matchmaker* then the *Matchmaker* with eventually reply with the set of matching providers".

assert $MM\_RESPONSE(r \in \mathcal{R}) =$
  $\mathbf{G}(request(r) \rightarrow (MM\_PROC\_REQ(r) \mathbf{U} \vee_{P \subseteq M(r)} tell(r, P)))$

We would be tempted to define a response property of the system with *Matchmaker* similar to the *Front-agent*: "If there is a matching *Provider* offer already registered with the *Matchmaker* when a request is issued by a *Requester* we would be tempted to require the system to guarantee that the request will be served by a matching *Provider*".

assert $MM\_MATCHING\_RESPONSE\_BAD(r \in \mathcal{R}) =$
  $\mathbf{G}((request(r) \wedge MATCHES(r)) \rightarrow (PROC\_REQ(r) \mathbf{U} receive\_reply(r, p \in \mathcal{P})))$

Note however that this property does NOT hold ! Between the time the matching offers are provided to the *Requester* by the *Matchmaker* and the time when the *Requester* decides to contact a matching *Provider*s it may happen that the matching *Provider* decides to cancel her offer by deregistering with the *Matchmaker*.

The correct reformulation of this property would require to guarantee that the system will either perform the request and reply accordingly (action *receive_reply*) or will indicate explicitly that fulfilling the request was refused (action *refuse_request*).

$$\textbf{assert } MM\_MATCHING\_RESPONSE(r \in \mathcal{R}) =$$
$$\mathbf{G}\left((request(r) \land MATCHES(r)) \rightarrow \right.$$
$$\left.(PROC\_REQ(r) \, \mathbf{U} \, \{receive\_reply(r, p \in \mathcal{P}), refuse\_request(r, p \in \mathcal{P})\})\right)$$

4 Conclusions and Future Work

In this paper we shown how FSP models of *Matchmaker* and *Front-agent* can be formally verified using FLTL temporal logic. For each model we defined an initial set of qualitative properties and we checked them with the help of LTSA analysis tool. As future work we intend to extend and apply the verification framework to more types of middle-agents and more complex models.

References

1. Bădică, A., Bădică, C.: Formal Specification of Matchmakers, Front-agents, and Brokers in Agent Environments Using FSP. In: Ultes-Nitsche, U., Moldt, D., Augusto, J.C. (eds.) Proc. MSVVEIS 2008 – 6th International Workshop on Modelling, Simulation, Verification and Validation of Enterprise Information Systems, pp. 9–18. INSTICC Press (2008)
2. Bădică, C., Ganzha, M., Paprzycki, M.: Developing a Model Agent-based E-Commerce System. In: E-Service Intelligence: Methodologies, Technologies and Applications. Studies in Computational Intelligence, vol. 37, pp. 555–578. Springer, Heidelberg (2007)
3. Bădică, A., Bădică, C., Lițoiu, L.: Middle-Agents Interactions as Finite State Processes: Overview and Example. In: Proc. 16th IEEE International Workshops on Enabling Technologies: Infrastructure for Collaborative Enterprises (WETICE 2007), pp. 12–17 (2007)
4. Bădică, A., Bădică, C.: Formal modeling of agent-based english auctions using finite state process algebra. In: Nguyen, N.T., Grzech, A., Howlett, R.J., Jain, L.C. (eds.) KES-AMSTA 2007. LNCS (LNAI), vol. 4496, pp. 248–257. Springer, Heidelberg (2007)
5. Decker, K., Sycara, K.P., Williamson, M.: Middle-agents for the internet. In: Proc.of 15th International Joint Conference on Artificial Intelligence IJCAI 1997, vol. 1, pp. 578–583. Morgan Kaufmann, San Francisco (1997)
6. Dwyer, M.B., Avrunin, G.S., Corbett, J.C.: Patterns in property specifications for finite-state verification. In: ICSE 1999: Proc. 21st International Conference on Software Engineering, pp. 411–420. IEEE Computer Society Press, Los Alamitos (1999)
7. Fasli, M.: Agent Technology For E-Commerce. Wiley, Chichester (2007)
8. FIPA: Foundation for Physical Agents, http://www.fipa.org
9. JADE: Java Agent Development Framework, http://jade.cselt.it
10. Klusch., M., Sycara, K.P.: Brokering and matchmaking for coordination of agent societies: A survey. In: Omicini, A., Zambonelli, F., Klusch, M., Tolksdorf, R. (eds.) Coordination of Internet Agents. Models, Technologies, and Applications, pp. 197–224. Springer, Heidelberg (2001)
11. Magee, J., Kramer, J.: Concurrency. State Models and Java Programs, 2nd edn. John Wiley & Sons, Chichester (2006)

A Social and Emotional Model for Obtaining Believable Emergent Behaviors

Javier Asensio, Marta Jiménez, Susana Fernández, and Daniel Borrajo

Planning and Learning research Group
Dept. of Computer Science, University Carlos III de Madrid
Av. de la Universidad 30. 28911 Leganés (Madrid), Spain
http://www.plg.inf.uc3m.es

Abstract. This paper attempts to define an emotional model for virtual agents that behave autonomously in social worlds. We adopt shallow modeling based on the decomposition of the emotional state in two qualities: valence (pleasantness or hedonic value) and arousal (bodily activation) and, also, for the agent personality based on the *five factors* model (openness, conscientiousness, extroversion, agreeableness and neuroticism). The proposed model aims to endow agents with a satisfactory emotional state achieved through the social actions, i.e. the development of social abilities. Psychology characterizes these social abilities for: using the language as a tool (verbal and nonverbal communication), being learned, producing reciprocal reward among the individuals involved in the communication and for depending on the individual features. We have implemented our model in the framework of a computer game, AI-LIVE, to show its validity.

Keywords: autonomous agents, behavior models, emotions and personality.

1 Introduction

Artificial Intelligence, as the branch of computer science that is concerned with the automation of intelligence behavior, has traditionally been interested in human behavior and its cognitive capabilities to emulate them or build intelligent systems. But, very few works have addressed the explicit representation and reasoning about important characteristics of humans that make them prefer some decisions over others: their emotion states and behaviors. Humans and emotions are highly intertwined, as well as the social and emotional behaviors. Emotions have been studied by different disciplines, such as psychology, neurobiology or philosophy (see [3] for an overview) without having reached a consensus on the emotional phenomena. However, many AI researchers agree with the functional hypothesis of Fridja [5] that emotions have the adaptive value of serving a purpose, called the *functional view of* emotions. Thus, some researchers want to endow artificial agents with emotions and thence with sociality, giving that emotions influence and shape the development of sociality as much as sociality influences and shaped the development of emotions.

D. Dochev, M. Pistore, and P. Traverso (Eds.): AIMSA 2008, LNAI 5253, pp. 395–399, 2008.
© Springer-Verlag Berlin Heidelberg 2008

The latter years have witnessed some work in the field like the ones by: Cañamero discussing the ideas relative to the construction of emotional artifacts that have to interact in a social world [2]; or Silverman and his collaborators that focus on challenges to improve the realism of socially intelligent agents and on the attempt to reflect the state of the art in human behavior modeling [9]. And also, there are several interesting applications employing animated agents to enhance their effectiveness as tutors, sales persons, or actors, among other roles, pointed out in [8].

This poses the need of designing and implementing emotional models for social interactions between artifacts, interacting with both, humans and other artifacts. The ability to provide formal approaches to emotions is a key aspect of current and future applications, that range from assistive technology, intelligent user interfaces, educational tools, automatic generation of scripts for films or tv shows, or games. This paper attempts to define such a social and emotional model. We have implemented it in the framework of a computer game to show its validity. The computer game is AI-LIVE [4], that is oriented towards the intensive use of AI controlled Bots. The game borrows the idea from the popular THE SIMS, where the player creates individual characters (units) that have significant autonomy, with their own drives, goals, and strategies for satisfying those goals, but where the human player can come in and stir things up by managing both the individual characters and their environment. Since our goal is to develop autonomous AI-driven for this kind of games, currently AI-LIVE only has automated players. The emotional model we propose here is based on the assumption that any emotional state can be decomposed into two qualities: valence (pleasantness or hedonic value) and arousal (bodily activation), borrowing the idea from [1]. We prefer implementing such a shallow model, rather than a deep one as in [7]. We also adopt shallow modeling for the agent personality based on the *five factors* model (openness, conscientiousness, extroversion, agreeableness and neuroticism) [6].

The proposed model aims to endow agents with a satisfactory emotional state achieved through social actions, i.e. the development of social abilities. In a first approach, the basic components of a model that complies with this idea will be: i) the **agent psyche model** that comprises agent's personality (the `five factors`), emotional state (`valence` and `arousal`) and social state (`age`, `gender` and place in the `social hierarchy`), ii) the **social model** for defining how every agent relates to the rest of agents and iii) the **emotional engine** that allows agents achieve *satisfactory* emotional states through communicative actions. In our design every personal relation is unidirectional and it involves only two actors (source and destination). This social link between agents is featured by an **emotional state**, being this state different from the actor's emotional state, but always under its influence; a **kind** of social relationship, can be based on age, blood or social status; the **strength** of the relationship, the stronger is the relation between two agents the more it will be affected by changes in the actors emotions and the **start time** for locating the beginning of the relationship. All these components are represented with numeric variables. The emotional engine defines how these relationships and emotions evolve over time. It consists of a set of formulae that compute the variation on the numeric variables involved in the model. In this paper, emotions as well as relations can only evolve by performing communicative actions, as *talking* or *shouting*. So far, we have only implemented the action *tellLikeness* where actors talk about their likeness of an specific object.

The remainder of the paper shows the empirical results that validate the model and the conclusions derived from the work, together with future research lines.

2 Experiments and Results

The experiments aim to assess the emotional model through the communicative action *tellLikeness* where an actor tells another his/her likeness of a particular object, L_{SO}, defined by a valence and arousal values. All the experiments were performed running the AI-LIVE system [4] with two rule-based AI clients: the sender (As), is the actor who talks and the receiver (Ar), the actor who listens. We let As fulfils a certain amount of *tellLikeness* actions (20), always talking about the likeness of the same object, and we let Ar only listen to him/her (we ignore Ar actions when s/he gets the turn). When As tells his likeness to Ar we observe the following: i) As changes slightly the values valence and arousal of L_{SO} transmitted to Ar, depending on his/her emotional state and his/her personality. ii) Ar also changes the values L_{SO}^{t} received from As, depending on his/her personality. iii) The relationship between Ar and As, R, changes its parameters strength, valence and arousal depending on the likeness difference, $L_{SO}^{t} - L_{RO}$, and on the social distance between both agents. iv) Ar changes his/her emotional state depending on the relationship changes above. And finally, Ar can change his/her likeness L_{RO} depending on the new strength.

The independent variables in all the experiments were: **1.** Agents (As and Ar) personality: (agreeableness, conscientiousness, extroversion, openness, neuroticism) all with values between 0 and 1. We used three kinds of actors according with their personality: positive extreme (1, 1, 1, 1, 0); negative extreme (0, 0, 0, 0, 1); and, neutral (0.5, 0.5, 0.5, 0.5, 0.5). **2.** Agents emotional state: (valence, arousal) (values between -1 and +1). We used three emotional states: positive extreme (1, 1), negative extreme (-1, -1), and neutral (0, 0). **3.** The relationship between Ar and As, R: kind, can be based on age, blood or social status, though we used age in the experiments; strength (values between 0, if they are stranger, and 1); valence and arousal to represent the emotions that Ar arouse to As and startTime, we used 0. **4.** Agents social role: age (values between 0 and 100), though we used 20 in the experiments; sex (0 male or 1 female), we only used male in the experiments; and social status (values between 0 and 100), we used 5. **5.** As likeness of the object O, L_{SO}. **6.** Ar likeness of the object O.

And we measured the transmitted likeness, as well as the variation on the strength of the relationship between the actors. Figure 1 shows the results of some of the experiments. We only plot the most significant ones. For example, the first graph reflects how As personality and his/her emotional state influence the communication. Normally, when an actor express his/her likeness of an object, the user never says the exact value of the `valence` and `arousal`, but changes them slightly. The closer his/her personality and his/her emotional state are to the positive extremes, the more he strengthens his/her likes. Something similar happens when `Ar` listens the likeness `As` is telling to him/her. Depending on his/her personality, s/he receives slight variations on the values of the `valence` and `arousal` transmitted by `As`. This variation only depends on the personality and emotional state of `As` but we have introduced a random value (between 0.5 and 1.3) that multiplies `As` emotional state and hence when $As.x = 0$ the likeness transmitted is always the same: 1.01 when `As` personality is the positive extreme, 0.97 when is neutral and 0.94 when is the negative extreme. The `As` likeness to the object is fixed to 1 in all the cases.

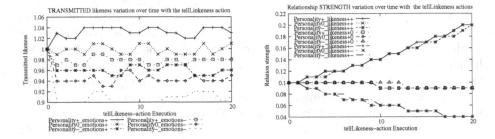

Fig. 1. Results for the *tellLikeness* action

The other graph reflects how the strength of the relationships changes depending on the initial affinity for the object. Initially, $R.strength = 0.1$, $R.valence = 0$, $R.arousal = 0$ and the personality and emotional state of `As` are fixed to the positive extreme for all the experiments. We varied `Ar` personality from the positive extreme (*Personality+*), neutral (*Personality0*) and negative extreme (*Personality–*) and the `As` and `Ar` likeness of the object. For example, *likeness++* means that the likeness emotions of both agents are the positive extreme, *likeness+0* means that `As` likeness is the positive extreme and `Ar` likeness is neutral and *likeness–* means that both agents' likeness are the negative extreme. If the likeness for the object is similar for both agents, each communicative action strengthens the relationship. On the contrary, the strength decreases when their likeness are different, being the change proportional to the likeness unsimilarity, so the *likeness+-* graphs get lower strength values than the *likeness+0* ones. The variations of `Ar` emotional state and likeness are proportional to this strength variation and we do not plot them for space reasons.

3   Conclusions and Future Work

This paper proposes an emotional and social model for obtaining believable emergent behaviors in autonomous agents. The initial hypothesis is that any agent can affect any other agent emotions by performing a communicative action as talking. A communicative action involves two agents and depends on the agents personality, their emotional states and the relationship between them. Therefore, the model must include a representation for all these objects, an implementation of the communicative actions and an emotional engine for evolving agents emotions. The emotional engine consists of a set of formulae that compute the variation on the numeric variables involved in the model. There are quite a few parameters that we have tuned experimentally. We have implemented all of this in the framework of a computer game inspired in the commercial game THE SIMS where each *sim* is an autonomous agent implemented with an AI technique. We have performed some experiments where an agent tells another its likeness about the same object and we have measured the variation of the emotions and relationship variables. The results show that the emotional engine evolves them in a coherent way.

In the future, we would like to augment the communication to include actions as *gossip* or *shouting*. Also, we would like to add actions where the agent not only communicates emotions but ideas and thoughts, as well. Furthermore, it would be interesting to generalize the emotional model to affect the whole decision making process of the agents.

References

1. Barrett, L.F.: Discrete emotions or dimensions? the role of valence focus and arousal focus. Cognition and Emotion 12, 579–599 (1998)
2. Cañamero, D.: Building emotional artifacts in social worlds: Challenges and perspectives. In: AAAI Fall Symposium on Emotional and intelligent II: The tangled knot of social cognition, pp. 22–30. AAAI Press, Menlo Park (2001)
3. Cornelius, R.: The Science of Emotion. Research and tradition in the psychology of emotion. prentice-hall, Upper Saddle River (1996)
4. Fernández, S., Adarve, R., Pérez, M., Rybarczyk, M., Borrajo, D.: Planning for an ai based virtual agents game. In: Proceedings of the Workshop AI Planning for Computer Games and Synthetic Characters in the ICAPS (2006)
5. Frijda, N.H.: Emotions in Robots. In: Comparative Approaches to Cognitive Science, pp. 501–516. MIT Press, Cambridge (1995)
6. McCrae, R., John, O.: An introduction to the five-factors model and its applications. Journal of Personality 60, 175–215 (1992)
7. Ortony, A., Clore, G.L., Collins, A.: The Cognitive Structure of Emotions. Syndicate of the University of Cambridge (1998) (in press)
8. Prendinger, H., Ishizuka, M.: Designing animated agents as social actors and dramatis personae for web-based interaction. Special Issue on Software Agent and its Applications (2002)
9. Silverman, B.G., Johns, M., Cornwell, J., O'Brien, K.: Human behavior models for agents in simulators and games: Part i: Enabling science with pmfserv. Departmental Papers (ESE) (2006)

A Learning Approach to Early Bug Prediction in Deployed Software

Saeed Parsa, Somaye Arabi, and Mojtaba Vahidi-Asl

Iran University of Science and Technology
parsa@iust.ac.ir, {atarabi,mojtaba_vahidi}@comp.iust.ac.ir

Abstract. In this paper the use of Support Vector Machines to build programs behavioral models predicting misbehaviors while executing the programs, is described. Misbehaviors can be detected more precisely if the model is built considering both the failing and passing runs. It is desirable to create a model which even after fixing the detected bugs is still applicable. To achieve this, the use of a bug seeding technique to test all different execution paths of the program in both failing and passing executions is suggested. Our experiments with a test suite, EXIF, demonstrate the applicability of our proposed approach.

1 Introduction

Early bug detection approaches [1-3] apply behavioral models to determine misbehaviors caused by any bug in the program. Behavioral models are mostly built within a learning process. In statistical learning approaches [1], [3], [4] the main intention is to build a reasonable model based on a limited number of test cases. The test cases should cover both passing and failing executions because, the resultant model have to determine the possibility of whether the program is malfunctioning or is on the right track. Applying a fault seeding technique, all different paths both in failing and passing executions can be tested [6]. Since the resultant model considers all different execution paths, minor corrections to the program will not affect the model.

The aim has been to construct a model that classifies the program behavior based on its termination status using an efficient learning method. Theoretically, most learning methods suffer from over fitting problem [8]. To avoid the problem, we have applied Support Vector Machine (SVM) learning approach [9-10]. SVM is one of the rare learning methods which avoids over-fitting by choosing a specific hyper-plane amongst those that can separate the data in the feature space.

The remaining parts of this paper are organized as follows: Section 2 introduces our proposed method for behavioral modeling. SVM models are described in Section 3. Section 4 evaluates functionality of the proposed method in a case study and includes experimental results. We conclude with final remarks in section 5.

2 Behavioral Modeling

Applying a SVM modelling technique, a statistical linear equation can be built for each predicate within a program. The equation can be applied to decide, in a relatively

D. Dochev, M. Pistore, and P. Traverso (Eds.): AIMSA 2008, LNAI 5253, pp. 400–404, 2008.
© Springer-Verlag Berlin Heidelberg 2008

small amount of time, whether the program is proceeding on the right track. In order to prepare the training data, the program is firstly instrumented. The instrumented code is executed several times with different input data. The training data collected at each instrumented point, represents the state of all the predicates from the program starting point to the predicate at the instrumented point, for a specific execution of the program.

The training data is kept in an array of pointers where each pointer addresses an array of the values of a different predicate in the program execution path. In each execution of the program zero or more values may be computed for a predicate, depending on the number of times the predicate is executed. Therefore, the values of the predicate have to be kept in a vector. In each run of the program the labels and the values of each predicate, p are kept in a vector, ExcPath (p,i), representing the execution path including all the predicates from the program starting point up to the ith appearance of the predicate, p, is constructed.

3 Modeling with Support Vector Machine

Support Vector Machine (SVM) is one of the binary classifiers based on maximum margin strategy introduced by Vapnik [9]. Consider the training data (x_i, y_i); $(1 \leq i \leq l)$, where x_i is a predicate vector in n dimensional program execution space and y_i is a Boolean number representing the program termination status. SVM finds a hyper-plane $w.x+b = 0$ to separate the program execution space into two regions of failing and passing executions. In this relation, the vector w is a normal vector which is perpendicular to the hyper-plane and the parameter b indicates the offset of the hyper-plane from the origin along the normal vector w. The hyper-plane should be selected in such a way that the two segments could be easily distinguished. As shown in Figure 1, the hyper-plane separates the program predicate vectors with class label '+1', for passing executions, from those with class label '-1', for failing execution [7].

The SVM model for each instrumented point of the program under test can be applied to predict bugs while executing the instrumented program, P*. In order to predict bugs at run time, a function f(x), representing the SVM hyper-plane is applied to the predicate vector x, constructed at each instrumentation point. The function computes the degree of the similarity of the vector x with the support vectors x_i, built at the training stage. Based upon the degree of the similarity of x to the support vector x_i, it is determined whether x belongs to a failing or a passing execution. The hyper-plane function, f(x), is as follows:

$$f(x) = sign \sum_{i}^{l} (a_i y_i k(x_i, x) + b) \tag{1}$$

Where a_i is the Lagrange multiplier and $k(x_i, x)$ is called a kernel function, it calculates similarity between two arguments x_i and x.

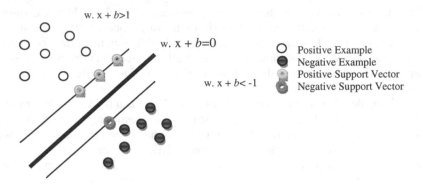

Fig. 1. A separating hyper-plane to classify training data

4   Evaluation

EXIF 0.6.9 is a program for image processing. There are three potential bugs in this program that crashes the program according to the Liblit reports [12]. Applying our proposed approach, the three reported bugs were found in the software. In addition, we could detect a 4th bug which was not reported before.

The ideas before our instrumentation framework are inspired from SOBER infra-structure [5]. The instrumentation procedure was applied to predicates in branch statements and return values. A fault seeding technique described in [6] could be applied to examine all different paths in a program flow graph. To create SVM models, a software tool called LIBSVM [11] may be used. In order to build SVM learning models for early bug prediction, the training data should be prepared in vectors each containing the values of a specific predicate in an execution of the program together with the program termination status. The following Kernel function could be applied to determine the status of a program execution at each instrumentation point:

$$K(x,y) = x^T \qquad (2)$$

The above function is applied by a SVM modeler called LIBLINEAR [8]. To evaluate optimal values for the parameters in the kernel functions a 10-fold cross validation technique can be applied. By applying a cross validation technique, the constructed model can predict the classes of the predicate vectors built dynamically at each instrumentation point, more precisely. Our experiments were conducted on a 2.81 GHz Pentium(R)-4PC with 1G Ram running Linux Ubuntu 7 with gcc 4.2.0. In order to build a SVM behavioral model for EXIF, 50 different fault seeded versions of the programs were made. The test data was collected in 1523 runs including 1416 successful and 107 failing runs. In addition, the seeded versions were executed 1523 times.

After building the SVM model for the EXIF test suite, the model was applied to predict bugs in 100 failing runs of the programs. In about 85 failing executions, the constructed model could predict all the three reported bugs. In addition a fourth bug was predicted. Figure 2, represents the predicates manifesting the bugs.

Predicate#	Predicate	Function
P₁	*strlen (val)==NULL*	*exif_entry_get_value()*
P₂	*i<0*	*Jpeg_data_set_exif_data()*
P₃	o + s > buf_size is TRUE	exif-mnote-data-canon
P₄	*(data->ifd[4])->count is* FALSE	exif_data_save_data_content()

Fig. 2. predicates in EXIF program code classified as faulty by the model

In EXIF's machine readable output mode, the following function returns a NULL pointer, by mistake:

```
0288: exif_entry_get_value(ExifEntry *e, char *val,  unsigned int maxlen)   {
0677:     case EXIF_FORMAT_SSHORT:
0678:         snprintf (val, maxlen, "%i",v_short);
0680:         maxlen -= strlen (val); ...
0691:     case EXIF_FORMAT_LONG: ...
0766:     return strlen (val) ? val : NULL; }
```

Our SVM model, reports the predicate at line 0766 as a main cause of the EXIF program failure when executing the instruction:

```
0111      fprintf (stdout, "%s", exif_entry_get_value (e, v, sizeof (v)));
```

Our model also, for the first time, reported the predicate at line 0677 as the cause of the failure. This line has not been already reported as a cause of the failure by LIBLIT [12]. Our model also indicated the following predicates as the causes of the failure, in addition to the ones already reported by LIBLIT:

```
0358:   if (data->ifd[EXIF_IFD_INTEROPERABILITY]->count) ...
0645:   for (i = 0; list[i].tag && (list[i].tag != e->tag); i++); ...
```

These two predicates are the main causes of the program failure at the following line:
```
0189:   memcpy(*buf + doff,n->entries[i].data, s);
```

We also injected 20 different faults, each in a distinct copy of the EXIF program, by applying the fault seeding technique described in [6]. Our proposed model could detect 17 out of the 20 bugs in the 20 faulty versions.

5 Conclusions

The aim is to prevent damages due to software failures by alerting anomalous behaviors before any failure. In order to predict failures in a program execution, a learning approach may be applied to build a behavioral model of the program. Constructing a separate model for each predicate within the program code, it will be possible to highlight faulty path ending up at that predicate. A comprehensive model should consider all possible execution paths both in faulty and successful executions of the program under test. A fault seeding technique may be best applied to consider any execution path when collecting training data for building the model. The Support Vector Machine learning technique provides a precise behavioral model to analyze a program behavior, in a relatively small amount of time. The model can be further improved by applying a cross validation technique.

References

1. Baah, G.K.: On-line anomaly detection of deployed software: a statistical machine learning approach. In: Proceedings of the 3rd international workshop on Software quality assurance, Portland, pp. 70–77 (2006)
2. Hangal, S., Lam, M.: Tracking down software bugs using automatic anomaly detection. In: 24th international conference on software engineering, Orlando, pp. 291–301 (2002)
3. Bowring, J., Rehg, J.M., Harrold, M.J.: Active Learning for automatic classification of software behavior. In: International symposium on software testing and analysis, Boston, pp. 195–205 (2004)
4. Brun, Y., Ernst, M.: Finding Latent Code Errors via Machine Learning over Program Executions. In: 26th Int'l Conference on software Engineering (ICSE 2004), Edinburgh, pp. 480–490 (2004)
5. Liu, C., Yan, X., Fei, L., Han, J., Midkiff, S.P.: Sober: Statistical model-based bug localization. In: 10th European Software Engineering Conference/13th ACM SIGSOFT Int'l Symposium Foundations of Software Eng (ESEC/FSE 2005), Lisbon, pp. 286–295 (2005)
6. Harrold, M.J., Jefferson Offutt, A., Tewary, K.: An approach to fault modeling and fault seeding using the program dependence graph. Journal of Systems and Software 36(3), 273–295 (1997)
7. Burges, C.J.C.: A Tutorial on Support Vector Machines for Pattern Recognition, Data Mining and Knowledge Discovery, 121–167 (1998)
8. Hsu, C.-W., Chang, C.-C., Lin, C.-J.: A Practical Guide to Support Vector Classification, http://www.csie.ntu.edu.tw/~cjlin/papers/guide/guide.pdf
9. Vapnik, V.: The Nature of Statistical Learning Theory. Springer, New York (1995)
10. Gunn, S.R.: Support Vector Machines for Classification and Regression. Technical Report, Faculty of Engineering, Science and Mathematics School of Electronics and Computer Science, University Of Southampton (1998)
11. Chang, C.-C., Lin, C.-J.: LIBSVM - A Library for Support Vector Machines, http://www.csie.ntu.edu.tw/~cjlin/libsvmtools
12. Liblit, B.: Cooperative Bug Isolation. PhD thesis, University of California, Berkeley (2004)

Evolutionary Decision Support System for Stock Market Trading

Piotr Lipinski

Institute of Computer Science,
University of Wroclaw, Wroclaw, Poland
lipinski@ii.uni.wroc.pl

Abstract. This paper proposes a decision support system for stock market trading, which is based on an evolution strategy algorithm applied to construct an efficient stock market trading expert built as a weighted average of a number of specific stock market trading rules analysing financial time series of recent price quotations. Although applying separately, such trading rules, which come from practictioner knowledge of financial analysts and market investors, give average results, combining them into one trading expert leads to a significant improvement and efficient investment strategies. Experiments on real data from the Paris Stock Exchange confirm the financial relevance of investment strategies based on such trading experts.

1 Introduction

Computational technologies, such as neural networks or evolutionary algorithms, find more and more applications in financial data analysis. Recent research focuses on application of computational intelligence to financial modelling [1], evolutionary portfolio optimization [2] or financial expert systems [4], [5].

This paper refers to the problem of supporting the decision making process for stock market trading. It proposes a decision support system based on trading experts built as weighted averages of a number of specific stock market trading rules. Such trading rules come from practictioner knowledge and are usually applied by financial analysts to identify particular market phenomena or detect trends in stock prices [6]. Single trading rules, applied separately, give average results, but combining them into one trading expert leads to relevant investment strategies. Such trading experts extend previous research on studing trading rules [3] and subsets of trading rules [4].

This paper is structured in the following manner: Section 2 introduces the decision making process. Section 3 describes an evolutionary algorithm to solve the optimization problem of finding efficient trading experts. In Section 5, some experiments and their results are discussed. Finally, Section 5 concludes the paper.

2 Decision Making Process

In our approach, the first component of the decision support system is the concept of *a stock market trading rule*, which is a function $f : \mathcal{K} \mapsto y \in \mathbf{R}$ mapping

D. Dochev, M. Pistore, and P. Traverso (Eds.): AIMSA 2008, LNAI 5253, pp. 405–409, 2008.

a factual financial knowledge $\mathcal{K}$, representing available trading information, to a trading signal y, where values y lower than a certain threshold α_1 correspond to a sell signal, values y greater than a certain threshold α_2 correspond to a buy signal, and remaining values y correspond to no signal (in our experiments, $\alpha_1 = -0.5$ and $\alpha_2 = 0.5$). In our approach, the factual financial knowledge $\mathcal{K}$ represents historical stock prices.

Technical analysis [6] provides a large number of such trading rules. However, in practice, trading rules are often discordant producing opposing trading advices, so investors must spend some effort to transform these advices into a final decision. A solution may be to follow the majority of trading rules or to define a subset of favorite rules and consider only advices produced by this subset [4].

Our approach focuses on a more complex solution, where trading rules form *a stock market trading expert*, the second component of the decision support system, which is a weighted average of trading rules, and the final decision is a weighted average of single advices. Formally, let $\mathcal{R} = \{f_1, f_2, \ldots, f_d\}$ be the entire set of available trading rules (in our experiments, there are $d = 350$ trading rules defined by technical analysis indicators described in details in [6]) and $w_1, w_2, \ldots, w_d \in \mathbf{R}$ be a set of weights. For a factual financial knowledge $\mathcal{K}$, the final trading signal is $y = w_1 \cdot f_1(\mathcal{K}) + w_2 \cdot f_2(\mathcal{K}) + \ldots + w_d \cdot f_d(\mathcal{K})$. The function $E : \mathcal{K} \mapsto y \in \mathbf{R}$ is the trading expert.

In order to compare trading experts, our approach proposes a kind of a simulation, which defines a performance of a trading expert. Consider a specific time period and a hypothetical investor with an initial capital, composed of an initial amount of cach and an initial quantity of stocks. At each successive date in the time period considered, the trading expert proposes a trading signal. If it is a sell signal, the investor sells 50% of stocks. If it is a buy signal, he invests 50% of money in stocks. Each transaction charges the investor 0.2% transaction cost. Finally, the simulation gives a sequence of cash volumes and quantities of stocks for succesive dates.

Afterwards, in order to assess a trading expert, our approach uses one of three performance measures, introduced by financial analysts and market traders, which consider not only the future return rate, but also the risk related to achieving it [5], namely, the Sharpe ratio ϱ_{Sh}, the Sortino ratio ϱ_{So} and the Sterling ratio ϱ_{St}, defined by the following formulas:

$$\varrho_{Sh} = \frac{\mathbf{E}[R] - r_0}{\mathbf{Std}[R]}, \quad \varrho_{So} = \frac{\mathbf{E}[R] - T}{\mathbf{Std}[R|R < T]}, \quad \varrho_{St} = \frac{\mathbf{E}[R] - r_0}{\mathbf{MDD}[R]},$$

where R denotes the return rate, i.e. the ratio of the next-day capital to the previous-day capital, r_0 denotes the return rate of the risk-free asset (in our experiments, it equals 3% annually), T denotes a target return rate (in our experiments, it is the value of the stock market index) and $\mathbf{MDD}$ denotes the maximum drawn down of the return rate, i.e. the maximum loss during the time period considered.

3   Discovering Trading Expertise

Discovering efficient trading experts is an optimization problem of finding weights of trading experts leading to high values of the performance measure. Naturally, the problem is considered in the context of a given set $\mathcal{R}$ of trading rules, a given performance measure ϱ, a given stock and a given training period.

Our approach solves such an optimization problem using an evolutionary algorithm, based on Evolution Strategies (ES) [7], which evolves a population of N trading experts E represented by vectors $\mathbf{w} = (w_1, w_2, \ldots, w_d) \in \mathbf{R}^d$. Each trading expert is evaluated according to the objective function, which is the performance measure ϱ.

Figure 1 shows the framework of the evolutionary algorithm designed to optimize the objective function ϱ with a population $\mathcal{P}$ composed of N individuals, where M denotes the number of children generated by the parent population $\mathcal{P}$, τ and τ_0 are constant algorithm parameters (in our experiments, $\tau = 0.50$ and $\tau_0 = 0.25$). As in classic Evolution Strategies [7], each individual consists not only of the main chromosome $\mathbf{w} = (w_1, w_2, \ldots, w_d) \in \mathbf{R}^d$ representing the trading expert E, but also of an additional chromosome $\boldsymbol{\sigma} = (\sigma_1, \sigma_2, \ldots, \sigma_d) \in \mathbf{R}^d$ containing additional parameters affecting the unique evolution of each trading expert.

EVOLUTION-STRATEGY($\varrho, N, M, \tau, \tau_0$)
1   $\mathcal{P} \leftarrow$ RANDOM-POPULATION(N);
2   POPULATION-EVALUATION($\mathcal{P}, \varrho$);
3   **while not** TERMINATION-CONDITION($\mathcal{P}$)
4   **do**
5       $\mathcal{P}^{(P)} \leftarrow$ PARENT-SELECTION($\mathcal{P}, M$);
6       $\mathcal{P}^{(C)} \leftarrow$ CROSSOVER($\mathcal{P}^{(P)}$);
7   MUTATION($\mathcal{P}^{(C)}, \tau, \tau_0$);
8   REPLACEMENT($\mathcal{P}, \mathcal{P}^{(C)}$);
9   POPULATION-EVALUATION($\mathcal{P}, \varrho$);

Fig. 1. The Evolution Strategy designed to optimize an objective function ϱ with a population $\mathcal{P}$ composed of N individuals, where M, τ, τ_0 are algorithm parameters

First, the algorithm creates an initial population $\mathcal{P}$ at random. For $i = 1, 2, \ldots, N$, each gene w_{ij} of each chromosome $\mathbf{w}_i$ is generated randomly with standard normal probability distribution $\mathcal{N}(0, 1)$, each gene σ_{ij} of each chromosome $\boldsymbol{\sigma}_i$ as well. Afterwards, the population evolves under the influence of evolutionary operators, namely *parent selection, crossover, mutation,* and *replacement,* until a termination condition is satisfied.

In *parent selection,* the algorithm creates a parent population $\mathcal{P}^{(P)}$ with M individuals randomly chosen from the population $\mathcal{P}$ in such a way that the probability of choosing an individual is proportional to its value of the objective function (each individual may be chosen zero, one or more times).

In *crossover*, each pair of parent individuals $(\mathbf{w}_1, \boldsymbol{\sigma}_1)$ and $(\mathbf{w}_2, \boldsymbol{\sigma}_2)$ produces a pair of children $(\tilde{\mathbf{w}}_1, \tilde{\boldsymbol{\sigma}}_1)$ and $(\tilde{\mathbf{w}}_2, \tilde{\boldsymbol{\sigma}}_2)$. The first child is created by *global intermediary recombination*, i.e. its chromosomes are averages of parent chromosomes

$$\tilde{\mathbf{w}}_1 = (\mathbf{w}_1 + \mathbf{w}_2)/2, \quad \tilde{\boldsymbol{\sigma}}_1 = (\boldsymbol{\sigma}_1 + \boldsymbol{\sigma}_2)/2.$$

The second child is created by *uniform crossover*, i.e. its chromosomes inherit each gene from a randomly chosen parent (drawn for each gene separately).

In *mutation*, which is the most important part of the algorithm, each child individual $(\mathbf{w}, \boldsymbol{\sigma})$ is modified by adding a random noise. First, the algorithm modifies the chromosome $\boldsymbol{\sigma}$,

$$\sigma_i \leftarrow \sigma_i \cdot \exp(\varepsilon_i + \varepsilon_0),$$

for $i = 1, 2, \ldots, d$, where ε_i is generated randomly with normal distribution $\mathcal{N}(0, \tau^2)$ (drawn for each gene separately), and ε_0 is generated randomly with normal distribution $\mathcal{N}(0, \tau_0^2)$. Next, the algorithm modifies the chromosome $\mathbf{w}$, using the vector $\boldsymbol{\sigma}$ already processed,

$$w_i \leftarrow w_i + \varepsilon_i,$$

for $i = 1, 2, \ldots, d$, where ε_i is generated randomly with normal distribution $\mathcal{N}(0, \sigma_i^2)$ (drawn for each gene separately).

In *replacement*, our approach uses *a tournament selection*, which consists in randomly choosing a number of individuals from the union of the parent and the children population, the probability of choosing an individual is the same for each individual, selecting the best of them to the new population and repeating the process N times.

Normally, the algorithm terminates when it completes a specific number of iterations or when there is no increases in objective function values over a specific number of recent iterations.

4   Experiments

In order to validate our approach, some experiments were performed on real-life data from the Paris Stock Exchange.

Table 1 shows the financial relevance of investment strategies based on trading experts. Experiments concern the stock Renault, the time period starting on September 3, 2007 and lasting 60 days. The first column describes the algorithm configuration. The next columns reports the performance $\varrho(E)$ of the trading expert, its return rate, the return rate of the benchmark B&H strategy, which consists in investing all the capital in stocks at the start of the period and keeping it until the end of the period. Results show that, in this case, the best strategies are with the Sterling ratio, the population size $N = 200$ and the offspring population size $M = 300$.

In order to test the capabilities of the different algorithm configurations, 100 optimization problems were prepared with randomly selected stocks and randomly selected training and test periods. Each algorithm configuration was applied with each problem. Results confirmed preliminary conclusions.

Table 1. Characteristics of solutions to the optimization problem of discovering an optimal investment strategy with different algorithm configurations for the stock Renault

Settings	Performance	Return	B&H
Sharpe, $N = 100$, $M = 150$	0.0195	0.0936%	0.0632%
Sharpe, $N = 200$, $M = 300$	0.0243	0.1149%	0.0632%
Sharpe, $N = 500$, $M = 750$	0.0213	0.0984%	0.0632%
Sortino, $N = 100$, $M = 150$	0.0398	0.1132%	0.0632%
Sortino, $N = 200$, $M = 300$	0.0587	0.1424%	0.0632%
Sortino, $N = 500$, $M = 750$	0.0436	0.1012%	0.0632%
Sterling, $N = 100$, $M = 150$	0.0332	0.1328%	0.0632%
Sterling, $N = 200$, $M = 300$	0.0403	0.1734%	0.0632%
Sterling, $N = 500$, $M = 750$	0.0401	0.0923%	0.0632%

5 Conclusions

This paper proposes an evolutionary algorithm for discovering efficient stock market trading experts built as weighted averages of a number of specific practictioner trading rules. Preliminary experiments on time series from the Paris Stock Exchange confirm the financial relevance of investment strategies based on such trading experts. However, some remaining problems concern the performance of the optimization algorithm and computing time.

References

1. Brabazon, A., O'Neill, M.: Biologically Inspired Algorithms for Financial Modelling. Springer, Heidelberg (2006)
2. Korczak, J., Lipinski, P., Roger, P.: Evolution Strategy in Portfolio Optimization. In: Collet, P., Fonlupt, C., Hao, J.-K., Lutton, E., Schoenauer, M. (eds.) EA 2001. LNCS, vol. 2310, pp. 156–167. Springer, Heidelberg (2002)
3. Lipinski, P.: Dependency Mining in Large Sets of Stock Market Trading Rules. In: Pejas, J., Piegat, A. (eds.) Enhanced Methods in Computer Security, Biometric and Intelligent Systems, pp. 329–336. Kluwer Academic Publishers, Dordrecht (2005)
4. Lipinski, P.: ECGA vs. BOA in Discoverying Stock Market Trading Experts. In: Proceedings of Genetic and Evolutionary Computation Conference, GECCO 2007, pp. 531–538. ACM, New York (2007)
5. Lipinski, P., Korczak, J.: Performance Measures in an Evolutionary Stock Trading Expert System. In: Bubak, M., van Albada, G.D., Sloot, P.M.A., Dongarra, J. (eds.) ICCS 2004. LNCS, vol. 3039, pp. 835–842. Springer, Heidelberg (2004)
6. Murphy, J.: Technical Analysis of the Financial Markets, NUIF (1998)
7. Schwefel, H.-P.: Evolution and Optimum Seeking. John Wiley and Sons, Chichester (1995)

Suboptimal Nonlinear Predictive Control Based on Neural Wiener Models

Maciej Ławryńczuk

Institute of Control and Computation Engineering, Warsaw University of Technology
ul. Nowowiejska 15/19, 00-665 Warsaw, Poland, tel. +48 22 234-76-73
M.Lawrynczuk@ia.pw.edu.pl

Abstract. This paper is concerned with a computationally efficient (sub-optimal) nonlinear Model Predictive Control (MPC) algorithm based on neural Wiener models. The model contains a linear dynamic part in series with a steady-state nonlinear part which is realised by a neural network. The model is linearised on-line, as a result the nonlinear MPC algorithm needs solving a quadratic programming problem. The algorithm gives control performance similar to that obtained in nonlinear MPC, which hinges on non-convex optimisation. In order to demonstrate accuracy and computational efficiency of the considered MPC algorithm, a polymerisation reactor is studied.

Keywords: Process control, Model Predictive Control, Wiener models, neural networks, optimisation, linearisation, quadratic programming.

1 Introduction

In Model Predictive Control (MPC) algorithms [6,11] at each consecutive sampling instant k a set of future control increments is calculated

$$\Delta u(k) = [\Delta u(k|k) \; \Delta u(k+1|k) \ldots \Delta u(k+N_u-1|k)]^T \qquad (1)$$

It is assumed that $\Delta u(k+p|k) = 0$ for $p \geq N_u$, where N_u is the control horizon. The objective is to minimise the differences between the reference trajectory $y^{ref}(k+p|k)$ and the predicted outputs values $\hat{y}(k+p|k)$ over the prediction horizon $N \geq N_u$. The following quadratic cost function is usually used

$$J(k) = \sum_{p=1}^{N}(y^{ref}(k+p|k) - \hat{y}(k+p|k))^2 + \sum_{p=0}^{N_u-1} \lambda_p(\Delta u(k+p|k))^2 \qquad (2)$$

where $\lambda_p > 0$. Only the first element of the determined sequence (1) is applied to the process, the control law is $u(k) = \Delta u(k|k) + u(k-1)$. At the next sampling instant the prediction is shifted one step forward and the procedure is repeated.

MPC algorithms based on linear models have been successfully used for years in advanced industrial applications [6,10,11], for example distillation columns or chemical reactors. It is mainly because they can take into account constraints

D. Dochev, M. Pistore, and P. Traverso (Eds.): AIMSA 2008, LNAI 5253, pp. 410–414, 2008.
© Springer-Verlag Berlin Heidelberg 2008

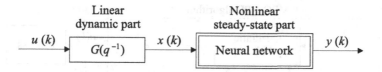

Fig. 1. The structure of the neural Wiener model

imposed on both process inputs (manipulated variables) and outputs (controlled variables), which usually decide on quality, economic efficiency and safety. Moreover, MPC techniques are very efficient in multivariable process control. Because properties of many processes are nonlinear, different nonlinear MPC approaches have been developed [8,11]. In particular, MPC algorithms based on neural models [2] are recommended [5,9,11]. Neural models can be efficiently used on-line in MPC because they have excellent approximation abilities, small numbers of parameters and simple structures. Furthermore, neural models directly describe input-output relations of process variables, complicated systems of algebraic and differential equations do not have to be solved on-line as it is the case in MPC based on fundamental models.

Neural models are trained using recorded data sets. Obtained models are of black-box type, their structure has nothing to do with the physical nature of the process. As an alternative, one can consider the Wiener model which contains a linear dynamic part in series with a steady-state nonlinear part. Wiener models can be efficiently used for modelling and control of various processes [1,3,4].

For identification of Wiener models different approaches have been proposed: correlation methods, parametric regression methods and non-parametric regression methods [3]. In the simplest case polynomials are used in the steady-state nonlinear part of the model. Unfortunately, the accuracy of polynomial approximation is limited. That is why Wiener models with steady-state neural part are recommended [3]. Because multilayer perceptron neural networks are universal approximators, a neural network with at least one hidden layer can approximate any smooth function to an arbitrary degree of accuracy.

This paper describes a computationally efficient (suboptimal) nonlinear Model Predictive Control (MPC) algorithm based on neural Wiener models. The model is linearised on-line taking into account the current state of the process, as a result the algorithm needs solving only a quadratic programming problem. The algorithm gives control performance similar to that obtained in nonlinear MPC with full nonlinear optimisation repeated at each sampling instant.

2   Suboptimal MPC Based on Neural Wiener Models

The structure of the neural Wiener model is shown in Fig. 1. It contains a linear dynamic part described by a transfer function $G(q^{-1})$ in series with a nonlinear steady-state part $y(k) = g(x(k))$ realised by a MultiLayer Perceptron (MLP) feedforward neural network with one hidden layer and a linear output [2].

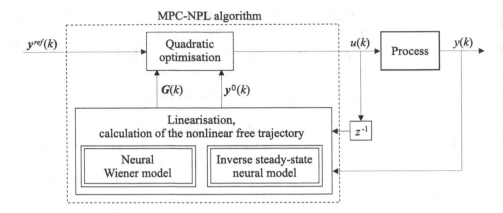

Fig. 2. The structure of the MPC-NPL algorithm

If for prediction in MPC a nonlinear model is used without any simplifications, a nonlinear optimisation problem has to be solved on-line at each sampling instant. Such an approach has limited applicability because computational burden of nonlinear MPC is enormous and it may terminate in local minima. That is why the MPC scheme with Nonlinear Prediction and Linearisation (MPC-NPL) [5,11] is adopted here. The structure of the algorithm is depicted in Fig. 2. At each sampling instant k the neural Wiener model is used on-line twice: to find a local linearisation and a nonlinear free trajectory. Thanks to the linearisation, the nonlinear output prediction can be approximated by the sum of a forced trajectory, which depends only on the future (on future control moves) and a free trajectory $\boldsymbol{y}^0(k)$, which depends only on the past

$$\hat{\boldsymbol{y}}(k) = \boldsymbol{G}(k)\Delta\boldsymbol{u}(k) + \boldsymbol{y}^0(k) \tag{3}$$

where the dynamic matrix $\boldsymbol{G}(k)$ contains step-response coefficients of the local linear approximation of the nonlinear Wiener model [5,11].

Using (3), the cost function (2) becomes a quadratic function of decision variables $\Delta\boldsymbol{u}(k)$

$$J(k) = \left\| \boldsymbol{y}^{ref}(k) - \boldsymbol{G}(k)\Delta\boldsymbol{u}(k) - \boldsymbol{y}^0(k) \right\|^2 + \left\| \Delta\boldsymbol{u}(k) \right\|_\Lambda^2 \tag{4}$$

As a result, the suboptimal MPC-NPL algorithm needs solving on-line a quadratic programming problem, the necessity of nonlinear optimisation is avoided. In the discussed algorithm the inverse steady-state model $x(k) = g^{inv}(y(k))$ is also used. It is realised by the second MLP feedforward neural network.

At each sampling instant k of the algorithm the following steps are repeated:

1. Linearisation of the neural Wiener model: obtain the dynamic matrix $\boldsymbol{G}(k)$.
2. Find the nonlinear free trajectory $\boldsymbol{y}^0(k)$ using the neural Wiener model.
3. Solve the quadratic programming problem to determine $\Delta\boldsymbol{u}(k)$.
4. Apply $u(k) = \Delta u(k|k) + u(k-1)$.
5. Set $k := k+1$, go to step 1.

3   Simulation Results

The considered process is a polymerisation reaction taking place in a jacketed continuous stirred tank reactor [7]. The reaction is the free-radical polymerisation of methyl methacrylate with azo-bis-isobutyronitrile as initiator and toluene as solvent. The output $NAMW$ (Number Average Molecular Weight) is controlled by manipulating the inlet initiator flow rate F_I, i.e. $u = F_I$, $y = NAMW$.

Steady-state properties of the considered process are highly nonlinear as shown in Fig. 3. The manipulated variable changes from $F_I^{\min} = 0.003\ m^3/h$ (which corresponds to $NAMW^{\max} = 4.5524 \cdot 10^4\ kg/kmol$) to $F_I^{\max} = 0.06$ m^3/h (which corresponds to $NAMW^{\min} = 1.4069 \cdot 10^4\ kg/kmol$). The steady-state gain of the system changes approximately 37 times.

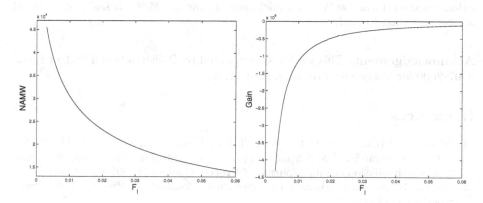

Fig. 3. The steady-state characteristic (*left*) and the steady-state gain (*right*)

Three models are used. The fundamental model is used as the real process during simulations. It is solved using the Runge-Kutta RK45 method. The model is simulated open-loop in order to generate data for identification as a result of which a linear model and a neural Wiener model are obtained. Both empirical models are of the second order. The steady-state neural part the Wiener model has 4 hidden nodes, the inverse model has 5 hidden nodes.

Parameters of MPC are $N = 10$, $N_u = 3$, $\lambda_p = 0.2$, constraints are: $F_I^{\min} = 0.003\ m^3/h$, $F_I^{\max} = 0.06\ m^3/h$, the sampling time is 1.8 min. Simulation results are shown in Fig. 4 for a given reference trajectory ($NAMW^{ref}$). Due to process nonlinearities, the MPC algorithm based on a linear model is unstable.

Two nonlinear MPC schemes are compared with the same neural Wiener model: the MPC algorithm with Nonlinear Optimisation (MPC-NO) and the MPC-NPL algorithm. Closed-loop performance obtained in the suboptimal MPC-NPL algorithm with quadratic programming is practically the same as in the computationally prohibitive MPC-NO approach with nonlinear optimisation repeated at each sampling instant on-line. The computational cost of MPC-NPL is 0.6201 MFLOPS while in the case of MPC-NO it soars to 7.4681 MFLOPS.

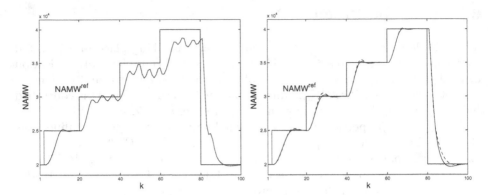

Fig. 4. *Left*: the MPC algorithm based on the linear model; *right*: the MPC-NO algorithm based on the neural Wiener model (*dashed*) and the MPC-NPL algorithm based on the same model (*solid*)

Acknowledgement. This work was supported by Polish national budget funds 2007-2009 for science as a research project.

References

1. Bloemen, H.H.J., Chou, C.T., Boom, T.J.J., Verdult, V., Verhaegen, M., Backx, T.C.: Wiener model identification and predictive control for dual composition control of a distillation column. Journal of Process Control 11, 601–620 (2001)
2. Haykin, S.: Neural networks – a comprehensive foundation. Prentice Hall, Englewood Cliffs (1999)
3. Janczak, A.: Identification of nonlinear systems using neural networks and polynomial models. In: A block-oriented approach. Lecture Notes in Control and Information Sciences, vol. 310 (2005)
4. Kalafatis, A.D., Wang, L., Cluett, W.R.: Linearizing feedforward-feedback control of pH processes based on Wiener model. Journal of Process Control 15, 103–112 (2005)
5. Ławryńczuk, M.: A family of model predictive control algorithms with artificial neural networks. International Journal of Applied Mathematics and Computer Science 17, 217–232 (2007)
6. Maciejowski, J.M.: Predictive control with constraints. Prentice Hall, Englewood Cliffs (2002)
7. Maner, B.R., Doyle, F.J., Ogunnaike, B.A., Pearson, R.K.: Nonlinear model predictive control of a simulated multivariable polymerization reactor using second-order Volterra models. Automatica 32, 1285–1301 (1996)
8. Morari, M., Lee, J.H.: Model predictive control: past, present and future. Computers and Chemical Engineering 23, 667–682 (1999)
9. Nørgaard, M., Ravn, O., Poulsen, N.K., Hansen, L.K.: Neural networks for modelling and control of dynamic systems. Springer, London (2000)
10. Qin, S.J., Badgwell, T.A.: A survey of industrial model predictive control technology. Control Engineering Practice 11, 733–764 (2003)
11. Tatjewski, P.: Advanced control of industrial processes, Structures and algorithms. Springer, London (2007)

Author Index

Lecture Notes in Artificial Intelligence (LNAI)

Vol. 4929: M. Helmert, Understanding Planning Tasks. XIV, 270 pages. 2008.

Vol. 4924: D. Riaño (Ed.), Knowledge Management for Health Care Procedures. X, 161 pages. 2008.

Vol. 4923: S.B. Yahia, E.M. Nguifo, R. Belohlavek (Eds.), Concept Lattices and Their Applications. XII, 283 pages. 2008.

Vol. 4914: K. Satoh, A. Inokuchi, K. Nagao, T. Kawamura (Eds.), New Frontiers in Artificial Intelligence. X, 404 pages. 2008.

Vol. 4911: L. De Raedt, P. Frasconi, K. Kersting, S. Muggleton (Eds.), Probabilistic Inductive Logic Programming. VIII, 341 pages. 2008.

Vol. 4908: M. Dastani, A. El Fallah Seghrouchni, A. Ricci, M. Winikoff (Eds.), Programming Multi-Agent Systems. XII, 267 pages. 2008.

Vol. 4898: M. Kolp, B. Henderson-Sellers, H. Mouratidis, A. Garcia, A.K. Ghose, P. Bresciani (Eds.), Agent-Oriented Information Systems IV. X, 292 pages. 2008.

Vol. 4897: M. Baldoni, T.C. Son, M.B. van Riemsdijk, M. Winikoff (Eds.), Declarative Agent Languages and Technologies V. X, 245 pages. 2008.

Vol. 4894: H. Blockeel, J. Ramon, J. Shavlik, P. Tadepalli (Eds.), Inductive Logic Programming. XI, 307 pages. 2008.

Vol. 4885: M. Chetouani, A. Hussain, B. Gas, M. Milgram, J.-L. Zarader (Eds.), Advances in Nonlinear Speech Processing. XI, 284 pages. 2007.

Vol. 4874: J. Neves, M.F. Santos, J.M. Machado (Eds.), Progress in Artificial Intelligence. XVIII, 704 pages. 2007.

Vol. 4870: J.S. Sichman, J. Padget, S. Ossowski, P. Noriega (Eds.), Coordination, Organizations, Institutions, and Norms in Agent Systems III. XII, 331 pages. 2008.

Vol. 4869: F. Botana, T. Recio (Eds.), Automated Deduction in Geometry. X, 213 pages. 2007.

Vol. 4865: K. Tuyls, A. Nowe, Z. Guessoum, D. Kudenko (Eds.), Adaptive Agents and Multi-Agent Systems III. VIII, 255 pages. 2008.

Vol. 4850: M. Lungarella, F. Iida, J.C. Bongard, R. Pfeifer (Eds.), 50 Years of Artificial Intelligence. X, 399 pages. 2007.

Vol. 4845: N. Zhong, J. Liu, Y. Yao, J. Wu, S. Lu, K. Li (Eds.), Web Intelligence Meets Brain Informatics. XI, 516 pages. 2007.

Vol. 4840: L. Paletta, E. Rome (Eds.), Attention in Cognitive Systems. XI, 497 pages. 2007.

Vol. 4830: M.A. Orgun, J. Thornton (Eds.), AI 2007: Advances in Artificial Intelligence. XIX, 841 pages. 2007.

Vol. 4828: M. Randall, H.A. Abbass, J. Wiles (Eds.), Progress in Artificial Life. XII, 402 pages. 2007.

Vol. 4827: A. Gelbukh, Á.F. Kuri Morales (Eds.), MICAI 2007: Advances in Artificial Intelligence. XXIV, 1234 pages. 2007.

Vol. 4826: P. Perner, O. Salvetti (Eds.), Advances in Mass Data Analysis of Signals and Images in Medicine, Biotechnology and Chemistry. X, 183 pages. 2007.

Vol. 4819: T. Washio, Z.-H. Zhou, J.Z. Huang, X. Hu, J. Li, C. Xie, J. He, D. Zou, K.-C. Li, M.M. Freire (Eds.), Emerging Technologies in Knowledge Discovery and Data Mining. XIV, 675 pages. 2007.

Vol. 4811: O. Nasraoui, M. Spiliopoulou, J. Srivastava, B. Mobasher, B. Masand (Eds.), Advances in Web Mining and Web Usage Analysis. XII, 247 pages. 2007.

Vol. 4798: Z. Zhang, J.H. Siekmann (Eds.), Knowledge Science, Engineering and Management. XVI, 669 pages. 2007.

Vol. 4795: F. Schilder, G. Katz, J. Pustejovsky (Eds.), Annotating, Extracting and Reasoning about Time and Events. VII, 141 pages. 2007.

Vol. 4790: N. Dershowitz, A. Voronkov (Eds.), Logic for Programming, Artificial Intelligence, and Reasoning. XIII, 562 pages. 2007.

Vol. 4788: D. Borrajo, L. Castillo, J.M. Corchado (Eds.), Current Topics in Artificial Intelligence. XI, 280 pages. 2007.

Vol. 4775: A. Esposito, M. Faundez-Zanuy, E. Keller, M. Marinaro (Eds.), Verbal and Nonverbal Communication Behaviours. XII, 325 pages. 2007.

Vol. 4772: H. Prade, V.S. Subrahmanian (Eds.), Scalable Uncertainty Management. X, 277 pages. 2007.

Vol. 4766: N. Maudet, S. Parsons, I. Rahwan (Eds.), Argumentation in Multi-Agent Systems. XII, 211 pages. 2007.

Vol. 4760: E. Rome, J. Hertzberg, G. Dorffner (Eds.), Towards Affordance-Based Robot Control. IX, 211 pages. 2008.

Vol. 4755: V. Corruble, M. Takeda, E. Suzuki (Eds.), Discovery Science. XI, 298 pages. 2007.

Vol. 4754: M. Hutter, R.A. Servedio, E. Takimoto (Eds.), Algorithmic Learning Theory. XI, 403 pages. 2007.

Vol. 4737: B. Berendt, A. Hotho, D. Mladenic, G. Semeraro (Eds.), From Web to Social Web: Discovering and Deploying User and Content Profiles. XI, 161 pages. 2007.

Vol. 4733: R. Basili, M.T. Pazienza (Eds.), AI*IA 2007: Artificial Intelligence and Human-Oriented Computing. XVII, 858 pages. 2007.

Vol. 4724: K. Mellouli (Ed.), Symbolic and Quantitative Approaches to Reasoning with Uncertainty. XV, 914 pages. 2007.

Vol. 4722: C. Pelachaud, J.-C. Martin, E. André, G. Chollet, K. Karpouzis, D. Pelé (Eds.), Intelligent Virtual Agents. XV, 425 pages. 2007.

Vol. 4720: B. Konev, F. Wolter (Eds.), Frontiers of Combining Systems. X, 283 pages. 2007.

Vol. 4702: J.N. Kok, J. Koronacki, R. Lopez de Mantaras, S. Matwin, D. Mladenič, A. Skowron (Eds.), Knowledge Discovery in Databases: PKDD 2007. XXIV, 640 pages. 2007.

Vol. 4701: J.N. Kok, J. Koronacki, R. Lopez de Mantaras, S. Matwin, D. Mladenič, A. Skowron (Eds.), Machine Learning: ECML 2007. XXII, 809 pages. 2007.

Vol. 4696: H.-D. Burkhard, G. Lindemann, R. Verbrugge, L.Z. Varga (Eds.), Multi-Agent Systems and Applications V. XIII, 350 pages. 2007.